CONTROL AND DYNAMIC SYSTEMS

Advances in Theory and Applications

Volume 57

CONTRIBUTORS TO THIS VOLUME

FRANK F. ABDI
ERNEST S. ARMSTRONG
JEAN-NOEL AUBRUN
CHRISTINA L. BLOEBAUM
ALAIN C. CARRIER
CHUNG-WEN CHEN
EMMANUEL G. COLLINS, JR.
W. GAWRONSKI
BERNARD GROSSMAN
RAPHAEL T. HAFTKA
PRABHAT HAJELA
JEN-KUANG HUANG
DAVID C. HYLAND
SURESH M. JOSHI
JER-NAN JUANG
JAMES A. KING
KATHRYN LENZ
K. B. LIM
RAFAEL LIVNEH
TRENT R. LOGAN
KENNETH R. LORELL
PEIMAN G. MAGHAMI
WILLIAM H. MASON
HITAY ÖZBAY
T. S. PAN
DOUGLAS J. PHILLIPS
S. S. RAO
GARY L. SLATER
JAROSLAW SOBIESZCZANSKI-SOBIESKI
ERIC R. UNGER
V. B. VENKAYYA

CONTROL AND DYNAMIC SYSTEMS

ADVANCES IN THEORY AND APPLICATIONS

Edited by

C. T. LEONDES

School of Engineering and Applied Science
University of California, Los Angeles
Los Angeles, California
and
Department of Electrical Engineering
and Computer Science
University of California, San Diego
La Jolla, California

VOLUME 57: MULTIDISCIPLINARY ENGINEERING SYSTEMS: DESIGN AND OPTIMIZATION TECHNIQUES AND THEIR APPLICATION

ACADEMIC PRESS, INC.
A Division of Harcourt Brace & Company
San Diego New York Boston London Sydney Tokyo Toronto

ACADEMIC PRESS RAPID MANUSCRIPT REPRODUCTION

This book is printed on acid-free paper. ∞

Academic Press, Inc.
1250 Sixth Avenue, San Diego, California 92101-4311

United Kingdom Edition published by
Academic Press Limited
24–28 Oval Road, London NW1 7DX

International Standard Serial Number: 0090-5267

International Standard Book Number: 0-12-012757-1

PRINTED IN THE UNITED STATES OF AMERICA
93 94 95 96 97 98 QW 9 8 7 6 5 4 3 2 1

CONTENTS

CONTRIBUTORS

Numbers in parentheses indicate the pages on which the authors' contributions begin.

Frank F. Abdi (25), *North American Aircraft, Rockwell International Corporation, Los Angeles, California 90009*

Ernest S. Armstrong (281), *NASA Langley Research Center, Hampton, Virginia 23665*

Jean-Noel Aubrun (181), *Research & Development Division, Lockheed Missiles & Space Company, Inc., Palo Alto, California 94304*

Christina L. Bloebaum (1), *Department of Mechanical and Aerospace Engineering, State University of New York at Buffalo, Buffalo, New York 14260*

Alain C. Carrier (181), *Research & Development Division, Lockheed Missiles & Space Company, Inc., Palo Alto, California 94304*

Chung-Wen Chen (331), *Mars Mission Research Center, North Carolina State University, Raleigh, North Carolina 27695*

Emmanuel G. Collins, Jr. (233), *Harris Corporation, Government Aerospace Systems Division, Melbourne, Florida 32902*

W. Gawronski (109), *Jet Propulsion Laboratory, California Institute of Technology, Pasadena, California 91109*

Bernard Grossman (55), *Department of Aerospace and Ocean Engineering, Virginia Polytechnic Institute and State University, Blacksburg, Virginia 24061*

Raphael T. Haftka (55), *Department of Aerospace and Ocean Engineering, Virginia Polytechnic Institute and State University, Blacksburg, Virginia 24061*

Prabhat Hajela (1), *Department of Mechanical Engineering, Aeronautical Engineering and Mechanics, Rensselaer Polytechnic Institute, Troy, New York 12180*

Jen-Kuang Huang (331), *Department of Mechanical Engineering and Mechanics, Old Dominion University, Norfolk, Virginia 23529*

David C. Hyland (233), *Harris Corporation, Government Aerospace Systems Division, Melbourne, Florida 32902*

Suresh M. Joshi (281), *NASA Langley Research Center, Hampton, Virginia 23665*

Jer-Nan Juang (331), *Spacecraft Dynamics Branch, NASA Langley Research Center, Hampton, Virginia 23665*

James A. King (233), *Harris Corporation, Government Aerospace Systems Division, Melbourne, Florida 32902*

Kathryn Lenz (369), *Department of Mathematics and Statistics, University of Minnesota at Duluth, Duluth, Minnesota 55812*

K. B. Lim (109), *NASA Langley Research Center, Hampton, Virginia 23665*

Rafael Livneh (423), *Aerospace Research Center, College of Engineering and Applied Sciences, Arizona State University, Tempe, Arizona 85287*

Trent R. Logan (25), *North American Aircraft, Rockwell International Corporation, Los Angeles, California 90009*

Kenneth R. Lorell (181), *Research & Development Division, Lockheed Missiles & Space Company, Inc., Palo Alto, California 94304*

Peiman G. Maghami (281), *NASA Langley Research Center, Hampton, Virginia 23665*

William H. Mason (55), *Department of Aerospace and Ocean Engineering, Virginia Polytechnic Institute and State University, Blacksburg, Virginia 24061*

Hitay Özbay (369), *Department of Electrical Engineering, The Ohio State University, Columbus, Ohio 43210*

T. S. Pan (153), *IBM, Poughkeepsie, New York 12602*

Douglas J. Phillips (233), *Harris Corporation, Government Aerospace Systems Division, Melbourne, Florida 32902*

S. S. Rao (153), *School of Mechanical Engineering, Purdue University, West Lafayette, Indiana 47907*

Gary L. Slater (423), *Department of Aerospace Engineering and Engineering Mechanics, University of Cincinnati, Cincinnati, Ohio 45221*

Jaroslaw Sobieszczanski-Sobieski (1, 25), *Interdisciplinary Research Office, NASA Langley Research Center, Hampton, Virginia 23665*

Eric R. Unger (55), *Department of Aerospace and Ocean Engineering, Virginia Polytechnic Institute and State University, Blacksburg, Virginia 24061*

V. B. Venkayya (153), *Wright Laboratory, Wright-Patterson Air Force Base, Ohio 45433*

PREFACE

The modern era of engineering systems began with the end of World War II and accelerated with the introduction of digital computers, integrated electronics, advances in the modeling of complex systems, increased understanding of complex interactive systems issues and techniques, the introduction of multidisciplinary engineering design teams, improvements in techniques for systems design optimization, and many other advances that will continue in the future. In particular, the past decade has seen the use of the multidisciplinary approach to engineering systems design. This approach will continue to grow and expand, in increasingly effective ways, in the design of future engineering systems. The implications are many—including increased system reliability, safety, and efficiency, improved system maintainability, cost effectiveness, and expanded systems capability. As a result, it is appropriate that we present multidisciplinary engineering systems design in this international series. Thus this volume is devoted to the most timely theme of "Multidisciplinary Engineering Systems Design and Optimization Techniques and Their Application."

The first contribution to this volume is "Decomposition Methods for Multidisciplinary Synthesis," by Christina L. Bloebaum, Prabhat Hajela, and Jaroslaw Sobieszczanski-Sobieski. This contribution speaks directly to, and presents important and powerful techniques for, the multidisciplinary system's synthesis process. As a result, it is a particularly appropriate contribution with which to begin this volume.

The next contribution is "Multilevel Optimization Techniques for Aircraft," by Trent R. Logan, Frank F. Abdi, and Jaroslaw Sobieszczanski-Sobieski. Among the most challenging of modern engineering systems, for which the multidisciplinary design approach is absolutely essential, is that of aircraft systems. In fact, it is from this field that so much of the modern methodology for multidisciplinary engineering systems design is being developed, and this is to be expected. In this contribution many important aspects of multidisciplinary engineering design methodology are presented and illustrated by example.

The next contribution is "Integrated Aerodynamic–Structural Design Optimization of Aircraft Wings," by Eric R. Unger, Raphael T. Hatka, Bernard Grossman, and William H. Mason. In this contribution a generalized treatment of the integrated aerodynamic–structural design problem is described. The methods presented are illustrated by a number of examples that provide an important appreciation for the requirements and the challenges associated with multidisciplinary design problems.

The next contribution is "Actuator and Sensor Placement for Control of Flexible Structures," by K. B. Lim and W. Gawronski. Ultimately, a multidisciplinary engineering design problem will involve a control process. This, particularly for so many of today's rather complex engineering systems, involves the effective placement or distribution of sensor and actuator systems. This contribution presents general techniques for doing this — motivated by the rather challenging problem of the control of flexible structures.

The next contribution is "Structural Modification Techniques for Robust Actively Controlled Structures," by S. S. Rao, T. S. Pan, and V. B. Venkayya. One of the major challenges in the design of modern complex engineering systems is the development of effective system models (i.e., the equations that are relevant to the specific disciplines and that adequately represent the system). From an adequate system model, or set of equations, the design and system control processes can proceed. Because of inevitable model uncertainties, robust techniques must be utilized in the system design and control process. This contribution provides effective means for doing this by considering the specific area of controlled structures.

The next contribution is "Active Control for Segmented Mirror Optical Systems," by Jean-Noel Aubrun, Alain C. Carrier, and Kenneth R. Lorell. In modern multidisciplinary engineering systems the concept of active control can result in some essential benefits. For example, the structural lifetime of an aircraft can be extended by using a heavier, more rigid, aircraft structure or by the utilization of active control for structural load alleviation. In this process, aircraft sensors are used to sense structural displacements and provide information to aircraft control surface actuators to generate countervailing forces on the aircraft structure. Through the use of active control, structural load alleviation is achieved that can, in general, be more cost effective than increasing aircraft structural weight. In this contribution the same technique of active control is utilized to achieve space telescope performance, which previously was achieved by means of structural design.

The next contribution is "Robust Control Experiments for Two Highly Elastic Space Structures," by Emmanuel G. Collins, Jr., Douglas J. Phillips, David C. Hyland, and James A. King. This contribution is an important treatment that confirms the performance improvement achievable in a complex multidisciplinary system — in this case, two examples of complex space structures and the iterative design procedures that are necessary to achieve the desired result.

The next contribution is "Applications of an Optimization-Based Integrated Controls–Structures Design Methodology for Flexible Space Structures," by Peiman G. Maghami, Suresh M. Joshi, and Ernest S. Armstrong. Current design procedures in multidisciplinary engineering systems often treat the design process iteratively, wherein the system design is first conducted by considering various factors such as structural loads and then the controller design is considered. The process is then iterated until the final design is achieved. This contribution presents important techniques wherein the system design and the control system design are treated in a completely integrated sense, and the techniques presented are exemplified by the particular case of space structures.

The next contribution is "Adaptive Linear System Identification and State Estimation," by Chung-Wen Chen, Jer-Nan Juang, Jen-Kuang Huang. In the interdisciplinary system design process there are usually two different ways to obtain a model for a dynamic system. The first is the analytic approach: it starts from the derivation of governing equations based on physical laws, and from this the system model is developed. One example technique used in this approach is the finite element method. The second is an experimental approach that uses the measured input and output data obtained from experiments or operations to find a system model directly, such as some modal testing techniques. In general, the system models obtained by analytical means need further confirmation and refinement using experimental results before the model can be used for control system design. System identification is a term used for the technique that deals with developing mathematical models for dynamic systems based on their input and output data. This contribution is an in-depth treatment of the system identification technique, which is then exemplified by means of structural system problems.

The next contribution is "Analysis and Robust Control Techniques for an Ideal Flexible Beam," by Kathryn Lenz and Hitay Özbay. In this contribution the process of developing system models by the analytic approach is presented and exemplified by a flexible beam. In addition, this contribution presents techniques which are easily generalized, for developing simplified models and for developing robust control systems.

The final contribution is "Simple Adaptive Controller with Applications to Wing Flutter," by Rafael Livneh and Gary L. Slater. As mentioned in the previous contributions, once the system model is developed by either analytical or experimental means, the system controller can be developed by the practical and effective or robust means presented in the previous contributions. Another effective approach is a reference model approach based on the analytically or experimentally derived model for the development of the system control. That is, the control is based on the reference model and the actual system is made to be effectively controlled according to this reference model by adaptive control techniques. The principles involved here, which are generally applicable, are exemplified by a rather complex 23rd order wing flutter model control problem.

This volume clearly manifests the significance and power of the multidisciplinary systems engineering approach that is absolutely essential in modern complex engineering systems. The co-authors are all to be commended for their contributions to this volume, which will provide a significant reference source for workers on the international scene for years to come.

Decomposition Methods for Multidisciplinary Synthesis

Christina L. Bloebaum

Department of Mechanical and Aerospace Engineering
State University of New York at Buffalo
Buffalo, New York 14260

Prabhat Hajela

Department of Mechanical Engineering, Aeronautical
Engineering and Mechanics
Rensselaer Polytechnic Institute
Troy, New York 12180

Jaroslaw Sobieszczanski-Sobieski

Interdisciplinary Research Office
NASA Langley Research Center
Hampton, Virginia 23665

I. INTRODUCTION

Recent interest in improving product design quality while minimizing manufacturing costs has resulted in research initiatives involving the incorporation of all design criteria in a simultaneous treatment, known as 'Concurrent Engineering'. However, design synthesis for multidisciplinary systems is complicated by often unwieldy numbers of design variables and constraints, as well as the inherent interactions between disciplinary analyses. The computational demands associated with the optimization of such large systems may exceed the capabilities of even the most advanced computers today. Even if the size of the problem can be accommodated, any interactions between disciplinary analyses would present further difficulties. Recent experiences have demonstrated the importance of accurately representing these interactions during the design process in order to achieve a holistic design strategy. In some cases, such as the design of hypersonic aircraft, failure to represent the discipline couplings realistically could result in a design that looks good on paper, but in actuality would not get off the ground.

Although it may be acknowledged that use of an integrated design approach is desirable and often essential, the traditional synthesis approach is not a feasible mechanism of implementation. The synergistic nature of integrated design can be realized, however, if a decomposition technique is employed.

Decomposition approaches reduce the complexity associated with multidisciplinary optimization by modeling the large coupled engineering system by smaller, more tractable, but coordinated subsystems. The advantage of subsystem modularity allows for implementation of specialized methods for analysis, computational efficiency, and the ability to incorporate human intervention and decision making in the form of an expert systems capability. This type of approach not only makes sense from a computational standpoint, but also from a managerial one, as design organizations generally are composed of distinct engineering groups that correspond to disciplinary topics (i.e., structural dynamics group, high speed aerodynamics group, etc.). Not only would these groups of engineers be in physically separate locations, each would also have a computational environment specific to its needs. In this type of situation, decomposition methods would lend themselves to distributed and parallel processing implementations, which would further maximize operational efficiency.

An increased number of conferences, journals, and publications devoted to multidisciplinary optimization demonstrates recent interest in the topic. Numerous papers have been published recently which deal specifically with multidisciplinary optimization applications in such diverse areas as naval structure design [1,2], spacecraft design [3], rotorcraft design [4], automobile design [5], and aircraft design [6]. The proposed methodologies to deal with the multidisciplinary design problem have been almost as diverse as the applications and have mostly proved disappointing. Two review papers on the field of multidisciplinary synthesis are of particular noteworthiness. The requirements and opportunities available in multidisciplinary analysis and synthesis applications are reviewed in Ref. 7. The potentials and achievements of multidisciplinary optimization are reviewed in Ref. 8.

In the present chapter, the different hierarchical and non-hierarchical architectures found in multidisciplinary environments will be presented. Various decomposition approaches will be developed and discussed for different multidisciplinary applications. Among these is a hierarchic linear decomposition approach, and two methods for use in non-hierarchic environments, wherein lateral couplings between subsystems exist. A brief discussion of method performance will be presented for each. An in-depth development of the non-hierarhic, system sensitivity based approach can be found in the chapter "Techniques for Sensitivity Analysis in Aircraft Design", by Bloebaum, Hajela, and Sobieszczanski-Sobieski, while application of the approach to aircraft design optimization can be found in the chapter, "Multilevel Optimization Techniques for Aircraft", by Logan, Abdi, and Sobieszczanski-Sobieski.

II. Hierarchical Decomposition Architectures

Three types of architectures exist which can be used to classify complex engineering systems. These are hierarchic, non-hierarchic, and hybrid-hierarchic. A brief discussion of each architecture is presented. An understanding of the differences in decomposition architectures is essential in accurately implementing a synthesis strategy.

A. Hierarchic Decomposition

Hierarchic decomposition is applicable in a complex engineering system in which information flow and interactions are strictly top-down (or can be modeled top-down). The system sends information to the appropriate subsystems that relate further information to its own subsystems following an analysis. This information flow is demonstrated in Figure 1.

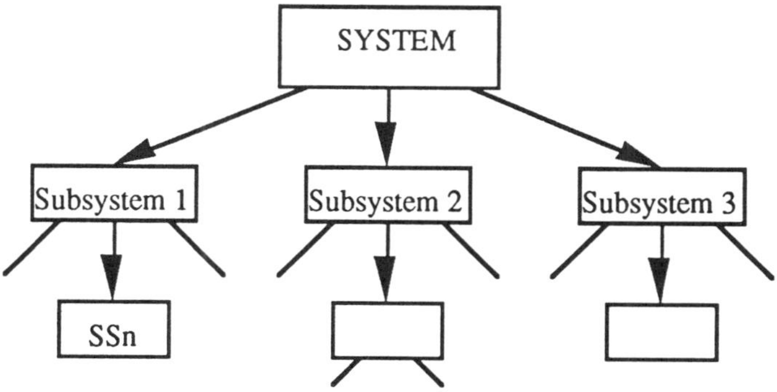

Figure 1. Hierarchical decomposition of an engineering system.

The flow of information continues through the levels, from system to subsystem to sub-subsystem, until the analyses at all levels have been exhausted. At this point, when the bottom level has been reached, the solution associated with each analysis is available for use in an optimization strategy.

An example of such a hierarchic system can be found in the structural analysis and design of a built-up structure comprised of components. As long as the appropriate boundary conditions and loadings are translated from one level to another, the analysis can be decomposed into a series of component analyses.

B. Non-hierarchic Decomposition

In the non-hierarchic architecture, information flow is not limited to top-down, but can be multi-directional, so that lateral couplings between subsystems can be accommodated. This type of architecture is often found in multidisciplinary applications in which each disciplinary analysis is dependent upon output from the other disciplines.

In Figure 2 it can be seen that subsystem 1 (ss1) interacts with all other subsystems through the transference of output data. If subsystem 1 represents a structures analysis, for example, and subsystem 2 (ss2) represents a controls analysis, the structural output may include the modal frequencies and mode shapes which are required for the optimal control analysis. Similarly, the structures analysis may require actuator mass data as input from the controls analysis.

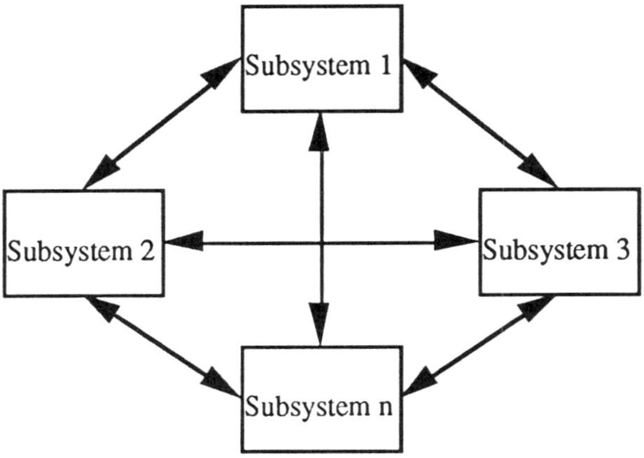

Figure 2. Non-hierarchic system architecture.

C. Hybrid-hierarchic Decomposition

A hybrid-hierarchic architecture consists of, as the name implies, both hierarchic and non-hierarchic components. For example, it can be seen that em-

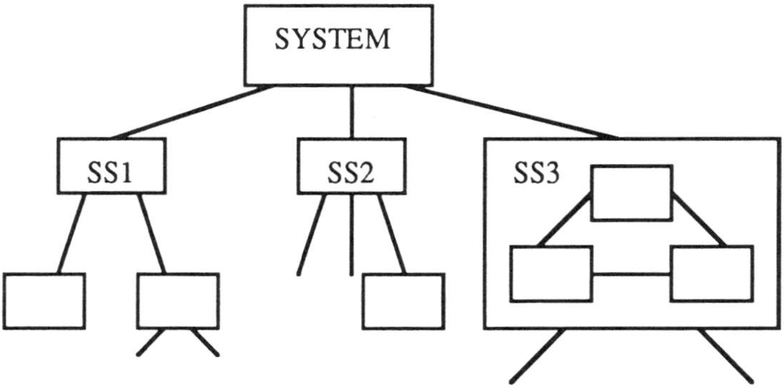

Figure 3. Hybrid-hierarchic representation of non-hierarchic system embedded within hierarchic architecture.

bedded in a subsystem box associated with a hierarchic level is a non-hierarchic system. The other situation might occur as well, wherein a hierarchy is embedded within a non-hierarchic architecture. For example, a multidisciplinary application in which structures is a contributing discipline could have a hierarchy representing structural analysis established within the structures subsystem box.

Such an architecture can also be represented in an N-Square Matrix [9] (Figure 4) where forward information flows (such as that from SS1 to SS2, and from SS2 to SS3) are top-down flows, and SS2, SS3, and SS4 represent a non-hierarchic circuit in which lateral dependencies and information feedbacks exist.

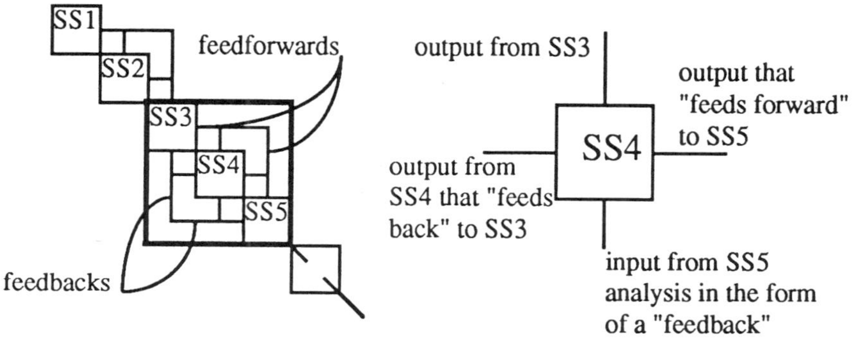

Figure 4. N-Square matrix representation of complex system.

Such a representation is helpful in visually quantifying discipline and subsystem dependencies.

III. OPTIMAL DESIGN PROBLEM STATEMENT

Optimal design is concerned with achieving the best design according to some prescribed criteria while satisfying certain associated restrictions. Wilde [10] defines optimal design as being "the best feasible design according to a preselected quantitative measure of effectiveness". The motivation behind optimization applications is to exploit the available limited resources in such a way as to maximize output [11]. The importance of minimum weight design of structures is especially crucial to the aerospace industry where aircraft designs are controlled more by weight considerations than by cost.

A. Design Problem Statement

The concept of optimizing a structure implicitly suggests that there is some freedom to change the structure. This change is accomplished by means of design variables which are parameters that typically have a range of potential movement associated with them. The objective function is essentially a merit function that has some explicit or implicit relation to at least a subset of the design variables and can be improved through manipulation of those variables. Typically, in realistic optimization applications, limits exist on design variables as well as certain functions (constraints) that are dependent on at least a subset of the design variables.

The following optimization statement demonstrates the notation used for the objective function, constraints and design variables.

$$\begin{array}{llll} \textit{Minimize} & F(X) & & \\ \textit{Subject to} & g_j(X) \leq 0 & j = 1,\ldots l & \\ & h_k(X) = 0 & k = 1,\ldots m & (1) \\ & X_i^L \leq X_i \leq X_i^U & i = 1,\ldots n & \end{array}$$

where (X) represents a vector of design variables, g_j and h_k are inequality and equality constraints, respectively, and F is the objective function.

In the complex engineering design problem associated with a multidisciplinary application, contributions to the design variables, constraints and the objective function are made from all the participating disciplines. The design variable and constraint vectors can then be described in terms of partitioned vectors where partitioned subsets are associated with each discipline's contributions.

B. Synthesis Approach

The general solution process for a gradient-based optimization problem can be seen in Figure 5. The process begins with an initialization of design variables and problem parameters from which an analysis is performed. A sensitivity analysis is then carried out to find derivative information of the output response quantities, such as the objective function and constraints. This sensitivity information is then used in the gradient-based optimizer which results in an improved value of the objective function. Typically, a piecewise linear approximation scheme is implemented to reduce computational costs. The process is terminated when no further improvement in the objective function can be made without violating the constraints.

For non-hierarchic systems, the intrinsically linked subsystem analyses must first be performed within an iterative framework to obtain a converged initial point. Figure 6 demonstrates a system comprised of four fully coupled subsystems. For this system, a converged initial point is defined as that point which satisfies the equations

$$SS1=0 \quad , \quad SS2=0 \quad , \quad SS3=0 \quad , \quad SS4=0 \qquad (2)$$

where SSi corresponds to the analysis associated with the ith subsystem. Once a converged initial point is obtained, the sensitivity analysis can be performed. The synthesis approach from Figure 5 can then be followed.

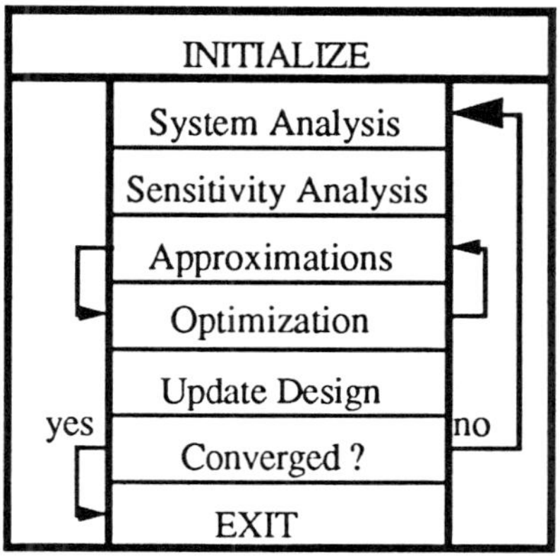

Figure 5. Flowchart of synthesis scheme for engineering system.

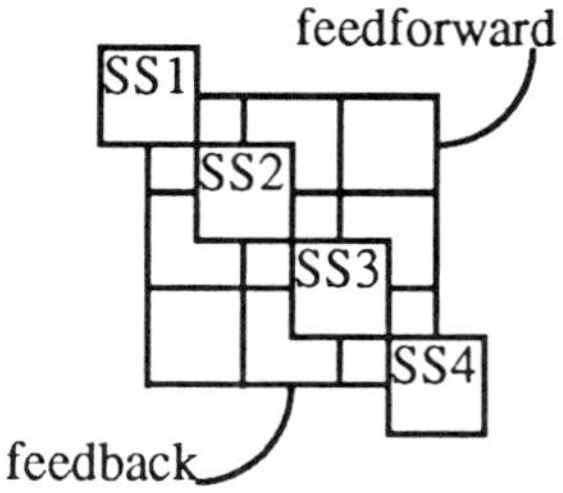

Figure 6. Fully coupled non-hierarchic system.

However, due to the large number of analyses required in the process, computational expenses are often exorbitant. The available computer tools used to perform analysis in such complex environments, such as structural or aerodynamic analyses, are inevitably computationally expensive. The piecewise linear optimization approach, or method of approximate programming, is extremely useful in reducing these computational expenses.

In the method of approximate programming [12], gradient information is used to create an approximate optimization problem that is solved instead of the fully nonlinear problem, thus reducing repeated costly analyses [13]. The optimization is then carried out in the neighborhood of the current design point. Move limits are imposed on the design variables during the optimization process to maintain the integrity of the linear approximations made on the output response quantities. Determination of move limit values is generally made based on problem-dependent heuristics and user experience.

Numerous approximation techniques have been investigated for application in approximate programming [14-16]. One of the most popular and easiest to implement is based on a first order Taylor series expansion of the objective function and constraints. Because this information is already available for use in the gradient-based optimizer, no additional effort is required. The linearized optimization problem based on this type of approximation is of the form

$$\begin{aligned} &\textit{Minimize} && \left\{F(\hat{X}) + (\mathrm{X}-\hat{\mathrm{X}})^T \nabla F(\hat{X})\right\} \\ &\textit{Subject to} && \left\{g_j(\hat{X}) + (\mathrm{X}-\hat{\mathrm{X}})^T \nabla g_j(\hat{X})\right\} \le 0 \\ & && \left\{h_k(\hat{X}) + (\mathrm{X}-\hat{\mathrm{X}})^T \nabla h_k(\hat{X})\right\} = 0 \\ &\textit{and} && \hat{X}_i - \alpha_i \le X_i \le \hat{X}_i + \beta_i \end{aligned} \tag{3}$$

where α and β are prescribed positive constants called move limits and $\hat{\mathbf{X}}$ is the design point where the objective function and constraints are linearized. These variables effectively serve to limit the variation domain of the design variables. The optimal design resulting from the approximate optimization problem with the additional move limit constraints then forms the initial point for the next cycle. The process is terminated when prescribed convergence criteria are met.

For a highly nonlinear problem, it is essential that appropriate move limits be established [17]. By allowing the design variables to change only within some percentage of the initial point, the inaccuracies introduced due to the linear approximations are effectively controlled.

IV. DECOMPOSITION METHODS FOR MULTI-DISCIPLINARY SYNTHESIS

The intuitive practice of breaking a large task into smaller more manageable tasks was applied by Sobieski in Ref. 18 in which a linear decomposition method was applied for hierarchic environments only. Early attempts to solve the non-hierarchic problem involved wrapping an optimization loop around the contributing disciplinary analyses [19]. Unfortunately, the approach was computationally prohibitive and tended to exclude human intervention and decision-making. The Global Sensitivity Equation Method [20-22] extended the modularity concept of Ref. 18 to include applications in the non-hierarchic environments existing in multidisciplinary problems. The Concurrent Subspace Optimization approach [23] extended the capabilities of the Global Sensitivity Equation method by performing concurrent optimizations within prescribed subsystems. These approaches will be discussed in detail in the following sections.

A. Multilevel Optimization Methods for Hierarchic Decomposition

The multilevel linear decomposition approach [18] is based on Bellman's Dynamic Programming method [24], with modifications to account for dimensionality problems. Bellman's method involved analyzing each subsystem in prescribed order from first to last, and then performing an optimization within each subsystem, working from last to first, recursively carrying the optimum design variables for each subsystem forward. Sobieski's linear decomposition method modifies Bellman's approach by using the concept of optimum sensitivity analysis to reduce dimensionality problems.

1. Linear Decomposition Algorithm

The method can best be introduced using a two level optimization problem, with the system level branching into distinct subsystems. The objective of the optimization problem is to minimize the objective function subject to constraints at both system and subsystem levels. Design variables are defined at both the system (X_i) and subsystem levels (Y_i). The system level design variables are functions of the subsystem variables, so that

$$\{X_i\} = f(\{Y_i\}) \tag{4}$$

Following a system analysis, each subsystem is optimized as a separate problem. The system variables are held constant during the subsystem optimizations, with an objective function corresponding to the level of constraint violation within that subsystem. A typical measure of constraint unsatisfaction is represented by an exterior penalty function [25]. More recently, however, a cumulative constraint representation (Kresselmeier-Steinhauser function) [26] has been used, which approximates the most critical constraint. The KS function can be written

$$C \equiv K.S. = \frac{1}{\rho} \ln \left[\sum_{j=1}^{m} \exp(\rho \cdot g_j) \right] \tag{5}$$

where m is the number of constraints being represented in the cumulative constraint (C) formulation and ρ is a user-prescribed constant. A smaller value of ρ allows more constraints to participate in the KS function while a larger value of ρ allows the most critical constraint to dominate the cumulative constraint representation.

The subsystem optimization can be written

$$\begin{aligned} &\textit{Minimize} && C^k(\{Y^k\}, \{X^k\}) \\ &\textit{Subject to} && \{X^k\} - f(\{Y^k\}) = 0 \\ & && \{Y^k\}_L \le \{Y^k\} \le \{Y^k\}_U \end{aligned} \tag{6}$$

where the first set of constraints ensures that the system level design variables remain constant and the second set are side constraints on the subsystem design variables.

Since the constrained minimum of a subsystem objective function, C^k, is a function of the parameters within vector $\{X^k\}$, an optimum sensitivity analysis can be performed to determine the sensitivity of C and Y to these parameters. Once these derivatives are known, linear extrapolations can be made to represent these quantities in the system level optimization, through the expressions

$$C^k = C^k + \sum_i \frac{dC^k}{dX_i^k}\Delta X_i^k \quad and \quad Y_j^k = Y_{j,o}^k + \sum_i \frac{dY_j^k}{dX_i^k}\Delta X_i^k \tag{7}$$

where k refers to a particular subsystem.

At the system level, therefore, the optimization problem becomes

$$\begin{array}{lll} Minimize & F(\{X\}) & \\ Subject\ to & g^S{}_j \leq 0 & j = 1,m(s) \\ & C^k \leq 0 & k = 1,nss \\ & X_L \leq X \leq X_U \ and \ \ Y^k{}_L \leq Y^k \leq Y^k{}_U & \end{array} \tag{8}$$

where the objective function is the system objective function (i.e. minimum weight), g^S are system level constraints, C^k are cumulative constraints associated with each subsystem k, and side constraints are placed on both system and subsystem design variables.

2. Performance of Method

The method was implemented in Ref. 27 using the framework structure seen in Figure 7. The entire framework represents the system level the with subsystems corresponding to each member (I-beam) of the structure. The design objective involved minimizing weight subject to static displacement and stress constraints. Subsystem design variables were comprised of cross-sectional dimensions of each I-beam (i.e. Y^{1T}=[b1,b2,h,t1,t2,t3]), while system level variables were cross-sectional area and moment of inertia (i.e. X^{1T}=[A1,I1]). Constraints associated with the subsystem level were stresses, while those at the system level corresponded to displacements.

Comparisons were made on solutions obtained by a traditional optimization approach and those found by application of the linear decomposition method. Both methods yielded the same optimum solution when starting from either feasible or infeasible initial designs. Minor discrepancies of less than 2% were attributed to use of the cumulative constraint representation, which is typically a conservative envelope function.

Application of the method to a three level structural problem was also investigated with equally satisfactory results. An extension of the portal frame was investigated in which the beam cross-section was comprised of stiffened

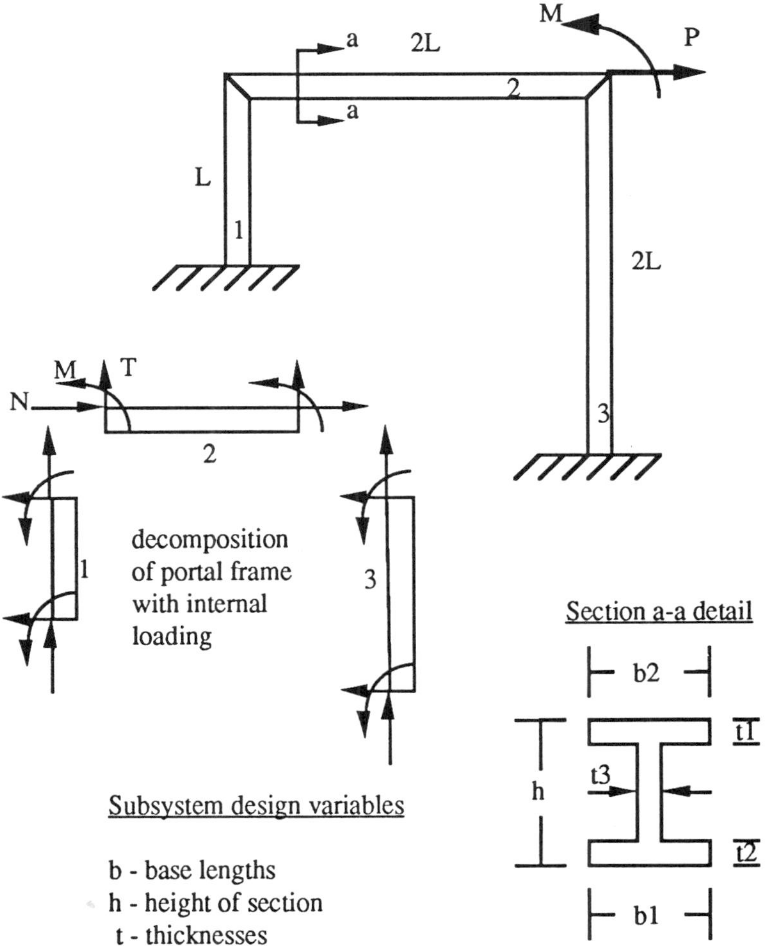

Figure 7. Portal framework as test case for linear decomposition method.

panels [28]. It was observed that one of the largest advantages to the approach is that subsystems can be processed concurrently, thus permitting exploitation of parallelism.

B. Optimization Methods for Non-hierarchic Decomposition

Two basic solution strategies have been proposed for highly coupled engineering design problems. The first involves an ad hoc decomposition in which the participating analyses of the various subsystems are performed in some prescribed order. In such an approach, the resulting design is dependent upon the order in which the analyses are implemented. The more desirable strategy is one which embraces parallel processing, in which each subsystem is examined simultaneously and with due consideration of all subsystem interactions [29].

Two methods are introduced that provide a vehicle for automated, simultaneous design of complex, coupled engineering systems. Both methods are strictly algorithmic in nature, making use of problem-dependent heuristics only to the extent of initializing parameters prior to implementation of the design process. A development of these two formal approaches to system decomposition for design and optimization is presented.

1. Global Sensitivity Equation Approach

The Global Sensitivity Equation approach is a methodology for obtaining the total sensitivity of the output response quantities with respect to the system design variables. The total derivative information thus obtained is used in constructing the approximate optimization problem described in Section III.B. The design variables and constraints from each subsystem are considered at the system level in an 'all-in-one' optimization.

a. Determination of Global Sensitivities

The underlying concepts in this formal approach for decomposition are simple and make use of the fact that the first derivative of a nonlinear function at a point is equal to the first derivative of the function linear approximation at that point. Consider the case of two disciplines A and B, the interactions between which are illustrated schematically in Figure 8.

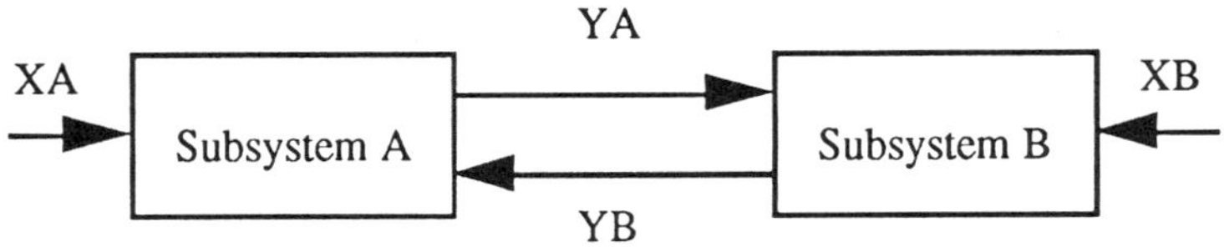

Figure 8. Interactions between two coupled subsystems.

The analysis equations for the two disciplines can be expressed in a symbolic form as follows.

$$A((X_A, Y_B), Y_A) = 0$$
$$B((X_B, Y_A), Y_B) = 0 \tag{9}$$

Here, X_A and X_B are the variables local to the system A and B, respectively. Y_A is the output vector for the system A and, in the most general form of coupling, this vector acts as a set of auxiliary input variables for system B. Similarly, Y_B is an auxiliary set of input variables for system A. Thus, the variables Y_A and Y_B provide the coupling links between the two systems. It is possible to rewrite the above expressions in an explicit form as follows.

$$Y_A = (X_{A,} Y_B)$$
$$Y_B = (X_B, Y_A) \tag{10}$$

A first order Taylor series representation and a simple application of the chain rule can be made to obtain the Global Sensitivity Equations. These are represented in matrix notation as,

$$\begin{bmatrix} I & -\dfrac{\partial Y_A}{\partial Y_B} \\ -\dfrac{\partial Y_B}{\partial Y_A} & I \end{bmatrix} \begin{Bmatrix} \dfrac{dY_A}{dX_A} \\ \dfrac{dY_B}{dX_A} \end{Bmatrix} = \begin{Bmatrix} \dfrac{\partial Y_A}{\partial X_A} \\ 0 \end{Bmatrix}$$

$$\begin{bmatrix} I & -\dfrac{\partial Y_A}{\partial Y_B} \\ -\dfrac{\partial Y_B}{\partial Y_A} & I \end{bmatrix} \begin{Bmatrix} \dfrac{dY_A}{dX_B} \\ \dfrac{dY_B}{dX_B} \end{Bmatrix} = \begin{Bmatrix} 0 \\ \dfrac{\partial Y_B}{\partial X_B} \end{Bmatrix} \tag{11}$$

This approach is described in more detail in Ref. 20-22, as well as the chapter entitled "Techniques for Sensitivity Analysis in Aircraft Design". The partial sensitivities within the Global Sensitivity Matrix can be computed locally within each subsystem, eliminating the need to perform computationally expensive interdisciplinary iterations. This also diminishes the possibility of errors associated with round-off and truncation in the iterative process, from having adverse effects on the quality of the sensitivity results. Another major advantage is the computational gains that can be obtained from exploiting the modularity of the method in a distributed or parallel processing format.

Since the Global Sensitivity Matrix (GSM) is a function of local sensitivities of outputs to inputs that can be obtained within each discipline independently, the approach essentially permits a decoupling of large systems into smaller subsystems. The sensitivities obtained from the above analysis can be used to develop linear approximations to the output response of each subsystem, which can be subsequently employed in the gradient-based piecewise linear optimization process.

b. Method Performance

Due to the complexities of large engineering problems, the dimensionality of the local sensitivity matrices may be prohibitively large for repetitive decomposition in an optimization sequence, and may contribute to substantial reductions in the numerical efficiency and accuracy.

It was demonstrated in Ref. 30 that the condition number associated with the GSM increases with increased dimensionality and is unacceptably large if a normalization scheme is not used. Details of the normalization process can be found in "Tehniques for Sensitivity Analysis in Aircraft Design".

In Ref. 30, the GSE was applied in a multidisciplinary aircraft design problem, wherein disciplines of aerodynamics, structures, flight dynamics, and performance contributed to the system design objectives. Results demonstrated the feasibility and versatility of the approach. Ref. 31 demonstrates the method's applications for industrial use. The chapter "Multilevel Optimization Techniques for Aircraft" further demonstrates the applications of the GSE approach.

2. Concurrent Subspace Optimization Method

The Concurrent Subspace Optimization (CSSO) method, originally proposed by Sobieski in Ref. 23, permits the decoupling of a large engineering system into smaller subsystem modules to achieve concurrent optimizations in each of these subspaces. This method essentially takes the concept behind the Global Sensitivity Equation Method one step farther, performing not only the sensitivity analyses within each subsystem, but the optimizations as well. Unlike the conventional method of subspace optimizations, however, the CSSO method eliminates the need for a full analysis in each subspace, thereby providing potential substantial computational savings. The method is particularly well suited to applications in a design organization setting in which tasks are distributed among groups of specialists representing physical subsystems and disciplines.

Recent technological and computer developments in the areas of cumulative constraint representations [26], sensitivity analysis for non-hierarchic systems [13,32], optimal sensitivity analysis [33], and distributed computing capabilities [34], provide the necessary components to create a decomposition methodology that allows for truly integrated synthesis and has the advantage of subsystem modularity. Such an advantage allows for implementation of specialized methods for analysis, computational efficiency, and the ability to incorporate human intervention and decision making in the form of an expert systems capability.

a. The CSSO Algorithm

As seen in Figure 9, a system analysis is first performed followed by a sensitivity analysis. Due to the modularity of the subsystems, the Global Sensitivity Equation Method previously described can be used. Constraints in each subspace are represented by a single cumulative constraint measure by means of a Kresselmeier-Steinhauser (KS) function (Eq. 5 in Section IV.A.1.).

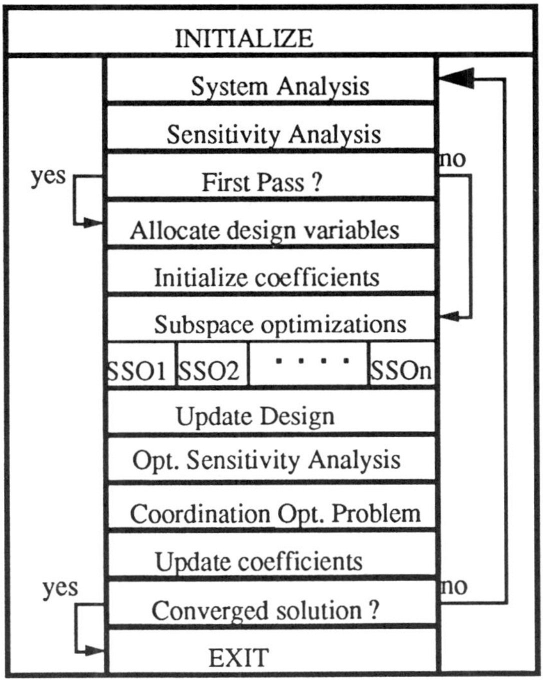

Figure 9. Flowchart of Concurrent Subspace Optimization Method.

The derivative of the cumulative constraint function with respect to the design variable X_i is determined analytically and combined in a chain differentiation with dg_j/dX_i. The derivative of the cumulative constraint, C, with respect to the design variables is written,

$$\frac{dC}{dX_i} = \left[\sum_{j=1}^{m} \exp(\rho \cdot g_j)\right]^{-1} \left[\sum_{j=1}^{m} \left\{\frac{dg_j}{dX_i} \exp(\rho \cdot g_j)\right\}\right] \qquad (12)$$

The design variables are then allocated to the subspaces on which they have the greatest effect. This allocation is based on objective function and constraint sensitivity information. A rank-ordering procedure is used to determine the subspaces for which design variables have the greatest impact.

Following design variable allocation, temporarily decoupled optimizations are performed in each subspace concurrently. The goal of these optimizations is to reduce the violation of the cumulative constraint with the least increase of the system objective function or greatest decrease if the cumulative constraint is already satisfied. The optimization problem can be stated as follows:

$$\begin{aligned} &\textit{Minimize} && F(X^k) && \\ &\textit{Subject to} && C^p \leq C^{po}\left[s^p\left(1-r_k^p\right)+\left(1-s^p\right)t_k^p\right] && p=1,NSS \\ & && X_k^L \leq X_k \leq X_k^U && \end{aligned} \tag{13}$$

where s^p, $r^p{}_k$, and $t^p{}_k$ are coefficients representing cross influences of one subspace on another. Since the subspace optimizations are decoupled, with only subsets of the system design variables in each subspace, it is essential that some form of coordination exist between subspaces. The coordination coefficients perform this duty. The distribution of these coefficients for a three subsystem architecture is demonstrated in Figures 10 and 11.

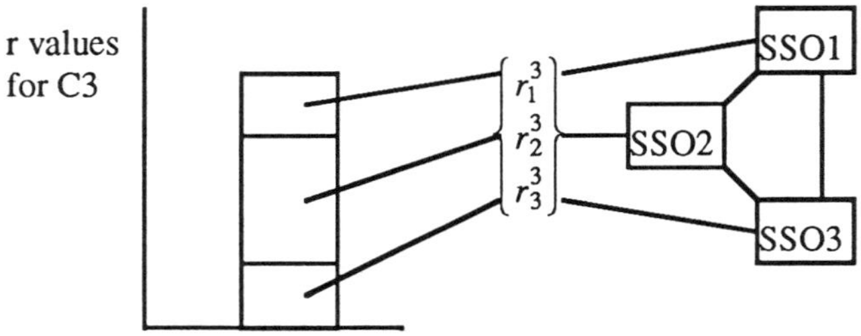

Figure 10. Distribution of r coefficient values associated with C3.

The $r^p{}_k$ coefficient represents the "responsibility" assigned to the kth SSO for reducing the violation of the cumulative constraint in the pth SSO. Even though design variables have been allocated to the subspace on which they have the greatest impact, it is easy to imagine how these variables would still affect constraints of other subspaces due to the couplings which exist in the non-hierarchic system. It then becomes necessary to account for this effect during the subspace optimizations. The $r^p{}_k$ coefficients essentially divide the responsibility for satisfying constraints amongst the subspaces. The coefficients represent the impact of the design variables within each subspace on the cumulative constraint. Figure 10 demonstrates the responsibilities assigned to Subspaces 1, 2, and 3 for satisfying the cumulative constraint associated with Subspace 3. The responsibility coefficients for a particular 'p' constraint must equal 1.0 over all subspaces 'k'. This essentially means that 100% of the responsibility for satisfying a subspace's cumulative constraint must be accounted for.

The $t^p{}_k$ coefficient represents the "trade-off" associated with each subspace that allows for the violation of a constraint in the pth SSO to obtain a reduction of the objective function, provided that the constraint will be over-satisfied in the kth SSO. Such a trade-off can occur only when the present and previous optimization cycles have produced satisfied constraints. Figure 11 demonstrates the trade-off in constraint satisfaction that might occur amongst three subspaces. It is essential that any violation that is permitted be compensated in other subspaces so that the sum of all trade-offs across the subspaces is zero. The "switch" parameter, s^p, is responsible for enabling or disabling the $r^p{}_k$ or $t^p{}_k$ coefficients depending on whether the constraints are initially violated.

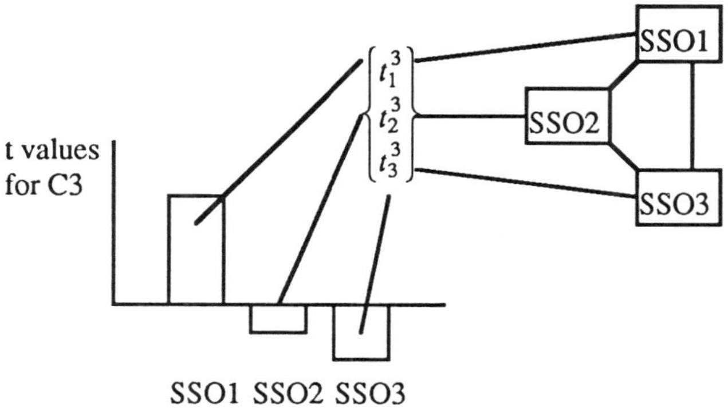

Figure 11. Distribution of t coefficient values associated with C3.

The constrained minimum of the system objective function, F, following the subspace optimizations is now a function of the r and t coefficients. An optimum sensitivity analysis (OSA) can therefore be performed to determine the sensitivity of F to these coefficients. For the case in which no equality constraints exist, it is possible to obtain a relation for Lagrange multipliers as a function of constraint and objective function gradients from as follows,

$$\lambda_k = -\left[\left[\frac{dC^k}{dX^k}\right]^T \left[\frac{dC^k}{dX^k}\right]\right]^{-1} \left[\frac{dC^k}{dX^k}\right]^T \left[\frac{dF}{dX^k}\right] \tag{14}$$

where C^k is the cumulative constraint measure for the kth subspace. Once the Lagrange multipliers are obtained, the optimum sensitivity derivative of F simplifies to,

$$\frac{dF}{dz_i} = \lambda_k \frac{dC^k}{dz_i} \tag{15}$$

where z is a variable representing either r or t. A complete discussion of optimum sensitivity analysis can be found in Ref. 33 as well as the chapter in this series entitled "Techniques for Sensitivity Analysis in Aircraft Design".

The derivative information obtained in the OSA is now used in the coordination optimization problem (COP) in which the system objective function is minimized with respect to the r^p_k and t^p_k coefficients. Completion of the COP yields new coefficients for use in the next SSO. The coordination optimization problem is defined in the following manner.

$$
\begin{aligned}
&\textit{Minimize} && F(r^p_k, t^p_k) \\
&\textit{Subject to} && \sum_k r^p_k = 1 \\
& && \sum_k t^p_k = 0 \\
& && 0 \le r^p_k \le 1 \\
& r^p_{k\,L} \le r^p_k \le r^p_{k\,U} && \text{and} \quad t^p_{k\,L} \le t^p_k \le t^p_{k\,U}
\end{aligned}
\tag{16}
$$

The coefficients are updated and the entire process is repeated until prescribed convergence requirements are met.

Perhaps the most important feature of the CSSO is that the modularity of the method allows for the efficient decoupling of the system to permit concurrent sensitivity analyses and subspace optimizations that can correspond to specialty groups within an organization. Further, the approach is particularly amenable to human judgment and intervention or the application of an artificial intelligence based system. These advantages and disadvantages are discussed in more details in following section.

b. Method Performance

The method was applied to a simple test case for verification purposes. The objective of the synthesis process was to find the minimum weight design of a truss structure subject to constraints derived from requirements of structural integrity and topology in the form of stress and static displacement limitations. Design variables contributing to the fulfillment of the optimization objectives were sizing variables and a structural geometry variable.

The ten-bar truss structure in Figure 12 was used to demonstrate the feasibility of the CSSO method. The structure was subjected to static loadings as shown in the figure.

The design variables for Subsystem 1 were the cross-sectional areas of the ten truss members. The output vector for the analysis associated with the subsystem contained the sizing variables, the objective function value, and a cumulative constraint measure representing the static stress constraints. The design variable for the topology subsystem was a geometry variable, D, which

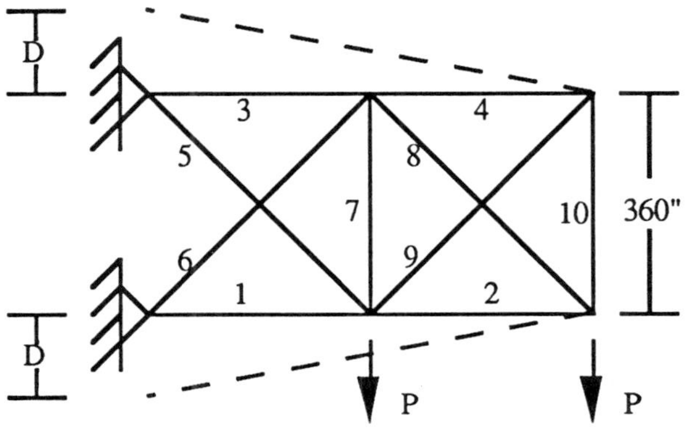

Figure 12. Application model for CSSO verification.

controlled the width of the truss structure at the wall. The output vector from the analysis contained the geometry variable, the objective function value, and a cumulative constraint measure representing the static lateral displacement constraints.

To simplify the initial method application, the design variables were permanently assigned to the subspaces at the outset of the design procedure. Optimization results for this structural problem were obtained by three different synthesis procedures for comparison purposes, one of which was the CSSO Method. An all-in-one optimization was performed for the other two cases in which the sensitivity information was obtained by the finite difference method and by the GSE method. The results from the application demonstrated the feasibility of the CSSO for a small problem. It also demonstrated the uses of a distributed processing environment as well as the incorporation of heuristics by means of an embedded expert system capability. A more detailed discussion of the CSSO and its heuristic counterpart can be found in Ref. 35.

V. CONCLUDING REMARKS

The recent thrust to improve quality of products as well as productivity in the United States has led to the realization that traditional design practices are inefficient and outmoded. It is in the design phase that the most potential exists to take advantage of the synergism of the inherently coupled disciplines to increase the overall quality of the product as well as to reduce the time and effort required for development and design of the product. The multidisciplinary interactions existing in large scale engineering design problems provide a unique set of difficulties associated with unwieldy numbers of design variables and constraints. Such obstacles require design techniques that take advantage of the extensive couplings of the disciplines in the analyses and optimizations, producing an efficient methodology to perform multidisciplinary synthesis.

The development of a design framework appropriate for large engineering systems (in which a distinct system hierarchy is difficult to identify) is essential. The concept of decoupling large complex problems into smaller, more tractable subsystems has been presented. Decomposition methods permit the concurrent consideration of the design criteria within all disciplines, thus providing an environment particularly amenable to parallel processing.

Although the methods investigated proved applicable for the optimal design of large engineering systems, room for improvement exists. The concept of multidisciplinary design optimization has gained substantial recognition in the engineering community in the last ten years. The importance of developing methodologies appropriate for non-hierarchic environments has been emphasized as a key function of future design techniques. Such techniques require a necessary organizational as well as technical modification to achieve their full potential. The design process itself has undergone major change, as the concept of incorporating all the design criteria in a simultaneous treatment has emerged as not only desirable, but necessary. Recent initiatives have focused on consideration of manufacturability and supportability as well as performance objectives in a "Concurrent Engineering" (CE) approach. An investigation of the applicability of these decomposition approaches in a CE problem is an obvious next step, as these methods allow for the use of specialized methods for analysis and optimization. Such innovative concepts as artificial neural networks, fuzzy logic, simulated annealing, and genetic algorithms are all representative of the types of techniques that could be incorporated into a decomposition approach to gain maximum flexibility within each subspace.

VI. REFERENCES

1. Dhir, S. K. and Hurwitz, M. M., "Role of Optimization in Interdisciplinary Analyses of Naval Structures", *Recent Experiences in Multidisciplinary Analysis and Optimization*, NASA CP 2327 (1984).

2. Hughes, O. F., "Structural Optimization of Large Ocean-Going Structures", *Recent Experiences in Multidisciplinary Analysis and Optimization*, NASA CP 2327 (1984).

3. Ferebee, M. J., "IDEAS, A Multidisciplinary Computer-Aided Conceptual Design System for Spacecraft", *Recent Experiences in Multidisciplinary Analysis and Optimization*, NASA CP 2327 (1984).

4. Miura, H., "Overview: Applications of Numerical Optimization Methods to Helicopter Design Problems", *Recent Experiences in Multidisciplinary Analysis and Optimization*, NASA CP 2327 (1984).

5. Prasad, B. and Magee, C. L., "Application of Optimization Techniques to Vehicle Design - A Review", *Recent Experiences in Multidisciplinary Analysis and Optimization*, NASA CP 2327 (1984).

6. Sensburg, O., Fulhas, K., and Schimdinger, G., "Interdisciplinary Design of Aircraft Structures for Minimum Weight", AIAA Paper No. 88-2302 (1988).

7. Tolson, R. H. and Sobieszczanski-Sobieski, J., "Multidisciplinary Analysis and Synthesis: Needs and Opportunities", AIAA Paper No. 85-0584 (1985).

8. Sobieszczanski-Sobieski, J., "Multidisciplinary Optimization for Engineering Systems: Achievements and Potentials", NASA TM 101566 (1989).

9. Rogers, J. L., "A Knowledge-based Tool for Decomposition of a Complex Design Problem", NASA TP 2903 (1989).

10. Wilde, D. J., *Globally Optimal Design*, John Wiley and Sons, New York, N.Y. (1978).

11. Haftka, R. T., Gurdal, Z. and Kamat, M. P., *Elements of Structural Optimization*, Kluwer Academic Publishers, Boston, Mass. (1990).

12. Griffith, R. E., Stewart, R. A., "A Nonlinear Programming Technique for the Optimization of Continuous Processing Systems", Management Science, Vol. 7 (1961).

13. Schmit, L. A., Miura, H., "Approximation Concepts for Efficient Structural Synthesis", NASA CR 2552 (1976).

14. Storaasli, O. O., Sobieszczanski-Sobieski, J., On the Accuracy of the Taylor Approximation for Structure Resizing, *AIAA Journal*, Vol. 12 (1974).

15. Starnes, J.H. Jr., Haftka, R.T., Preliminary Design of Composite Wings for Buckling, Stress, and Displacement Constraints, *Journal of Aircraft*, Vol. 16 (1979).

16. Fadel, G. M., Riley, M. F., Barthelemy, J.-F., "Two point Exponential Approximation Method for Structural Optimization", *Structural Optimization*, Vol. 2 (1990).

17. Morris, A. J., ed., *Foundations of Structural Optimization: A Unified Approach*, John Wiley and Sons, New York, N.Y. (1982).

18. Sobieszczanski-Sobieski, J., "A Linear Decomposition Method for Large Optimization Problems - Blueprint for Development", NASA TM 8324 (1982).

19. Kroo, I. and Takai, M., "A Quasi-Procedural, Knowledge-Based System for Aircraft Design", *AIAA/AHS/ASEE Aircraft Design, Systems, and Operations Meeting*, Atlanta Georgia, AIAA Paper No. 88-4428 (1988).

20. Sobieszczanski-Sobieski, J., "On the Sensitivity of Complex, Internally Coupled Systems", *AIAA Journal,* Vol. 28, No. 1 (1990).

21. Bloebaum, C. L., "Implementation of Global Sensitivity Analysis in Structural/Control Optimization", Master's Thesis, University of Florida (1987).

22. Sobieszczanski-Sobieski, J., Bloebaum, C. L., Hajela, P., "Sensitivity of Control-Augmented Structure Obtained by a System Decomposition Method", *AIAA Journal*, Volume 29, Number 2, pp. 264-270 (1991).

23. Sobieszczanski-Sobieski, J., "Optimization by Decomposition: A Step from Hierarchic to Non-Hierarchic Systems", *Recent Advances in Multidisciplinary Analysis and Optimization,* NASA CP-3031 (1988).

24. Bellman, R., *Adaptive Control Processes: A Guided Tour*, Princeton University Press (1961).

25. Vanderplaats, G. N., *Numerical Optimization Techniques for Engineering Design: With Applications*, McGraw-Hill Book Company, New York, N.Y. (1984).

26. Kresselmeier, G. and Steinhauser, R., "Application of Vector Performance Optimization to a Robust Control Loop Design for a Fighter Aircraft", *International Journal of Control*, Vol. 37, No. 2, pp. 251-284 (1983).

27. Sobieszczanski-Sobieski, J., James, B. B., and Dovi, A. R., "Structural Optimization by Multilevel Decomposition", *AIAA Journal,* Vol. 23, No. 11, pp. 1775-1782 (1985).

28. Sobieszczanski-Sobieski, J., James, B. B., and Riley, M. F., "Structural Optimization by Generalized Multilevel Optimization", AIAA Paper No. 85-0698 (1985).

29. Weisshaar, T. A., Newsom, J. R., Zeiler, T. A., and Gilbert, M. G., "Integrated Structure/Control Design - Present Methodology and Future Opportunities", ICAS Paper No. 86-4.8.1 (1986).

30. Hajela, P. and Bloebaum, C. L., "Application of Global Sensitivity Equations in Multidisciplinary Aircraft Synthesis", *Journal of Aircraft*, Vol. 27, No. 12, pp. 1002-1010 (1990).

31. Abdi, F. F., Ide, H., Shankar, V. J., and Sobieszczanski-Sobieski, J., "Optimization for Nonlinear Aeroelastic Tailoring Criteria", *International Council for Aeronautical Sciences, 16th Congress*, Jerusalem, pp. 1083-1091 (1988).

32. Haftka, R., "Sensitivity Analysis", *Proceedings of the NASA-ASI Session on Modern Structural Optimization*, Liege, Belgium (1980).

33. Sobieszczanski-Sobieski, J., Barthelemy, J.-F., and Riley, K. M., "Sensitivity of Optimum Solutions of Problem Parameters", *AIAA Journal*, Vol. 20, No. 9 (1982).

34. Rogers, J. L., Dovi, A. R., and Riley, K. M., "Distributing Structural Optimization Software Between a Mainframe and a Minicomputer", *Proceedings of the Engineering Software Second International Conference and Exhibition*, London, England (1981).

35. Bloebaum, C. L., "Formal and Heuristic System Decomposition Methods in Multidisciplinary Synthesis", Ph.D. Dissertation, University of Florida, Gainesville, Florida and NASA CP 4413 (1991).

Multilevel Optimization Techniques for Aircraft

Trent R. Logan
Frank F. Abdi

North American Aircraft
Rockwell International Corporation
Los Angeles, California 90009

Jaroslaw Sobieszczanski-Sobieski

Interdisciplinary Research Office
NASA Langley Research Center
Hampton, Virginia 23665

A multilevel decomposition methodology for which design variables have relatively equal impact at each level is discussed as an approach to large-scale system optimization. Multidisciplinary optimization techniques utilizing sensitivity derivatives and a global sensitivity matrix to combine the levels are used to develop a total optimum design. A hypothetical hypersonic vehicle is used to illustrate the process when considering trajectory issues as well. An advanced fighter wing example is used to illustrate the potential of improved design optimization in a multidisciplinary environment as compared to conventional sequential design development.

I. INTRODUCTION

Contemporary flight vehicle design requires consideration of an extensive parameter array for successful configuration development. For example, weight sensitivity of hydrogen fueled single-stage-to-orbit vehicles, as measured by the growth factor, is in the range of ten-to-one and demands highly sophisticated and optimized airframe structure. Furthermore, hypersonic vehicle material selection, which strongly influences structural concepts and manufacturing techniques, is driven by service temperatures, which are determined in turn by flight path parameters, particularly speed and altitude. Optimum structural design for this case requires concurrent consideration of the structure, its materials, and the mission itself.

Further, emerging military aircraft design concepts require integration of an increasing number of conflicting design features. For example, high angle-of-attack and high-acceleration maneuver wing flexibility and planform solutions conflict with efficient high-speed cruise solutions. These maneuvers place higher than normal demands on control authority and low airframe weight.

In this environment, multidisciplinary optimization methods provide a powerful design technique and are expected to provide vehicle designs having the best balance of structure, vehicle concept, mission capability, and cost. Payoff can be decreased design cycle time or an improved design while providing a natural base for concurrent engineering approaches for the total design process. These compromises are not new issues, but new tools are available which make the trades easier and more accurate, and which can yield a design that is truly the best for all issues considered. The tools are formal optimization capabilities which handle complex problems based on a variety of mathematical approaches, large-scale computing capabilities, and interdisciplinary interaction based on simulation and system sensitivities. Implementation of these tools for complex vehicle system design is made possible by advanced computing capabilities which allow economical simulation of complex aerodynamic, structural and material performance, and mission analysis. These computational tools form an accurate model simulating the vehicle being designed, and avoid compromises previously required when using approximating mathematical equations to describe desired behavior.

Vehicle optimization based on this sensitivity analysis allows mapping of interactions which occur among vehicle components and engineering disciplines in the real vehicle onto the data flow among the computational modules in the system mathematical model. This mapping, in turn, yields equations quantifying design variable effects on the vehicle. These data support both engineering judgement and formal optimization to control and amplify interdisciplinary design trade-offs.

This chapter addresses a multidisciplinary design approach based on sensitivity derivatives and global sensitivity equations which account for interdisciplinary effects. A primary motivation for exploiting these issues is that desired very high weight savings as noted previously must go beyond development of advanced structural and material applications only, and exploit active controls and new flight vehicle configurations. Furthermore, application of multidisciplinary design optimization techniques can provide new designs that would not necessarily be developed using conventional approaches.

As an example, a conceptual design study for a fighter aircraft, Vanderplaats [1] developed a result for a wing planform that was different when optimized for the entire mission than would have been expected based on past experience. The optimization process and tools that are now in place allow a larger number of variables to be considered than can be accounted for in the traditional trade-study environment, providing more degrees of freedom and more opportunities to save weight or otherwise improve a design. As another example, Abdi et al [2] and Ide et al [3] performed a preliminary study of a hypersonic vehicle forebody combining aerodynamics and structural effects to optimize inlet mass flow and developed a significantly different shape than the baseline .

During the past several years, mathematical optimization tools and structured approaches to multidisciplinary design have been developed to address complex aircraft design problems. Sobieszczanski-Sobieski [4] presented a systematic methodology for decomposing large complex systems into a series of smaller subsystems, using sensitivity derivatives between these subsystems as the basis of generating a global sensitivity equation for total system optimization. This approach has been subsequently expanded and demonstrated by additional applications, e.g., Refs. [5-7]. Other authors have considered the interaction of aerodynamics, structures, and controls on aircraft design. For example, Grossman et al [8] focused on computational aspects of the aerodynamics-structures problem for a transport wing, and showed that a modular (black box) method for computing sensitivity derivatives was significantly superior to finite-difference methods. Simos and Jenkinson [9] presented conceptual design results for a shorthaul aircraft, obtained by optimizing operational and mission-performance factors. Wrenn and Augustine [10] developed a multilevel decomposition of a transport wing, which was optimized for mission performance, structural weight, and detailed structure efficiency. In that study, design data and related sensitivity derivatives were passed between levels for sequential optimization at the mission level.

The approach taken here is a multilevel breakdown of the total vehicle system, with levels linked by key parameter sensitivity derivatives, using a global sensitivity approach for optimization at the highest (flight-mission) level of interest. An issue is which variables are to be considered in the design process. Variables considered at each level of detail during the design process should have similar relative effect on results being sought at each specific level. Note that, for some problems, an "all in one" approach can be used with a high degree of success, particularly at the conceptual level, or for a system in which only a few parameters are available for optimization.

The purposes of this chapter are to present a strategy for multilevel, multidisciplinary design of an aircraft that allows focusing of effort on various levels of design complexity as the design progresses, and to demonstrate the advantage of total system optimization as compared to conventional sequential design. The following sections discuss a general strategy, detailing the approach to level definition and level interconnection; a background on optimization formulation; and, finally, several examples to illustrate application of the concepts.

II. OVERALL STRATEGY

Sobieszczanski-Sobieski et al [6] present a summary of approaches to system decomposition, which includes formal and intuitive methods, and which has more recently been developed in further depth by Bloebaum [11]. Formal methods could be based on a mathematical structure from which the necessary relationships may be derived, whereas intuitive methods are based on detailed physical understanding of the problem. Although an adequate formal structure could be developed using a technique such as a detailed work-breakdown structure, from implementation and flexibility points of view for the conceptual design process, a heuristic approach has more flexibility and will be used for developments in this chapter. Further discussion of decomposition methods is given in this book by Bloebaum, et al in the chapter "Decomposition Methods for Multidisciplinary Synthesis".

A. SYSTEM STRUCTURE

An entire airplane conveniently can be broken into hierarchic subsets with reasonably well-defined interfaces. Thus, for example the hardware interface between a wing box and the fuselage or a control surface are natural divisions. At the conceptual/preliminary design level, these physical breakdowns still make sense, but rather than being strictly hierarchical, must instead be represented by a complex network with many-to-many relations among the elements since they are not well defined at this stage. The cross linking would in some cases be quite strong, dominating the design at that level, whereas in other cases it would have little or no effect. For complex problems, consideration of all variables at once would lead to a massive computational problem for which a clear optimum would be difficult to determine because of convergence issues and masking of important detail parameters by more dominant ones.

The best approach to keep the number of variables within practical limits will be to reduce as many as possible of these new relationships to simple ones, preferably to eliminate many. This can be achieved by considering various levels of design-parameter influence, for which parameters have similar influence at each level, and by not considering weak linking. System leveling or decomposition will use heuristic reasoning, experience, and sensitivity studies to display only those parameters that have significant effect on measures of merit at each level.

The highest level for a vehicle would be the system level, where a measure of total system effectiveness could be, for example, lifecycle cost, or a selected mission parameter. At this level, takeoff gross weight is significant, as is airframe cost, but parameters such as composite panel fiber orientation or panel stiffening concept have minor measurable effects. These latter factors are, however, of key importance to structural component weight and cost, and would therefore be evaluated at a lower level.

As an example, Fig.1 illustrates the multilevel breakdown for a wing. Motivation for the breakdown is ease and rapidity of design by focusing on the primary parameters that influence each level. At the first level, the total wing would be considered, with primary aerodynamic performance and control parameters as shown. At the second level, major wing subsystems would be evaluated, which for this example will be the wing box and the control surfaces, the latter considered as assemblies. The third level would evaluate detailed

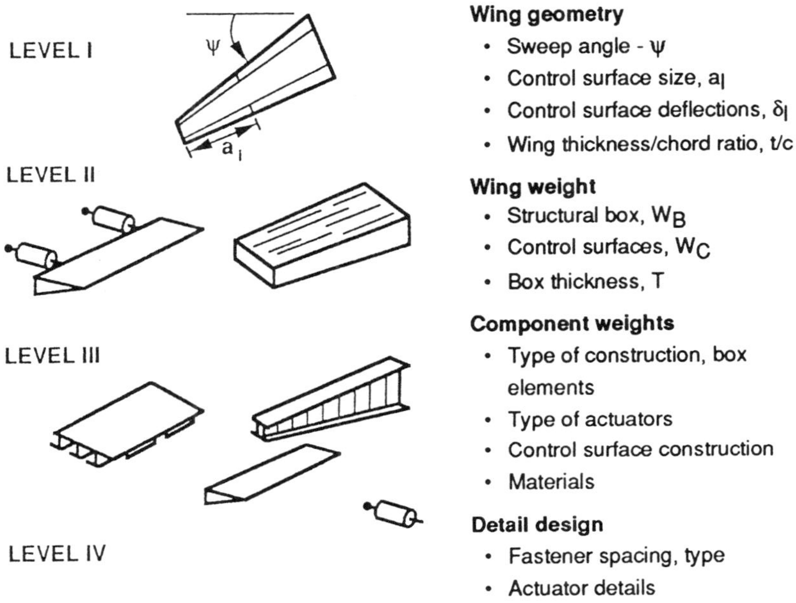

Figure 1. Wing Multi-Level Definition

components such as wing skin panels and actuators. This is the level at which most conventional optimization is performed and there is a wide variety of computerized and manual tools available for analysis. For structural optimization, as an example, an in-depth review is given by Venkayya [12], and a recent example is given by Sensburg [13]. The fourth level would address very detailed optimization consisting of issues such as fastener type and spacing, actuator details, and so forth, and are largely calculations which accommodate practical and rule-of-thumb design issues. At the conceptual stage of the design cycle, primary concentration would be at the first and second levels. Support data from lower levels would typically be supplied by equations and data developed from past designs. Preliminary design would consider the first two levels, possibly the third level to some extent. Detailed design would focus primarily on the third and fourth levels, with only minor redevelopment of the second level.

B. MULTIDISCIPLINARY OPTIMIZATION

To describe the approach, it is convenient to consider a system limited to only three contributing analysis, or "black boxes" which is a number small enough to simplify the discussion and yet large enough to develop a pattern that readily generalizes to a system composed of n black boxes. To give the three black boxes a physical meaning, they are named A, S, and M and represent the aerodynamics, structures and mission, respectively. It is understood that each black box contains an appropriate mathematical model, e.g., a finite element program with the pertinent pre- and post-processors in S. However, the very purpose of the black box concept is to hide the internal detail and to consider the black box as an input-to-output transformer that is, simply, a set of simultaneous equations whose solution vector is the output. The equations may be written in a compact form as a vector functions set to zero and operating on a set of arguments. The list is ordered so that the last on the list is the output vector while the input vectors are grouped in the inner parentheses. The input group begins with the vector of design variables (independent variables of the problem) X followed by the Y-vectors output from the other black boxes in the system:

$$\begin{aligned} &A((X, Y\underline{s}, Y\underline{m}), Y\underline{a}) = 0 \\ &S((X, Y\underline{m}, Y\underline{a}), Y\underline{s}) = 0 \\ &M((X, Y\underline{a}, Y\underline{s}), Y\underline{m}) = 0 \end{aligned} \tag{1}$$

In the above, $Y\underline{a}$ contains the aerodynamic pressure coefficients, $Y\underline{s}$ includes the elastic deformations, and $Y\underline{m}$ comprises the mission performance definition. The design variables in X may represent major configuration parameters such as wing aspect ratio and sweep angle, cross-sectional dimensions, and specific impulse.

Of course, not every element of each of the Y and X input vectors is actually used in each of the black boxes. The usage is selective, for instance, S uses the aerodynamic pressure data in $Y\underline{a}$ but may ignore the local Mach number values. Similarly, S uses the wing geometry and cross-sectional dimensions data in X, while only the former is used in A.

The system analysis produces an output vector $Y' = [\underline{Ya}: \underline{Ys}: \underline{Ym}]$ as an implicit function of X. Owing to the output-to-input transmissions of the Y vectors, the system equations have to be treated as simultaneous even though they are partitioned into the A, S, and M sets. If any of the constituent black boxes contains nonlinear analysis, the system becomes iterative.

When the system analysis is completed, the design process requires answers to the "what if" questions regarding the influence of perturbations in X. These questions may be posed directly by engineers in an interactive mode or they may be generated by a formal optimization procedure. Derivatives of Y with respect to X quantify answers to these questions and may be approximated by a finite difference technique that requires repetition of the system analysis for an increment of one $X\underline{k}$ at a time. This finite difference technique, or a parametric study approach similar to it, have well-known disadvantages of the high cost and often-inadequate accuracy, however, an alternative free of these disadvantages was proposed by Sobieszczanski-Sobieski [14] in the form of of an algorithm for a system sensitivity analysis that yields the derivatives of Y with respect to X directly.

That algorithm requires the system analysis to be completed as a prerequisite, that is a solution Y for a given X must be available. Then, for the given X and the corresponding, known Y, the algorithm begins with generation of the derivatives of the output with respect to the input for each black box. These derivatives are by definition the partial derivatives so that they can be computed for each black box taken in isolation, and treating all the X and Y inputs as independent variables. Computation of the partial derivatives amounts to a disciplinary sensitivity analysis and may be executed using any technique available for a particular black box at hand. Preferably, a quasi-analytical method should be used (a survey of such methods for various disciplines is provided by Adelman and Haftka [15]) but a finite difference method may have to be used wherever a quasi-analytical method has not yet been implemented. For the three-part system, the partial derivatives are collected in the Jacobian matrices J_{as}, J_{am}, J_{sa}, J_{sm}, J_{ma}, J_{ms}, and in the vectors R - one R for each design variable. Each matrix J corresponds to the output-to-input transmission that links the black box identified by the second subscript to the one identified by the first subscript. For example, the i-th column of J_{as} contains the partial derivatives of the elements of Y_a with respect to the i-th element of Y_s received in A. A k-th R vector contains the partial derivatives of Y with respect to the k-th design variable, X_k.

As shown in Ref. [14] on the basis of the implicit function theorem, the above partial derivatives may be used to assemble a set of simultaneous, linear, algebraic equations that yield a vector of derivatives of Y with respect to X_k:

$$\begin{vmatrix} J & -Jas & -Jam \\ -Jsa & I & -Jsm \\ -Jma & -Jms & I \end{vmatrix} \{dY / dX_k\} = \{R\}_k \tag{2}$$

The derivatives in $\{dY/dX_k\}$ have the meaning of total derivatives and reflect the interactions among the black boxes, unlike the partial derivatives that appear as coefficients in the above equations and represent only the sensitivity of the isolated black boxes. For example, if X_k is the wing sweep angle, then the partition $\{dY_a/dX_k\}$ in $\{dY/dX_k\}$ will indicate the rate of change of the body aerodynamic pressure accounting for vehicle angle of attack influence on the aerodynamics directly and, indirectly, through elastic deformation changes..

The derivatives $\{dY/dX_k\}$ may be used to guide a formal optimization algorithm and for extrapolation purposes to evaluate the effects of small but finite increments of the design variables. In strongly nonlinear problems, it may be cost-effective to enhance the extrapolation by the second order terms and the above system sensitivity analysis may be formally extended to calculate the second derivatives as shown by Sobieszczanski-Sobieski [16]. The improvement of the extrapolation accuracy gained by using at least the diagonal terms in the matrix of the second derivatives was demonstrated by Abdi, et al [7].

III. HYPERSONIC VEHICLE OPTIMIZATION EXAMPLE

A generic hypersonic vehicle, a one-half model of which is shown in Fig. 2, will be used to illustrate how the method can be applied and to demonstrate computation formulation and approaches to numerical solution. This vehicle configuration has a long forebody and a small wing, sized by takeoff and landing requirements, and is typical of a class of manned horizontal

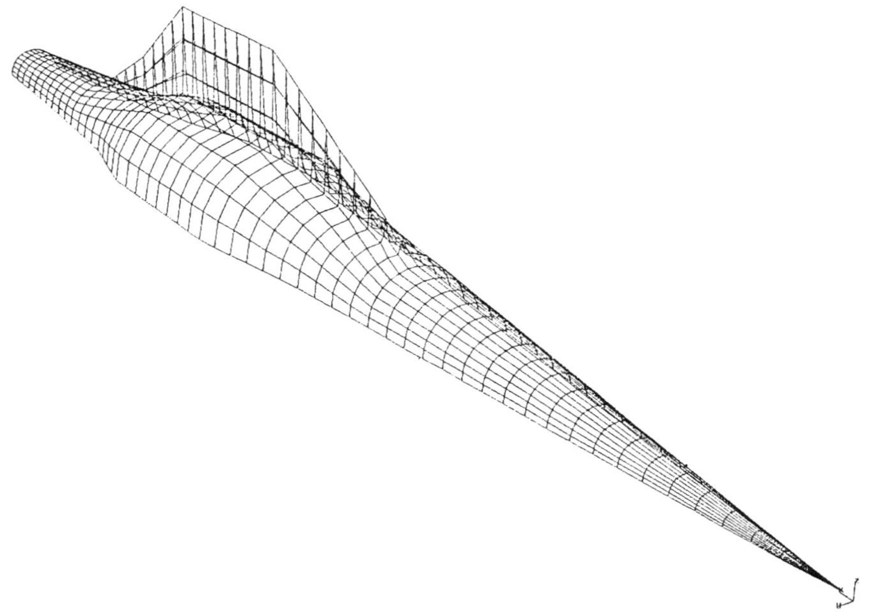

Figure 2. Baseline Vehicle

to earth, with a conventional landing. For the orbital case, cross-rangerequirements and orbit entry parameters have critical effects on material selection, primarily because of temperature considerations, and consequent airframe weight and cost. For a cruise flight, cruise altitude and speed effect vehicle temperature, engine efficiency, and therefore airframe weight.

Figure 3 illustrates the overall strategy used, with primary emphasis on structural behavior and material application. Aerodynamic performance and mission capability are included a part of the trade-off to optimize vehicle weight and overall performance. A through review of conceptual hypersonic vehicle design is given by Hunt and Martin [17], which includes all design aspects at the conceptual / preliminary level. The design study considered here, while only a subset of the total problem, illustrates an approach to global optimization for the overall design process. Actual ascent, orbit, and descent strategies are far more complex than discussed here, and are detailed for example, by Nguyen [18], and by Beltracchi [19].

A. OPTIMIZATION STRATEGY

Figure 4 summarizes the levels of equivalent design parameter influence used in this example. The design optimization process proceeds from level i to level iv with local optimization taking place at each level, related to the next level by sensitivity derivatives of parameters common to interfacing levels. The lowest level provides evaluation of structural details and material effects, leading to eventual mission analysis as the integrator of weight and performance

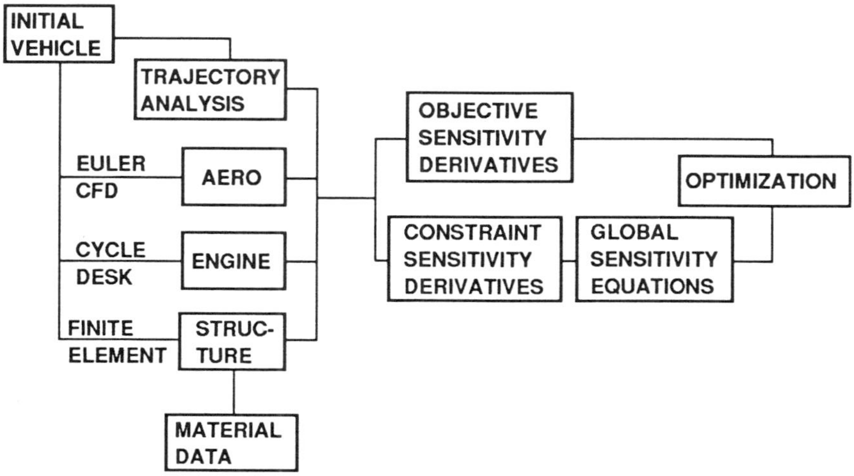

Figure 3. Design Study Strategy

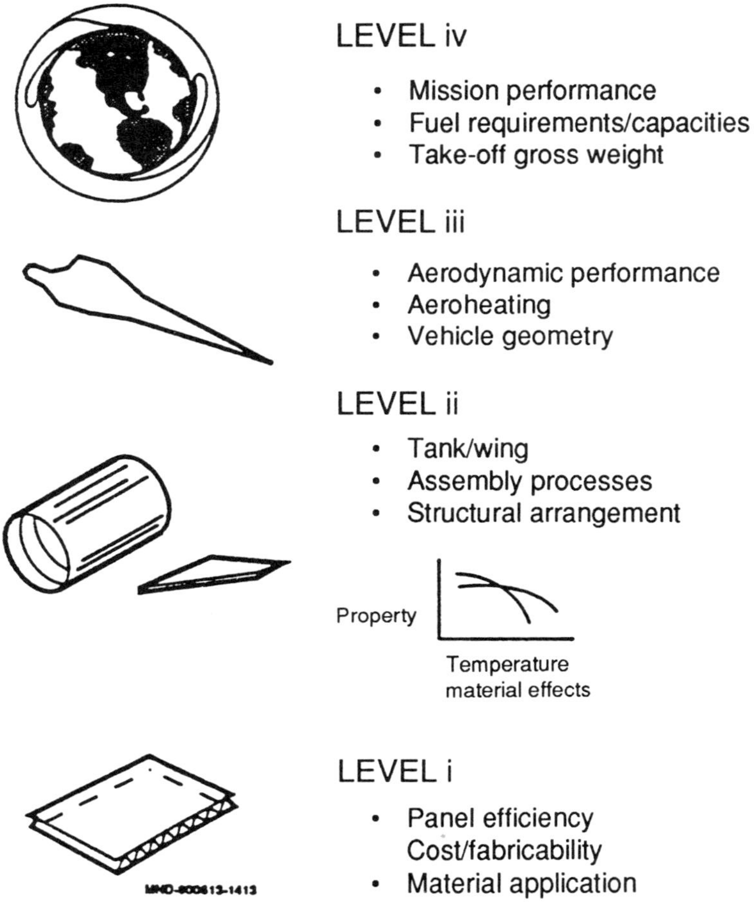

Figure 4. System Multi-Level Definition

trades. The approach illustrated here uses Euler CFD to compute aerodynamic sensitivities, an FEM-based structure code for sub-optimization of the detail structure weight, and mission simulation to perform the multilevel design objective synthesis. Take-off and landing segments are allowed to vary to account for variations in vehicle weight. The cruise range is held constant, with the altitude and speed varied to explore effects on vehicle weight.

Structural optimization focuses on material selection and typical structure configuration, driven in turn by the temperature capability of the material and requirements of cruise parameters. For a typical cruise leg, a higher cruise altitude will tend to produce lower temperatures at the same speed, or speed can be increased to that allowed by the material, possibly increasing engine efficiency for a lower total weight. The advantage in this example would be a flight profile tailored to a specific material set that is lower cost or more producible than another.

The principal structural trade is between structures using advanced high fabrication cost materials with high service temperature capabilities (intermetallic titaniums for example) and structures using near-term more easily fabricated materials (conventional titaniums) that may require more insulation (weight) or supplementary cooling.

B. OPTIMIZATION FORMULATION

Following the previous discussion,the problem is formulated as a constrained optimization for which minimization of the function F,

$$\frac{\min F(X)}{X} \tag{3}$$

subject to the constraints

$$g(X) = 0 \tag{4}$$

is to be obtained. The vector X contains the independent design variables of interest. Variables held constant or not explicitly evaluated during the optimization are not displayed. The formulation minimizes weight considering different materials, with aerodynamic performance as the primary constraint.

Because it dominates total structural weight for the type of vehicles being considered, concentration is on the fuselage structure which is also the primary fuel tank, McQuilkin and Logan [20]. Basic structural configuration trades, illustrated as level i in Fig. 5, may be accomplished using standard engineering calculations, test data, regression equations, or optimization codes. For new vehicle design, there is usually inadequate historical data to develop the necessary relations, and lower level structural analysis and optimization are used to generate required data.

DESIGN CONCEPTS	CRITERIA	MAT'L	DESIGN TEMPT	t (in.)	pt lb/ft^2	REMARKS
SPF/DB BREADED STIFFENED PANEL	UNIXIAL $N_x = 750$	Ti_3 Al RSR T1 SIC/Ti_3Al SIC/RSR T1	1200°F 1800°F 1200°F 1800°F	0.52 0.40 0.50 0.40	1.23 1.21 1.10 1.17	• Low thermal stress
	BI-AXIAL $N_x = 750$ $N_x = 300$	Ti_3 Al RSR T1 SIC/Ti_3Al SIC/RSR T1	1200°F 1800°F 1200°F 1800°F	0.063 0.070 0.057 0.072	1.48 1.40 1.08 1.30	
CORRUGATE D PANEL	UNIAXIAL $N_X = 750$	Ti_3 Al RSR T1 SIC/Ti_3Al SIC/RSR T1	1200°F 1800°F 1200°F 1800°F	0.035 0.045 0.034 0.045	0.85 0.90 0.78 0.87	• Minimum thermal stress • Requires external surface
HONEYCOMB	UNIAXIAL $N_x = 750$	Ti_3 Al RSR T1 RENE°41	1200°F 1800°F 1600°F	0.030 0.033 0.30	0.75 0.87 1.31	• Closeout weight to be added

Figure 5. Wing Panel Concepts Evaluation

At level iii of the fuselage breakdown the objective function is structural weight, related to aerodynamic performance at the next level through selected shape parameters. Representing the structural component of Eq. (1), fuselage weight is expressed as

$$W = f_1((r, l, T, Ym), Yw) \tag{5}$$

where r is the fuselage radius, l the nominal fuselage length, and T a measure of structural temperature for critical areas. Here, T is the variable that will be used to relate structural weight to component design at the lower level and to aerodynamic heating at the next level and is the explicit part of the Y_a matrix in the structural component of Eq. (1). Note that there is an implicit relation between T and W through material properties as related to temperature. In this formulation, T is an array of skin temperatures related to specific structural areas of interest. The vector Ym contains mission effects on weight, and Yw is the vector of vehicle weights. For this example the notation will be limited to

$$W = f_1(r, l, T) \tag{6}$$

Only the fuselage diameter and length are displayed since these variables will be explicitly varied, and will be used to tie the weight to aerodynamic performance at the next level. Structural weights and required sensitivities are developed from a finite element based structural optimization code. This code is used to optimize the fuselage structure for variations in length and diameter, using criteria for strength, stiffness, panel buckling, and local crippling. This process for weight sensitivities considered optimization of the structural shell at level iii, supported by trade studies and detailed component optimization at level iv.

Using these results, overall structural optimization is performed as previously discussed. The basic model, used for both aerodynamic and structural analyses, consisted of ten stations as shown in Fig. 6. At each station a number of curve segments, which in turn are defined by a set of points, defines the vehicle cross section. This baseline geometrical information is input to a program which performs a spline fit to the data points. The geometry is then discretized according to a user-specified number of segments in the traverse direction. Improved definition is developed in the longitudinal direction by using curve fitting techniques with specified longitudinal stations, resulting in the 57 station model shown in Fig. 7.

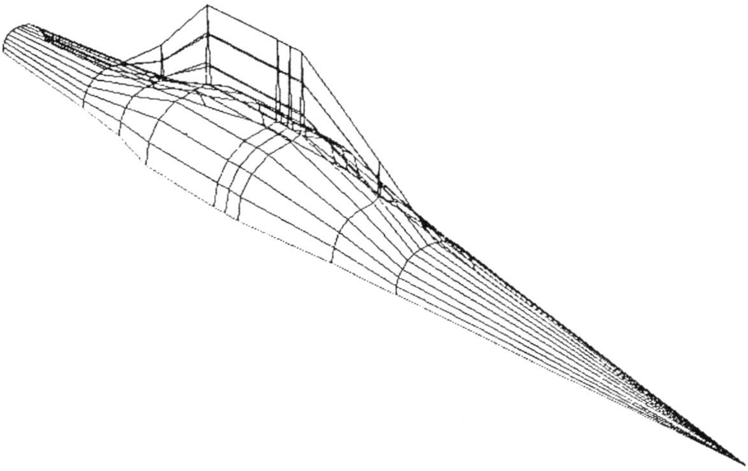

Figure 6. Base 10 Station Model

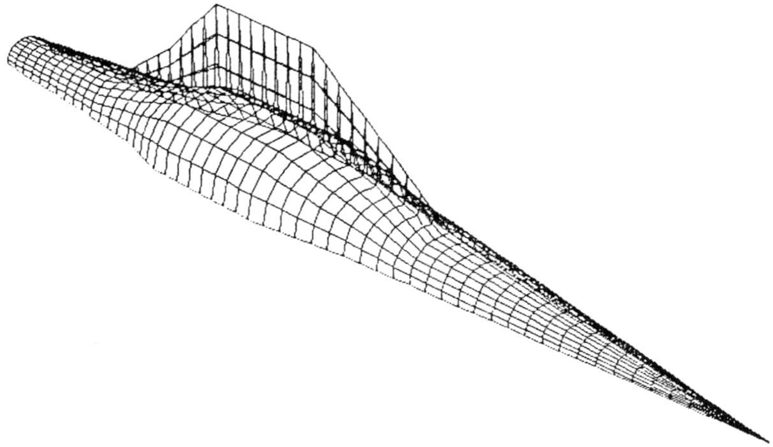

Figure 7. Enhanced 57 Station Model

A finite element model using membrane elements with sample substructures is shown in Fig. 8. Model generation included different component concepts and materials to facilitate trade studies, and it is at this step where material effects are integrated into overall structural performance. Kinematic constraints were assigned to the two aft nodes to prevent rigid body

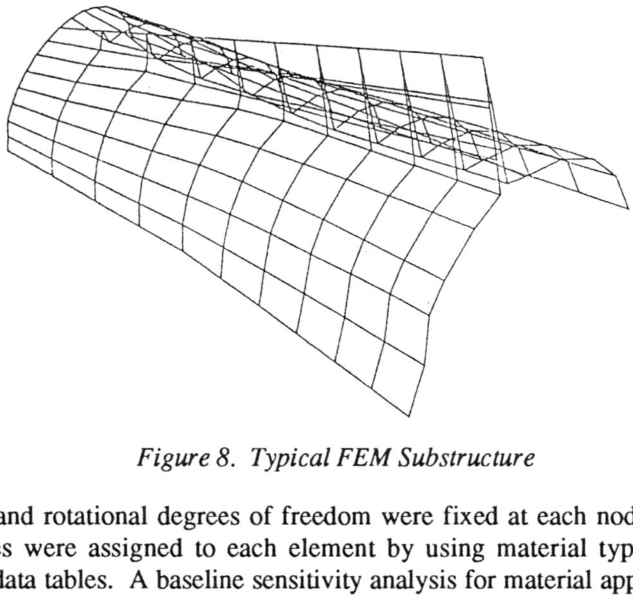

Figure 8. Typical FEM Substructure

motion, and rotational degrees of freedom were fixed at each node. Material properties were assigned to each element by using material type codes and internal data tables. A baseline sensitivity analysis for material application was run using a series of point design loads and material data. Since material properties are generally nonlinear with respect to temperature, a series of linear analyses was used to represent nonlinear sensitivity of the vehicle temperature deformation through temperature dependent material properties. Because of strong effects on engine inlet performance, nose tip deflections were used as a guide to vehicle stiffness. Figure 9 shows tip deflection sensitivity to temperature, and Fig. 10 shows tip deflection as a function of applied loads.

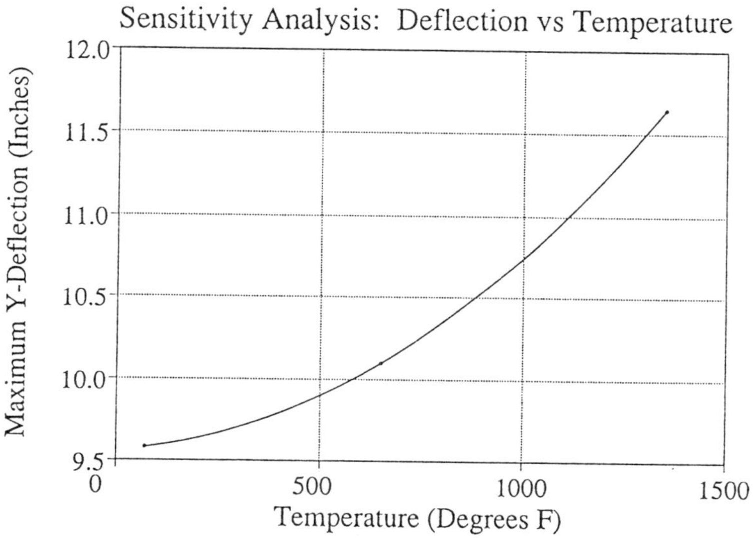

Figure 9. Sensitivity Analysis—Effect of Temperature on Nose Tip Deflection

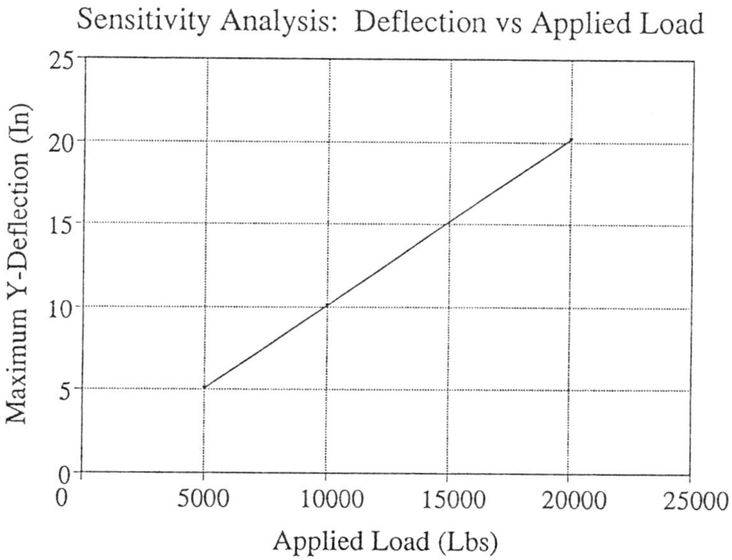

Figure 10. Sensitivity Analysis—Effect of Point Load on Nose Tip Deflection

Aerodynamic computations at level ii utilized a three dimensional unsteady Euler CFD code developed by Chakravarthy and Szema [21] with results for Mach 5.0 and $\alpha=0°$ shown in Fig. 11. This code solves the steady/unsteady Euler equations based on time and space marching modes. The space marching mode is a Gauss-Seidel relaxation method which includes forward sweeps, backwards sweeps, or both. This code utilized an internal grid generator which created a two dimensional grid in the Y-Z plane at consecutive planes along the X direction. Following notation of Eqs. (5) and (6), the aerodynamic segment of Eq. (1) is expressed as :

$$A= f_2(l,d,t_i,T) \tag{7}$$

where t_i are trajectory parameters. Aeroheating, represented by the vector T, is a function of velocity, altitude, surface geometry, angle of attack, and material characteristics such as roughness, emissivity, and catalytic activity.

Trajectory analysis and influence on vehicle weight is computed using a program such as POST (Program to Optimize Simulated Trajectories), Brauer, et al [22,23] with the aircraft performance mission option. For purposes of this case study, mission analysis is represented by :

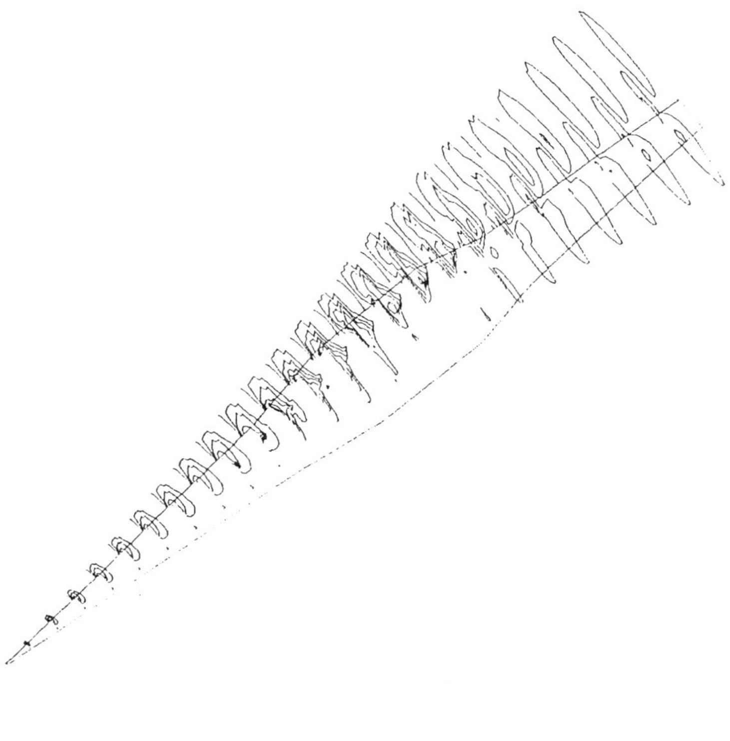

Figure 11. Aerodynamic Pressure Distribution for $M=5,\alpha=0°$

$$M = f_3 (D, h, v, W) \tag{8}$$

where D is vehicle drag, h the cruise altitude, v the cruise velocity. Fuselage shape is varied to optimize performance (weight for a specified mission leg) as speeds are varied. For purposes of this example, propulsion efficiency for the baseline mission was computed by a propulsion performance code, and was held constant. A propulsion extension to the optimization would be handled by including an appropriate equation to the set in Eq. (1), and including a fourth vector, Y_P.

The objective is minimum take-off gross weight (TOGW), based on varying material application and flight conditions, constrained by aerodynamic heating and performance. Specific Impulse (I_{SP}) in this formulation is a component of Y_m in Eq. (5) and is maximized as a constraint. Then, the following matrix equation of sensitivity derivatives is the governing equation:

$$\begin{vmatrix} I & 0 & -\dfrac{\partial f(m)}{\partial M} \\ -\dfrac{\partial f(A)}{\partial m} & I & -\dfrac{\partial f(A)}{\partial M} \\ -\dfrac{\partial f(M)}{\partial m} & -\dfrac{\partial f(M)}{\partial A} & I \end{vmatrix} \begin{Vmatrix} \dfrac{dm}{dx} \\ \dfrac{dA}{dx} \\ \dfrac{dM}{dx} \end{Vmatrix} = \begin{vmatrix} \dfrac{\partial f(m)}{\partial x} \\ \dfrac{\partial f(A)}{\partial x} \\ \dfrac{\partial f(M)}{\partial x} \end{vmatrix} \tag{9}$$

where the notation m has been introduced to represent material behavior as a subset of structures in the case to be studied, and are engineering properties as a function of temperature. The functions f(m), f(A) and f(M) represent materials, aerodynamic behavior, and mission, respectively and are, in general, matrices. Note that material behavior was selected as a subset of structures for specific display as compared to the initial Eq. (1). The sensitivity derivatives are as follows:

$\partial f(m)/\partial M$- material property change as affected by changes in mission parameters.
$\partial f(A)/\partial m$- aerodynamic changes as affected by material properties.
$\partial f(A)/\partial M$- aerodynamic changes as affected by mission parameter changes.
$\partial f(M)/\partial m$- mission parameters changes as affected by material properties.
$\partial f(M)/\partial A$- mission parameter changes as affected by aerodynamics.
dm/dx, dA/dx, dM/dx- global sensitivity derivatives with respect to explicit design variables, x_i.

The right hand vector is the set of partial derivatives of material properties, aerodynamic forces and temperatures, and mission parameters with respect to the explicit design variables, x_i.

Solution of Eq. (9), yields the global sensitivity equations needed to establish interdisciplinary effects. Sensitivities are represented as linear Taylor series of expansions about the baseline design, which using Eq. (6) as an example, becomes

$$W = W_0 + C_1 \frac{\partial f_1}{\partial r} \Delta r + C_2 \frac{\partial f_1}{\partial l} \Delta l + C_3 \frac{\partial f_1}{\partial T} \Delta T \tag{10}$$

where the constants are coefficients of the expansion. Sensitivity derivatives for this example were calculated simply by plotting results of several computer runs and computing the slope of a smooth curve through the data points at the point of interest. An extensive literature exists on sensitivity derivative computation techniques, for example Ref. [15], and an appropriate method may be used for a specific problem. Further mathematical details for the development are given in Refs. [2,7].

To achieve minimum TOGW, baseline vehicle parameters were perturbed. Fig. 12 shows the effect of TOGW on mission fuel weight, including effects of structural optimization, materials application studies, and aerodynamic performance. In this study, fuel weight was used as a common link between TOGW and I_{SP}. Table I summarizes optimization results. Take-off Gross Weight decreased by 1.6 percent, with an I_{SP} increase of 5.6 percent. The optimum Mach Number also increased, with a slight increase in cruise altitude. Further details are given in Ref. [24].

In Ref. [25], Levine, et al considered a different approach, for which I_{SP} was maximized by considering geometry changes to the vehicle forebody. For that study, fuel volume was held constant, cruise speed was set at M=16, and dynamic pressure was held constant. This approach yielded a 12.5 percent I_{SP} increase, shown in Table II, when considering forebody flexibility.

These results indicate the potential of optimization techniques when applied to hypersonic vehicle design. Once the work is invested in setting up the problem, a variety of approaches can be considered to allow a designer to fully understand parameter perturbation effects on various performance measures, and on vehicle configuration.

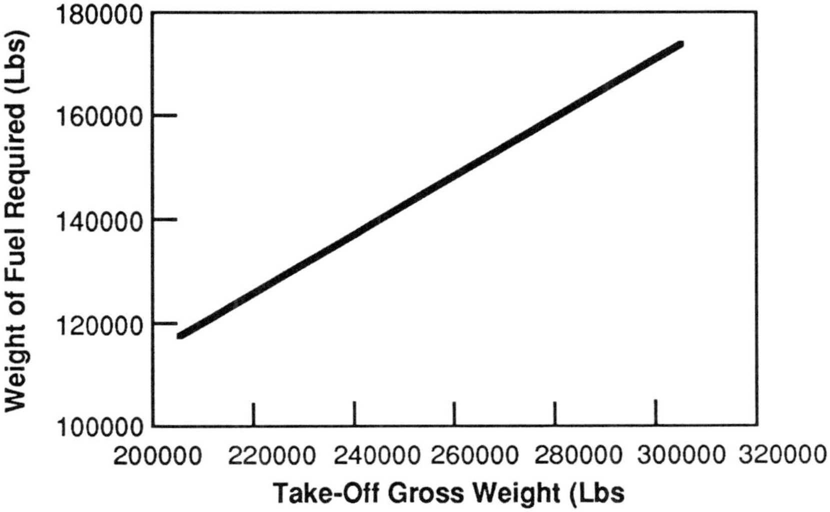

Figure 12. Effect of TOGW on Fuel Weight

Table I. Hypersonic Vehicle Optimization Results

Parameter	Baseline Value	Optimized Value
Mach Number	15	16
TOGW (lbs)	193,000	190,000
ISP (sec)	11,800	12,500
Fuel Flow (lbs/sec)	75.6	76.0

These results indicate the potential of optimization techniques when applied to hypersonic vehicle design. Once the work is invested in setting up the problem, a variety of approaches can be considered to allow a designer to fully understand parameter perturbation effects on various performance measures, and on vehicle configuration.

Table II. Forebody Optimization Results

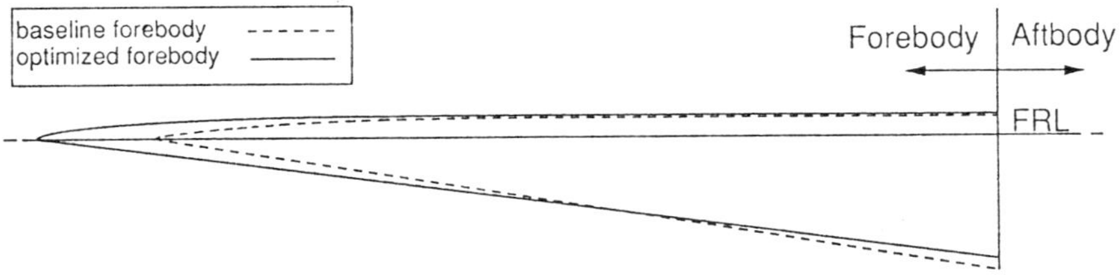

Optimization Parameter	Optimization Results
Design Vanable	
1. Forebody length	1.0209
2. Cone angle	0.9693
3. Upper surface height	1.0029
4. Geometric transition length	1.0760
5. Elevon deflection	0.8620
6. Bodyflap deflection	1.0320
Objective	
Effective trimmed Isp	1.1259

MPD-920716-1186

IV. ADVANCED FIGHTER WING EXAMPLE

A simple case study from Ref. [26] will be shown to illustrate the previous discussions. An example of a fighter wing is used to further explore leveling concepts and to illustrate ideas that have been presented. The example will consider a simplified thread through the entire process.

The wing chosen, Fig. 13, has been used in previous studies, for example Ide et al [27] and part of that data base will be used here. First, formulation of the problem for the wing itself will be developed, followed by system level consideration of mission performance. The objective function, or functions for the case of a multiobjective optimization, is the measure of merit used to guide the optimization process. In the case of the entire airplane,takeoff gross weight (TOGW), or any mission performance parameter such as mission radius can be selected. For the case at hand, the single objective function of TOGW is used for the airplane as a whole (the parent system) with weight of each subsystem used as objective functions for suboptimization at each level.

A. FORMULATION

The problem will be formulated as a constrained optimization in accordance with Eq. (3) and using a similar approach as in the previous section. The example will minimize weight for a given mission, using aerodynamic performance criteria as constraints. At the first level of the wing breakdown, the

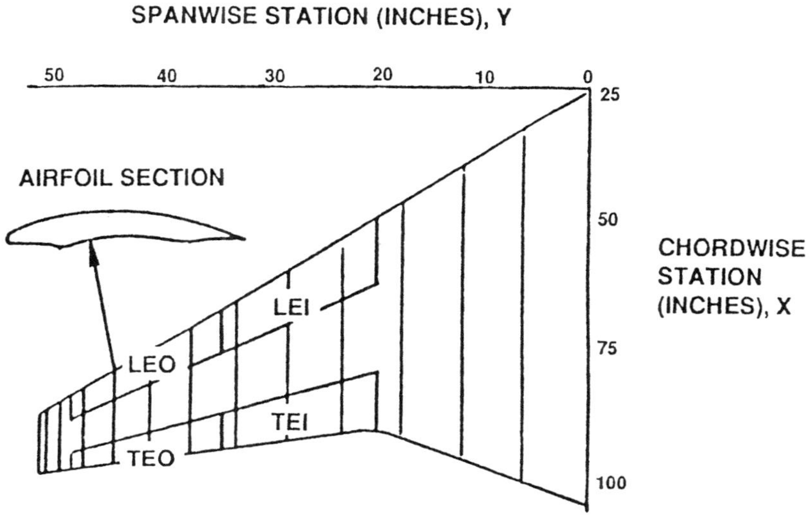

Figure 13. Advanced Fighter Wing Planform

objective function is weight, related to aerodynamic performance at the next level. Wing weight W is expressed as

$$W = W(T) \tag{11}$$

where T is the wing root thickness. A change in notation using W in this case instead of f has been made for convenience and clarity of primary variables. In this formulation, only the root thickness is displayed since it is the only variable that will be explicitly varied and will be used to tie the weight to aerodynamic performance at the next level. Wing structural weights will be developed from a suboptimization problem based on a finite element model of the wing, as discussed later. As with other parameters to be studied, the function is an implicit one. The actual relations and sensitivities are derived from a computer simulation that accounts for all other variables required to fully describe the contributing analysis.

The ratio of lift to drag L/D and the roll rate Θ are used as simple measures of aircraft performance and are defined as follows:

$$L/D = L(\delta_i, T, \psi) \tag{12}$$

$$\dot{\Theta} = \dot{\Theta}(\delta_i) \tag{13}$$

$$\begin{aligned}&\delta_i = \text{control-surface deflections, i = 1-4}\\&T = \text{wing root thickness}\\&\psi = \text{sweep angle}\end{aligned} \tag{13}$$

Side constraints on lift and roll rate are used to assure minimum performance. Other variables, such as aspect ratio, are held constant for this study. Wing root thickness is used as the wing-thickness measure for aerodynamic performance as well as a primary parameter for structural-weight sensitivity as discussed above. At this level, overall wing geometry to be studied includes thickness and sweep-angle changes.

B. SENSITIVITY DERIVATIVES

Sensitivities are developed as linear Taylor series expansions of these relations as follows:

$$W = W_o + A\frac{\partial W}{\partial T}\Delta T \tag{14}$$

$$L = L_o + B\sum_{i=1}^{4}\frac{\partial L}{\partial \delta}\Delta\delta_i + C\frac{\partial L}{\partial T}\Delta T + D\frac{\partial L}{\partial \psi}\Delta\psi \tag{15}$$

$$\dot{\Theta} = \dot{\Theta}_o + E\sum_{i=1}^{4}\frac{\partial\dot{\Theta}}{\partial\delta_i}\Delta\delta_i \tag{16}$$

where the constants are coefficients of the expansion. Sensitivities are obtained by simulation of structural and aerodynamics behavior.

Structural-weight sensitivities were obtained from a finite element based structural optimization code with the model illustrated in Fig. 14. This code was used to optimize the wing structure for the baseline sweep angle and root thickness using criteria for strength, stiffness, panel buckling, and local crippling and wrinkling as well as minimum gauges. This design-optimization step considered a variety of panel configurations, including stiffener type and spacing as a function of spar spacing. The wing is constructed of advanced composite skins and metallic substructure.

This inner optimization was performed because of the complexity of the structural optimization process and the definition detail desired. Weight sensitivities could also be derived from regression curves, should an adequate data base exist. The process for weight sensitivities thus considered optimization of the structural box at level ii, Fig. 1, supported by a data base of detailed component optimization at level iii. Buckling-efficiency studies were based on the NASA OPCOM code and, since execution times were very rapid, were used

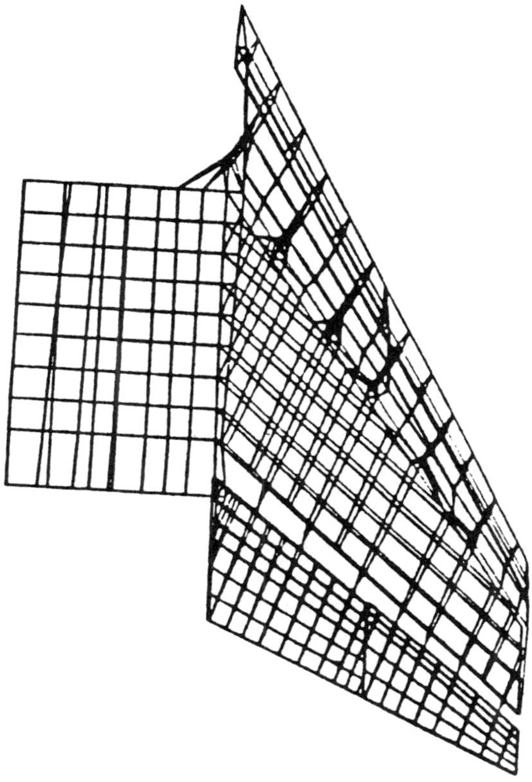

Figure 14. Wing Finite Element Model

to provide a data base for subsequent finite element definition. Additional details of the model are given in Ref. [7]. Panel efficiency studies yielded a corrugated stringer as the optimum shape and this configuration was used for further model development.

Wing box weight sensitivities to root thickness variations were computed by varying root thickness and recomputing box weight. To avoid complexities in calculating aerodynamic sensitivities for this illustration, a straight taper to a constant thickness was used. Resulting sensitivities are shown in Fig. 15, where a 10% root thickness variation was used. Weight of fixed leading and trailing edges, tips, and other secondary structure weight, as well as weight of non-optimum material such as fasteners, and sealants were assumed constant but could be included if desired.

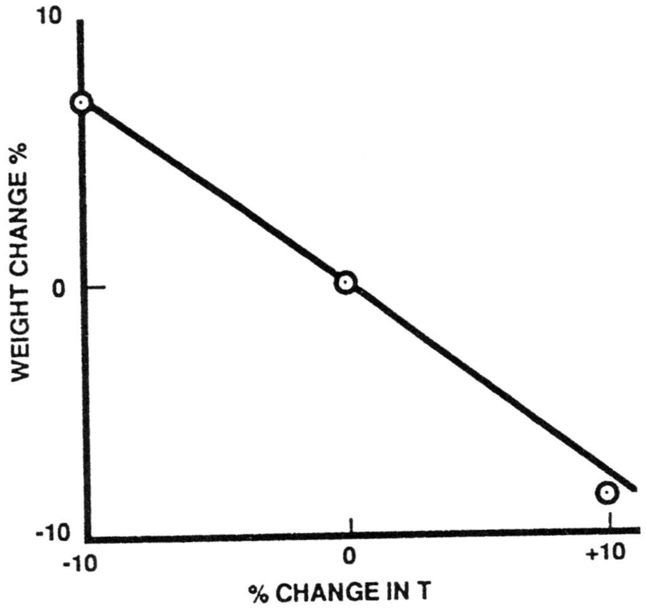

Figure 15. Wing Weight/Thickness Sensitivity

Aerodynamic sensitivities were computed using a three-dimensional full-potential computational fluid dynamics (CFD) code described by Ide and Shankar [28]. Parameters varied were baseline sweep ± 10 degrees; trimmed control-surface length, and deflections +5 degrees and +10 degrees; and baseline root thickness ± 10%. Sensitivities were calculated manually, and are presented in Table III for M = 0.9.

Resolution between performance and weight can be made several ways. Multiobjective optimization can be developed by assigning ratios to separate objectives and performing the optimization cycle. Alternately, a variable linking process using historical data or some other relation can be used to associate weight and performance. Here, the approach will be to use a mission-effectiveness simulation to resolve weight and aerodynamics conflicts. For purposes of this illustration, range at constant speed will be used, in the form

$$R = R(L/D, W) \tag{17}$$

Table III. Primary Parameter Sensitivities

Sensitivity With Respect To	L/D	CM_{roll}	Useful Range
$\frac{\partial}{\partial \psi}$	-.06218 deg^{-1}	-.005211 deg^{-1}	ψ = 50 deg → 60 deg
$\frac{\partial}{\partial a}$	+.06095 $in.^{-1}$	+.01826 $in.^{-1}$	a = 7.3 → 22.0 in.
$\frac{\partial}{\partial \delta}$	+.02920 deg^{-1}	+.1123 deg^{-1}	δ = 0 deg → 2 deg
$\frac{\partial}{\partial (\%T)}$	-.09003 $(\% \text{ change})^{-1}$	+.003556 $(\% \text{ change})^{-1}$	ΔT = 10% → + 10%

Following previous notation, the first-order Taylor expansion is developed as

$$R = R_o + F\frac{\partial R}{\partial L}\Delta L + G\frac{\partial R}{\partial W}\Delta W \tag{18}$$

where the notation *L is* used to represent *L/D* as before.

For steady cruise conditions, the Breguet equation can be used to compute appropriate derivatives. The problem has then been formulated to minimize aircraft weight for a constant range, with aerodynamic performance adequate to meet minimum requirements for lift and roll rate. For the fighter being studied, the sensitivities required in Eq. (18) are given in Table IV for M = 0.9.

Using these sensitivity parameters, a constrained optimization for minimum weight was performed using the modified method of feasible directions option of the Automated Design Synthesis program. Convergence was typically obtained in several cycles, with results summarized in Table V. Baseline wing sweep angle was 50 degrees aft. As can be seen, the optimum design converged to the maximum allowed wing thickness to provide minimum wing structural weight ratio while maintaining lift, but sacrificing drag. Wing sweep was increased slightly. Constant engine efficiency was assumed, with fuel weight allowed to vary with L/D changes. Roll rate was held to be the same as the baseline design as a constraint, with resized control surfaces used for trimming.

Table IV. Range Sensitivities

$\frac{dR}{d\,L/D}$ (miles)	$\frac{dR}{dW}$ (miles/lb)
127.6	-0.28

Table V. Fighter Wing Optimization Results

Root Thickness, T	+10%
Sweep Angle, Ψ	52.4 deg
Wing Box Weight	-8.3%
TOGW	-0.8%

The significant wing structural weight reduction of 8.3% is due to sacrificing overall aerodynamic efficiency for this single condition. Further analysis of these results as well as comparison of different approaches to local optimization is presented by Ide, et al [29]. Although the tradeoff is a valid one, more complete mission simulation would be expected to alter specific results, and would be required for a complete study. For example, sensitivities of the objective functions to variations in key mission summarized was shown by Simos and Jenkinson [9] to have potentially significant effects. Furthermore, studies by Johnson [30] showed a wide variation in optimum commercial aircraft wing planform as a result of changing objective functions. These results indicate the potential for significant vehicle-design improvement when the entire system is considered, with sufficient detail at the various levels to adequately model each discipline. Significant extension is now needed to fully develop an adequate computer system and data base to take full advantage of these developments.

V. SUMMARY

This chapter has described a decomposition algorithm and multidisciplinary optimization method that integrated a hypersonic vehicle cruise mission, aerodynamic performance, and structural design. The methodology is based on a multi-level decomposition which is arranged such that design variables have relatively similar impact at each level, facilitating suboptimization of primary design drivers at each level, and on simulation of principal vehicle behavior through use of computer codes representing aerodynamics, (CFD), structures (FEM), and mission performance. An example illustrated implementations of key features of the methodology and formulation approach.

Following this discussion, an additional example of an aircraft wing was used to show potential of this process to develop an improved design compared to one developed by conventional sequential approaches and, hopefully, the value of a system view of the design process.

For the complex integration required for contemporary advanced vehicles, the power of multidisciplinary optimization methods is readily apparent with continued use assured, although significant future development remains. In particular, as study fidelity is increased to account for representation of more design variables, integrated analysis approaches such as described by Wieting [31], and overall process management and automated sensitivity derivative computation will become necessary. To manage the anticipated even larger matrix of global sensitivity equations, formal suboptimization control methods and cascading of results via subsets of sensitivity equations which can be manipulated at the designer's direction will be needed. An initial approach for mechanization of such a process is presented by Levine, et al [25]. Automated sensitivity derivative computation, including those of higher order, again under the designer's direction, will provide needed accuracy and design speed. These further developments will provide a major boost to acceptance and utilization of multidisciplinary optimization methods.

VI REFERENCES

1. Vanderplaats, G. N., "Numerical Optimization Techniques for Engineering Design", McGraw-Hill, New York.

2. Abdi, F., Ide, H., Levine, M., and Austel, L., "The Art of Spacecraft Design: A Multidisciplinary Challenge", in "Proceedings of the Second NASA/Air Force

Symposium on Recent Experiences in Multidisciplinary Analysis and Optimization", NASA Langley Research Center, Hampton, VA, 289-308, September 28, 1988.

3. Ide, H., et al, "Hypersonic Vehicle Forebody Design Studies and Aerodynamic Trends", AIAA Paper 89-2182, July, 1989.

4. Sobieszczanski-Sobieski, J., "A Linear Decomposition Method for Large Optimization Problems - Blueprint for Development", NASA TM-83248, February, 1982.

5. Sobieszczanski-Sobieski, J., "Sensitivity Analysis and Multidisciplinary Optimization for Aircraft Design: Recent Advances and Results", *Proceedings of the 16th Congress of the International Council of he Aeronautical Sciences*, AIAA, Washington, DC, 1820 - 1826, 1988.

6. Sobieszczanski-Sobieski, J., Barthelemy, J. M., and Giles, G. L., "Aerospace Engineering Design by Systematic Decomposition and Multilevel Optimization," *Proceedings of the 14th Congress of the Internationall Council of the Aeronautical Sciences,* AIAA, New York, 828-840, 1984.

7. Abdi, F., Ide, H., Shankar, V. J., and Sobieski, J., "Optimization of Nonlinear Aeroelastic Tailoring Criteria", *Proceedings of the 16th Congress of the Internationall Council of the Aeronautical Sciences,* AIAA, Washington, DC, 1083 - 1091 , 1988.

8. Grossman, B. et al, "Integrated Aerodynamic—Structural Design of a Transport Wing", AIAA Paper 89-2129, July 1988.

9. Simos, D., and Jenkinson, L. R., "Optimization of the Conceptual Design and Mission Profiles of Short-Haul Aircraft", *J.of Aircraft* **25** , 618-624, July 1988.

10. Wrenn, G. A., and Augustine, R. D., "Multilevel Decomposition Approach to the Preliminary Sizing of a Transport Aircraft Wing", *J.of Aircraft,* **25**, 632-638, July, 1988.

11. Bloebaum, C. L. , "Formal and Heuristic System Decomposition Methods in Multidisciplinary Synthesis", NASA CR 4413, December, 1991.

12. Venkayya, V. B., "Recent Developments in Large-Scale Structural Optimization", Second NASA/Air Force Symposium on Recent Advances in Multidisciplinary Analysis and Optimization, Hampton, VA, September 28-30, 1988.

13. Sensburg, O., Schweiger, J., Godel, H., and Lotze, A., "The Integration of Structural Optimization in the General Design Process for Aircraft", 17th Congress, International Council of the Aeronautical Sciences, Stockholm, Sweden, September, 1990.

14. Sobieszczanski-Sobieski, J., "Sensitivity of Complex, Internally Coupled Systems", *AIAA Journal* **28**, 153-160 January, 1990.

15. Adelman, H. A., and Haftka, R. T., "Sensitivity Analysis of Discrete Structural Systems", *AIAA Journal* **24**, 823-832, May, 1986.

16. Sobieszczanski-Sobieski, J., "High Order Sensitivity Analysis of Complex, Coupled Systems", *AIAA Journal* **28**, 756-758, April, 1990.

17. Hunt, J.L and Martin, J.G., "Hypersonic Airbreathing Vehicle Conceptual Design (Focus on AeroSpace Planes)", in "Second NASA/Air Force Symposium on Recent Experiences in Multidisciplinary Analysis and Optimization", Hampton VA., September 28, 1988.

18. Nguyen, H. N., "Optimal Ascent Trajectories of the Horizontal Takeoff Single-Stage and Two-Stage-to-Orbit Launchers", AIAA Guidance, Navigation, and Control Conference, New Orleans, LA, August 12-14, 1991.

19. Beltrachi, T., "A Decomposition Approach to Solving the Allup Trajectory Optimization Problem", 28th Aerospace Sciences Meeting, Reno, NV, January 8-11, 1990.

20. McQuilkin, F., and Logan, T., "Optimization and Validation of a Fuselage Fuel Tank Structural Concept for the NASP", AIAA Paper 90-5262, AIAA Second International Aerospace Planes Conference, Orlando, FL, October, 1990.

21. Chakravarthy, S. R., and Szema, K., "Euler Solver for Three Dimensional Supersonic Flows with Subsonic Pockets", *J.of Aircraft* **24**, 73-82, February, 1987.

22. Brauer, G. C., et al, "Program to Optimize Simulated Trajectories (POST)", Martin Marietta Corp, NASA CR-132689, April, 1975.

23. Brauer, G.C., Cornick, D.E., and Stevenson, R., "Capabilities and Applications of the Program to Optimize Simulated Trajectories (POST)", Martin Marietta Corp, NASA CR-2770, February, 1977.

24. Logan, T. R., Abdi, F., and Sobieszczanski-Sobieski, J., "Optimization of Aircraft Configurations in a Multidisciplinary Environment", Paper ICAS-90-

2.3.2, 17th Congress, International Council of the Aeronautical Sciences, Stockholm, Sweden, September, 1990.

25. Levine, M., Ide, H., and Hollowell, S., "Multidisciplinary Hypersonic Configuration Optimization", Third Air Force/NASA Symposium on Recent Advances in Multidisciplinary Analysis and Optimization, San Francisco, CA., September 24-26, 1989

26. Logan, T. R., "A Strategy for Multilevel Optimization of Aircraft", *J.of Aircraft* **27** 1068-1072, October, 1989.

27. Ide, H., Abdi, F., and Shankar, V. J., "CFD Sensitivity Study for Aerodynamic/Control Optimization Problems", AIAA Paper 88-2336, April 1988.

28. Ide, H., and Shankar, V. J., "Application of a Fast, Time-Accurate Full-Potential Scheme to a Statically Flexible Wing in the Transonic Regime", AIAA Paper 87-0707, April 1987.

29. Ide, H., Levine, M., and Ominsky, D., "Evaluation of Current Multiobjective Optimization Methods for Aerodynamic Problems Using CFD Codes", AIAA Paper 90-0955, AIAA 31st Structures, Structural Dynamics and Materials Conference, Long Beach, CA., April, 1990.

30. Johnson, V. S., "Optimizing Conceptual Aircraft Design for Minimum Life Cycle Cost", *Proceedings of the Second NASA/Air Force Symposium on Recent Experiences in Multidisciplinary Analysis and Optimization,* NASA Langley Research Center, Hampton, VA, 309-332, September, 1988.

31. Wieting, A. R., Dechaumphai, P., Bey, K., Thornton, E. A., and Morgan, K., "Application of Integrated Fluid-Thermal-Structural Analysis Methods", 16th Congress, International Council of the Aeronautical Sciences, Jeruselem, Israel, September 28 - August 2, 1988.

INTEGRATED AERODYNAMIC – STRUCTURAL DESIGN OPTIMIZATION OF AIRCRAFT WINGS

Eric R. Unger
Raphael T. Haftka
Bernard Grossman
William H. Mason

Department of Aerospace and Ocean Engineering
Virginia Polytechnic Institute and State University
Blacksburg, Virginia 24061

I. INTRODUCTION

Aircraft design requires the integration of several disciplines, including aerodynamics, structures, controls, and propulsion. Traditionally, the design process would proceed sequentially with feedback limited to the conceptual design level. In the last few years there has been more interest in integrating the design procedure. This was motivated by the introduction of composite materials which permit extensive aeroelastic tailoring, as well as by the requirements of advanced vehicles, such as the aerospace plane, where because of extreme flight conditions, the interdisciplinary interactions are particularly important.

One of the more important interdisciplinary interactions in modern aircraft design is that of aerodynamics and structures. One aspect of that interaction is the use of composite materials to tailor the deformation of aircraft wings so as to affect aerodynamic loads. This so-called aeroelastic tailoring is the focus of many publications (see [1,2] for recent reviews of the state of the art). Aeroelastic tailoring is particularly important for forward-swept-wing aircraft, *e.g.* the Grumman X-29A, because of the aeroelastic instabilities that tend to plague forward-swept wings.

CONTROL AND DYNAMIC SYSTEMS, VOL. 57

Aeroelastic tailoring is a limited form of integrated design, as the structure is designed to affect aerodynamic properties, but the aerodynamic design does not consider structural constraints except in a very broad sense (such as in limits on wing aspect ratio and thickness). In [3] it has been shown that in an integrated design process, it may be reasonable to use a suboptimal aerodynamic design if it results in reduced structural weight, because the weight savings may be converted into improved aerodynamic efficiency.

Past work from the authors and others has considered the integrated aerodynamic-structural design optimization problem. Initial work at Virginia Tech [3,4], began with the integrated design of a sailplane wing utilizing simple lifting-line aerodynamics in conjunction with beam theory for the structural analysis. Subsequent work [5,6], extended the model complexity in a transport example to include vortex-lattice theory for aerodynamics and finite-element theory for the structural analysis. Similar work by Gallman, Kroo, and Smith [7], considered the design of a joined-wing transport utilizing vortex-lattice and beam theory for the aerodynamic and structural analysis respectively. The addition of active control systems has also been considered in work by Livne, Friedmann, and Schmitt [8], with kernel-function aerodynamics and equivalent plate theory for the structure. Additional design work with these types of design problems can be found in [9–11].

This chapter provides a generalized treatment to the integrated aerodynamic-structural design problem. The systematic approach addresses the prohibitive computational costs of multi-disciplinary design with some of the analysis and optimization techniques available today. The first of these methods involves a dramatic reduction to the cost sensitivity analysis through the use of modular sensitivity analysis, [9,12]. An example of the modular technique with aerodynamic, structural, and active control influences was considered in [13]. Additional savings in computational effort are achieved by making use of both detailed numerical models and simple algebraic expressions traditionally employed at the conceptual design phase. This method of a dual level of complexity in the analysis is called here *variable-complexity modeling.*

The remainder of this chapter demonstrates the practical application of the methods presented. The first of these design cases is the combined aerodynamic-structural design optimization of a subsonic transport aircraft. This case provides a detailed example of modular-sensitivity analysis and an introduction to variable-complexity modeling. The second design example illustrates the aerodynamic design of a High-Speed Civil Transport (HSCT) utilizing a substantial amount of variable-complexity modeling. At the conclusion of this chapter some closing remarks are made and lists of nomenclature and references are presented.

II. ANALYSIS

The ability to effectively represent the coupled structural and aerodynamic response of the wing is essential to the design process. The method presented below provides such a representation in a generalized formulation to allow for a broad range of analysis methodology. Additional information on the solution of aeroelastic systems can be found in Refs. [14] and [15].

II-A. Aeroelastic Formulation

Performing structural analysis on a finite-element-based wing model yields a relatively large system of equations:

$$Ku = F_a + F_I \, , \tag{1}$$

where K is the stiffness matrix, u is the displacement vector, F_a represents the applied aerodynamic loads, and F_I gives the inertial loads. To reduce the cost of the aeroelastic analysis, (in particular the cost of derivatives of aerodynamic loads with respect to structural deformations) it is advantageous to decrease the number of degrees of freedom in the structural model. Such a reduction can be realized by utilizing a reduced-basis structural analysis. To employ this approach the displacement vector is approximated as

$$u = \Phi\theta \, , \tag{2}$$

where Φ the mode-shape matrix composed of n_m deformation shapes and θ is a vector of modal magnitudes defined as the generalized displacements. Since a system that may have hundreds or even thousands of degrees of freedom may be adequately represented with perhaps twenty or thirty mode shapes (vibration modes are commonly utilized), a clear advantage can be obtained using this approximation. Here we assume that the modes are orthonormal with respect to a matrix M (mass matrix in the case of vibration modes), that is

$$\Phi^T M \Phi = I . \tag{3}$$

Substituting this approximation into Eq. (1) and premultiplying by Φ^T yields the reduced structural system as

$$\Phi^T K \Phi \theta \equiv K_G \theta = \Phi^T (F_a + F_I) \equiv F_{aG} + F_{IG} , \tag{4}$$

where $K_G = \Phi^T K \Phi$ is the generalized stiffness matrix and $\Phi^T F_a$ is the generalized aerodynamic force vector denoted as F_{aG}, with F_{aG_i} being the work done by the aerodynamic loads on the ith mode.

The loads, F_a, are determined from an aerodynamic analysis procedure. It is assumed that these aerodynamic forces are dependent on some set of design parameters, p, the root angle of attack, α, (which we assume to represent the only body contribution), and the vector of generalized displacements, θ. The load vector F_a is then obtained as

$$F_{aG} = f_1(p, \alpha, \theta) . \tag{5}$$

The angle of attack, α, is determined for a given flight condition from the requirement that α is sufficient to provide an F_{aG} that provides the required lift

$$f_2(p, F_{aG}) = nW - N^T F_a = 0 , \tag{6}$$

where N is a summation vector ($N^T F_a \equiv \sum_{i=1}^{m} F_{a_i}$), n is the load factor, and W is the weight of the aircraft. To transform this equation to generalized coordinates we approximate F_a as

$$F_a \cong M \Phi F_{aG} . \tag{7}$$

This approximation is consistent with that of Eq. (3) in that if we multiply Eq. (7) by Φ^T, we find that it is satisfied exactly. Then Eq. (6) can be written as

$$f_2(p, F_{aG}) = nW - \hat{N}^T F_{aG} = 0\,, \tag{8}$$

where

$$\hat{N}^T = N^T M \Phi\,. \tag{9}$$

The vector of generalized displacements, θ, can also be represented functionally as

$$\theta = f_3(p, F_{aG})\,, \tag{10}$$

with f_3 representing the solution of Eq. (4).

II-B. Solution Procedure

With the structural and aerodynamic response of the wing in functional form it is now necessary to solve the static system for a given flight condition. Inspection of Eqs. (5), (8), and (10) reveals a set of coupled, generally nonlinear equations for the vector of generalized aerodynamic loads, F_{aG}, the wing-root angle of attack, α, and the vector of generalized displacements, θ. For the analysis problem, the vector of design parameters, p, may be considered to be specified. In [9] and [12], Sobieski presents a modular approach for the evaluation of sensitivity derivatives in such coupled interdisciplinary equations. This modular approach allows the individual discipline analysis procedures to be treated as *black boxes* that do not need to be changed during the coupling procedure. Here we are interested in solving the coupled system with a Newton's Method that will require cross-disciplinary sensitivities. Therefore a modular approach similar to [9] and [12] may also be applied to the analysis problem with f_1 representing an *aerodynamic black box* and f_3 a *structural black box*.

Given an initial estimate for the solution F^0_{aG}, α^0, θ^0 we can use Newton's method to improve that estimate. The iterative process may be written as

$$J \Delta Y = \Delta f\,, \tag{11}$$

where

$$\Delta Y = \begin{Bmatrix} \Delta F_{aG} \\ \Delta \alpha\,, \\ \Delta \theta \end{Bmatrix}, \tag{12}$$

and

$$\Delta f = \left\{ \begin{array}{c} f_1(p, \alpha^0, \theta^0) - F_{aG}^0, \\ f_2(p, F_{aG}^0) \\ f_3(p, F_{aG}^0) - \theta^0 \end{array} \right\}, \tag{13}$$

and the Jacobian J is given, (for the balanced flight condition), as

$$\begin{aligned} J &= \begin{bmatrix} I & -\partial f_1/\partial \alpha & -\partial f_1/\partial \theta \\ -\partial f_2/\partial F_{aG} & 0 & 0 \\ -\partial f_3/\partial F_{aG} & 0 & I \end{bmatrix}, \\ &= \begin{bmatrix} I & -qR & -qA \\ \hat{N}^T & 0 & 0 \\ -S_G & 0 & I \end{bmatrix}. \end{aligned} \tag{14}$$

The Jacobian contains the dynamic pressure q, the generalized incremental aerodynamic force vector, qR, the generalized aerodynamic influence coefficient matrix, qA, and the generalized flexibility matrix S_G. The incremental aerodynamic force vector is defined such that its component qr_i represents the *ith* generalized aerodynamic force due to a unit change in α, and the aerodynamic influence coefficient matrix, is defined such that its component qa_{ij} represents the *ith* generalized force to do a unit change in θ_j. Similarly, the generalized flexibility matrix, is such that s_{ij} is the change in θ_i due to a unit change in F_{aG_j}, and from Eq.(3) $S_G = K_G^{-1}$. In many cases the matrix A has to be calculated by using finite differences, that is perturbing each θ_i and recalculating f_1. This makes the calculation of A the most expensive part of the Jacobian evaluation.

Partial solution of Eq. (11) yields the following three equations for the increments $\Delta\theta$, $\Delta\alpha$ and ΔF_{aG}:

$$(I - qS_G A^x)\Delta\theta = S_G B \Delta f_1 + \frac{S_G R}{\hat{N}^T R} \Delta f_2 + \Delta f_3, \tag{15}$$

$$\Delta\alpha = \frac{\Delta f_2 - \hat{N}^T \Delta f_1 - q\hat{N}^T A \Delta\theta}{q\hat{N}^T R}, \tag{16}$$

$$\Delta F_{aG} = \Delta f_1 + qR\Delta\alpha + qA\Delta\theta, \tag{17}$$

where we define

$$B \equiv I - \frac{R\hat{N}^T}{\hat{N}^T R}, \tag{18}$$

and

$$A^x \equiv B\,A\,. \tag{19}$$

It is also possible to solve the above system using fixed point iteration. In this approach we use the initial guesses for α and θ to generate the next approximation for F_{aG} with equation f_1. These generalized forces are then used to get the next approximation for θ with equation f_3. The next approximation for α is obtained from an approximation to Eq. (8) where

$$\hat{N}^T f_1(p, \alpha + \Delta\alpha, \theta) \cong \hat{N}^T F_{aG} + q\hat{N}^T R \Delta\alpha = nW\,, \tag{20}$$

where F_{aG} and θ represent the current approximation to the generalized forces and displacements and $\Delta\alpha$ gives the approximate change in the angle of attack.

The choice between the fixed-point iteration and the Newton iteration depends on the rate of convergence of the fixed-point iteration, on the cost of calculating the Jacobian, and on the order of the system, [16]. In most cases where *black-box* analysis codes are involved, derivatives have to be calculated by finite differences. In such cases, some of the terms of the Jacobian can be quite expensive to evaluate (especially the A matrix which requires n_m separate aerodynamic analyses). Thus, unless the convergence of the fixed-point iteration is extremely slow, the fixed-point iteration would be preferred to Newton's method. However, as shown in the next section, the Jacobian needs to be evaluated for the sensitivity analysis. In situations where sensitivities are needed, it may be advantageous to combine the fixed-point iteration with one or two Newton iterations near convergence.

Also important to the analysis problem is the evaluation of the aeroelastic divergence instability. This condition is evaluated at a fixed angle of attack, because it can be assumed the pilot does not react fast enough to change the angle of attack as the wing diverges. The instability is characterized by a homogeneous solution to Eq. (11) not including $\Delta\alpha$ and the weight balance equation, that is

$$\begin{bmatrix} I & -qA \\ -S_G & I \end{bmatrix} \begin{Bmatrix} \Delta F_{aG} \\ \Delta\theta \end{Bmatrix} = 0\,. \tag{21}$$

Equation (21) is an eigenvalue problem for q. The lowest eigenvalue is the divergence dynamic pressure q_D. We denote the corresponding eigenvector as $[F_{aGD}, \theta_D]^T$. Equation (21) can be reduced to a standard linear eigenproblem by substituting for $\Delta\theta$ in terms of ΔF_{aG} to obtain

$$(AS_G - \frac{1}{q}I)\Delta F_{aG} = 0 . \tag{22}$$

In the present aeroelastic divergence analysis only the wing is modeled, neglecting the effect of interaction with the fuselage. This effect may be of consequence for forward-swept wings, *e.g.* Rodden, [17,18].

III. SENSITIVITY CALCULATION

During a design procedure it is necessary to evaluate the derivatives (or sensitivities) of the system response with respect to vector of design design parameters p. These sensitivities could be evaluated by differentiating Eqs. (15), (16), and (17) (*e.g.*, [4]). However, this approach requires the calculation of derivatives of the matrices A and S_G which can be very costly. Here, instead, we follow [9] and [12] and utilize a modular approach by differentiating Eqs. (5), (8) and (10) with respect to p to obtain

$$JY' = f' , \tag{23}$$

where a prime denotes differentiation with respect to p and where

$$Y' = [\, F'_{aG} \quad \alpha' \quad \theta' \,]^T , \tag{24}$$

and

$$f' = [\, f'_1 \quad f'_2 \quad f'_3 \,]^T , \tag{25}$$

along with the definition $f'_i \equiv \partial f_i / \partial p$, for $i = 1, 2, 3$. The Jacobian J appearing in Eq. (23) is the same as that used in the analysis procedure as defined by Eq. (14). Equation (23) can be partially solved to yield

$$(I - qS_G A^x)\theta' = S_G B f'_1 + \frac{S_G R}{\hat{N}^T R} f'_2 + f'_3 , \tag{26}$$

$$\alpha' = \frac{f_2' - \hat{N}^T f_1' - q\hat{N}^T A\theta'}{q\hat{N}^T R}, \tag{27}$$

$$F_a' = f_1' + qR\alpha' + qA\theta'. \tag{28}$$

where B and A^x are defined in Eqs. (18) and (19). This approach does not require any derivatives of A and S_G but only partial derivatives of f_1, f_2 and f_3. For example, f_1' denotes the derivative of F_{aG} with respect to a design variable when α and θ are fixed.

To find the derivative of the divergence dynamic pressure q_D with respect to a design parameter p, we differentiate Eq. (21) at $q = q_D$ with respect to p

$$\begin{bmatrix} I & -q_D A \\ -S_G & I \end{bmatrix} \begin{Bmatrix} F_{aGD}' \\ \theta_D' \end{Bmatrix} + \begin{bmatrix} 0 & -(q_D A)' \\ -S_G' & 0 \end{bmatrix} \begin{Bmatrix} F_{aGD} \\ \theta_D \end{Bmatrix} = 0. \tag{29}$$

We premultiply Eq. (29) by the left eigenvector of Eq. (21), $[F_{aGL}^T, \theta_L^T]$, defined by

$$[F_{aGL}^T, \theta_L^T] \begin{bmatrix} I & -q_D A \\ -S_G & I \end{bmatrix} = 0, \tag{30}$$

and obtain

$$[F_{aGL}^T, \theta_L^T] \begin{bmatrix} 0 & -(q_D A)' \\ -S_G' & 0 \end{bmatrix} \begin{Bmatrix} F_{aGD} \\ \theta_D \end{Bmatrix} = 0, \tag{31}$$

or

$$q_D' = -\frac{q_D F_{aGL}^T A' \theta_D + \theta_L^T S_G' F_{aGD}}{F_{aGL}^T A \theta_D}. \tag{32}$$

Equation (32) contains derivatives of A and S_G with respect to p which we have previously managed to avoid. However, the corresponding terms can be simplified. Using the definition of A, Eq. (14), we note that

$$qA'\theta_D = \frac{\partial}{\partial p}\left(\frac{\partial f_1}{\partial \theta}\right)\theta_D. \tag{33}$$

To see how $A'\theta_D$ can be calculated without obtaining A' consider a more generic case. Let f be a function of a vector X, and let D be a given unit vector. Let X_0 be a particular choice for X, then the scalar product of the gradient $\partial f/\partial X$ at X_0 and the vector D, is the directional derivative of f in the direction D, that is

$$\left.\frac{\partial f}{\partial X}\right|_{X_0} D = \lim_{\epsilon\to 0}\frac{1}{\epsilon}\left[f(X_0+\epsilon D)-f(X_0)\right]$$
$$= \frac{d}{d\epsilon}[f(X_0+\epsilon D)]_{\epsilon=0}\,. \tag{34}$$

It is easy to check that Eq. (34) holds even if D is not a unit vector, but has arbitrary magnitude. Equation (34) provides us with a way of calculating the product $\partial f/\partial X$ times D without calculating the individual components of $\partial f/\partial X$. If we now consider Eq. (33) and use θ as the vector X and θ_D as the vector D we can write for each component f_{1_i}

$$\left(\frac{\partial f_{1_i}}{\partial \theta}\right)\theta_D = \frac{d}{d\epsilon}[f_{1_i}(\theta_0+\epsilon\theta_D)]_{\epsilon=0}\,, \tag{35}$$

where θ_0 is the nominal value of θ, and the index i, varies from 1 to n_m. Note that $\partial f_{1_i}/\partial\theta$ is a row vector with elements composed of the derivatives of f_{1_i} with respect to the individual components of θ. Equation (35) can also be written as

$$\left(\frac{\partial f_1}{\partial \theta}\right)\theta_D = \frac{d}{d\epsilon}[f_1(\theta_0+\epsilon\theta_D)]_{\epsilon=0}\,, \tag{36}$$

where $\partial f_1/\partial\theta$ is a matrix whose ith row is $\partial f_{1i}/\partial\theta$. The right-hand side is typically calculated by finite differences which involves two calculations of f_1. Then a finite-difference derivative with respect to p involves n_p more calculations of f_1 at perturbed values of p (n_p is the number of design variables). This compares with $n_m \times n_p + 1$ calculations of f_1 to obtain A' by finite differences.

The term $S'F_{aGD}$ is easier for two reasons. First, it is not very expensive to calculate K'_G by finite differences and also since $S_G = K_G^{-1}$

$$S'_G = -S_G K'_G S_G\,. \tag{37}$$

IV. OPTIMIZATION PROCEDURE

The multi-disciplinary wing optimization can be formulated as

$$\textit{minimize} \quad f(p)\,, \tag{38}$$

$$\begin{aligned} \textit{such that} \quad & g_s(p) \geq 0\,, \\ & g_a(p) \geq 0\,, \\ & g_p(p) \geq 0\,, \end{aligned}$$

where p is the vector of design parameters and g_s represents the structural constraints, g_a the aerodynamic constraints, and g_p the performance constraints. The objective function, $f(p)$, could represent any property of the aircraft, although it is common to minimize weight or cost.

While it would be possible to provide an optimization algorithm with the relations required to solve the aeroelastic problem (which would provide the information needed for the constraint analysis), the cost of the several hundred necessary system solutions would be prohibitive. This difficulty is avoided with the use of approximate optimization procedures outlined in the following sections.

IV-A. Sequential Approximate Optimization

In this approach, a sequence of approximate optimization problems or cycles is solved. At the beginning of each of these cycles, expensive constraints are approximated and an approximate set of constraints is generated. The most common form of this approximation is a simple linearization where the approximate set of constraints takes the form

$$\tilde{g}(p) = g(p_0) + \frac{\mathrm{d}g}{\mathrm{d}p}(p_0)\,(p - p_0)\;, \tag{39}$$

where $\tilde{g}(p)$ represents a set of approximate constraint functions and $g(p_0)$ represents the exact set at the initial design point p_0. The expense of this frequently used approach lies in the derivative evaluations. However, if Newton's method is used to solve the aeroelastic system, the Jacobian containing the required sensitivity terms is available.

In many circumstances a constraint is a function of several terms, some may be trivial to calculate while others may be expensive. For such constraints, it makes sense to linearize the expensive terms as opposed to the entire constraint to yield a better approximation. As an example consider the calculation of aircraft range. Range is dependent on the design parameters, (which presumably define the fuel load, wing area, altitude, etc.), the aircraft weight, and the aircraft drag. Functionally the range is given as

$$R_c = R_c(p, W, D)\,, \tag{40}$$

where W is the weight and D is the drag. The drag is considerably more expensive to calculate than the weight and should be linearized. The approximate range is given as

$$\tilde{R}_c = R_c(p, W, \tilde{D})\,, \tag{41}$$

where

$$\tilde{D}(p) = D(p_0) + \frac{\mathrm{d}D}{\mathrm{d}p}(p_0)\,(p - p_0)\,. \tag{42}$$

The approximate optimization problem is solved to produce an approximate optimum, and the process is called an optimization cycle. It is assumed that at design points far away from the initial design, the approximations will become inaccurate. Thus it is necessary to apply *move limits* on the design variables. That is, the optimization procedure will not be allowed to change design variables beyond prescribed upper and lower bounds,

$$p_l^i \leq p^* \leq p_u^i\,, \tag{43}$$

where p^* represents the cycle optimum, p_l^i is the lower design bound for the ith cycle, and p_u^i give the upper bound on the same cycle.

The result of one optimization cycle is used as the initial design point for another approximate optimization and the cycle is repeated until convergence. The number of cycles required for convergence and the size of the move limits is highly dependent on the individual problem and experimentation is often necessary.

IV-B. Variable-Complexity Modeling

Variable-complexity modeling involves combining generally expensive detailed or complex numerical analysis with simple algebraic expressions that capture the behavior of the detailed models. This dual-model approach is utilized during the optimization sequence in two methods. The first of these methods is illustrated in §V. Design Case 1, and involves the coupling of detailed analysis for one discipline and simple analysis for the other. In this example, a simple equation model of structural weight is first used in a numerical optimization sequence to obtain an aerodynamic design which approximately accounts for structural effects. Then, this design is refined with integrated aerodynamic-structural optimization using detailed models for both aerodynamics and structures.

The second method of variable-complexity modeling involves the use of simple models to approximate the behavior of the detailed models during the optimization cycles. This type of modeling is demonstrated extensively in §VI. Design Case 2, where algebraic models for wave drag and drag due to lift are utilized during optimization in addition to the detailed numerical methods.

IV-B.1. Scaled approximation

With scaled approximation, linearized approximations are replaced with simple models which capture the behavior of the more exact detailed calculations over the global design space. While such simple models capture correct behavior in the global sense, they may not be particularly accurate at a local design point and require a suitable scaling factor, β, which is defined as the ratio between the detailed and simple models:

$$\beta(p_0) = \frac{f_d(p_0)}{f_s(p_0)}, \tag{44}$$

where p_0 is the initial design point, f_d represents the result from the complex or detailed model and f_s gives the result from the simple or global-approximation model. Note that this scaling factor is held constant during a given optimization cycle.

The advantage of such an approach is twofold. First, these simple models generally do not catastrophically break down far from the linearization point like locally linearized approaches can. The consequence of this behavior is the ability to use larger move limits during a given optimization cycle. The second advantage of this method is that sensitivity terms may be computed very inexpensively from the algebraic models.

IV-B.2. Global-Local Approximation

Another form of Variable Complexity Modeling which shows promise is the combination of the local and global approximations. The following is a brief description of one such method called here *Global-Local* Approximation (GLA).

As previously mentioned the scaling factor β is held constant during a optimization cycle in the scaled procedure. It is possible to use sensitivity information from the complex model to vary the scaling factor, [19]. In this case the scaling factor is given as

$$\beta(p) = \beta(p_0) + \beta'(p_0)(p - p_0)\,, \tag{45}$$

where β' is the derivative of the scaling factor with respect to the design variables. Equation (45) can be written as

$$\beta(p) = \frac{f_d(p_0)}{f_s(p_0)} \left[1 + (p - p_0) \left(\frac{f'_d(p_0)}{f_d(p_0)} - \frac{f'_s(p_0)}{f_s(p_0)} \right) \right] , \tag{46}$$

where f'_d represents the detailed model sensitivity, and f'_s the simple model sensitivity. The advantage of this method is that it allows a further increase in move limits if the derivatives of the detailed model can be accurately obtained. A disadvantage of the method is that it requires the use of costly detailed model sensitivities.

V. DESIGN CASE 1: SUBSONIC TRANSPORT

First we consider the design of a forward-swept composite wing for a subsonic transport aircraft. The design of the wing was formulated so as to reduce the weight of the aircraft while maintaining a given range. The wing-design problem requires properties for the rest of the aircraft, which are obtained from a reference transonic transport design, described in Table

I and Ref. [20]. Due to the overwhelming computational costs associated with using non-linear transonic aerodynamics in numerical optimization procedures, the cruise Mach number was reduced from 0.78 to 0.68 to allow for the use of a compressibility-corrected vortex-lattice method, as in [5]. Furthermore, the aft-swept wing of the reference aircraft was replaced with a forward-swept wing to allow for natural laminar flow over a substantial portion of the wing, (which can be realized for moderate forward sweep).

Table I. Reference Subsonic Transport Specifications

Weight, N:		
	W_{gw}	449,400
	W_{rs}	285,200
	W_{uf}	27,380
	W_p	136,800
	W_{rw}	30,200
Wing:		
	Aspect ratio	14
	Area, m^2	83.98
	Span, m	34.29
	Thickness	12%
	Sweep	15°
	Taper Ratio	.25
Av. Cruise:		
	Mach No.	.78
	C_L	.672
	L/D	20.7
	Specific Fuel Consumption	.430
Range, m:		2.34×10^6

V-A. Subsonic Analysis

For this example, a given set of 48 points, called here the *load set nodes*, were selected for the aeroelastic analysis. These nodes were used to define mode shapes comprising a value of unity at one of the load-set nodes and zero elsewhere. With this particular set of mode shapes, the generalized displacements give the value of the vertical deflection at each of load set nodes, and the generalized forces give the value of the forces at each of these nodes. F_{aG} was evaluated by interpolating the wing deflections to the aerodynamic control points, evaluating the aerodynamic loads, and transferring these forces to the load set nodes. The generalized flexibility matrix, S_G, was calculated by applying a unit load at one load-set node and solving for the displacements at the other nodes. The vertical aerodynamic loads at the load set, are determined from a vortex-lattice method (*e.g.*, [21]), which is utilized to compute the lift and induced drag. The wing is discretized into panels, with each panel containing an element of a horseshoe vortex of strength γ_j. By enforcing flow tangency at each panel, a vector of circulation strengths Γ is computed from the wing geometry, which depends upon p, the vector of design parameters, along with the vector of vertical displacements, θ, and the angle of attack, α. The aerodynamic forces are computed from a local application oı the Kutta-Joukowski theorem, and compressibility effects are included through a Göthert transformation. The profile drag for each wing section is calculated from the measured airfoil drag polar.

The vertical displacements at the load set are calculated by finite-element analysis using a modification of the WIDOWAC program, [22]. First the nodal-displacement vector U is calculated from the aerodynamic loads and the gravitational and inertial loads. The vertical displacements at the load set θ are then extracted from U.

The interpolation of forces and deflections was accomplished with the use of an aerodynamic-structural interface. This interface provided the means to transfer forces acting on the aerodynamic panels to the load-set nodes and to transfer elastic deflections from the load-set nodes to the aerodynamic panels. Linear isoparametric shape functions were used with the grid defined by the load set nodes to unambiguously define aerodynamic points and interpolate deflections to the aerodynamic nodes or forces to the load set thus conserving work and energy. With this type of interpolation, a deflection at the *ith* node only effects the wing locally near that given node.

V-B. Induced Drag Calculation

The induced drag was calculated in the *far-field* or *Trefftz* plane, an approach commonly utilized in industry. In the far field, the negligible streamwise-flow variation yields a two-dimensional, cross-plane flow induced by the trailing vortex sheet. The circulation distribution across the wing can be represented by a Fourier series, [23], as

$$\Gamma(\psi) = 2bU_\infty \sum_{n=1}^{\infty} A_n \sin n\psi \,, \tag{47}$$

where b is the span and $\cos\psi = 2y/b$. Using the circulation distribution determined from the vortex-lattice method, the coefficients A_n can be found using a Fourier transform. Then the induced drag is calculated as

$$D_i = -\rho_\infty U_\infty^2 b^2 \sum_{n=1}^{\infty} nA_n^2 \sin^2 n\psi \,, \tag{48}$$

which may be then integrated to yield

$$C_{D_i} = \frac{C_L^2}{\pi AR}\left[1 + \sum_{n=2}^{\infty} n \left(\frac{A_n}{A_1}\right)^2\right] . \tag{49}$$

The lift coefficient is determined from the total lift, $C_L = L/qS$, with the lift force obtained by an integration of the lift per unit span as $L = \int l\, dy$ and $l = \rho_\infty U_\infty \Gamma$. The computations were performed with 32 terms in the series.

V-C. Solution Procedure

The coupled aerodynamic and structural system for this case was solved with a single Newton iteration, which was found to provide accurate results without prohibitive cost. As an initial guess for the single iteration we started with a rigid-wing approximation $F_{aG}^0 = F_{aGr} = \Phi^T F_{ar}$, $\alpha^0 = \alpha_r$, $\theta^0 = 0$, where

$$F_{aGr} = f_1(p,0,0) + q\alpha_r R \,, \tag{50}$$

$$\alpha_r = \frac{nW - \hat{N}^T f_1(p,0,0)}{q\hat{N}^T R} . \tag{51}$$

V-D. Forward-Swept-Wing Design Problem

The objective function to be minimized is the gross weight of the aircraft W given as

$$W = W_s + W_{uf} + W_p\,, \tag{52}$$

where W_s is the aircraft standard empty weight, W_{uf} is the usable fuel weight, and W_p is the payload weight. The payload weight is taken to be the same as for the reference aircraft. The usable fuel weight is a design variable adjusted by the optimization procedure so as to satisfy the range requirement. The standard empty weight of the aircraft, W_s, is calculated from the standard empty weight of the reference aircraft, W_{rs}, by assuming that structural weight savings in the wing are amplified by a growth factor η due to corresponding savings in non-structural weight and in the tail and fuselage. That is

$$W_s = W_{rs} - \eta\,(W_{rw} - W_w)\,, \tag{53}$$

where W_{rw} and W_w are the structural weight of the wings of the reference and design aircraft, respectively and η was selected to be 2.8. This value is arrived at by assuming that 50% of a wing's total weight is non-structural and that every pound of total wing weight contributes an additional 0.4 lbs. of fuselage and tail weight [24]. For this calculation, the structural wing weight of the design aircraft was calculated by summing up the weights of the structural members.

The wing structure was designed to withstand a 2.5g pull-up maneuver with a 1.5 factor of safety. This maneuver was assumed to follow an altitude loss, and occurs at 2.5 times the cruise dynamic pressure. The wing skin was made of 0-deg, ±45-deg, and 90-deg graphite-epoxy laminate with the zero direction being a design parameter used to create favorable bending-twist coupling to prevent aeroelastic divergence. Maximum strain constraints of 0.012 in the fiber and normal direction and in shear were imposed on each ply. These strain allowables are very high and are based on expected properties of future advanced composite materials. Spar caps were made of unidirectional material with a maximum stress capacity of 0.262 GPa. This flexible structure was modeled by a finite-element model shown in Fig. 1.

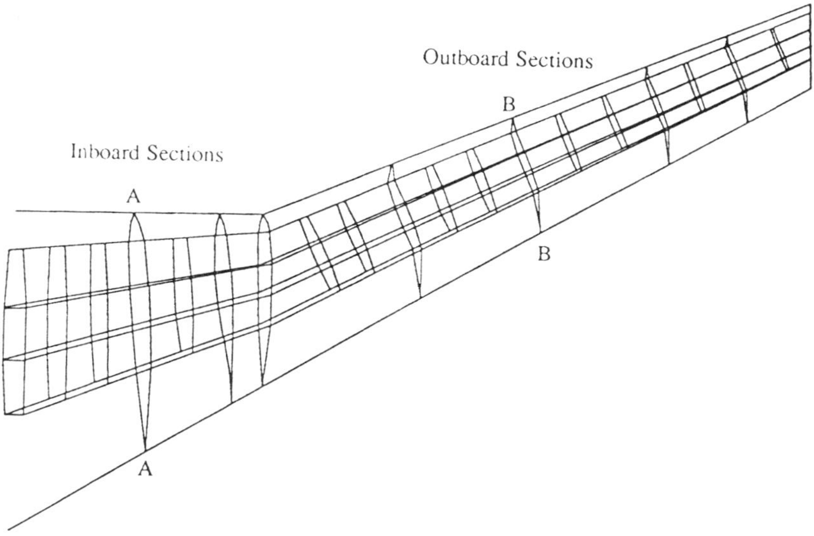

Figure 1. Structural Layout and finite-element model of the Subsonic Design Wing

The wing was required to have an aeroelastic divergence dynamic pressure larger than 1.2 times the dynamic pressure at the pull-up maneuver.

The aerodynamic design was primarily controlled by the requirement for the range to be equal to that of the reference aircraft (2340 Km). The aerodynamic loads were calculated using a vortex lattice-model with 120 panels. The airfoil section used for all spanwise stations was a natural laminar-flow airfoil designated as $HSNLF(1) - 0213$, described in [25]. The viscous drag was computed using section lift coefficients and a drag polar obtained from data in [25]. The details of the calculation of the range are given in [5].

Table II lists the design variables used for the optimization procedure. The first 6 variables are planform design variables shown in Figure 2. The next two variables are twist angle variables that define the wing twist before structural deformations are included, thus defining a *jig shape*.

The two performance variables listed in Table II affect mostly the range. The cruise dynamic pressure is a surrogate for the cruise altitude and, together with the usable fuel, is selected so as to achieve the desired range at minimum weight.

Table II. Design Variables for Subsonic Transport

8 Geometric Design Variables	1. Root chord 2. Break chord 3. Tip chord 4. Root to break 5. Break to tip 6. Sweep angle 7. Twist angle at break 8. Twist angle at tip
2 Performance Design Variables	9. Cruise dynamic pressure 10. Usable fuel weight
29 Structural Design Variables	11.–34. Ply thicknesses 35.–38. Spar-cap areas 39. Ply orientation

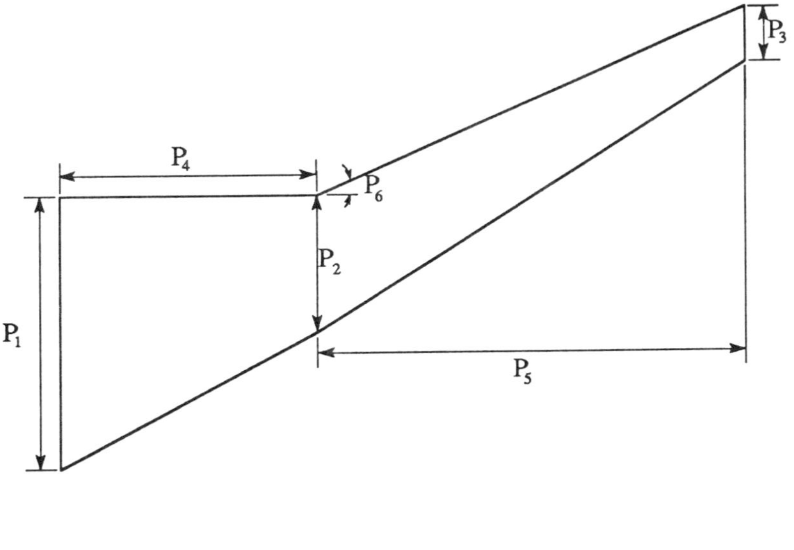

Figure 2. Subsonic Transport Planform Design Variables

There are 24 structural design variables which define the thickness of the zero, ninety, and forty-five degree plies. The wing is divided into four regions—two in the inboard box, (see Fig. 1) and two in the outboard box. The line A–A in Figure 1 marks the division of the inboard box and the line B–B marks the division of the outboard box. The ply thicknesses are taken to be constant in each region, making for 12 variables for the lower skin and 12 variables for the upper skin. Four design variables control the cross-sectional areas of the two front and two rear spar caps. Finally, one design variable is used to control the zero direction for the laminate for both lower and upper skins.

The constraints first used in the present optimization are listed in Table III and consist of the strain, stress, aeroelastic divergence, fuel volume, and range constraints discussed above along with a landing speed and a local stall constraint imposed on the design problem. The two additional design constraints in Table III were added when preliminary optimization results showed that the wing area was becoming very small. The first additional constraint is a landing speed maximum of 150 knots at 5000 ft. altitude, with an assumed maximum lift coefficient of 1.5 (flaps deployed). This new constraint significantly affected the design by imposing a lower bound on the wing area for a given gross weight. The second constraint limits the lift coefficient of the outboard wing section to 1.0 during cruise. In these sections the chords tend to get very small thus forcing the local lift coefficients to enter a stall regime. The addition of the constraint effectively prevented the outboard wing chords from becoming too small. The application of a negative twist by the optimizer was also effective in limiting the local lift coefficients.

Table III. Design Constraints for Subsonic Transport

305 Structural Constraints	1.–228. Max. skin strain
	229.–304. Max. spar-cap stress
	305. Divergence dynamic pressure
4 Performance Constraints	306. Range
	307. Fuel volume
	308. Landing speed
	309. Outboard section C_L

V-E. Approximate Optimization Procedure

The wing optimization was formulated as a sequential approximate optimization algorithm with move limits. Each approximate optimization problem starting from an initial design p^0 is formulated as:

$$\text{minimize} \quad W(p) = W(p^0) + \sum_{i=1}^{n_p} \frac{\partial W}{\partial p_i} \Delta p_i \tag{54}$$

$$\begin{aligned}
\text{such that} \quad & g_s(p^0) + \sum_{i=1}^{n_p} \frac{\partial g_s}{\partial p_i} \Delta p_i \geq 0, \\
& V(p) \geq V_r, \\
& L_s(p, W) \leq L_{s_r}, \\
& R_c(p, D) \geq R_r, \\
\text{where} \quad & D = D(p^0) + \sum_{i}^{n_p} \frac{\partial D}{\partial p_i} \Delta p_i,
\end{aligned}$$

and where the vector g_s represents the structural constraints on the stresses, strains and aeroelastic stability and n_p is the number of design variables. The quantities V and V_r are the available wing volume and the required fuel volume, respectively. The quantities L_s and L_{s_r} are the landing speed and the required landing speed, respectively. The calculated range, R_c, depends upon the total drag of the aircraft and is required to be greater than the range of the reference aircraft. The range constraint is calculated using the exact range relationships with a linear approximation to the drag.

The optimizer used is the NEWSUMT-A program [26], which is based on an extended interior penalty function procedure, and allows for various levels of constraint and objective function approximations.

V-F. Optimization Results

The optimization process was begun from a nominal design without regard to the initial quality. The design problem required approximately 57 cycles to converge, with each iteration cycle taking approximately 1100 CPU seconds on an IBM 3090. The initial and final designs are tabulated in Table IV and the initial and final planforms are sketched in Fig. 3.

We see that the final wing configuration is dramatically different from the initial design and that of the reference aircraft. The final design has a relatively low aspect ratio of 10.61 and a span of 38.8 meters; (double the total distance from the root to the tip in Table IV to obtain the total span for the aircraft). Additionally, the root chord is very large, 8.8 meters, giving the design a delta-wing appearance. The combination of this delta-type wing in conjunction with the tip extension provided a wing with low bending moments at the root to minimize the necessary structural weight.

Table IV. Initial and Final Designs of Subsonic Transport Wing

	Initial Design*	**Final Design**
Gross Weight (N)	501,100	444,600
Total Wing Weight (N)	60,290	18,490
Usable Fuel Weight (N)	40,000	36,020
Range Margin	0.9%	0.0%
Chord Lengths (m)		
Root	6.000	8.800
Break	4.000	1.727
Tip	1.000	0.523
Distance from Root to Break (m)	8.000	11.900
Distance from Break to Tip (m)	14.000	7.523
Total Wing Area (m^2)	150.0	142.19
Aspect Ratio	12.91	10.61
Leading Edge Sweep Angle (deg)	-20.05	0.0
Ply Orientation (deg)	26.1	21.3

* Some constraints violated

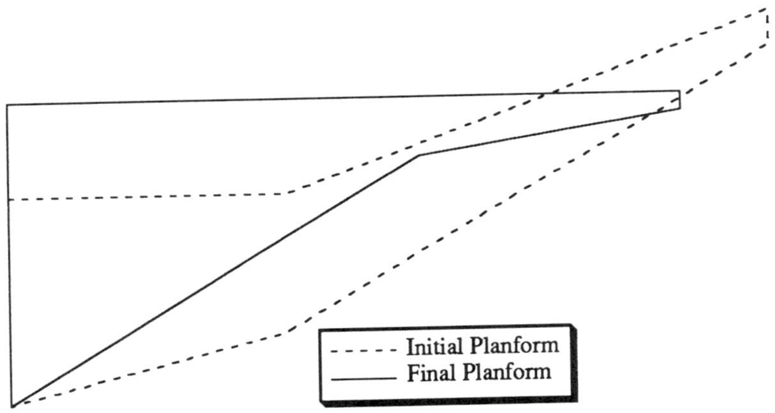

Figure 3. Initial and Final Wing Planforms – Initial Combined Design, Subsonic Transport

Of interest is how the design wing differs from the that of the reference aircraft. With so much of the wing loading for the design wing near the root, its structural weight is considerablely lower than that of the reference aircraft. While the dramatic difference between the two weights may be more pronounced than is realistic, it is clear that the design wing offers distinct weight advantages over a more conventional design. The penalty for this light, delta-type wing is in fuel costs. The design wing uses 32% more fuel than the reference aircraft. This difference can be attributed to the fact that the reference aircraft flies at a more efficient Mach number, and has better aerodynamic efficiency. However, this increase in fuel weight was more than offset by the structural-weight savings; (recall that no accounting for fuel economy is made in the objective function).

It is instructive to consider the iteration history of the design optimization. In Fig. 4, the convergence history of the gross weight is presented. The initial decrease followed by an increase in the weight resulted from the omission of a landing-speed constraint at the beginning of the optimization. When it became apparent that the wing was shrinking to the point that the aircraft could not land at a high-altitude airport, a landing-speed constraint was gradually introduced. Consequently, the wing area increased to allow for such landings with an attendant weight increase. However, once this constraint was satisfied, the wing area changed only slightly and the weight decreased again with more efficient planforms and structural thicknesses. The landing-speed constraint was active at all times, and effectively imposed a lower bound on the wing area for a given gross weight.

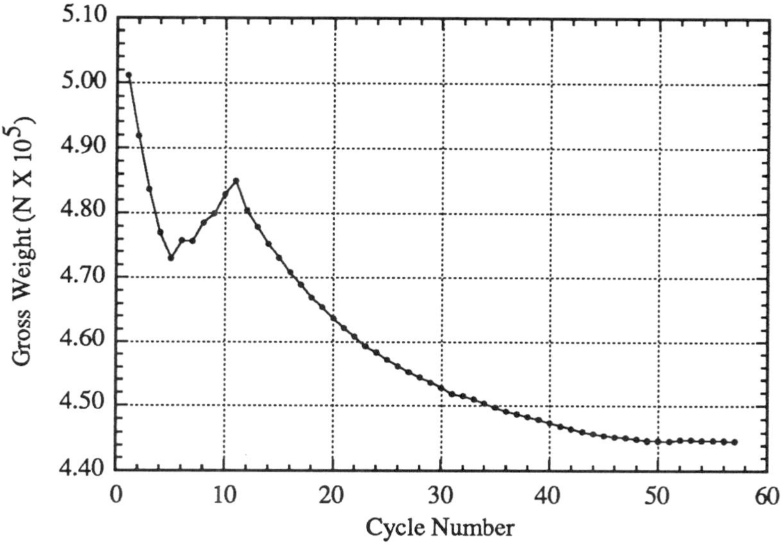

Figure 4. Weight History – Initial Combined Design, Subsonic Transport

The iteration history of the wing aspect ratio is shown in Fig. 5. An inspection of this figure reveals the inefficiency of the optimization, which first increased the aspect ratio, and then lowered it. This inefficient convergence is partly due to surplus structural material in the initial design and to the poor initial guess of the planform.

As expected the subcritical Mach number of 0.68 did not require any sweep and the optimization completely removed the sweep of the outboard-section leading edge, as illustrated in the planform design in Fig. 3. The iteration history of the leading-edge sweep angle in shown in Fig. 6. Due to the way the design variables have been selected, there was necessarily some forward sweep of the quarter chord.

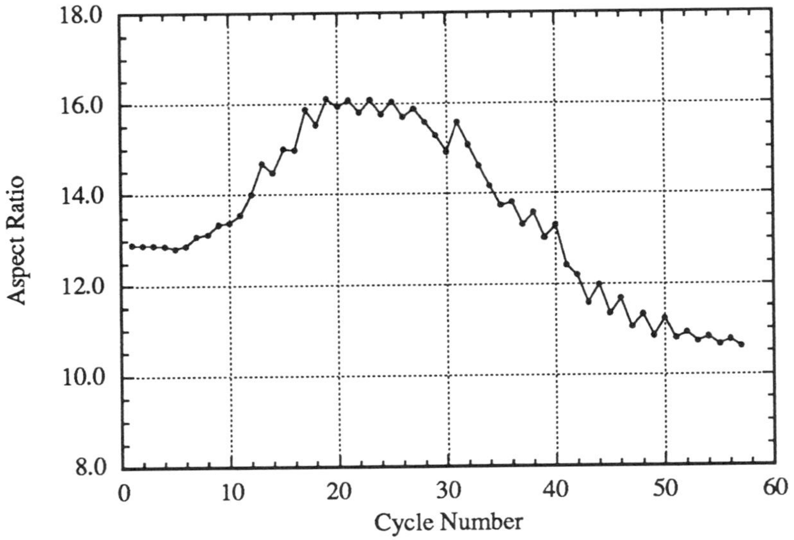

Figure 5. Aspect Ratio History – Initial Combined Design, Subsonic Transport

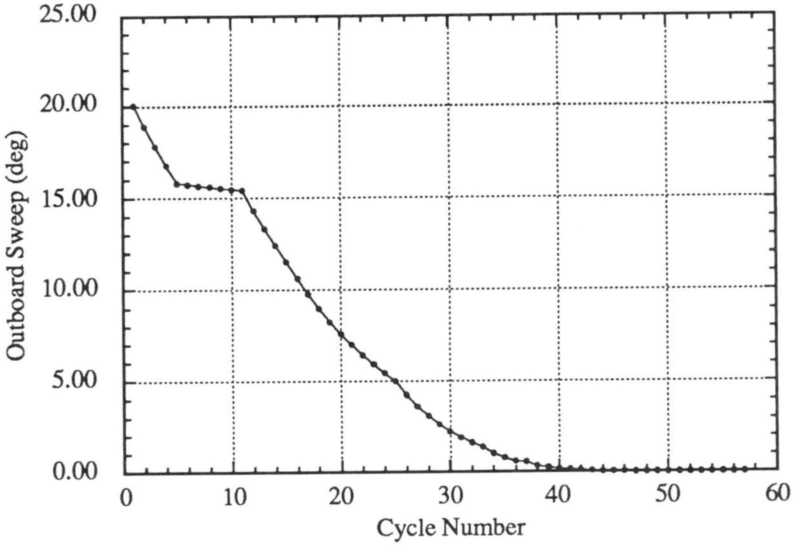

Figure 6. Outboard Leading Edge Sweep History – Initial Combined Design, Subsonic Transport

Other important lessons were learned from the convergence studies. It was found that tight move limits on the order of 4–6% were needed for the approximate optimization cycle (except at the last four cycles where the move limits were reduced to 2%). The jaggedness of the convergence history of the aspect ratio, Fig. 5, indicates that move limits could not be increased. Tighter move limits would result in smoother but slower convergence. Also, convergence required a nearly feasible initial design with only small constraint violations. Since the number of optimization cycles required for convergence was large, we believed that a better initial design would reduce CPU costs. Such an initial design could be generated by using a less complex structural model. This approach is a form of variable-complexity modeling and is discussed below.

V-G. Aerodynamic design with algebraic weight equation

Since the number of iterations and required CPU time mentioned above are quite large it seems logical that one could reduce them with a better initial design. This initial design can be provided by the use of variable-complexity modeling as discussed previously in §IV-B. For this case the simpler problem of aerodynamic design with wing weight estimated by an algebraic weight equation taken from the conceptual design code FLOPS [27], was solved. These weight estimation routines rely on correlations developed using wing weights of actual aircraft, as well as structural optimization studies with the TSO program, [28].

The calculation of the vertical loads for this problem is exactly the same as that of the previous example, while the calculation of the vertical displacements is replaced by algebraic weight relations that approximately account for the effects of geometry on wing weight. The solution to the coupled system of equations is trivial. The resulting solution is merely the rigid-wing condition given by

$$F_{aGr} = f_1(p,0) + q\alpha_r R\,, \quad \text{and} \quad \alpha_r = \frac{\frac{1}{2}nW - \hat{N}^T f_1(p,0)}{q\hat{N}^T R}\,. \tag{55}$$

These relations are very inexpensive as the generalized aerodynamic influence matrix, qA, and the generalized flexibility matrix, S_G, are unnecessary.

For the sensitivity derivatives of the rigid wing we differentiate equations (5) and (8) (ignoring θ) with respect to p to obtain

$$JY' = f' , \tag{56}$$

with

$$Y' = [\, F'_{aG} \quad \alpha' \,]^T \ , \quad \text{and} \quad f' = [\, f'_1 \quad f'_2 \,]^T \, . \tag{57}$$

and the Jacobian as

$$J = \begin{bmatrix} I & -\partial f_1/\partial \alpha \\ -\partial f_2/\partial F_{aG} & 0 \end{bmatrix} = \begin{bmatrix} I & -qR \\ \hat{N}^T & 0 \end{bmatrix} . \tag{58}$$

Note that the Jacobian for this problem contains no terms which are computationally expensive so we again have substantial computational saving over the complex model.

The design formulation and optimization procedures for the simple-model problem is almost exactly the same as that of the previous example, except that the structural constraints and design variables are unnecessary. The reduced set of design variables consist of the six planform variables, the two twist variables, and the performance variables. Note that the twist variables for this analysis represent the shape of the wing during cruise, and do not define a *jig shape* as with the flexible wing design.

In the optimization, it was not necessary to linearize the weight relations as the cost of the calculation was minimal. Typically the weight predicted by the simple-model analysis was much larger than the weight obtained from the complex model. Part of the difference is due to the high allowables corresponding to future advance materials used here and not included in the wing weight correlations. Another contribution to the difference is that the flexible analysis does not include all the load cases on the wing and all the constraints (such as buckling).

V-H. Aerodynamic design optimization results

The optimization results from the aerodynamic design with algebraic weight equations described in §V-G, is now considered. A scaled weight that corresponds to the same advanced material properties and the same load conditions and structural constraints as in the detailed finite-element model was used in the optimization. The scaling factor was chosen so that the gross weight of the simple-model and the detailed finite-element analysis would match for one planform. The planform selected for this was the final design of the combined problem described previously. With this planform the scaling factor was found to be 0.4, which represents a drastic weight difference.

The aerodynamic design problem was optimized using these scaled-down weights starting with the same initial design as that of the initial combined aerodynamic-structural design problem (see Fig. 3). The initial and final designs are presented in Table V with a sketch of the two planforms given in Fig. 7. The aerodynamic optimum obtained with the scaled weight equation closely approximates the planform for the combined aerodynamic-structural design optimum. This is an encouraging result since the goal of the aerodynamic design is an improved initial design point. Additionally, the final weight of the wing is much closer to that of the combined design. Figures 8 and 9 give the iteration history of gross weight and aspect ratio of the aerodynamic optimization with each cycle using approximately 100 CPU seconds on an IBM 3090 which is less than 10% of the combined design CPU time.

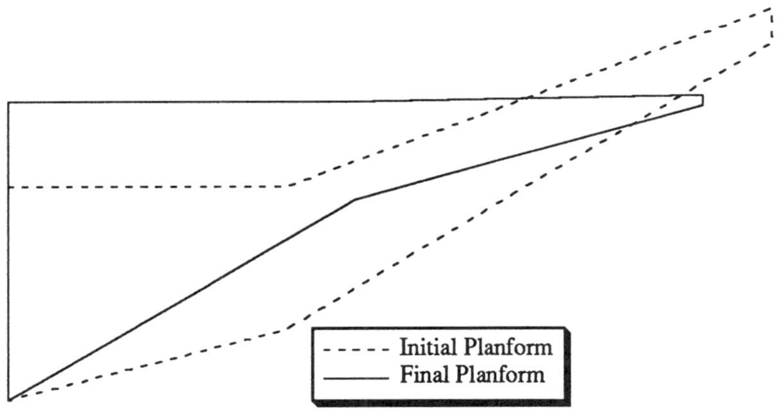

Figure 7. Initial and Final Planforms – Aerodynamic Design, Subsonic Transport

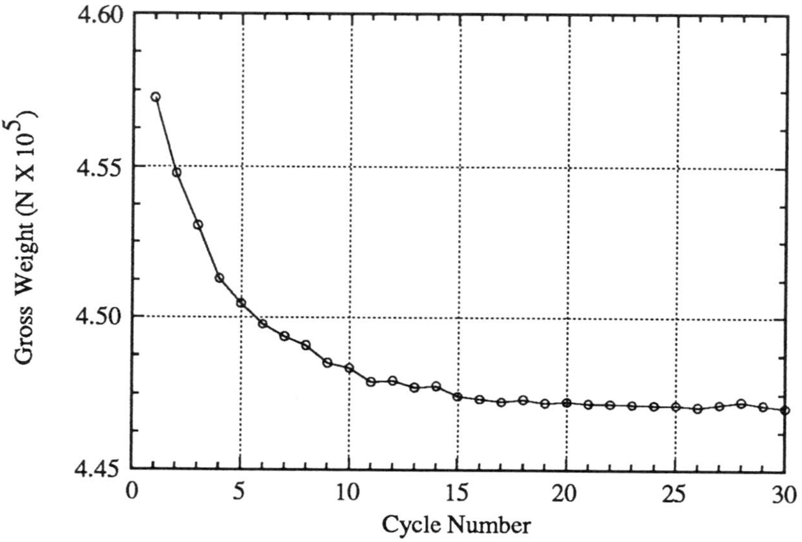

Figure 8. Weight History – Aerodynamic Design, Subsonic Transport

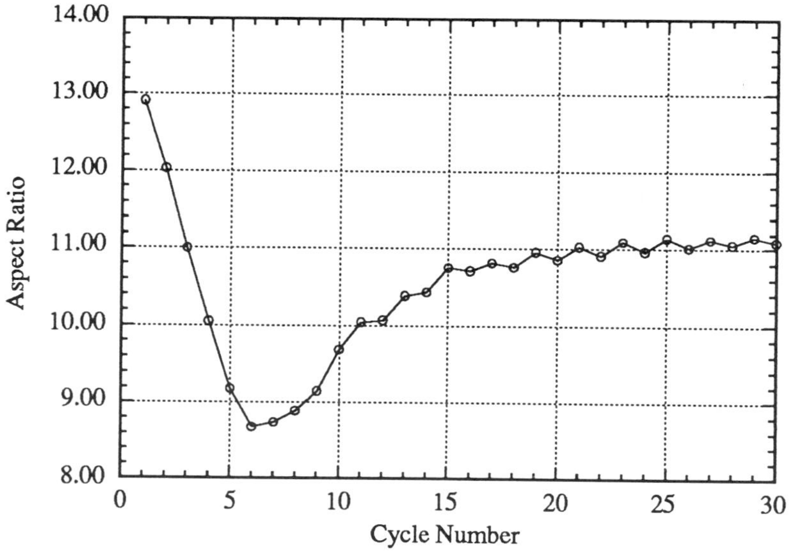

Figure 9. Aspect Ratio History – Aerodynamic Design, Subsonic Transport

Table V. Initial and Final Designs for the Aerodynamic Design Problem, Subsonic Transport

	Initial Design*	Final Design
Gross Weight (N)	457,300	447,000
Total Wing Weight (N)	26,490	20,430
Usable Fuel Weight (N)	40,000	37,780
Range Margin	1.71%	0.0%
Chord Lengths (m)		
Root	6.000	8.434
Break	4.000	2.784
Tip	1.000	0.294
Distance from Root to Break (m)	8.000	10.005
Distance from Break to Tip (m)	14.000	10.000
Total Wing Area (m^2)	150.0	141.0
Aspect Ratio	12.91	11.19
Leading Edge Sweep Angle (deg)	-20.05	-1.05

* Some constraints violated

V-I. Combined design from aerodynamic optimum

With the aerodynamic optimum in hand, the combined design problem was rerun using this design as an initial condition. Figure 10 gives a flowchart of this variable-complexity approach. The optimization was restarted from the aerodynamic optimum. Since the aerodynamic design yields only a planform and performance variables, it was necessary to estimate the structural design variables. The initial value for these structural variables was selected to be the optimum structure for the nominal combined design case. Alternatively, a relatively inexpensive structural optimization (for fixed aerodynamic design) may be used at this point to get an initial structural design. Such an optimization may be also used to determine a good weight scale factor.

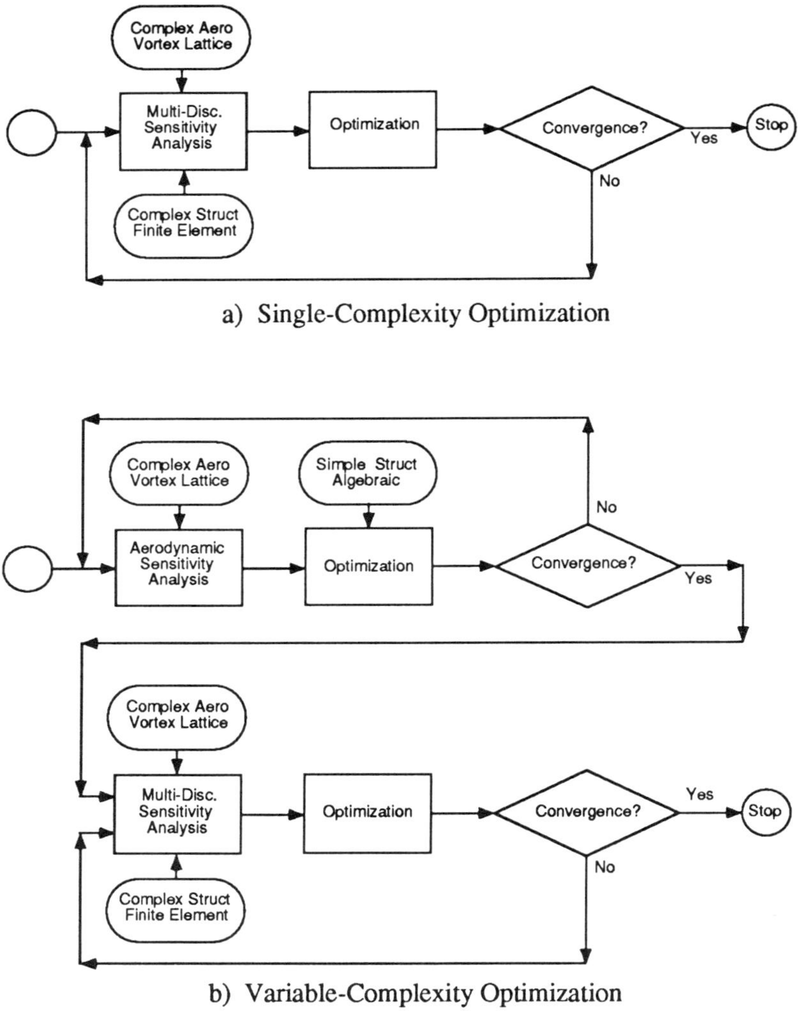

Figure 10. Variable-Comlexity Optimization Flowchart for Subsonic Transport

The initial and final designs are given in Table VI with a sketch of the planform given in Fig. 11. It is clear that this initial guess is excellent and only a minimal amount of combined optimization was required. Indeed, Fig. 12 reveals that the aspect ratio varied only slightly, and history of the gross weight presented in Fig. 13 demonstrates sufficient convergence in only 11 design cycles. Figure 14 presents the CPU times for the optimization runs and reveals a reduction of over 75% for this case over the nominal optimization. This figure also demonstrates the almost negligible cost of the aerodynamic design.

Table VI. Optimum for Combined Design from Aerodynamic Optimum for the Subsonic Transport

	Initial Design*	**Final Design**
Gross Weight (N)	446,300	443,600
Usable Fuel Weight (N)	37,780	35,440
Range Margin	0.0%	0.0%
Chord Lengths (m)		
Root	8.434	8.550
Break	2.784	2.061
Tip	0.294	0.266
Distance from Root to Break (m)	10.005	11.501
Distance from Break to Tip (m)	10.00	8.514
Total Wing Area (m^2)	141.00	141.85
Aspect Ratio	11.19	11.30
Leading Edge Sweep Angle (deg)	-1.5	0.0
Ply Orientation (deg)	21.2	21.4

* Some constraints violated

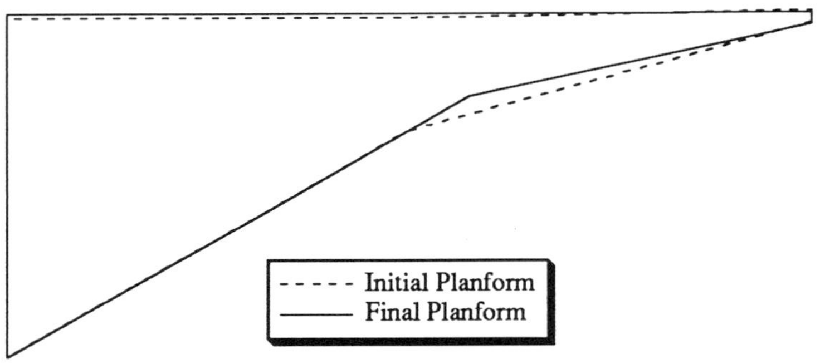

Figure 11. Initial and Final Planforms – Combined Design from Aerodynamic Final Design for Subsonic Transport

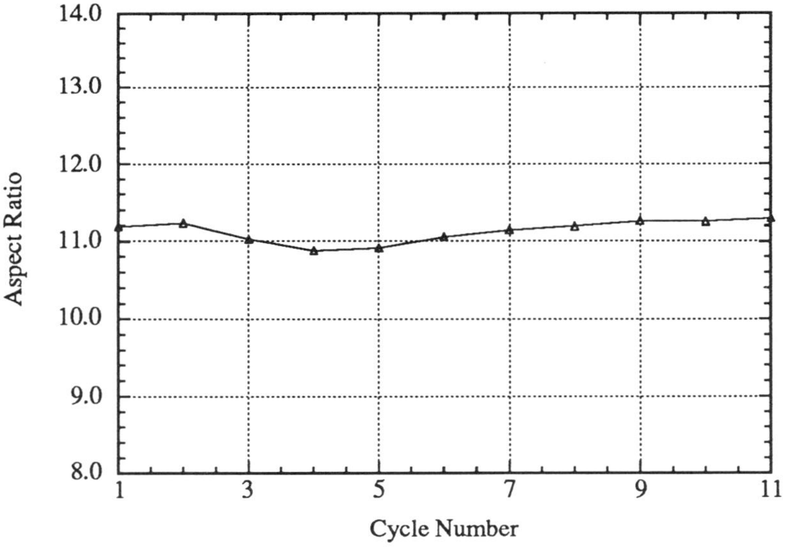

Figure 12. Aspect Ratio History – Combined Design from Aerodynamic Final Design for Subsonic Transport

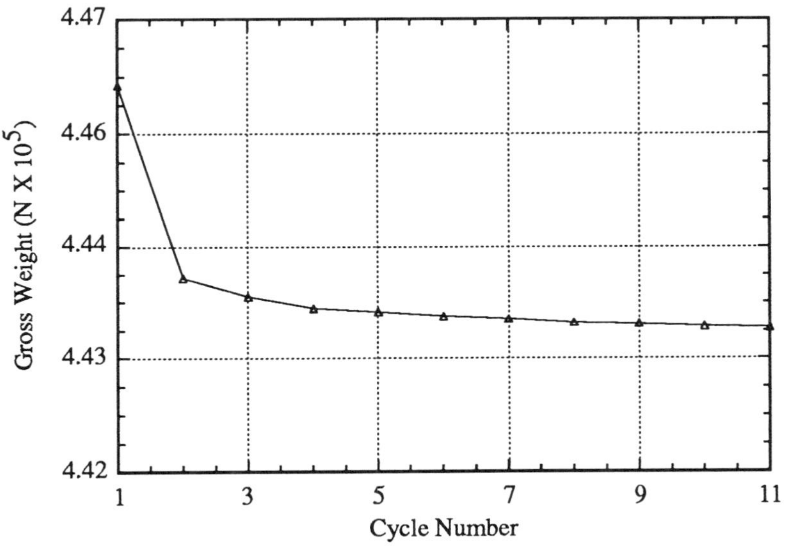

Figure 13. Gross Weight History – Combined Design from Aerodynamic Final Design for Subsonic Transport

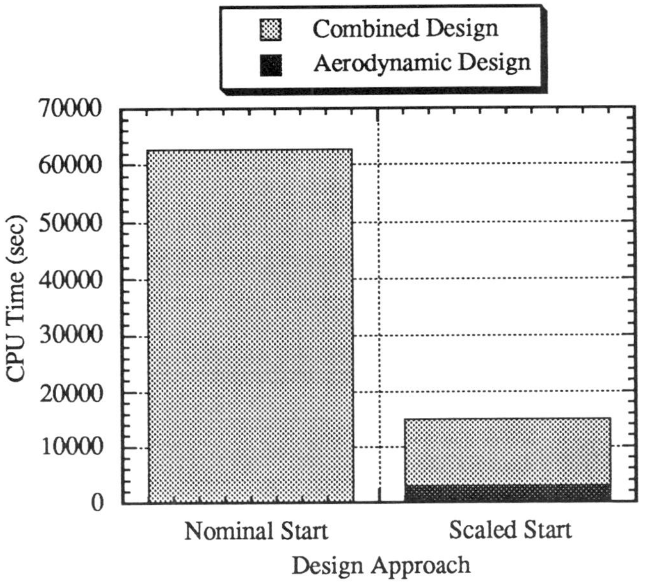

Figure 14. Computational Times for Variable-Complexity Strategies for Subsonic Transport

Comparing Table VI to Table IV, we see a gross weight reduction of 1000N, even though the final planform is not very different than the nominal case. It appears that the nominal and new design represented two local minima, but investigation reveals that this is not the case. It is now believed that numerical errors in the calculation of the drag sensitivities prevent the nominal design from reaching this lighter design. In any event, the difference in gross weight between the two designs is only 0.25%.

VI. DESIGN CASE 2: HIGH SPEED CIVIL TRANSPORT

For this case we consider the aerodynamic design of a Mach 3.0 transport aircraft with a 6500 mi. range. As with the subsonic design case, this problem was formulated to reduce the take-off gross weight while maintaining the given range. In this problem we consider only aerodynamic design using a simplified algebraic equation for the weights. This design can be considered a prelude to a combined aerodynamic-structural wing design using detailed structural models. The weights for the analysis are provided by a weight equation taken from FLOPS, as described for the subsonic aerodynamic design problem, see §V. The baseline concept for this aircraft is provided by NASA's AST3I given in [29]. Some specifications of the AST3I are given in Table VII and a sketch of the vehicle appears in Fig. 15.

VI-A. Supersonic aerodynamic analysis

As in the previous design case, drag provides the major aerodynamic consideration for the design. The drag consists of the volumetric wave drag, the drag due to lift, and the friction drag. Evaluation of these terms represents a considerable cost increase over drag evaluation for the subsonic design case (for a similar level of accuracy). This is the motivation for the use of variable-complexity modeling in the aerodynamic analysis during optimization. This modeling is specifically applied to the wave and drag to lift terms. A brief description of simple and detailed models for each of these terms follows.

An estimation of the wave drag of the wing is provided by an algebraic model given in [30]. This model is based on theoretical and empirical

Table VII. AST3I Specifications

Weights, lbs:		
	Wing	56,258
	Total structures	153,381
	Empty weight	263,658
	Operating weight	272,981
	Passenger weight(250)	41,250
	Passenger baggage weight	11,000
	Zero-fuel weight	327,796
	Mission fuel weight	385,900
	Gross weight	713,696
Wing:		
	Area, ft^2	12,185
	MAC, ft	129.65
	Span, ft	150.0
	Aspect ratio	1.85
	Root t/c (%)	2.0
	Tip t/c (%)	3.5

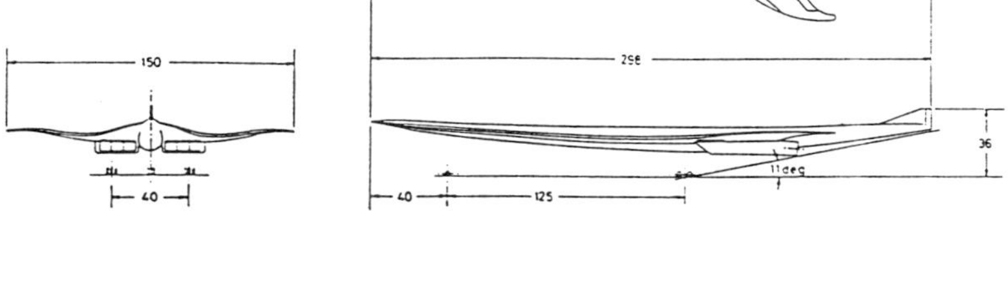

Figure 15. AST3I Configuration

studies. The algebraic model is based upon a trapezoidal wing, where the planform description is limited to the leading and trailing edge sweeps, the taper ratio, and the aspect ratio. Since the exact description of the aircraft planform is more complex then this simple description, an equivalent trapezoidal wing is utilized with the algebraic model. The details of this approximation can be found in [31].

The detailed wave-drag estimates are calculated using the classic Harris wave-drag program, [32]. This program numerically computes the value of the far-field integral arising from slender-body theory, and is known to adequately predict the wave drag of supersonic transport-class aircraft.

A simple model for the drag-due-to-lift term is provided by utilizing analytical results obtained for linearized supersonic thin-airfoil theory, [33]. As with the wave-drag approximation, these analytic estimates are only available for a simplified planform. In this case the simplified planform is an arrow wing where the geometrical description consists of the leading and trailing-edge angles. Once again it is necessary to provide an equivalent wing to the more general geometry for approximation purposes (see [31] for details).

The more detailed drag due to lift analysis is provided by the techniques developed by Carlson *et. al.*, [34–36]. In this approach, a panel method is utilized for the integral solution to the linearized supersonic potential equation, including the effects of attainable leading edge suction. The approach can be applied to wings of arbitrary planform.

The friction-drag component utilizes a single model approach with standard algebraic estimates. The flow is assumed to be turbulent, and the so-called van Driest II method is used, as recommended by Hopkins and Inouye [37], based on an extensive set of evaluations.

VI-B. Supersonic transport design problem

An analytic wing definition was created to provide a practical representation of the complex wing geometry utilizing a minimum number of design variables, [31]. These variables are listed in Table VIII and consist of 11 planform variables, 7 airfoil variables, and 3 performance variables.

The first ten planform variables define the coordinates of break points for the leading and trailing edge. These break points define three straight

Table VIII. Design Variables for HSCT Wing

11 Planform Design Variables	1. LE Break 1, x
	2. LE Break 1, y
	3. LE Break 2, x
	4. LE Break 2, y
	5. LE Wing tip, x
	6. TE Break 1, x
	7. TE Break 1, y
	8. TE Break 2, x
	9. TE Break 2, y
	10. Tip chord
	11. Wing semi-span
7 Airfoil Design Variables	12. t/c at wing root
	13. t/c at LE break 1
	14. t/c at LE break 2
	15. t/c at tip
	16. Max t/c location
	17. "Slope" of leading edge half angle and t/c
	18. Leading edge radius (subsonic sections)
3 Performance Design Variables	19. Flight fuel
	20. Initial cruise altitude
	21. Climb rate

line segments for the leading and trailing edges which are exponentially blended to provide a smooth transition between segments as in [38]. The final planform variable defines the wing semi-span as illustrated in Fig. 16.

The first four airfoil variables define a distribution of wing thickness across the span. These variables provide t/c at four wing stations and it is assumed that the actual wing thickness varies linearly between the points. The point of maximum thickness at each airfoil is assumed to be constant along the span and defined by the next design variable. Using the original AST3I aircraft as a basis, a provision for both supersonic and subsonic airfoil sections is maintained by the next two design variables. The first of

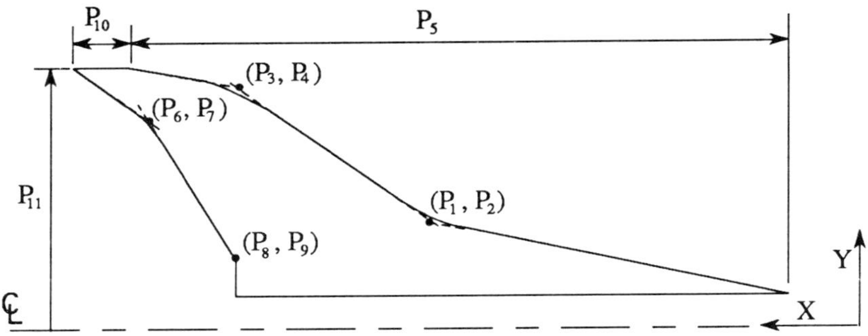

Figure 16. HSCT Planform Description Parameters

these gives the leading-edge half angle for supersonic sections by defining the linear relation between the wing t/c and this angle. The final airfoil variable provides the leading-edge radius for the subsonic sections.

The first performance variable gives the flight fuel for the aircraft which is utilized for cruise and loiter. The final two design variables provide information for the cruise profile giving the starting altitude and the climb rate. This rate is assumed to be a constant until the aircraft reaches a maximum altitude of 70,000 ft and levels off.

A total of 36 constraints, listed in Table IX, were used for this design problem. These constraints limit the wing's chords, thicknesses, TE angle, and breakpoint locations, the aircraft's approach lift coefficient, landing angle (to present tailscrape), and range. Additional constraints assure that adequate wing volume is available to hold the fuel and that lift coefficients are not too large.

VI-C. Sequential approximate optimization procedure

In this example an integrated variable-complexity approach was utilized in the optimization sequence. Since the algebraic model did an excellent job of approximating the response of the complex Harris model, the variable-complexity design approach discussed in §IV-B.1 was utilized for wave drag calculations. Unfortunately, the drag-due-to-lift term proved to

Table IX. Design Constraints for HSCT Wing

20 Geometric Constraints	1.–11.	Wing chords
	12.–15.	Wing thicknesses
	16.–19.	Break point locations
	20.	Root TE angle
16 Performance Constraints	21.	Range
	22.	Landing α
	23.	Fuel volume
	24.	Landing C_L
	25.–35.	Landing section C_l
	36.	Maximum altitude

be more difficult. The arrow-wing drag-due-to-lift approximations had difficultly capturing the the drag-due-to-lift for the more general planforms and it was necessary to utilize the more costly global-local approximation (§IV-C) approach during some optimization cycles. Another problem associated with the drag term was the *noisy* results obtained from the detailed analysis. This noise often resulted in erroneous derivative evaluations and produced erratic optimization convergence. The optimization for this problem was formulated as:

$$
\begin{aligned}
\textit{minimize}\quad & W(p) \\
\textit{such that}\quad & V_s(p) \geq V_r, \\
& C_{L_a}(p, W) \leq 1.0, \\
& \alpha_{L_a}(p, W) \leq 12.0^\circ, \\
& R_c(p, D) \geq 6500mi, \\
& c_l(p, y) \leq c_{l_{max}}, \\
& c(p, y) \geq c_{min}, \\
& g_w(p) \leq 0 \\
\textit{where}\quad & D = D_f + \beta_w(p_0)D_{w_s} + \beta_l(p)D_{l_s}, \\
\textit{and}\quad & \beta_l(p) = \beta_l(p_0) + \beta_l'(p_0)(p - p_0).
\end{aligned}
\tag{59}
$$

The quantities V_s and V_r are the available wing volume and the required fuel volume, respectively. The quantities C_{L_a} and α_{L_a} are the approach lift coefficient and angle respectively. The calculated range, R_c, depends upon the total drag of the aircraft and is required to be greater than 6500 miles. The range constraint is calculated using the exact range relationship with the approximation to the drag as where D_{w_s} and D_{l_s} give the wave and drag-due-lift results from the simple approximate models. The terms for c_l and c provide the limitations for the local lift coefficients and chords at each span location (hence the dependence on y). The final term g_w represents the wing geometric constraints which consist of minimum wing chord, thicknesses, and root trailing edge angle. Once again the optimizer utilized was NEWSUMT-A.

VI-D. Optimization results

The initial and final planforms for the optimization are shown in Fig. 17. The initial design represents an evaluation of the parametric model of the AST3I using the methods described above. Representative subsonic and supersonic airfoil sections for the initial and final configurations are shown in Figs. 18 and 19. The move limits and the drag approximation methods employed in the design sequence is shown in Table X. This sequence was the result of significant experimentation and was driven largely by the difficulties associated with approximation errors in the drag estimation. The difficulty associated with poor derivatives becomes more noticeable the closer a design approaches convergence. This explains the transition from the GLA approximations to the scaled approximations and the small move limits shown in Table X. Of all the constraints employed, only four were active at the conclusion of the optimization. They were the landing angle-of-attack limitation, the range constraint, the altitude limitation, and the root trailing-edge sweep limitation.

Figures 20 and 21 present convergence histories of the designs for aircraft gross weight and range. The oscillations in both range and gross weight due to the drag approximation errors are apparent in both figures.

Table XI presents some details of the initial and final designs. The final designs not only weigh less, but meet the range and landing angle-of-attack requirements. The principal changes made to the initial design were to decrease the inboard leading-edge sweep and to increase the outboard

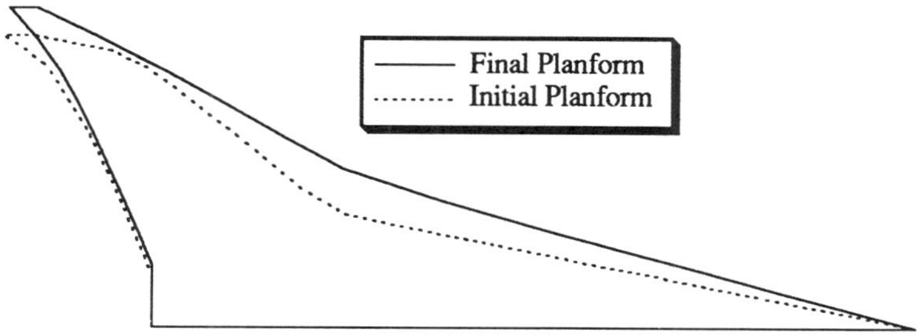

Figure 17. HSCT Initial and Final Planform

Table X. Optimization Schedules

Cycles	**Move Limits**	**Drag Approx Wave**	**Drag Approx Lift**
1–3	7%	Scaled	GLA
4–7	5%	Scaled	GLA
8–10	3%	Scaled	GLA
11–13	2%	Scaled	Scaled
14–15	1%	Scaled	Scaled

sweep and the reduction in wing thickness. For example, the wing area was redistributed inboard compared to the baseline configuration and was increased by 17.4%. The span was increased by 9.8%. The tendency in the designs was to trade increased wing weight (thinner wings) for aerodynamic efficiency (decreased wave drag) to save mission fuel. As may be seen in Table XI, the wing weight increased while fuel weight decreased for a net savings. Figure 22 presents a weight breakdown of the initial and final designs.

Aerodynamically, the final configuration shows improvement in wave drag and drag due to lift. Figure 23 shows drag breakdowns for the aircraft at the start of cruise. In Fig. 24, the lift-to-drag ratios are presented for the initial and final designs. There was an increase of 12.2% in the maximum L/D ratio from the initial to final design. Of this increase, approximately

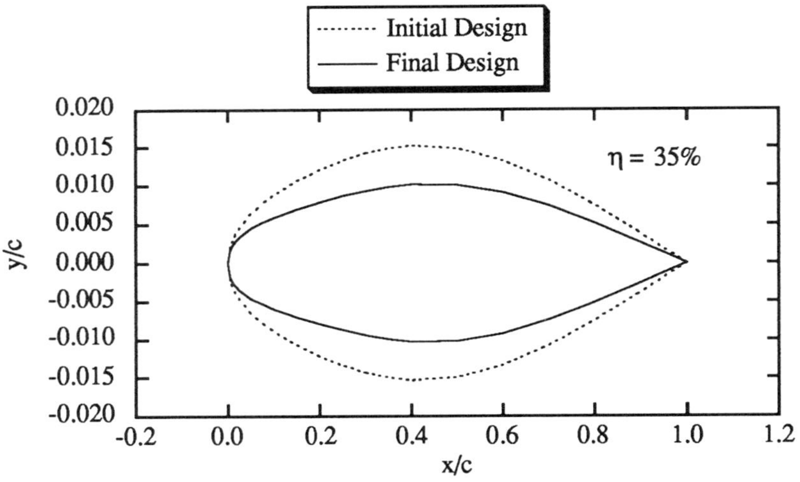

Figure 18. Subsonic LE Airfoil Thickness Distribution for HSCT ($\eta = 2y/b$)

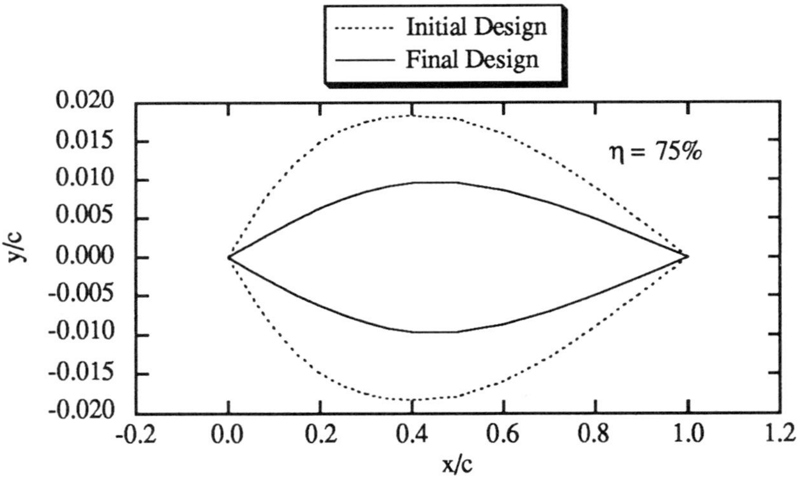

Figure 19. Supersonic LE Airfoil Thickness Distribution for HSCT ($\eta = 2y/b$)

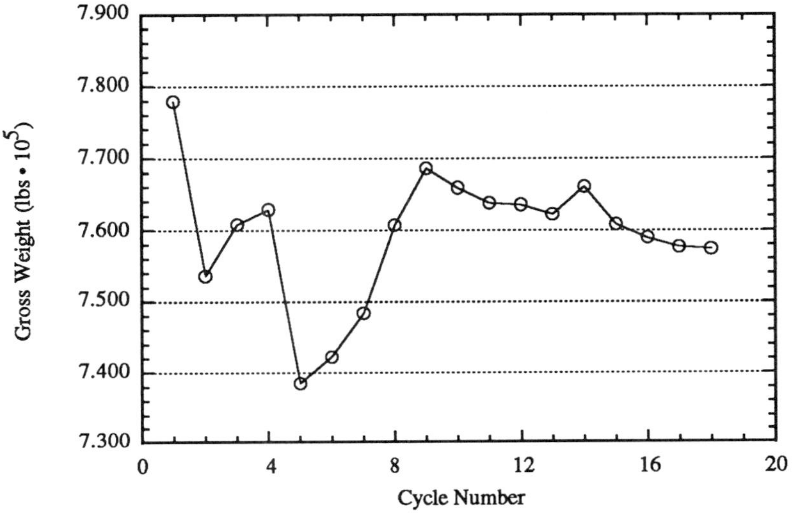

Figure 20. Weight Convergence History – HSCT Design

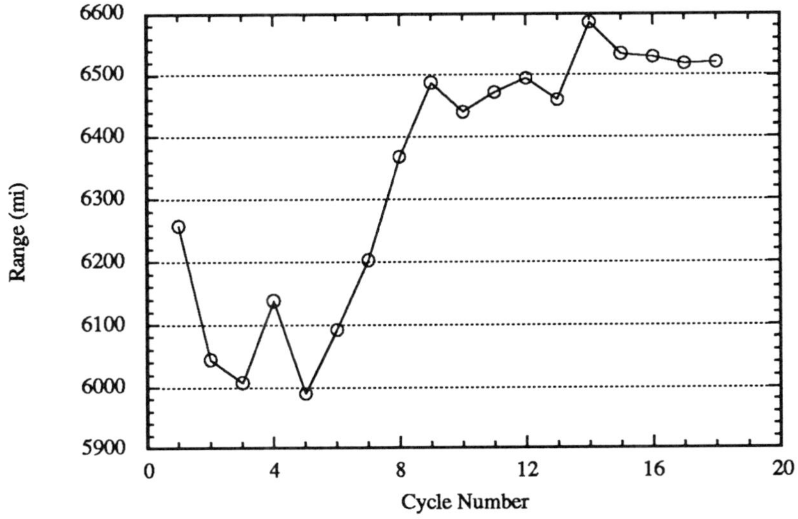

Figure 21. Range Convergence History – HSCT Design

Table XI. Initial and Final Design for HSCT Wing

	Initial Design*	**Optimum Design**
Weights		
Gross Weight (lbs)	788,600	757,400
Wing Weight (lbs)	92,400	104,900
Fuel Weight (lbs)	400,000	357,900
Range Margin (%)	-2.35	0.32
Landing Angle Margin (%)	-20.4	0.00
Sweep (deg)		
Inboard LE	79.56	76.02
Outboard LE	53.13	59.85
Inboard TE	0.00	0.00
Outboard TE	27.14	26.61
Thickness (% t/c)		
Root	2.000	1.903
LE Break 1	3.500	2.058
LE Break 2	3.500	1.902
Tip	3.500	1.894
Wing Area ($ft^2 \times 10^4$)	1.223	1.436
Aspect Ratio	1.854	1.869
L/D Max	7.900	8.865

* Some constraints violated

70% was achieved through reductions in C_{D_0} and the remaining 30% in drag due to lift. The points corresponding to start and end of cruise are indicated. While the aircraft flies close to the maximum L/D ratio, the altitude limitation restricts the ability to fly a more nearly optimal cruise profile.

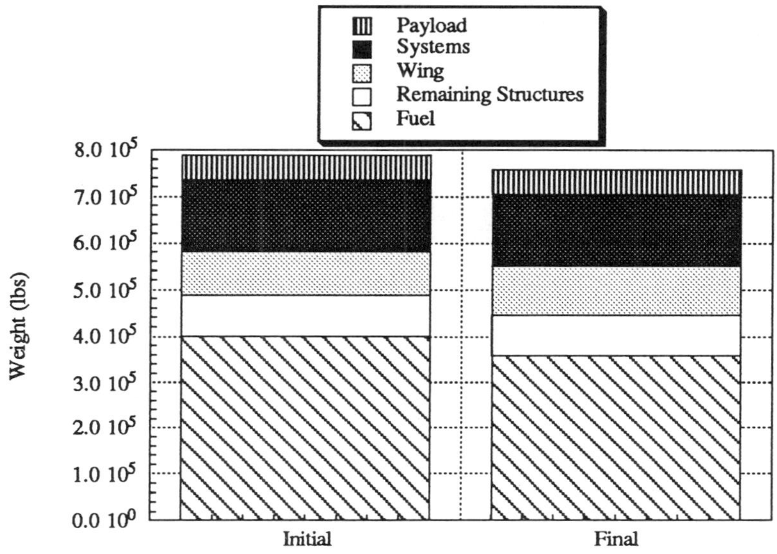

Figure 22. Weight Breakdowns – HSCT Design

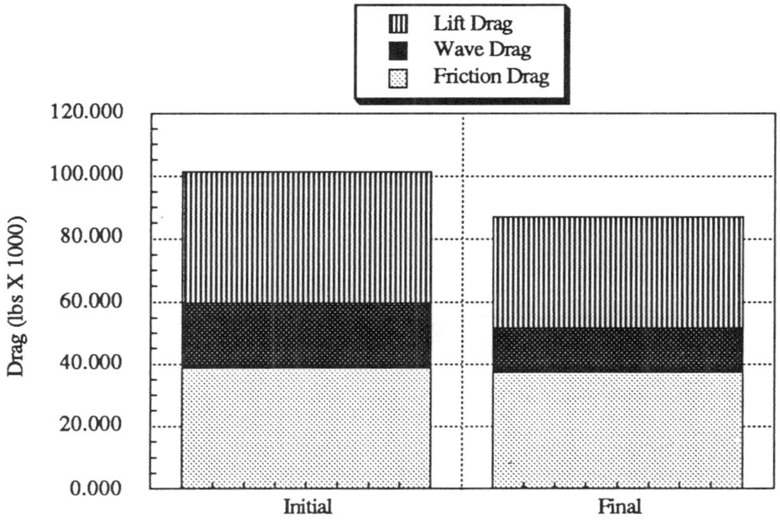

Figure 23. Drag Breakdowns at Start of Cruise – HSCT Design

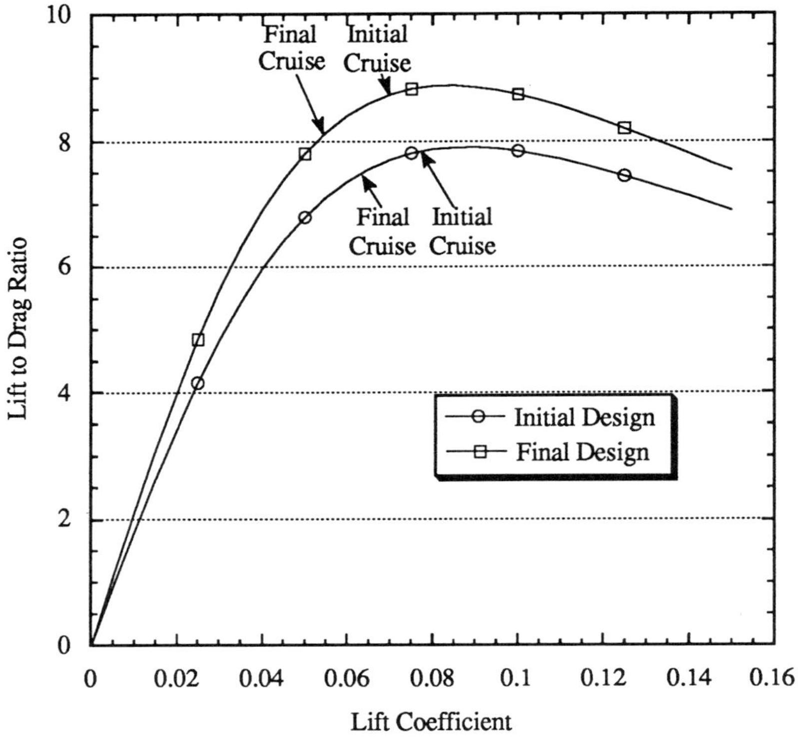

Figure 24. Lift to Drag Ratios – HSCT Design

VII. CONCLUDING REMARKS

A generalized treatment of the integrated aerodynamic-structural design problem has been described. The generality and modularity of the technique presented allows the utilization of any analysis technique desired without changing the aeroelastic formulation. This technique is also flexible enough to incorporate additional influences beyond that of the coupled aerodynamic-structural design problem.

It is hoped that the limited number of examples that have been presented will instill the reader with an appreciation for the requirements and difficulties associated with multi-disciplinary design problems. At this time, work is progressing to include more complex structural analysis into the supersonic transport example and further develop the variable-complexity techniques for the integrated design problem.

VIII. ACKNOWLEDGEMENTS

This work has been supported over the past several years by NASA Langley Research Center under grants NAG–1–505, NAG–1–603, NAG–1–1160, and by the National Science Foundation under grants DMC–8615336 and DDM–9008451. Thanks is given to these institutions for their support. The authors also wish to thank the following individuals for their extensive contributions to this work over the years: G.J. Strauch, W.M. Eppard, P.J. Kao, D. Polen, M. Rais-Rohani, and M.G. Hutchison of VPI & SU and J. Sobieski at NASA Langley. Finally, we must thank L.A. McCullers of ViGYAN Research Associates for his help in implementing the weight relations from FLOPS, and P. Coen of NASA Langley for providing us with help on the AST3I.

IX. NOMENCLATURE

A – Generalized aerodynamic influence matrix
AR – Aspect ratio
b – Total wing span
$c(y)$ – Local wing chord
c' – Thrust specific fuel consumption
C_D – Drag coefficient
C_{D_i} – Induced drag coefficient
$C_l(y)$ – Section lift coefficient
C_L – Lift coefficient
D – Drag
e – Span efficiency
f_1 – Aerodynamic *black box*
f_2 – Balanced flight equation
f_3 – Structural *black box*
F_a – Aerodynamic load vector
F_{aG} – Generalized force vector
F_{aGD} – Generalized force vector at divergence
F_{aGr} – Generalized forces at rigid wing conditions
F_I – Gravitational and inertial force vector
I – Identity matrix
J – Jacobian matrix
K_G – Generalized stiffness matrix
L – Lift
M – Mass matrix
n – Load factor
n_m – Number of mode shapes
n_p – Number of design variables

N	– Summation vector	V	– Wing volume
p	– Design variable Vector	W	– Total aircraft weight
q	– Dynamic pressure	W_p	– Payload weight
q_D	– Divergence dynamic pressure	W_s	– Structural wing weight
		W_{uf}	– Usable fuel weight
R	– Generalized incremental aerodynamic force vector	W_w	– Design wing weight
		α	– Angle of attack
R_c	– Calculated design aircraft range	α_r	– Rigid angle of attack
		Γ	– Vorticity vector
R_r	– Reference aircraft range	η	– Growth factor
S	– Total wing area	θ	– Generalized displacement
S_G	– Generalized flexibility matrix	Ψ	– Vector of mode shapes vector
u	– Vector of displacements		

X. REFERENCES

1. Haftka, R.T., "Structural Optimization with Aeroelastic Constraints – A Survey of US Applications," *International Journal of Vehicle Design*, **7**, pp. 381–392 (1986).
2. Zeiler, T.A., and Weisshaar, T.A., "Integrated Aeroservoelastic Tailoring of Lifting Surfaces," *Journal of Aircraft,* **25**, pp. 76–83 (1988).
3. Grossman, B., Strauch, G.J., Eppard, W.M., Gurdal, Z., and Haftka, R.T., "Integrated Aerodynamic/Structural Design of a Sailplane Wing", AIAA Paper 86-2623, *AIAA Aircraft Systems, Design & Technology Meeting*, Dayton, Ohio, (1986).
4. Haftka, R.T., Grossman, B., Eppard, W.M., Kao, P.J., and Polen, D., "Efficient Optimization of Integrated Aerodynamic-Structural Design," *International Journal for Numerical Methods in Engineering,* **28**, pp. 593–607 (1989).
5. Grossman, B., Haftka, R.T., Kao, P.-J., Polen, D.M., Rais-Rohani, M., and Sobieszczanski-Sobieski, J., "Integrated Aerodynamic-Structural Design of a Transport Wing", *Journal of Aircraft*, **27**, No. 12, pp. 1050–1056 (1990).

6. Unger, E.R., Rais-Rohani, M., Hutchison, M.G., Haftka, R.T., and Grossman, B., "Multidisciplinary design of a Subsonic Transport Wing", *Proceedings of the Third Air Force / NASA Symposium On Recent Advances in Multidisciplinary Analysis and Optimization*, San Francisco, Ca. (1990).
7. Gallman, J.W., Kroo, I.M., and Smith, S.C., "Design Synthesis and Optimization of Joined-Wing Transports," AIAA Paper 90-3197, *AIAA/AHS/ASEE Aircraft Design, Systems and Operations Conference*, Dayton, Ohio (1990).
8. Livne, E., Friedmann, P. P., and Schmit, L. A., "Studies in Integrated Aeroservoelastic Optimization of Actively Controlled Composite Wings," AIAA Paper 91–1098–CP (1991).
9. Sobieszczanski-Sobieski, J., "Sensitivity Analysis and Multidisciplinary Optimization for Aircraft Design: Recent Advances and Results," Paper ICAS–88-1.7.3, *16th Congress of the International Council of the Aeronautical Sciences*, Jerusalem Israel, pp. 953–964 (1988),
10. Wrenn, G.A., and Dovi, A.R., "Multilevel Decomposition Approach to the Preliminary Sizing of a Transport Aircraft Wing," *Journal of Aircraft*, **25**, 632–639 (1988).
11. Malone, B., and Mason W.H., "Multidisciplinary Optimization in Aircraft Design Using Analytic Technology Models", AIAA Paper 91–3187, *AIAA/AHS/ASEE Aircraft Design Systems and Operations Meeting*, Baltimore, MD. (1991)
12. Sobieszczanski-Sobieski, J., "On the Sensitivity of Complex Internally Coupled Systems," AIAA Paper No. CP–88–2378, and *AIAA Journal*, **28**, No. 1, pp. 153–160 (1990).
13. Rais-Rohani, M., "Integrated Aerodynamic-Structural-Control Wing Design," Ph.D. Dissertation, Virginia Polytechnic Institute and State University, August (1991).
14. Karpel, M., "Reduced Size First-Order Subsonic and Supersonic Aeroelastic Modeling," AIAA Paper 90-1154-CP (1990).
15. Dowell, E.H., Curtiss, H.C. Jr., Scanlan, R.H., and Sisto, F., *A Modern Course in Aeroelasticity*, Kluwer Academic Publishers, Boston (1989).
16. Haftka, R.T., Sobieszczanski-Sobieski, J., and Padula, S.L., "On Op-

tions for Interdisciplinary Analysis and Design Optimization", *Structural Optimization* (in press).

17. Rodden, W. P., "Aeroelastic Divergence of Unrestrained Vehicles", *Journal of Aircraft,* **21**, No. 1, pp. 94–96 (1984).
18. Rodden, W. P., "Comment on *General Formulation of the Aeroelastic Divergence of Composite Swept-Forward Wing Structures,*" *Journal of Aircraft*, **26**, No. 7, pp. 694–695 (1989).
19. Haftka, R. T., "Combining Global and Local Approximations," *AIAA Journal*, **29**, No. 9, pp. 1523-1525 (1991).
20. Turriziani, R.V., "Continuation of a Study of the Effects of Advanced Technologies on the Sizing and Fuel Efficiency of a 150 Passenger 1220 nmi Airplane," File: 3-92000/5LTR-007 (1985).
21. Moran, J., *An Introduction to Theoretical and Computational Aerodynamics*, John Wiley and Sons, Inc., New York, pp. 130–135 (1984).
22. Haftka, R.T. and Starnes, J.H. Jr., "WIDOWAC: Wing Design Optimization with Aeroelastic Constraints—Program Manual," NASA TM X-3071 (1974).
23. Karamcheti, K., *Principals of Ideal-Fluid Aerodynamics*, John Wiley and Sons, Inc., New York, pp. 562–563 (1966).
24. McCullers, L.A., *Private communication* (1990).
25. Sewall, W.G., McGhee, R. J., Viken, J. K., Waggoner, E. J., Walker, B. S. and Millard, B. F., "Wind-Tunnel Results for a High-Speed, Natural Laminar-Flow Airfoil Designed for General Aviation Aircraft," NASA TM-87602 (1987).
26. Grandhi, R.V., Thareja, R. and Haftka, R.T., "NEWSUMT-A: A General Purpose Program for Constrained Optimization using Constraint Approximations," *ASME Journal of Mechanisms, Transmissions and Automation in Design,* **107**, pp. 94–99 (1985).
27. McCullers, L.A., "FLOPS - Flight Optimization System, Version 2.0, User's Guide," PRC Systems Services Report, Hampton, VA (1986).
28. McCullers, L.A., Lynch, R.W., "Dynamic Characteristics of Advanced Filamentary Composites Structures," AFFDL-TR-73-111, Vol. II (1974).
29. Robins, A. Warner, Dollyhigh, S.M., Beissner, F.L., Jr., Geiselhart, K., Martin, G.L., Shields, E.W., Swanson, E.E., Coen, P.G., and Morris, S.,

"A Concept Development of a Mach 3.0 High-Speed Civil Transport", NASA TM-4058 (1988).

30. Schemensky, R. T., "Development of an Empirically Based Computer Program to Predict the Aerodynamic Characteristics of Aircraft," AFFDL-TR-73-144, Vol. 1 (1973).
31. Hutchison, M.G., Unger, E.R., Mason, W.H., Grossman, B., and Haftka, R.T., "Variable-Complexity Aerodynamic Optimization of an HSCT Wing Using Structural Wing-Weight Equations," *AIAA Paper 92-0212 presented at the 30th Aerospace Sciences Meeting & Exhibit,* Reno, NV (1992).
32. Harris, Roy V., Jr., "An Analysis and Correlation of Aircraft Wave Drag," NASA TM X-947 (1964).
33. Jones, R.T., and Cohen, D., *High Speed Wing Theory*, Princeton University Press, pp. 197-202 (1960).
34. Carlson, H. W., Miller, D. S., "Numerical Methods for the Design and Analysis of Wings at Supersonic Speeds," NASA TN D-7713 (1974).
35. Carlson, H. W., Mack, R. J., "Estimation of Leading-Edge Thrust for Supersonic Wings of Arbitrary Planforms," NASA TP-1270 (1978).
36. Carlson, H. W., Mack, R. J., Barger, R. L., "Estimation of Attainable Leading-Edge Thrust for Wings at Subsonic and Supersonic Speeds," NASA TP-1500 (1979).
37. Hopkins, E.J., and Inouye, M., "An Evaluation of Theories for Predicting Turbulent Skin Friction and Heat Transfer on Flat Plates at Supersonic and Hypersonic Mach Numbers," *AIAA Journal*, **9**, No. 6, pp. 993–1003 (1971).
38. Barnwell, R., "Approximate Method for Calculating Transonic Flow About Lifting Wing-Body Configurations", NASA TR R-452, pp. 58-61 (1976).

Actuator and Sensor Placement for Control of Flexible Structures

K.B. Lim

NASA Langley Research Center
Hampton, VA 23665

W. Gawronski

Jet Propulsion Laboratory
California Institute of Technology
Pasadena, CA 91109

I. INTRODUCTION

A. PROBLEM DESCRIPTION

The problem of actuator and sensor placement is closely connected with the controllability and observability properties of an open-loop system. This is particularly the case since during the early stages of a control system design, a precise specification of the closed-loop performance or control gains is typically not available. This fact motivates one to review and investigate the concept of state controllability and observability in the following.

The notion of state controllability was originally introduced by Kalman (1960) [20] for linear time-varying systems to define the conditions under which an input may arbitrarily influence the states. Although finite time and piecewise continuity of the input time histories are assumed in the definition of controllability, it is more useful for engineering applications that a measure of controllability would, in addition, involve the magnitude and direction of the controllable states, as well as control effort and its duration. The above need has been recognized early on and, subsequently, various definitions of the degree of controllability have been proposed (for example, [17, 18, 31, 43, 44]) to address the problem of optimal actuator and sensor location for control design.

CONTROL AND DYNAMIC SYSTEMS, VOL. 57

For linear time-invariant systems,

$$\dot{x} = Ax + Bu; \qquad y = Cx \tag{1}$$

the standard definition of controllability is based on the properties of the controllability matrix:

$$M_c = \begin{bmatrix} B & AB & \cdots & A^{n-1}B \end{bmatrix} \tag{2}$$

The set of states that can be reached from the zero state within a finite time is given by the range space of the above matrix, and the dimension of the controllable subspace is given by the rank of the controllability matrix. An important and useful property of the controllable subspace is that a system at any initial state within the subspace can be transferred to any final state within the subspace in finite time and control energy (see for example Chapter 1 of [24]).

In theory, the rank of matrices is well defined; in practice however, the rank of the controllability matrix is frequently a fuzzy number because indicators of the rank of a matrix (such as its singular values) often gradually decrease to machine accuracy. This means that although a system may in theory be full rank, i.e., fully controllable, the amount of control effort required may be physically unrealizable. In those cases for practical purposes, the system is considered not fully controllable. This dilemma, often encountered in actuator and sensor placement problems, is highlighted recently in [25] and involves an explicit or implicit (i.e., by default) decision that a control designer must eventually make.

Although it is well known (see for example page 61 of [24]) that the controllable subspace is identical to the subspace spanned by the set of eigenvectors corresponding to the set of controllable poles, no further geometrical properties have yet been investigated in the context of actuator placement problem. For instance, not all directions are equally controllable in the controllable subspace and indeed, the principal values and directions (see for example [38]) can be used to define the degree of controllability in state space. The set of principal controllable directions is not identical to the set of controllable eigenvectors, although they span the same subspace. Hence, a relationship between the two sets of directions is of interest. Furthermore, the relationship between modal controllability and observability (as in [15, 16, 19, 22]) and generic state controllability and observability (as in [4, 20, 24, 38]) is investigated. This relationship is important for a class of actuator and sensor placement techniques based on modal controllability and observability concepts (see for example [15, 16, 17, 22]).

To summarize, the goal of our investigation is basically twofold: (1) to define a useful metric to quantify the degree of goodness of an actuator (sensor) configuration, and (2) to develop an algorithm to compute the optimal configuration effectively.

B. ACTUATOR/SENSOR PLACEMENT MODEL

To define the actuator and sensor location model for flexible structures, consider the second-order system

$$M\ddot{\xi} + Q\dot{\xi} + K\xi = Eu \tag{3}$$

with the translational/rotational displacements (see [25] for the case with velocity and/or acceleration measurements)

$$y = F\xi \tag{4}$$

In the above equations, ξ denotes a large finite-element displacement vector (for example, 3828 degrees-of-freedom (DOF) for the experimental structure considered in §VII), and E is a matrix whose columns are formed from columns of the identity matrix. Each element of u corresponds to a candidate physical actuator whose DOF location is specified by the location of the nonzero entry in the corresponding column of E. For actuator candidate locations (for example, 50 candidate locations in §VII) the dimension of matrix E will be 3828 by 50. The displacement output matrix F is formed from rows of the identity matrix. The sensor locations are specified by the location of the nonzero entry in the corresponding rows of F. In the case when actuators or sensors are not parallel to a defined DOF, the corresponding column of E or row of F will consist of a vector of direction cosines. For possible sensor locations (for example, 100 candidate locations), the dimension of the output matrix will be 100 by 3828.

In modal form, the above equations can be written as

$$\ddot{\eta} + diag(2\zeta\omega)\dot{\eta} + diag(\omega^2)\eta = \Psi^T Eu \tag{5}$$

$$y = F\Psi\eta \tag{6}$$

where

$$\xi = \Psi\eta \tag{7}$$

Ψ denotes the structural mode shape vector corresponding to all significant modes (for example, 12 modes are considered in §VII so that the dimension of matrix Ψ is 3828 by 12). The significant modes are defined to be a subset of structural modes that are important in terms of their degree of participation in a dynamic environment, through an analysis independent of the sensors and actuators. As an example, a set of significant modes could be deduced from a combination of modes that significantly influence a line-of-sight motion [26], modes that are in the bandwidth of a known disturbance [23, 12], and modes excited by a tracking command [12].

C. ORGANIZATION OF PAPER

This paper is organized as follows. In §II, a survey of previous work in actuator and sensor placement is given. Due to the difference in the definition of what constitutes an optimal placement, several different approaches are presented in the literature and some connections between them are given. However, an underlying theme in the various definitions is the degree of controllability and observability, which is discussed in §III. In this section, controllability and observability grammians are analyzed in the context of actuator and sensor placement. The geometric description of grammians leads to physical interpretations of controllability and observability concepts. In §IV, the grammians are related to the performance of a linear optimal regulator and an optimal state estimator. In §V, tests and measures of Popov, Belevitch, and Hautus (PBH) [19] and their relationship to state space controllability and observability conditions are examined. In §VI, the explicit forms of grammians for flexible structures are used for actuator and sensor placement. *Modal grammian coefficients* are defined and used as a measure of modal controllability and observability. The simple form, and especially the uncoupled contribution of each actuator to a particular mode, leads naturally to optimal location and minimal number of actuators or sensors that maintain a significant level of controllability and observability. The relationship between the gross measure of controllability and observability and modal grammian coefficients, and an interpretation of energy optimal degree of controllability of Longman et al., are given. In §VII, a design example illustrating this approach for optimal actuator placement is given for a model of a space structure testbed. Comparisons of closed-loop performance based on several actuator configurations are given. Finally, §VIII gives concluding remarks.

II. SURVEY OF PREVIOUS WORK

Many different methods of optimal placement are given in the literature. These differences can be attributed to the objectives of the problem (see for example [40]). For instance, spillover constraints are addressed in [30], transmission zero locations are optimized in [34], while [21, 29, 41] address placement of actuators/sensors for modal identification purposes. In applications however, a closed-loop system must typically satisfy several requirements simultaneously. For instance, fine pointing under transient and steady-state disturbance while maintaining robust stability to unmodeled dynamics is required, see references [1, 12, 28, 27] for examples.

Actuator and sensor placement methods that are tailored for specific performance variables should be distinguished from methods that are based on controllability and observability measures which are dependent on all internal variables. Several different metrics proposed are related and their explicit relationships are given in §VI. In the following, several approaches are reviewed and a brief outline is given in chronological order.

A. JUANG AND RODRIGUEZ [18]

In this approach, the optimal actuator location is defined as a location which gives the absolute minimum of the steady-state expectation of the LQR cost function for a given initial condition covariance matrix. The optimal sensor location is determined as a location which gives the absolute minimum of the steady-state estimation error. Juang and Rodriguez show by a simple example (1 actuator and 1 mode) that the optimal actuator location coincides with the maximum deflection location of a single mode. Although the problem of optimal location for a single actuator and sensor for a single structural mode is heuristically obvious, it is shown in §VI-B that a direct relationship exists between the location of the maximum modeshape amplitude and a scaling of the controllability and observability grammians. Depending on one's viewpoint, the Juang and Rodriguez approach is advantageous or disadvantageous since the closed-loop performance metric is optimized directly, and the locations are optimal with respect to a set of weighting matrices.

B. LONGMAN, ET AL. [31, 33, 44, 45]

A novel definition of the degree of controllability is given in Longman, et al. and other references thereof. The cited works also provide a good discussion of the insufficiency of the standard rank test of the controllability matrix for actuator and sensor problems. A degree of controllability is defined with respect to the recovery region. The recovery region defines the set of initial conditions which can be returned to the origin in time T using bounded energy control. The recovery region for time T and energy E_c for the system given by Eq. (1) is a set

$$\Re_c = \{x(0) \mid \exists u(t), t \in [0,T] \ni J(u(\cdot),T,0) \leq E_c, x(T) = 0\} \tag{8}$$

where $J(u(\cdot),T,0) = \frac{1}{2}\int_0^T u^T Ru \, dt$. Note that depending on the type of control constraint assumed, the size and the shape of the recovery region may vary. For example, if a control saturation constraint is used for defining the recovery region, then the bounded energy control constraint is replaced by $|u_i(t)| \leq 1$ (as in [45]). The above definition of the recovery region is based on the energy constraint and leads to the definition of the energy optimal degree of controllability (DOC) [33]. The energy optimal degree of controllability for time T and input energy E_c for normalized state-space system is defined as

$$\rho_c = \inf_{x(0)\notin\Re_c} \|x(0)\| \tag{9}$$

where $\|\cdot\|$ is the Euclidean norm. Other types of recovery regions are defined but may be difficult to compute (see for example p.224 in [45]). The recovery region corresponding to the energy optimal degree of controllability is given by (Eq. 7 in [33]):

$$\Re_c = \{x_o \mid \frac{1}{2}x_o^T W^{-1}(T,0)x_o \leq E_c\} \tag{10}$$

where $W(t,0) = e^{-At}W_c(0,t)e^{-A^Tt}$ and $W_c(0,t)$ is the controllability grammian as defined in §III-A. The above recovery region is an n-dimensional ellipsoid and hence the energy optimal DOC is obtained as the minimum distance from the origin to this boundary:

$$\rho_c = \min_{i=1,\ldots,n} [2E_c\lambda_i(W(T,0))]^{\frac{1}{2}} \tag{11}$$

where $\lambda_i(W(T,0))$ denotes the i-th eigenvalue of $W(T,0)$. For undamped large flexible spacecraft, in modal coordinates, i.e., $\ddot{\eta}_i + \omega_i^2\eta_i = b_i u$; $i = 1,\ldots,n/2$, the above energy optimal DOC simplifies further if the state variables x_i, $i = 1,\ldots,n$, are defined as $x_{2i-1} = \eta_i/N_i$ and $x_{2i} = \dot{x}_{2i-1}/\omega_i$, where N_i is the normalization factor for mode i. The energy optimal DOC for the case of large T and for $R = I$ and $N_i = 1$ simplifies to

$$\rho_c \approx \sqrt{E_cT}\min_i \left[\frac{b_ib_i^T}{\omega_i}\right] \tag{12}$$

Time T and input energy E_c scale the energy DOC index, which is a measure of worst case controllability. Indeed, the energy DOC as given by Eq.(12) is closely related to the modal grammian coefficients as defined by Eq.(84) (see proposition 15 in §VI-B). Note that the above explicit formulas for the energy optimal DOC are for nonrigid body, undamped, second-order structures. Analogous relationships are obtained for observability (see [33]).

C. DeLORENZO AND SKELTON [6, 43, 7]

In this approach, a linear stochastic system is considered and a Linear Quadratic Guassian (LQG) controller is assumed for control purposes. A set of sensor and actuator locations (from a given candidate set) and a corresponding LQG controller is sought such that the control input variance is minimized while satisfying specific output variance constraints. This approach of optimizing closed loop performance is basically different from the class of methods based on optimizing controllability and observability properties. An outline of the method is as follows. For a given set of actuators and sensors, a measure of the effectiveness of each sensor and actuator (which depends on current controller weight matrices, Q and R) for a given LQG controller is defined. The sensors and actuators are ranked in the order of their importance with respect to the introduced measure. Based on a current ranking, an appropriate number of sensors and actuators is deleted and the controller weights are recomputed. The above two steps are iterated until a suitable number of actuators and sensors is obtained. A numerical example of a space telescope is given.

D. HAMDAN AND NAYFEH [15]

In this approach, novel measures of modal controllability and observability are proposed, which amount to a generalization of the modal controllability and observability tests of Popov, Belevitch, and Hautus (PBH) [19]. The generalization of the PBH test involves quantifying the degree of controllability and observability of each mode by the following modal controllability and observability measures

$$f_i = \frac{\phi_i^T B}{||\phi_i||}; \qquad h_i = \frac{C\psi_i}{||\psi_i||} \tag{13}$$

where ψ_i and ϕ_i denote the i-th right and left eigenvectors of A, respectively. In this definition, the i-th modal degree of controllability depends on the angles between the i-th left eigenvector and each column of the input matrix. Similarly, the i-th modal degree of observability depends on the angles between the i-th right eigenvector and each row of the output matrix. By taking the norms of f_i and h_i, the degree of gross controllability and observability, i.e., controllability and observability due to all inputs and outputs for the i-th mode, is defined. While the above measures are geometrically clear, their quantitative connection to the reachable or observable states is not so. Indeed, modal controllability or observability are in general not the same as state controllability and observability, although they are similar for flexible structures. In the latter case, the eigenvectors are approximately the same as the principal directions of controllability and observability grammians. In §V, the relationship between modal controllability and observability and the controllability and observability subspaces is examined. Although the modal measures have been specialized to second-order systems in [15], a design strategy to achieve optimal locations for flexible structures is not yet available.

E. KAMMAR [21]

This approach addresses sensor placement for modal identification. The optimal locations are found for a given number of sensors and their candidate locations. The optimal locations are defined as those locations that minimize various norms of the state error covariance matrix whose inverse is defined as the Fisher information matrix [37]. To facilitate the search for optimal locations, a metric which captures the contribution of a sensor location to the "linear independence of an output" is introduced. This approach is equivalent to maximizing various norms of the steady-state observability grammian because, for large measurement noise, this grammian is related to an optimal covariance matrix of the steady-state estimation error. This will be demonstrated in §IV.

F. KIM AND JUNKINS [22]

This approach highlights the necessity of accounting for both the degree of controllability (observability) of each mode and its relative degree of physical significance in defining a metric for selecting optimal locations of actuators (sensors). This approach combines the gross measures of modal controllability (observability) in [15] and the performance metric in modal cost analysis [42] to form a new index

$$\alpha = \sum_{i=1}^{n} \frac{V_i}{V} ||f_i||_2^2 \tag{14}$$

where V_i and V denote the i-th component modal and total cost, respectively, and f_i is the gross measure of controllability for mode i as defined in Eq. (13). The numerical experiment outlined in [22] shows that for an optimal symmetric output feedback controller, actuator configurations with smaller control gains and better closed-loop eigenvalue conditioning correspond to a larger value of the controllability index Eq.(14). A second numerical example based on LQR design shows that the level of correlation between α and the closed-loop performance can vary depending on the control objective. Note that modal cost, Eq.(14), which is based on a summation of open-loop responses due to independent unit impulse inputs, may not fully capture the performance metric of a closed-loop system. For example, internal stability is dependent on all state variables rather than on physically significant inputs or outputs chosen by the designer. The combined index in Eq. (14) is shown to be closely related to modal grammians as outlined in proposition 14.

G. LIM [25]

In this approach, the intersection space of the controllability and observability subspaces corresponding to the i-th actuator and j-th sensor is defined, which gives the set of all states both observable and controllable by the pair. Using the above intersection space, the combined degree of controllability and observability of the k-th mode, with respect to a single actuator-sensor pair, is defined as the weighted orthogonal projection of the modes into the (i, j) intersection subspace such that the following combined index is defined:

$$\alpha_k^{ij} = \left(\psi_k^{ijT} W_{c_i} \psi_k^{ij}\right) \left(\psi_k^{ijT} W_{o_j} \psi_k^{ij}\right) = \tilde{\psi}_k^{ijT} \tilde{\Sigma}^{ij^2} \tilde{\psi}_k^{ij} \tag{15}$$

where ψ_k^{ij} denotes the projection of the k-th mode into the (i, j) intersection subspace, while $\tilde{\psi}_k^{ij}$ and $\tilde{\Sigma}^{ij}$ denote the k-th mode and Hankel singular value corresponding to the (i, j)-th actuator/sensor pair in balanced coordinates. In balanced coordinates, the above index amounts to a weighted-norm of the modes with respect to (i, j)-th Hankel singular values. The design method involves selecting an actuator and sensor pair for each mode which gives a maximum weighted orthogonal projection and then combining all selected pairs for the final set. A justification for combining independently optimal pairs is given in §VI-A. The idea of effectiveness (controllability and observability of a particular mode) and versatility (controllability and observability of all remaining modes) of actuator-sensor pairs is also introduced and used for analyzing the number of actuator and sensors needed in addition to optimal locations. This approach is useful when the dynamics of sensors and actuators interact significantly with the system to be controlled or observed (see for example [35, 10, 32]).

H. GAWRONSKI AND HADAEGH [11]

In this approach, the locations of actuators and/or sensors are determined to obtain specified Hankel singular values of a system, i.e., specific degrees of controllability and observability for each state are sought. The solution is found for a class of assignable systems. Of immediate interest is a location of sensors and actuators which assures equal controllability and observability of all states (i.e., which assures all Hankel singular values to be equal). The system with all Hankel singular values equal is called a uniformly balanced system. A method of sensor and actuator placement to obtain a uniformly balanced system is presented. The method is also applied to flexible systems. The basic drawback of the method lies in that it sets no restrictions on possible actuator or sensor locations; all locations are assumed available for placement.

III. GRAMMIANS: GEOMETRICAL AND PHYSICAL ASPECTS

In this section, the analytical properties and physical interpretations are given for both observability and controllability grammians of time-invariant systems. Controllability and observability grammians have several advantages when compared to controllability and observability matrices. The superior numerical properties of grammians are due to their symmetry, semidefiniteness, and compactness. In addition, physical interpretations of grammians are useful in the solution of the actuator and sensor placement problems, as will be shown in the following sections. A particularly revealing interpretation is the geometry of principal components of an ideal sensor or actuator configurations.

A. CONTROLLABILITY GRAMMIAN

For a linear time-invariant system (1), the state transition solution is as follows

$$x(t_1) = e^{A(t_1-t_o)}x_o + \int_{t_o}^{t_1} e^{A(t_1-\tau)}Bu(\tau)d\tau \tag{16}$$

and, for the zero initial conditions, the states reachable by the input, $u(\tau)$, $\tau \in [t_o, t_1]$, are in the subspace spanned by the range space of the controllability grammian [4]

$$W_c(t_o, t_1) = \int_{t_o}^{t_1} e^{A(t_1-\tau)}BB^T e^{A^T(t_1-\tau)}d\tau = \int_{0}^{t_1-t_o} e^{A\tau}BB^T e^{A^T\tau}d\tau \tag{17}$$

The finite time grammian depends on the length of the time interval, with the controllable subspace as its range space. Any initial state in this subspace can be transferred to an arbitrary final state within this subspace in a finite time and a finite control energy.

Proposition 1 *Let x_o and $\hat{x}$ be states in the controllable subspace. The input*

$$u(t) = B^T e^{A^T(t_1-t)}q; \quad t \in [t_o, t_1] \tag{18}$$

is the minimum energy control effort that transfers x_o at time t_o to $\hat{x}$ at time t_1 where q satisfies

$$\hat{x} - e^{A(t_1-t_o)}x_o = W_c(t_o, t_1)q \tag{19}$$

Proof: The grammian spans the controllable subspace and both $\hat{x}$ and the matrix exponential in Eq.(19) lie in the controllable subspace, therefore q always exists. By substituting the control input in Eq.(18) into the state transition solution in Eq.(16) and using Eq.(19), one can show that the above control input will drive the initial state to $\hat{x}$ at time t_1. To show that u is the minimum energy controller, one follows the minimum energy control proof in [5]: Let $\tilde{u}$ be any input that transfers x_o to $\hat{x}$ at time t_1 so that $\tilde{u} = u + \Delta u$. From Eqs.(16), (18), and (19), the following orthogonality property holds:

$$\int_{t_o}^{t_1} \Delta u^T(t) u(t) dt = 0 \tag{20}$$

which proves that $||\tilde{u}|| \geq ||u||$ where

$$J_u \triangleq ||u|| = \int_{t_o}^{t_1} u^T u \, dt \tag{21}$$

and that $u(t)$ is the minimum energy control. □

The above proposition leads to the following useful result for a minimum energy controller.

Proposition 2 *Let S_x be the set of reachable states at time t_1 from zero initial state at time t_o by the constrained input of unit energy, $||u(t)|| \leq 1$. The set S_x is the n-dimensional ellipsoid*

$$S_x = \{x | x = U_c \Sigma_c p, \ ||p||_2 \leq 1\} \tag{22}$$

where U_c and $\Sigma_c = diag(\sigma_{c_1}, \ldots, \sigma_{c_n})$ are the singular value decomposition, $W_c(t_o, t_1) = U_c \Sigma_c^2 U_c^T$.

Figure 1 shows a geometrical interpretation of the singular values and the principal directions of the controllability grammian over the interval $[t_o, t_1]$. For clarity of interpretation, the singular values are assumed to be arranged in a decreasing order, $\sigma_{c_1} \geq \sigma_{c_2} \geq \cdots \geq \sigma_{c_n} > 0$, so that U_{c_1} and U_{c_n} correspond to the most and the least controllable directions, and their degrees of controllability are given by the maximum and minimum singular values. For the case of not fully-controllable states, the radii in the uncontrollable directions are equal to zero.

The previous result involves the set of all reachable states from the zero initial condition through bounded control energy. For flexible structures, it is of interest to determine the minimum control effort necessary to reach a set of mode shapes [36]. Since the control profile to reach a particular state is nonunique, we consider minimum rather than limited energy control. By substituting the minimum energy profile of Eq.(18) and the terminal state e into the control energy expression Eq. (21), we obtain the following result, which provides further insight into the properties of controllability grammians of flexible structures.

Proposition 3 *For the control effort, J_u, defined by Eq.(21), and the system given by Eq.(1), the minimum control effort that transfers a system from a zero state at time t_o to the state e at time t_1 is given by*

$$J_u(e) = e^T W_c^{-1}(t_o, t_1) e = \sum_{i=1}^{n} \left(\frac{U_{c_i}^T e}{\sigma_{c_i}} \right)^2 \tag{23}$$

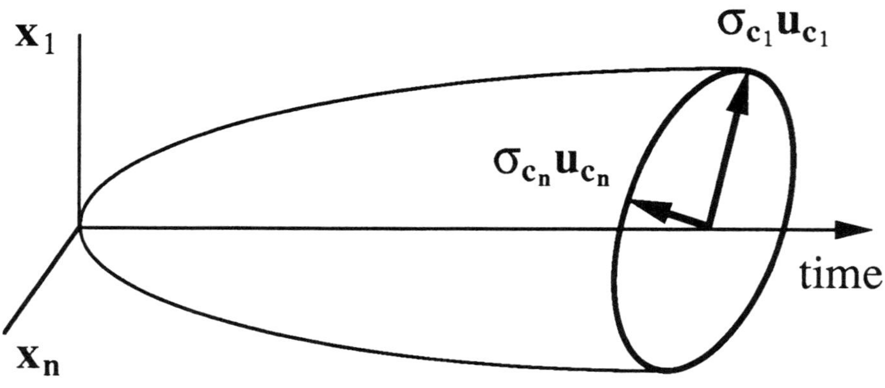

Fig. 1. Geometry of controllability grammian; states reachable by unit energy bounded input.

The above control energy is a sum of weighted projections of e onto the principal controllable directions. In particular, the inverse singular values, $1/\sigma_{c_1}^2$ and $1/\sigma_{c_n}^2$, denote the range of minimum and maximum control energies necessary to attain an arbitrary unit state vector at terminal time from the zero initial state. The corresponding vectors, U_{c_1} and U_{c_n}, indicate the most and least controllable directions. For a structural modeshape e, Eq.(23) shows the minimum control energy needed to reach this modeshape. The closer the modeshape is to the most controllable direction, the more controllable the mode is.

For actuator placement applications, an ideal actuator configuration should generate a set of principal controllable directions having the same degree of controllability and in the same directions as a set of significant modes. However, the principal directions form an orthogonal set while the target modes do not, in general. This and the fact that it is usually not reasonable to generate principal controllable directions with identical degrees of controllability in all directions (unless the uniformly balanced systems introduced by Gawronski and Hadaegh in §II-H are used), imply an inevitable tradeoff among the significant modes.

B. OBSERVABILITY GRAMMIAN

For a system given by Eqs.(1) with outputs $y(\tau)$, $\tau \in [t_o, t_1]$, the observed states lie in the range space of the observability grammian

$$W_o(t_o, t_1) = \int_{t_o}^{t_1} e^{A^T(\tau - t_o)} C^T C e^{A(\tau - t_o)} d\tau = \int_0^{t_1 - t_o} e^{A^T \tau} C^T C e^{A\tau} d\tau \tag{24}$$

If the system is not fully observable, the above range space is rank deficient. The null space of the observability grammian is the set of states that do not participate in the outputs. For systems which are not fully-observable, the states can only be reconstructed to within an element in this unobservable subspace [24]. Equivalent arguments can be made using the observability matrix

$$M_o = \begin{bmatrix} C \\ CA \\ \vdots \\ CA^{n-1} \end{bmatrix} \tag{25}$$

The following provides a geometrical interpretation of the observability grammian.

Proposition 4 *Let the total output energy be constrained as follows*

$$\int_{t_o}^{t_1} ||y(t)||^2 dt \leq 1 \tag{26}$$

The set of initial states that gives at most a unit output energy at final time t_1 is enclosed within the n-dimensional ellipsoid given by

$$S_{x_o} = \{x_o | x_o = U_o \Sigma_o^{-1} p, \; ||p||_2 \leq 1\} \tag{27}$$

where U_o and $\Sigma_o = diag(\sigma_{o_1}, \ldots, \sigma_{o_n})$ are the singular value decomposition, $W_o(t_o, t_1) = U_o \Sigma_o^2 U_o^T$.

Proof: The above result can be shown by substituting the output response due to initial conditions into the output energy inequality in Eq.(26). □

Figure 2 shows a geometrical interpretation of the singular values and the principal directions of the observability grammian over the interval $[t_o, t_1]$, where the singular values are arranged in a descending order, $\sigma_{o_0} \geq \sigma_{o_2} \geq \ldots \geq \sigma_{o_n} > 0$. The figure shows that the ellipsoid in the direction U_{o_1}, which corresponds to the largest singular value σ_1, has the shortest radius. This means that the unit initial state in this direction will produce the largest output energy. The complementary direction, U_{o_n}, with the largest radius of $1/\sigma_{o_n}$ corresponds to the least output energy direction; i.e., for small σ_{o_n}, a large initial state is necessary to generate a unit energy at the output. For the case of not fully observable states, the radii in the unobservable directions tend to infinity and the output energy will be zero for any finite initial vector lying in the unobservable subspace.

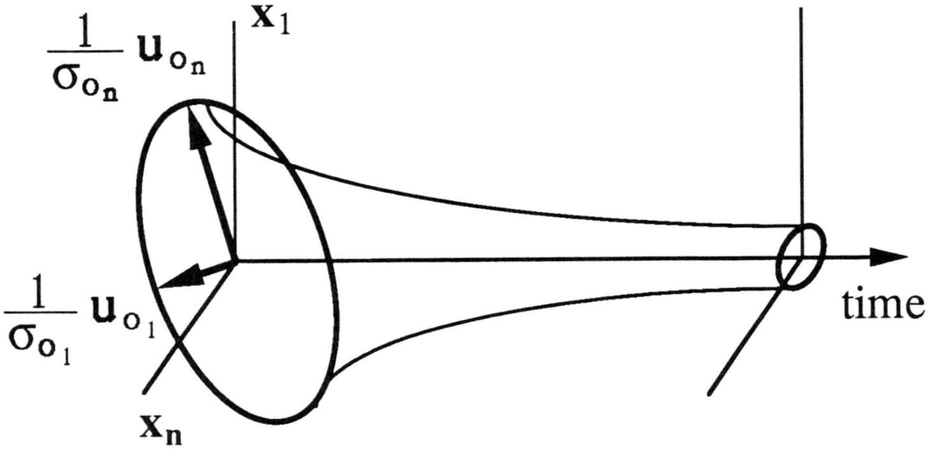

Fig. 2. Geometry of observability grammians; ellipsoid of initial conditions having unit output energy.

Analogous to the idea of controllability, it is of interest to consider the output energy corresponding to structural mode shapes. The following result, which follows directly from the previous results, provides further insight into the properties of observability grammians for flexible structures.

Proposition 5 *For the system given by Eq.(1), the output energy, J_y, as defined by Eq.(26) over the time interval t_o to t_1 corresponding to the initial state e is*

$$J_y = e^T W_o(t_o, t_1) e = \sum_{i=1}^{n} \left(\sigma_{o_i} U_{o_i}^T e\right)^2 \tag{28}$$

The above output energy is a sum of weighted projections e on to the principal observable directions. In particular, the singular values $\sigma_{o_1}^2$ and $\sigma_{o_n}^2$ of the grammian denote the range of maximum and minimum output energies for an arbitrary unit state vector at initial time. The corresponding vectors,

U_{o_1} and U_{o_n}, correspond to the most and least observable directions. The closer a modeshape is to the most observable direction, the more the output energy, i.e., more observable. For a structural mode e, Eq.(28) shows the output energy due to this mode.

An ideal configuration of the sensors should generate a set of principal observable directions with the same degree of observability, and must be in the same direction as a given set of significant modes. Since the principal directions form an orthogonal set and typically the set of significant modes do not, not all the modes lie in the principal directions. For sensor placement problems, this and the fact that it is usually not reasonable to generate principal observable directions with identical singular values in all directions, implies an inevitable tradeoff among the set of modes.

C. GRAMMIANS IN BALANCED COORDINATES

The geometry of the grammian previously discussed shows that the sets of principal controllability and observability directions are not identical and are not identically scaled. This means for instance that a state which is most controllable may not be colinear with the most observable state, and may have different scale. In the case of a cantilever beam undergoing both vertical and lateral motions with an actuator in one plane and a sensor in the other plane, this sensor-actuator configuration will not be able to either influence or detect motions corresponding to modes that do not have components in both vertical and lateral directions.

The geometry of the grammians simplifies in balanced coordinates as shown in the following.

Proposition 6 *For the system given by Eq.(1), let P be the coordinate transformation to the balanced coordinates, x_b, such that $x = Px_b$. Then the bounded energy controllability and observability ellipsoids defined by Eqs.(22) and (27) are given by*

$$\text{controllability:} \quad x_b^T \Sigma_b^{-2} x_b \leq 1 \tag{29}$$

$$\text{observability:} \quad x_b^T \Sigma_b^2 x_b \leq 1 \tag{30}$$

where $\Sigma_b^2 = diag(\sigma_{b_1}^2, \ldots, \sigma_{b_n}^2)$ denotes the Hankel singular values.

Proof: In balanced coordinates, the grammians are equal and diagonal so that $\Sigma_b^2 = P^{-1} W_c P^{-T} = P^T W_o P$. □

Figure 3 shows the controllability and observability ellipsoids in balanced coordinates where the i-th controllable and i-th observable directions are the same. The balanced state, δ_1, corresponds to the largest state that can be attained by a unit input energy and the smallest initial state which gives a unit output energy, i.e., δ_1 corresponds to the most controllable

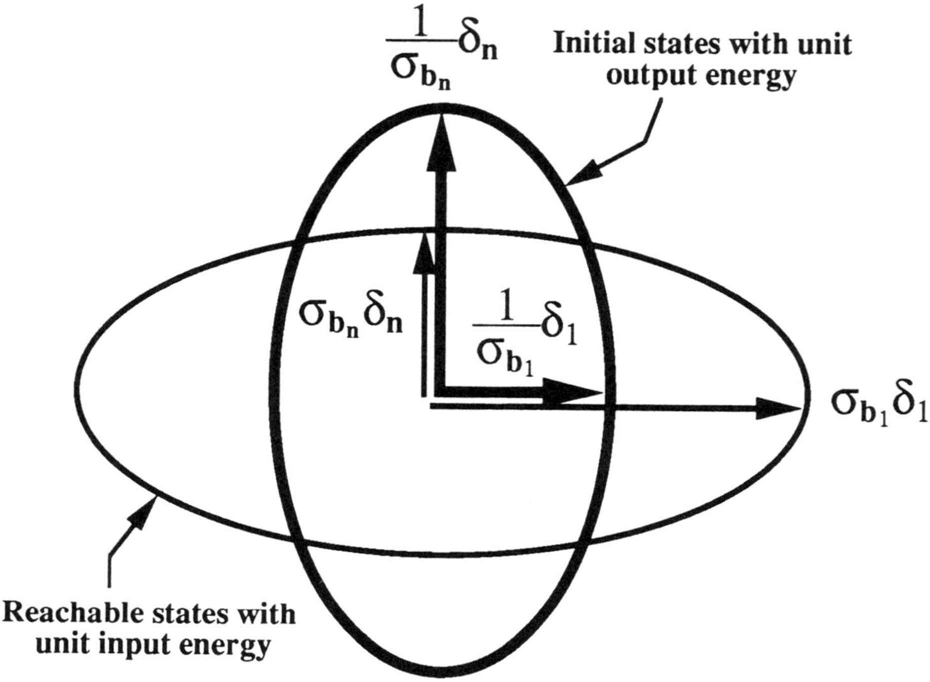

Fig. 3. Controllability and observability ellipsoids in balanced coordinates.

and observable state. The coincidence of controllability and observability ellipsoids is the basis for model reduction in balanced coordinates.

The i-th controllable and observable directions in the original coordinate system are given by the vector p_i which is the i-th column of the P matrix. Notice that the set of principal directions in balanced coordinates is orthogonal and nonorthogonal in the original coordinates. On the other hand, recall that the set of principal directions of controllability (or observability) grammians is orthogonal if determined separately (see Figs. 1 and 2). For a detailed discussion of balanced coordinates, the interested reader is referred to [38], and for applications to flexible spacecraft, [14], [8], and [9] are recommended.

D. STEADY-STATE GRAMMIANS

It is clear from Eqs.(17) and (24) that the existence of the grammian matrices in the limiting case, $t \to \infty$, depends on the stability of the system, since for unstable systems certain states will tend to infinity, thus the grammian ellipsoids explode with time even for bounded input. For stable systems however, both grammians exist and are referred to as steady-state grammians.

The steady-state controllability and observability grammians are given by

$$W_{c\infty} = \lim_{t_1\to\infty} W_c(t_o,t_1) = \int_0^\infty e^{A\tau}BB^Te^{A^T\tau}d\tau \quad (31)$$

$$W_{o\infty} = \lim_{t_1\to\infty} W_o(t_o,t_1) = \int_0^\infty e^{A^T\tau}C^TCe^{A\tau}d\tau \quad (32)$$

and can be computed from the following Lyapunov equations

$$AW_{c_\infty} + W_{c_\infty}A^T = -BB^T \quad (33)$$

$$A^TW_{o_\infty} + W_{o_\infty}A = -C^TC \quad (34)$$

In applications, control activity occurs over a finite time interval and the finite time grammian beyond a certain time threshold can be satisfactorily approximated by the steady-state grammian. The validity of the approximation is evident from the explicit relationship between the finite and infinite time grammian matrices [8]

$$W_c(t) = W_{c_\infty} - e^{At}W_{c_\infty}e^{A^Tt}; \qquad W_o(t) = W_{o_\infty} - e^{A^Tt}W_{o_\infty}e^{At} \quad (35)$$

The above relations depend on the decay of the matrix exponential, $e^{At} = \Phi e^{\Lambda t}\Phi^{-1}$, where Λ and Φ denote the eigenvalue and eigenvector matrix of A, respectively. Observe that the largest time constant of the system can be used as a threshold time for steady-state convergence.

E. THRESHOLD OF CONTROLLABILITY: AN EXAMPLE

The states that can be reached or observed are given by the range space of the steady-state grammians, which can be determined from the singular value decomposition of the grammians

$$W_{c_\infty} = \begin{bmatrix} U_c & U_c^\perp \end{bmatrix} \begin{bmatrix} \Sigma_c^2 & \\ & 0 \end{bmatrix} \begin{bmatrix} U_c^T \\ U_c^{\perp T} \end{bmatrix} \quad (36)$$

$$W_{o_\infty} = \begin{bmatrix} U_o & U_o^\perp \end{bmatrix} \begin{bmatrix} \Sigma_o^2 & \\ & 0 \end{bmatrix} \begin{bmatrix} U_o^T \\ U_o^{\perp T} \end{bmatrix} \quad (37)$$

where

$$\Sigma_c^2 = diag(\sigma_{c_1}^2, ..., \sigma_{c_{qc}}^2); \qquad \sigma_{c_1}^2 \geq \sigma_{c_2}^2 \geq ...\sigma_{c_{qc}}^2 > 0 \quad (38)$$

$$\Sigma_o^2 = diag(\sigma_{o_1}^2, ..., \sigma_{o_{qo}}^2); \qquad \sigma_{o_1}^2 \geq \sigma_{o_2}^2 \geq ...\sigma_{o_{qo}}^2 > 0 \quad (39)$$

The scalars q_c and q_o denote the dimension of the controllable and observable subspaces, respectively, and the columns of the U_c and U_o matrices form orthogonal bases for the range spaces of W_{c_∞} and W_{o_∞}, while the corresponding singular values indicate the degree of importance of the corresponding directions as discussed in more detail in §III-A and B.

In practice, some engineering judgment is necessary to determine the dimensions of controllability and observability subspaces. For example, consider the problem of defining the controllability for the experimental structure with 49 structural modes controlled by 8 actuators [2]. Figure 4 show the singular values of the controllability grammian of the experimental structure in a descending order. The figure shows that the singular values

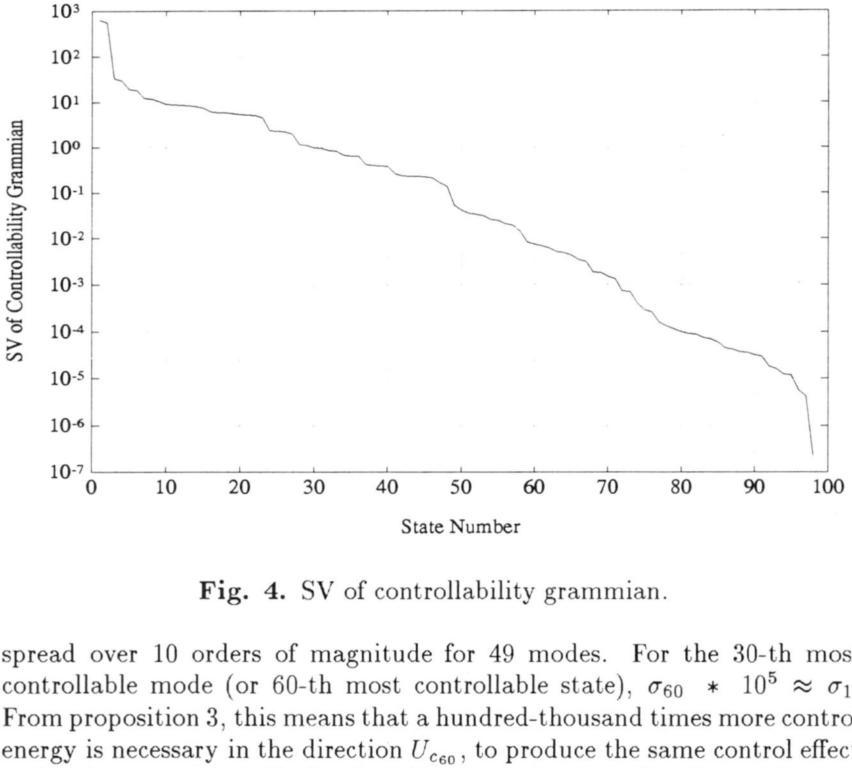

Fig. 4. SV of controllability grammian.

spread over 10 orders of magnitude for 49 modes. For the 30-th most controllable mode (or 60-th most controllable state), $\sigma_{60} \ * \ 10^5 \approx \sigma_1$. From proposition 3, this means that a hundred-thousand times more control energy is necessary in the direction $U_{c_{60}}$, to produce the same control effect as in the most controllable direction U_{c_1}. As a result, the 60-th state or the 30-th most controllable mode can be considered to be practically uncontrollable. Notice also from Figure 4 that the degree of controllability varies gradually.

For flexible structures, the above complicating aspects of gradual variation of the SV of the grammian and large discrepancy in the degree of controllability in different directions are basically unavoidable.

IV. ACTUATOR/SENSOR PLACEMENT AND OPTIMAL CONTROL

Better actuator and sensor configurations presumably lead to improved closed-loop performance. However, the actuator and sensor location problem is typically an open-loop problem since the closed loop performance and control gains are not usually available or fixed. In this section, the relationship between grammians and closed-loop performance for small gains is considered. This implies that by improving open-loop performance (via controllability and observability indices), the overall closed-loop control problem can be enhanced.

A closed-loop system whose compensator consists of a controller and a state estimator is considered. Let ν, w denote the process and measurement noise, respectively. For the system given by Eq.(1), consider the case of a full state feedback regulator with the control law $u = Kx$, and the steady-state quadratic performance J_r

$$J_r = \frac{1}{2}\int_0^\infty \left(x^T Q x + u^T R u\right) d\tau \tag{40}$$

For a given state feedback gain K, and a given initial condition $x(0) = x_o$, the closed-loop quadratic cost can be evaluated as follows:

$$J_r = \frac{1}{2}\int_0^\infty x_o^T e^{A_{reg}^T \tau} M e^{A_{reg}\tau} x_o \, d\tau = \frac{1}{2} x_o^T \Xi x_o \tag{41}$$

where

$$\Xi = \int_0^\infty e^{A_{reg}^T \tau} M e^{A_{reg}\tau} \, d\tau \tag{42}$$

where $A_{reg} = A + BK$ and $M = Q + K^T RK$. For a stable A_{reg}, Ξ satisfies

$$A_{reg}^T \Xi + \Xi A_{reg} = -M \tag{43}$$

The gain, $K = -R^{-1}B^T\Xi$, minimizes the performance J_r and in this case Eq.(43) becomes the controller algebraic Riccati equation (CARE) (see for example Chapter 3 of [24])

$$A^T\Xi + \Xi A - \Xi B R^{-1} B^T \Xi + Q = 0 \tag{44}$$

Proposition 7 *For a controllable system and weighting Q*

$$Q = W_c^{-1} B (I + R^{-1}) B^T W_c^{-1} \tag{45}$$

the solution Ξ of CARE is as follows

$$\Xi = W_c^{-1} \tag{46}$$

Proof: Introduction of Eq.(45) into (44) gives

$$A^T\Xi+\Xi A-\Xi BR^{-1}B^T\Xi+W_c^{-1}BB^TW_c^{-1}+W_c^{-1}BR^{-1}B^TW_c^{-1}=0 \quad (47)$$

By substituting Eq.(46) into the above equation one obtains

$$A^TW_c^{-1}+W_c^{-1}A+W_c^{-1}BB^TW_c^{-1}=0 \qquad (48)$$

or

$$W_cA^T+AW_c+BB^T=0 \qquad (49)$$

which is the controllability Lyapunov equation, satisfied for W_c. □

Consider a state estimator for an observable system given by Eq.(1) whose estimation error and corresponding covariance matrix are defined as

$$x_e(t)=x(t)-\hat{x}(t), \qquad \Theta(t)=E[x_e(t)x_e^T(t)] \qquad (50)$$

Define $d(t)=Gw(t)+\nu(t)$ as the combined external disturbance, where G is the estimator gain, and Z is the combined zero mean disturbance noise intensity such that $E[d(\tau_1)d^T(\tau_2)]=Z\delta(\tau_1-\tau_2)$, and define $A_{est}=A+GC$ as the estimator system matrix. Then the error covariance matrix Θ satisfies

$$\dot{\Theta}(t)=Z+A_{est}\Theta(t)+\Theta(t)A_{est}^T \qquad (51)$$

For a stable system with independent random vector processes $w(t)$ and $\nu(t)$, having covariance intensities W and V, respectively, the steady-state error covariance matrix satisfies the Lyapunov equation

$$A_{est}\Theta+\Theta A_{est}^T=-Z \qquad (52)$$

where $Z=GWG^T+V$. The optimal steady-state estimator which minimizes $trace(\Theta)$ is given by $G=-\Theta C^TW^{-1}$ and Eq.(52) becomes the estimator (or filter) algebraic Riccati equation (FARE)

$$A\Theta+\Theta A^T-\Theta C^TW^{-1}C\Theta+V=0 \qquad (53)$$

Analogous to the optimal regulator case, the result for the optimal estimator is stated as follows:

Proposition 8 *For a fully observable system and for the weight V as follows*

$$V=W_o^{-1}C^T(I+W^{-1})CW_o^{-1} \qquad (54)$$

the solution Θ of FARE is as follows

$$\Theta=W_o^{-1} \qquad (55)$$

The proof of the above proposition is analogous to the CARE case.

The previous two propositions state that the controllability and observability grammians are related to the inverse of the CARE and FARE matrices if a particular set of state and estimated state weights is chosen. Let us consider the special case when

$$\underline{\sigma}(R) \gg 1; \qquad \underline{\sigma}(W) \gg 1 \tag{56}$$

The above condition results in small gains for the optimal regulator and estimator. The above conditions are summarized as follows:

Proposition 9 (**small gain approximation**) *For a stabilizable and detectable system defined by Eq.(1) with controllability and observability grammian W_c, W_o, and the weighting matrices, R, W, of a steady-state optimal regulator and estimator*

$$Q = W_c^{-1} B B^T W_c^{-1}; \qquad V = W_o^{-1} C^T C W_o^{-1} \tag{57}$$

the following small gain approximation holds

$$\Xi^{-1} \approx W_c \tag{58}$$

$$\Theta^{-1} \approx W_o \tag{59}$$

provided that Eqs. (56) holds.

These particular forms of the weighting matrices in the small gain approximation have the following interpretations. For the regulator states, the above choice of weights penalizes the output $B^T W_c^{-1} x(t)$, which consists of weighted physical displacements and velocities at the actuator input locations. The states are weighted by the degree of controllability since

$$W_c^{-1} x(t) = \sum_{i=1}^{n} \frac{U_{c_i} U_{c_i}^T x(t)}{\sigma_{c_i}^2} \tag{60}$$

is the sum of weighted projections of the state on principal controllable directions. Less controllable states with smaller singular values are penalized more severely in order to have them more involved in the control process. This is reflected in the cost function

$$J_r = \frac{1}{2} \sum_{i=1}^{n} \frac{(U_{c_i}^T x_o)^2}{\sigma_{c_i}^2} \tag{61}$$

For example, an initial state of unit length in the least controllable direction (which is precisely the principal direction corresponding to the smallest σ_c)

will give a closed-loop cost of $1/\underline{\sigma}(W_c)$. Analogous interpretations hold for the measurement noise covariance matrix as defined in Eq. (57).

The above small gain approximations imply that the controllability grammian can be interpreted in terms of an optimal regulator performance. The small gain condition is not unexpected since grammians are open-loop properties. From Eq.(41), the regulator performance, $\overline{\sigma}(\Xi)$, gives the worst cost due to a unit initial state in the least controllable direction. From the small gain approximation, minimizing $\overline{\sigma}(\Xi)$ is equivalent to maximizing $\underline{\sigma}(W_c)$, i.e., maximize the minimum singular value of the controllability grammian. If the initial states are not well known and hence treated as independent random variables with unit variances, the quadratic cost reduces to $\frac{1}{2}trace(\Xi)$, which is approximately $\frac{1}{2}trace(W_c^{-1})$ or $\frac{1}{2}\sum_{i=1}^{n} 1/\sigma_{c_i}^2$. The latter cost is an average cost over all possible directions rather than the worst case cost. Clearly, any particular physical problem dictates the choice of the cost. In §VII-D, a comparison of actuator configurations for an experimental structure with different levels of minimum controllability is given.

Similarly, the observability grammian matrix can be interpreted as the optimal small gain estimator performance. The estimator performance $\overline{\sigma}(\Theta)$ follows from Eq.(50). From the small gain approximation, minimizing $\overline{\sigma}(\Theta)$ is equivalent to maximizing $\underline{\sigma}(W_o)$, i.e., maximize the minimum singular value of the observability grammian. Similar to the regulator case, the most suitable choice of the cost function should depend on whether the worst case state estimate variance needs to be optimized or an average performance of all states is desired.

In application, the regulator and estimator gains are bounded by constraints such as model uncertainties and process noise so that the performance of the final compensator design may be significantly influenced by small gain properties which are dependent on the grammians. In summary, optimizing the actuator and sensor locations to maximize the grammian matrices can directly improve the closed-loop regulator and estimator performance.

V. MODAL CONTROLLABILITY AND OBSERVABILITY

The measures of modal controllability and observability defined in Eq.(13) are geometrically clear and simple, but their quantitative connection to the reachable or observable states are not so. A closer examination reveals that the B and C matrices correspond to only subsets of controllability and observability spaces

$$R(B) \subseteq R(W_c); \qquad R(C^T) \subseteq R(W_o) \tag{62}$$

To investigate the connections between projections into $R(B)$, $R(C^T)$, $R(W_c)$, and $R(W_o)$ respectively, let S_c, $S_o^\perp$ represent the controllable subspace and its orthogonal complement

$$S_c = R([\psi_1, \ldots, \psi_r]); \qquad S_c^\perp = R([\phi_{r+1}, \ldots, \phi_n]) \tag{63}$$

where r denotes the dimension of the controllable subspace. Since $\Phi^T \Psi = I$, therefore $S_c \oplus S_c^\perp$ spans R^n. The controllable subspace is spanned by the right eigenvectors corresponding to the controllable poles (see p. 61 of [24]). The following result relates modal controllability to the state-space controllability.

Proposition 10

$$\phi_i^T B = 0 \Leftrightarrow \phi_i \in S_c^\perp; \qquad \phi_i^T B \neq 0 \Leftrightarrow \psi_i \in S_c \tag{64}$$

Proof: From the left eigenvector invariance property,

$$\begin{aligned} \phi_i^T M_c &= \phi_i^T \begin{bmatrix} B & AB & \cdots & A^{n-1}B \end{bmatrix} && (65) \\ &= \begin{bmatrix} \phi_i^T B & \lambda_i \phi_i^T B & \cdots & \lambda^{n-1} \phi_i^T B \end{bmatrix} && (66) \end{aligned}$$

it follows that

$$\phi_i^T B = 0 \Leftrightarrow \phi_i^T S_c = 0 \tag{67}$$

and as a corollary

$$\phi_i^T B \neq 0 \Leftrightarrow \phi_i^T S_c \neq 0 \tag{68}$$

Since $\phi_i^T S_c = 0 \Leftrightarrow \phi_i \in S_c^\perp$, the first equation is proved. For the second equation, $\phi_i^T S_c \neq 0 \Leftrightarrow \psi_i^T S_c^\perp = 0 \Leftrightarrow \psi_i \in S_c$ from bi-orthogonality of the eigenvectors and the proof follows from the corollary. □

The above proposition states that if the i-th mode is controllable, the left eigenvector has a nonzero component in the column space of B, and the right eigenvector is an element of the controllable subspace. When the mode is uncontrollable, the left eigenvector has a zero component in the column space of B and is an element in the uncontrollable subspace. Notice that degree of controllability derived from angular measurements between eigenvectors and columns of the B matrix (Eq.(13)) is quite different from the degree of controllability defined by the singular value of the controllability grammian (Eq.(22)).

For observability analysis, the unobservable subspace $S_o^\perp$ and its orthogonal complement subspace S_o are expressed as

$$S_o = R([\phi_1, \ldots, \phi_m]); \qquad S_o^\perp = R([\psi_{m+1}, \ldots, \psi_n]) \tag{69}$$

where m denotes the dimension of the observable subspace, and $S_o \oplus S_o^\perp$ spans R^n. The unobservable subspace is spanned by the right eigenvectors corresponding to the unobservable poles (see p. 75 of [24]). The following result relates modal observability to the state-space observability.

Proposition 11

$$C\psi_i = 0 \Leftrightarrow \psi_i \in S_o^\perp; \qquad C\psi_i \neq 0 \Leftrightarrow \phi_i \in S_o \tag{70}$$

Proof: From the right eigenvector invariance property,

$$M_o\psi_i = \begin{bmatrix} C \\ CA \\ \vdots \\ CA^{n-1} \end{bmatrix} \psi_i = \begin{bmatrix} C\psi_i \\ \lambda_i C\psi_i \\ \vdots \\ \lambda^{n-1} C\psi_i \end{bmatrix} \tag{71}$$

it follows that

$$C\psi_i = 0 \Leftrightarrow M_o\psi_i = 0 \tag{72}$$

and as a corollary

$$C\psi_i \neq 0 \Leftrightarrow M_o\psi_i \neq 0 \tag{73}$$

Since $M_o\psi_i = 0 \Leftrightarrow \psi_i \in S_o^\perp = 0$, the first equation is proved. For the second equation, $\psi_i^T S_o \neq 0 \Leftrightarrow \phi_i^T S_o^\perp = 0 \Leftrightarrow \phi_i \in S_o$ from bi-orthogonality and the proof follows from the corollary. □

VI. ACTUATOR/SENSOR PLACEMENT FOR FLEXIBLE STRUCTURES

This section focuses on the placement problem for flexible structures. We consider the relationship between subsets of actuators and sensors to their full sets, quantifying the more-is-better physical intuition. In addition, the special grammian structure for flexible structures is discussed in the context of actuator and sensor placement problems. Finally, a method of determining the actuator and sensor locations and their numbers is presented. The method is based on analyzing and exploiting the approximate singular values of grammians which are associated with individual modes, as a measure of the relative significance of each actuator.

A. COMBINATION OF ACTUATORS OR SENSORS

Consider two sets of inputs and corresponding input matrix $B = \begin{bmatrix} B_1 & B_2 \end{bmatrix}$. Define the controllability subspaces corresponding to each individual pair, (A, B_1) and (A, B_2), by

$$S_1^c = R\begin{bmatrix} B_1 & AB_1 & \cdots & A^{n-1}B_1 \end{bmatrix} = R[Q_1] \tag{74}$$

$$S_2^c = R\begin{bmatrix} B_2 & AB_2 & \cdots & A^{n-1}B_2 \end{bmatrix} = R[Q_2] \tag{75}$$

where $Q_i = \begin{bmatrix} B_i & AB_i & \cdots & A^{n-1}B_i \end{bmatrix}$. Similarly, the controllability subspace for the combined sets of inputs is given by

$$S^c = R\begin{bmatrix} B_1 & B_2 & AB_1 & AB_2 & \cdots & A^{n-1}B_1 & A^{n-1}B_2 \end{bmatrix} = R\begin{bmatrix} Q_1 & Q_2 \end{bmatrix} \tag{76}$$

Consider also two sets of outputs and corresponding output matrix $C = \begin{bmatrix} C_1 & C_2 \end{bmatrix}$. Define the observability subspaces corresponding to each individual pair, (A, C_1) and (A, C_2), by

$$S_1^o = R(P_1^T) = N(P_1)^\perp \tag{77}$$
$$S_2^o = R(P_2^T) = N(P_2)^\perp \tag{78}$$

where

$$P_i = \begin{bmatrix} C_i \\ C_iA \\ \vdots \\ C_iA^{n-1} \end{bmatrix} \tag{79}$$

The observability subspace for the combined output is given by $R(P^T) = S^o$ where $P = \begin{bmatrix} P_1 \\ P_2 \end{bmatrix}$.

Proposition 12

$$S_i^c \subseteq S^c, \quad S_i^o \subseteq S^o; \quad i \in [1, 2] \tag{80}$$

Proof: $R([Q_1Q_2]) = R(Q_1) \oplus R(Q_1)^\perp \cap R(Q_2) = R(Q_2) \oplus R(Q_2)^\perp \cap R(Q_1)$. Similarly, $R(P^T) = R(P_1^T) \oplus R(P_1^T)^\perp \cap R(P_2^T) = R(P_2^T) \oplus R(P_2^T)^\perp \cap R(P_1^T)$ □

Proposition 13

$$\sigma_j(W_c) \geq \sigma_j(W_{c_i}), \quad \sigma_j(W_o) \geq \sigma_j(W_{o_i}); \quad i \in [1, 2], \quad \forall j \tag{81}$$

Proof: Due to the linearity of Lyapunov equations, the combined grammians are $W_c = W_{c_1} + W_{c_2}$ and $W_o = W_{o_1} + W_{o_2}$. By assuming the singular values are arranged in a descending order, the singular value inequalities follow directly from Rayleigh's Principle [39]. □

For actuator and sensor placement applications, the above two propositions have useful consequences. First, any combination of controllable states corresponding to each set of actuators can be controlled by combining the actuators. Similarly, any combination of observable states from each set of sensors is observable from the combined set of the sensors. Second, any additional sensors or actuators will not degrade controllability or observability. While this fact is heuristically obvious, the result shows that the improvement is guaranteed in *all* principal directions. In general, additional sensors or actuators can only increase both the subspace dimension and the degrees of the controllability and observability and therefore enhance the system.

It is shown in reference [6] that it is possible for an LQG controller to perform better (maintain output variance at a lower input power) when a noisy actuator is removed. This may seem to contradict the above propositions, but it makes sense if one considers the case of an actuator which causes high controllability but is susceptible to significant noise; controllability is no longer the only determining factor in actuator selection. On the other hand, reference [6] shows that more sensors always yield an equal or better LQG controller under any measurement noise; this is consistent with the guaranteed equal or improved observability of the system.

B. GRAMMIANS OF FLEXIBLE STRUCTURES

Consider linear time-invariant flexible structure as given by Eq. (5). Defining the state vector as in [46, 9]

$$x = \begin{pmatrix} \dot{\eta}_1 & \omega_1\eta_1 & \cdots & \dot{\eta}_n & \omega_n\eta_n \end{pmatrix}^T \tag{82}$$

the corresponding state equations take the form of Eqs.(1) where $A = diag(A_1, \cdots, A_n)$, $B = (B_1^T, \cdots, B_n^T)^T$ and $C = (C_1, \cdots, C_n)$, with

$$A_i = \begin{bmatrix} -2\zeta_i\omega_i & -\omega_i \\ \omega_i & 0 \end{bmatrix}, \quad B_i = \begin{bmatrix} b_i \\ 0 \end{bmatrix}, \quad C_i = \begin{bmatrix} 0 & \frac{1}{\omega_i}c_i \end{bmatrix} \tag{83}$$

where $b_i = \psi_i^T E$ and $c_i = F\psi_i$. Notice that for small damping ratio the choice of the state vector in Eq.(82) gives normal state matrix and hence orthogonal eigenvectors. For flexible structures with distinct natural frequencies, the steady-state controllability and observability grammians asymptotically (as $\zeta \to 0$) approach 2-by-2 block diagonal matrices and are given by [46, 9]

$$W_{c_{ii}} \rightarrow I_2\frac{\beta_i^2}{4\zeta_i\omega_i} \tag{84}$$

$$W_{o_{ii}} \rightarrow I_2\frac{\theta_i^2}{4\zeta_i\omega_i^3}; \quad i \in M = \{1, \ldots, n\} \tag{85}$$

where $\beta_i^2 = b_i b_i^T$, $\theta_i^2 = c_i^T c_i$, I_2 is a 2 by 2 identity matrix, and n is the number of modes. The above approximate grammian relationships indicate that high frequency modes with large damping are among the least controllable and observable.

Note that for small damping and distinct frequencies, the 2-by-2 block diagonal grammians correspond to individual modes and the actuator and sensor locations affect the grammian matrices only through the scalars, β_i^2 and θ_i^2. The above scalars are related to the mode shapes and actuator locations since

$$\beta_i^2 = b_i b_i^T = \psi_i^T E E^T \psi_i = \sum_{j=1}^{r} \psi_{ij}^2 \tag{86}$$

$$\theta_i^2 = c_i^T c_i = \psi_i^T F F^T \psi_i = \sum_{k=1}^{m} \psi_{ik}^2 \tag{87}$$

Because of the significance of the β_i^2 and θ_i^2 terms, they will henceforth be called *modal grammian coefficients* (MGC). The terms ψ_{ij} and ψ_{ik} denote the normalized amplitudes of the i-th mode at the degrees of freedom corresponding to the j-th actuator and k-th sensor. Equations (84) to (87) show that the controllability and observability grammians are related to the amplitudes of the modes at all actuator and sensor locations. As a useful rule-of-thumb, *locations with larger modal amplitudes are better.* It also follows from the above equations that the addition of actuators or sensors can only increase the singular values of the grammian for each mode and hence improve the controllability or observability of every mode, which is consistent with the results in §VI-A. The only exception is when the actuator j or sensor k are located exactly at the node of mode i where $\psi_{ij} = 0$ or $\psi_{ik} = 0$, which will produce no improvement for that mode.

In the simplest case of a single actuator controlling a single structural mode, the location with the largest amplitude gives the largest observability and controllability grammians. The above grammian analysis helps to explain an earlier result in [18], which involves an optimal linear quadratic control of a single-mode simple beam model using a single actuator (see also §II-A). The numerical study in [18] shows that placing the actuator at the peak amplitude location of the mode shape produces the absolute minimum of the cost function of an optimal LQR.

For the case of single actuator (sensor) controlling (observing) multiple modes, observe that all the block diagonal components of grammians depend on a single actuator (sensor) parameter. This means that the controllability (observability) of all modes cannot be improved simultaneously by choosing the location of a single actuator (sensor). The problem becomes more complex for multiple sensors (actuators) and significant damping with closely spaced frequencies.

The modal grammians are closely related to the measure f_i of controllability as defined in Eq. (13) as follows:

Proposition 14

$$f_i = \frac{1}{\sqrt{2j}} b_i; \qquad \|f_i\|_2^2 = 2\zeta_i \omega_i W_{c_{ii}}. \tag{88}$$

The measure is proportional to the i-th mode displacements at the actuator locations. The modal grammians weighs the i-th mode by a factor inversely proportional to the i-th modal damping and frequency. This means that high frequency modes with significant damping are among the least controllable, a consequence consistent with physical intuition. However, for a given structure (i.e., for given frequency, damping, and mode shape), the optimization of actuator locations involves the $b_i b_i^T$ terms only so that both indices are effectively identical. Differences will occur only when the relative weighting of the modes is not identical and the term $4\zeta_i\omega_i$ in the modal grammian index may be interpreted as a relative weighting. For example, the performance metric as defined in Eq. (14) and proposed in [22] can be expressed in the form of a weighted sum of modal grammians:

$$\alpha = \sum_{i=1}^{n} \alpha_i W_{c_{ii}}; \qquad \alpha_i = 2\zeta_i\omega_i \frac{V_i}{V} \tag{89}$$

The modal grammians are also closely related to the energy optimal DOC as introduced in [31, 33, 44, 45]. By comparing Eq. (84) to Eq. (12) under the conditions assumed in the latter equation, it follows that for flexible structures:

Proposition 15

$$\rho_c \approx 4\sqrt{E_c T} \ \min_i [\zeta_i W_{c_{ii}}] \tag{90}$$

The energy optimal DOC is approximately the minimum weighted singular values of the controllability grammian. It is clear that the energy optimal DOC is closely related to a measure of the least controllable state as defined by the minimum singular value of the controllability grammian. The control energy and time interval appears only as a constant scale factor.

C. MODAL GRAMMIAN SYNTHESIS

Consider the problem of optimal actuator placement. For flexible structures, Eq. (86) gives the i-th MGC

$$\beta_i^2(r) = \sum_{j=1}^{r} \psi_{ij}^2; \quad i \in M \tag{91}$$

Let N denote the set of candidate actuator locations. Then $\beta_i^2(N)$ denotes the maximum achievable controllability, i.e. the i-th MGC when all N actuators are used at all candidate locations.

Typically, a designer does not know exactly how many actuators to use and where to locate them. However, a larger and redundant set of candidate locations is usually known. If all actuators corresponding to the candidate locations are used, an estimate of the maximum achievable controllability in terms of the MGC can be computed. Rather than selecting a particular number of actuators based on heuristic judgment and then seeking their optimal locations, we consider herein an analysis/design strategy based on sequential removal of the least significant actuators while approximately maintaining maximum achievable controllability.

First, let $\beta_i^2(N)$ and $\beta_i^2(N_s)$ denote the i-th MGC corresponding to N and a smaller subset N_s, respectively. Define the sum of normalized (with respect to N actuators) MGC for the set of N_s actuators, so that

$$\gamma^2(N_s, N) := \sum_{i \in M} \frac{\beta_i^2(N_s)}{\beta_i^2(N)} \tag{92}$$

$$= \sum_{i \in M} \frac{1}{\beta_i^2(N)} \sum_{j \in N_s} \psi_{ij}^2 \tag{93}$$

$$= \sum_{j \in N_s} \sum_{i \in M} \frac{\psi_{ij}^2}{\beta_i^2(N)} \tag{94}$$

$$= \sum_{j \in N_s} \gamma_j^2 \tag{95}$$

where γ_j is a normalized contribution index of j-th actuator, $\gamma_j = \|v_j\|$, and

$$v_j = \begin{pmatrix} \frac{\psi_{1j}}{\beta_{11}(N)} & \cdots & \frac{\psi_{nj}}{\beta_{nn}(N)} \end{pmatrix}^T \tag{96}$$

While the sum in the above definition indicates an average performance over all modes, the normalization indicates that all significant modes are equally important in the MGC sense. This choice of normalization results in the maximum achievable controllability of $\gamma^2(N, N) = n$. The index γ_j^2 is the MGC of j-th actuator; thus the sum of normalized MGC consists of non-negative contributions from individual actuators as in Eq. (95).

The sum of independent components motivates the following strategy of sequential removal of least significant actuators. The actuators are arranged in a sequence of ascending magnitudes of MGC. Actuators with small γ_j^2 are less significant and hence removed. The set of remaining actuators after a sequence of removal represents the globally optimal set with respect to all sets of actuators having the same actuator numbers. This global optimality is with respect to the maximum achievable controllability achieved by using all candidate locations.

Note that the above optimality is defined (by normalizing) with respect to a reference maximum achievable controllability value. However, when a large difference exists between the total number and a feasible number of actuators available for use, the initial maximum achievable controllability value may be an unrealistic reference. Therefore as a special case, the MGC can be locally normalized with respect to the available rather than a fixed number of actuators. In this case, N varies with each actuator removal such that $N = N_s + 1$ at each step of removal.

The above approach consists of searching for the least significant actuator whose removal causes a minimal change in $\gamma^2(N_s, N)$. Although a trace of the normalized grammian, i.e., an average controllability over all modes, is used above, other definitions may also be physically meaningful. In any case, the modal approach to a multi-mode problem makes it necessary to specify a relative weighting which reflects the physical problem as highlighted in [22]. Observe also that based on the above sequence of contribution indices, γ_j^2, a control engineer can trade off the degree of controllability and the number of actuators, thus the number of actuators and their locations are addressed simultaneously. Due to the similarity of the controllability and observability grammians for flexible structures, a method for selecting the number and locations of sensors analogous to the actuator problem can be devised.

VII. OPTIMAL ACTUATOR PLACEMENT: A DESIGN EXAMPLE

A. MODEL DESCRIPTION

In this section, the modal grammian approach is applied to an experimental structure illustrating the actuator placement procedure. The configuration of the Controls-Structures Interaction Evolutionary Model (CEM) is shown in Figure 5. A total of 50 candidate locations for the air thrusters are selected and shown in the figure. The structural model consists of 12 modes whose first six are suspension modes. The first column in Table I shows the open-loop eigenvalues of the structural modes. The frequencies are closely spaced and lightly damped, which is typical for flexible structures. The main purpose of the actuators is to supress the suspension and structural vibration motion due to external disturbance by using feedback control. Reference [2] gives a more detail description of the structural model. Based on the set of actuators 1 to 8, the results of several control designs are given in [3, 27, 28].

B. MODAL GRAMMIAN ANALYSIS

Four different strategies for actuator placement are considered. They correspond to four cases: (a) MGC normalized with respect to a fixed set of 50

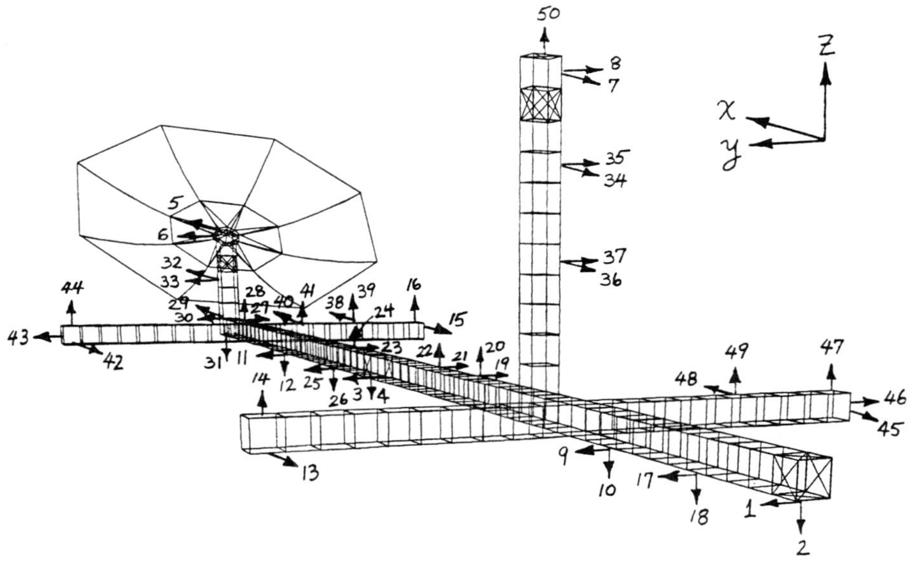

Fig. 5. Actuator configuration of CEM.

actuators, (b) MGC normalized with respect to the previous set of actuators ($N = N_s + 1$), (c) partial random order of removal where the last eight actuators correspond to the existing actuator configurations (locations 1-8) [2], and (d) random order of actuator removal. Only one case is shown from a sequence of random cases simulated. Figures 6(a) to 6(d) show the decrease in modal grammians, $W_{c_{ii}}$, as actuators are removed.

Table II show the corresponding order of actuator removal for the above four cases. Since cases (a) and (b) are based on a least significant actuator removal strategy, the optimal set of actuators for any given number can be deduced from the table.

In case (c) eight actuators (1-8) correspond to the existing configuration on the CEM [2] and for comparison purposes, the locations of eight actuators for the remaining three cases are given in Figure 7. Observe that six (1, 2, 8, 16, 39, 44) out of eight actuators are the same in case (a) and (b). On the other hand, only four actuators are the same (1, 2, 7, 8) in case

Table I. Open- and closed-loop eigenvalues of LQR

	Open-Loop		Case (a)		Case (b)		Case (c)		Case (d)	
Mode	$\omega(\frac{r}{s})$	ζ	$\omega(\frac{r}{s})$	ζ	$\omega(\frac{r}{s})$	ζ	$\omega(\frac{r}{s})$	ζ	$\omega(\frac{r}{s})$	ζ
1	.9297	.037	.9301	.033	.9696	.169	1.137	.611	1.046	.462
2	.9409	.001	.9718	.179	1.048	.478	1.231	.682	1.115	.550
3	.9732	.050	1.330	.771	1.456	.868	1.359	.793	1.320	.761
4	4.623	.005	4.818	.282	4.773	.192	4.750	.213	4.740	.064
5	4.752	.004	4.900	.241	4.857	.251	4.752	.089	4.924	.297
6	5.614	.002	6.067	.386	5.909	.316	5.725	.187	5.650	.175
7	9.410	.002	9.666	.223	9.565	.177	9.454	.109	9.448	.106
8	14.54	.001	14.63	.054	14.62	.057	14.56	.025	14.54	.011
9	15.29	.006	15.31	.051	15.33	.069	15.35	.084	15.32	.065
10	15.54	.001	15.55	.017	15.55	.017	15.54	.004	15.54	.005
11	16.27	.006	16.22	.089	16.25	.091	16.29	.065	16.28	.043
12	18.29	.014	18.29	.023	18.29	.027	18.28	.031	18.29	.027

(b) and (c) and only one (2) is the same in case (b) and (d) of the random sequence. Clearly, only cases (a) and (b) have similar configurations.

The modal grammian plots are initially flat even for the random order case, i.e., the indices are not initially sensitive to actuator removal. This is consistent with the physical intuition that actuator location is not an important issue when a redundant set of actuators is used. From another viewpoint, as more and more actuators are added, the additions will produce a smaller change in the indices.

Note that cases (a) and (b) which are based on MGC give more gradual drop in modal grammians than cases (c) and (d). However, when the number of remaining actuators approaches ten, the indices drop sharply. This break number is an indication that ten is the minimum number of actuators necessary to maintain the level of controllability generated by a large number of actuators. Notice also that cases (a) and (b) basically give larger indices than cases (c) and (d) for the most number of actuators and modes. However, because of the multi-mode nature of the problem, not every modal grammian is larger.

C. LQR PERFORMANCE

Using eight actuators, an optimal linear quadratic regulator (LQR) controller is designed for each case. The weighting matrices are chosen as identity matrices so that the quadratic integrand of the total energy (kinetic plus potential) of the system is the measure of the system performance. Thus the LQR control law is meant to minimize the time integral of the total energy of the system.

The closed-loop eigenvalues are shown in Table I. Most modes are well

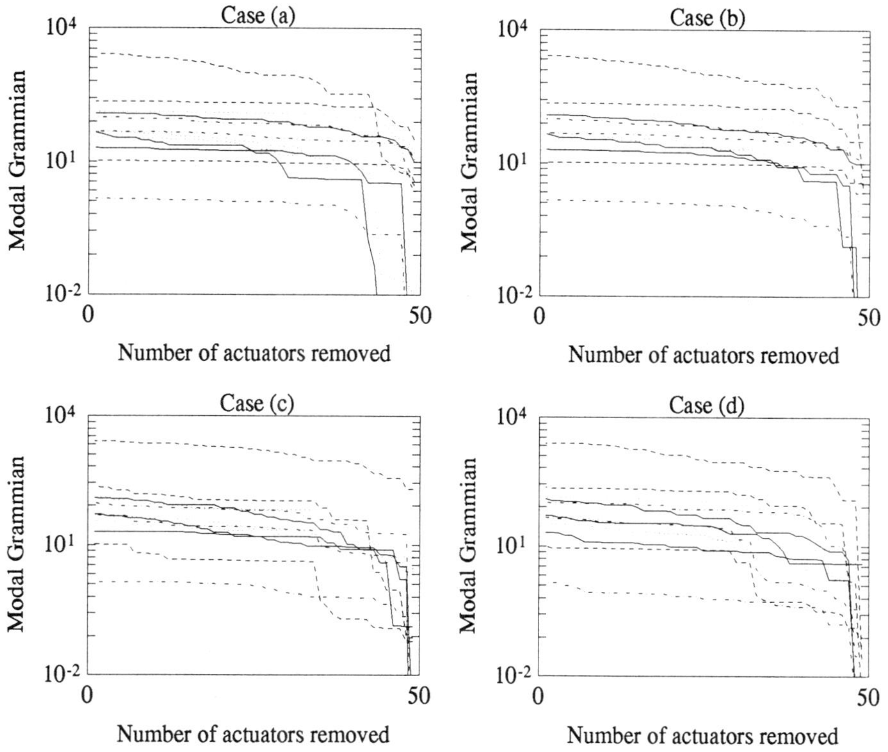

Fig. 6. Modal grammian history with sequential removal.

damped for all cases with two exceptions. In case (a) mode 1 is lightly damped and in cases (c) and (d) mode 10 is lightly damped. The lack of closed loop damping in mode 1 for case (a) is consistent with the observation that only case (a) does not have an actuator acting in the X-direction which is the primary motion of mode 1 (see Fig. 8(a)). Although this absence of actuator may initially appear counter intuitive, it turns out that most of the twelve modes selected involve motions along the Y and Z-directions so that X-direction actuators have low modal indices and do not contribute significantly to the sum of MGC. In fact, actuator number 7, which is an actuator in the X-direction, is the 9-th most important actuator for case (a) and was therefore left out when choosing a set of eight actuators (see case (a) in Table II). As for the second exception, mode number 10's small damping for cases (c) and (d) is caused by the fact that mode 10 is composed primarily of the X-axis rotation of the horizontal appendage near the antenna and antenna rotation (see Fig. 8(b)) and cases (c) and

Table II. Removal sequence of actuators.

Case (a):									
29	32	40	5	19	50	46	38	11	36
12	48	43	37	10	25	28	21	9	20
23	27	3	13	45	33	41	42	15	34
31	6	26	30	35	17	49	22	18	24
4	7	1	8	14	47	39	2	44	16
Case (b):									
29	32	40	19	50	46	5	12	11	43
10	38	37	28	25	36	21	20	48	9
41	23	27	26	3	31	13	49	33	34
35	22	42	6	18	45	24	30	14	15
4	47	17	39	8	7	44	2	1	16
Case (c):									
50	49	48	47	46	45	44	43	42	41
40	39	38	37	36	35	34	33	32	31
30	29	28	27	26	25	24	23	22	21
20	19	18	17	16	15	14	13	12	11
10	9	8	7	6	5	4	3	2	1
Case (d):									
3	39	5	24	48	7	4	1	45	12
30	6	21	33	28	31	17	35	25	18
41	29	11	9	26	23	13	40	42	16
20	46	44	15	8	34	32	36	50	37
49	14	10	2	19	43	47	22	27	38

(d) do not have actuators configured to actively influence this local mode (see Fig. 7).

Figure 9 shows the closed-loop performance of controllers measured by the magnitude of the eigenvalue of the matrix Ξ as given by Eq. (41). The eigenvalues are plotted in an ascending order. These eigenvalues are the principal costs corresponding to an orthogonal set of initial states. The closed-loop cost is dependent on the direction of the initial state. Case (b) has the best quadratic cost in all directions except in the direction of the smallest cost. Case (a) is the second best, followed by case (c) and (d). It is interesting that, although case (a) does not have an X-direction actuator, thus resulting in smaller damping for mode 1 when compared to cases (c) and (d), the closed-loop cost for case (a) is clearly better than in the latter two cases. This clearly demonstrate the strong dependence of actuator location on the modes considered for control and its control objective.

D. MAXIMIZING MINIMUM CONTROLLABILITY

Measures related to the least controllable state are frequently suggested for use as an indicator of actuator location optimization, for instance, the

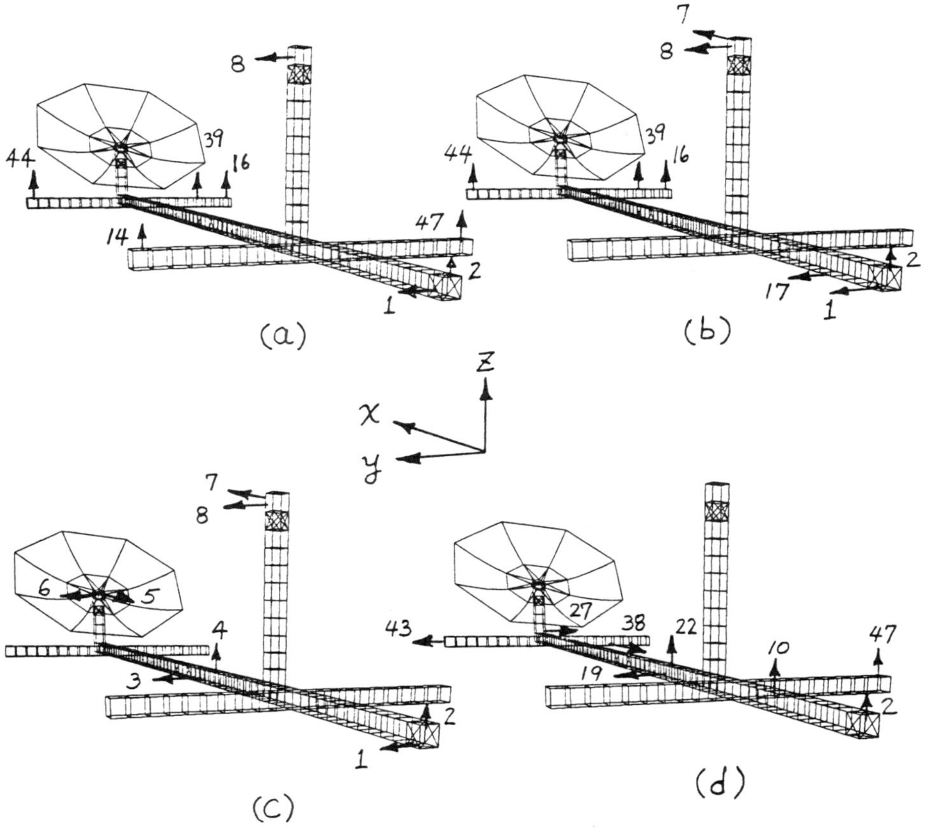

Fig. 7. Eight-actuator configurations

energy DOC in Eq.(11) proposed in [31, 33, 44, 45]. However, this may not always turn out to be the most suitable choice, as was discussed in §IV. Figure 10 shows the singular values of the controllability grammian corresponding to all four cases. Cases (a), (b) and (c) have larger singular values in the majority of principal directions than case (d). Hence, the random order case (d) is the worst configuration. In spite of this, the random order case (d) has relatively larger singular values than cases (a) and (c) in the least controllable principal direction. This means that the degree of minimal controllability may not always be a good indicator of actuator location. A metric based on minimum controllability will encourage the improvement in the worst controllable direction at the expense of all other directions, hence a possible loss of performance everywhere else.

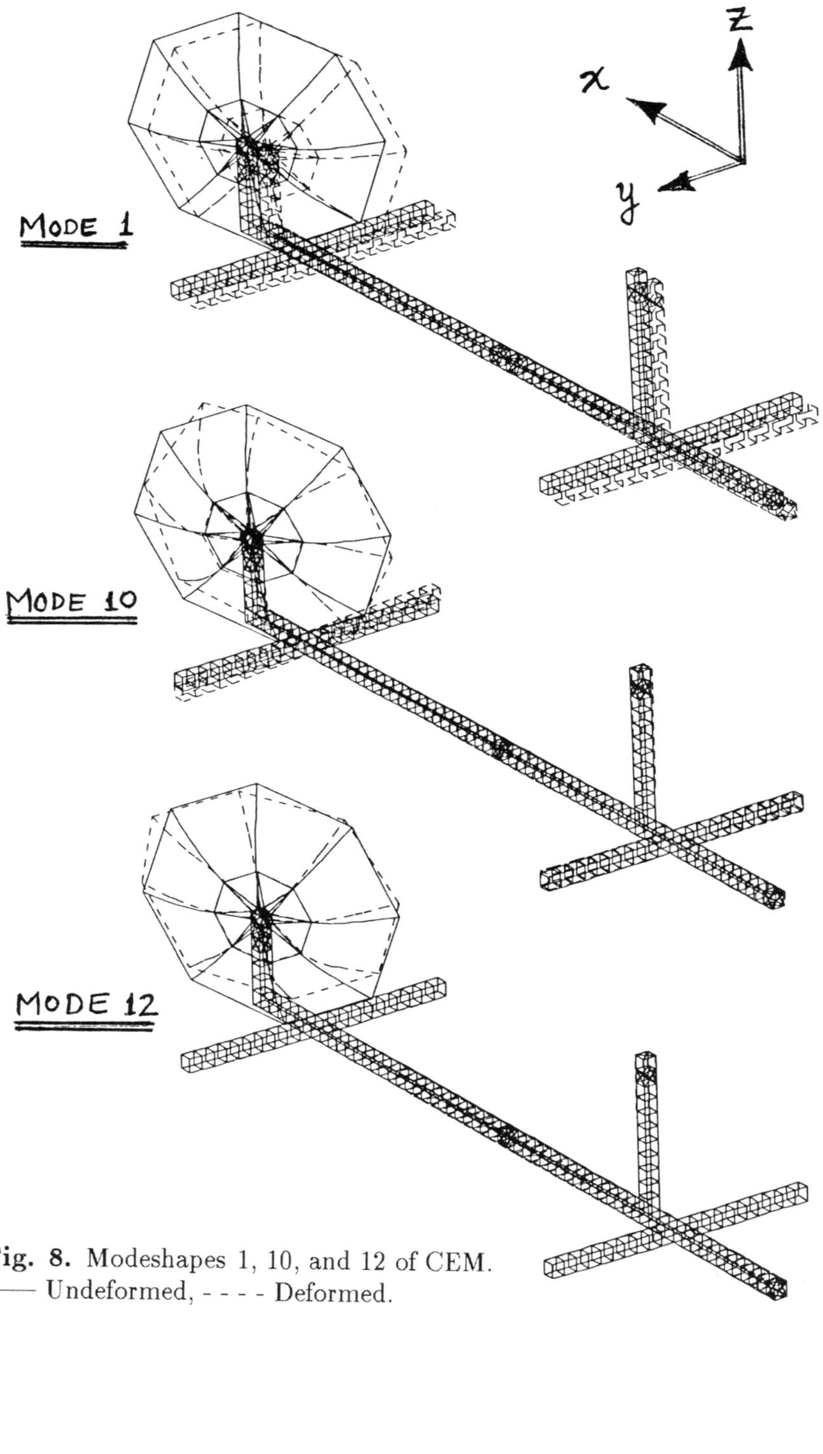

Fig. 8. Modeshapes 1, 10, and 12 of CEM.
—— Undeformed, - - - - Deformed.

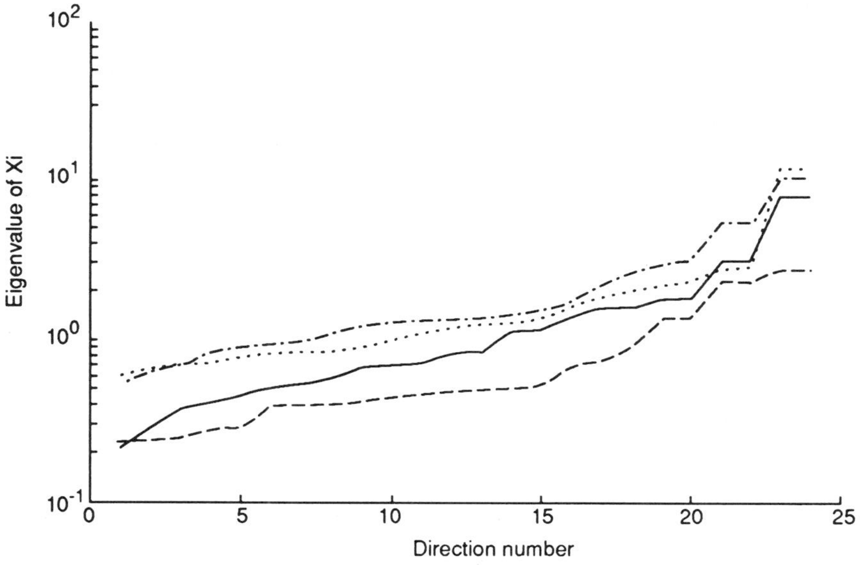

Fig. 9. Performance of LQR controller.
—— Case (a), - - - - Case (b), · · · · · · Case (c), — · — · —· Case (d).

E. LIMITATION OF CANDIDATE LOCATIONS

Figure 11 shows modal grammians for all twelve modes. Note that the degree of controllability for each mode is clear from Fig. 11, unlike Fig. 10. Again cases (a) and (b) are superior to (c) and (d). The low controllability of mode 1 in case (a) and the low controllability of mode 10 for cases (c) and (d) is also clear from these figures (cf. Fig. 10). In addition, Fig. 11 show small modal grammians for mode 12 for all four configurations. This indicates the ineffectiveness of the chosen configurations in controlling mode 12. This mode turns out to be an antenna-dominated mode (see Fig. 8(c)) and the set of 50 candidate locations does not include candidate locations on the antenna. This clearly demonstrates the strong dependence of actuator location performance on candidate location.

F. SUMMARY

The performances of the four actuator configurations discussed above can be put in the following order: cases (b), (a), (c), and (d). It follows from the above discussion that the overall performance depends on (1) type of

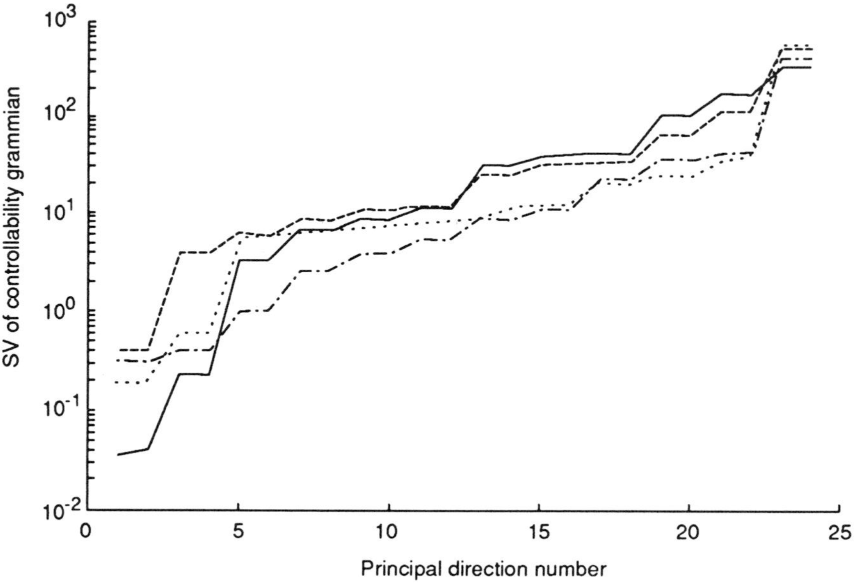

Fig. 10. Singular value of controllability grammian.
—— Case (a), - - - - Case (b), · · · · · · Case (c), — · — · —· Case (d).

normalization, (2) composition of the modes and their relative weightings, (3) definition of closed-loop performance, and (4) candidate set. The proposed approach provides a systematic and reasonable way for analysis and design of actuator and sensor placement such that the above characteristics can be taken into account. The design example also shows that (1) actuator location is not an important issue for heavily redundant sets of actuators; (2) a minimum number of presumably non-redundant sets of actuators can be determined; (3) an improved set of eight actuators can be used for controlling the selected twelve modes; (4) minimum controllability may not always be a good metric; and (5) heuristically motivated actuator configuration or closed-loop damping is not a reliable indicator of improved controllability.

VIII. CONCLUDING REMARKS

Various metrics for optimal placement have been examined and some connections have been made in a principal component framework. For flexible structures, a method for optimal placement via grammian synthesis has

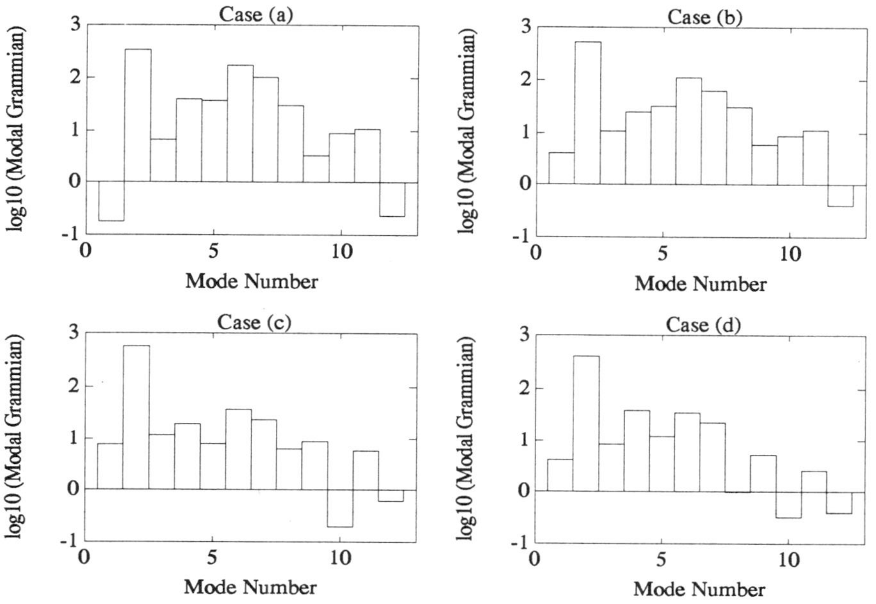

Fig. 11. Modal grammian.

been proposed. The method has desirable characteristics such as physical relevance, effectiveness, simplicity, and generality. For flexible structures, the singular values of controllability and observability grammians are directly related to the degrees of controllability and observability of individual modes. The small gain approximations imply that the controllability and observability grammians can be interpreted in terms of an optimal regulator and estimator performance. Consequently, optimizing the actuator and sensor locations through maximization of the grammian singular values can improve the closed-loop regulator and estimator performance. The numerical example for an existing flexible structure shows several interesting results. In the case where no other information is available, the following rule-of-thumb is recommended: actuators and sensors should be placed at locations with large motions. This information is easily available from structural mode shape data. Actuators and sensors should be configured such that principal directions corresponding to large singular values are parallel to the set of target modes.

IX. ACKNOWLEDGEMENTS

The authors would like to thank Mr. James P. Bailey for providing the finite element model of the CEM structure and the modeshape figures, Mr. Michael G. Powell for his help with graphics, and Drs. Lucas G. Horta and Kenny B. Elliott for carefully reading the manuscript and providing helpful comments.

X. REFERENCES

[1] Balas, G.J., and Doyle, J.C., "Robustness and Performance Tradeoffs in Control Designs for Flexible Structures," *IEEE 29-th CDC*, Hawaii, Dec. 1990.

[2] Belvin, W. K., et al., *Langley's CSI Evolutionary Model: Phase 0*, NASA TM 104165, Langley Research Center, Hampton, VA., Nov. 1991.

[3] Bruner, A., et al., "Active Vibration Absorber for the CSI Evolutionary Model: Design and Experimental Results," *Journal of Guidance, Control and Dynamics*, Vol. 15, No. 5, 1992, pp. 1253-57.

[4] Brockett, R.W., *Finite Dimensional Linear Space*, Wiley & Sons, Inc., New York, 1970.

[5] Chen, C.T., *Introduction to Linear System Theory*, Holt, Rinehart and Winston, Inc., 1970, pp. 401-402.

[6] DeLorenzo, M.L., "Selection of Noisy Sensors and Actuators for Regulation of Linear Systems," Ph.D. Thesis, School of Aeronautics and Astronautics, Purdue University, West Lafayette, IN, May 1983.

[7] DeLorenzo, M.L., "Sensor and Actuator Selection for Large Space Structure Control," *Journal of Guidance, Control and Dynamics*, Vol. 13, No. 2, 1990, pp. 249-257.

[8] Gawronski, W., and Juang, J-N., "Model Reduction for Flexible Structures," in *Control and Dynamic Systems, ed. C.T. Leondes*, Vol. 36, Academic Press, New York, 1990.

[9] Gawronski, W., and Williams, T., "Model Reduction for Flexible Space Structures," *Journal of Guidance, Control and Dynamics*, Vol. 14, No. 1, 1991, pp. 68-76.

[10] Gawronski, W., and Lim, K.B., "Controllability and Observability of Flexible Structures with Proof-Mass Actuators," *AIAA Dynamics Specialists Conference*, Dallas, TX., April 16-17, 1992.

[11] Gawronski, W., and Hadaegh, F.Y., "Balanced Input-Output Assignment," *International Journal for Systems Science*, Vol. 24, 1993.

[12] Gawronski, W., and Mellstrom, J., "Control and Dynamics of the Deep Space Network Antennas," in *Control and Dynamics Systems, ed. C.T. Leondes*, 1992.

[13] Golub, G.H., and VanLoan, C.F., *Matrix Computation*, John Hopkins University Press, Baltimore, MD, 1985.

[14] Gregory, C.Z. Jr., "Reduction of Large Flexible Spacecraft Models Using Internal Balancing Theory," *Journal of Guidance, Control and Dynamics*, Vol. 7, No. 6, 1984, pp. 725-732.

[15] Hamdan, A. M. A., and Nayfeh, A. H., "Measures of Modal Controllability and Observability for First- and Second-Order Linear Systems," *Journal of Guidance, Control and Dynamics*, Vol. 12, No. 3, 1989, pp. 421-428; and *Journal of Guidance, Control and Dynamics*, Vol. 12, No. 5, 1989, p. 768.

[16] Hughes, P.C., and Skelton, R.E., "Controllability and Observability of Linear Matrix Second-Order Systems," *Journal of Applied Mechanics*, Vol. 47, No. 6, 1980, pp. 415-420.

[17] Hughes, P.C., and Skelton, R.E., "Controllability and Observability for Flexible Spacecraft," *Journal of Guidance, Control and Dynamics*, Vol. 3, No. 5, 1980, pp. 452-459.

[18] Juang, J-N., and Rodriguez, G., "Formulations and Applications of Large Structure Actuator and Sensor Placements," *VPI/AIAA Symposium on Dynamics and Control of Large Flexible Spacecraft*, Ed. L. Meirovitch, Blacksburg, 1979, pp.247-262.

[19] Kailath, T., *Linear Systems*, Prentice-Hall, Englewood Cliffs, NJ, 1980.

[20] Kalman, R.E., "Contributions to the Theory of Optimal Control," *Bol. Soc. Mat. Mexicana*, Vol. 5, 1960, pp. 102-119.

[21] Kammer, D., "Sensor Placement for On-Orbit Modal Identification and Correlation of Large Space Structures," *Journal of Guidance, Control and Dynamics*, Vol. 14, No. 2, 1991, pp. 251-259.

[22] Kim, Y., and Junkins, J.L., "Measure of Controllability for Actuator Placement," *Journal of Guidance, Control and Dynamics*, Vol. 14, No. 5, 1991, pp. 895-902.

[23] Kullas, M.C., "Handbook on Astronaut Crew Motion Disturbances for Control System Design," NASA Reference Publication 1025, May 1979.

[24] Kwakernaak, H., and Sivan, R., *Linear Optimal Control Systems*, John Wiley & Sons, Inc., 1972.

[25] Lim, K.B., "Method for Optimal Actuator and Sensor Placement for Large Flexible Structures," *Journal of Guidance, Control and Dynamics*, Vol. 15, No. 1, 1992, pp. 49-57.

[26] Lim, K.B., and Horta, L.G., "A Line-of-Sight Performance Criterion for Controller Design of a Proposed Laboratory Model," *AIAA Dynamics Specialist Conference*, Long Beach, CA, April 1990, Paper No. 90-1226.

[27] Lim, K.B., Maghami, P.G., and Joshi, S.M., "Comparison of Controller Designs for an Experimental Flexible Structure," *IEEE Control Systems*, June 1992, pp. 108-118.

[28] Lim, K.B., and Balas, G.J., "Line-of-Sight Control of the CSI Evolutionary Model: μ Control," *American Control Conference*, Chicago, Ill, June 24-26, 1992.

[29] Lim, T.W., "Sensor Placement for On-Orbit Modal Testing," *32nd Structures, Structural Dynamics, and Materials Conference*, Baltimore, MD, April 1991.

[30] Lindberg, R.E., and Longman, R.W., "Optimization of Actuator Placement via Degree of Controllability Criteria Including Spillover Considerations," *AIAA/AAS Astrodynamics Conference*, San Diego, CA, Aug 9-11, 1982, AIAA 82-1435.

[31] Longman, R. W., Sirlin, S. W., Li, T., and Sevaston, G., " The Fundamental Structure of Degree of Controllability and Degree of Observability," *AIAA/AAS Astrodynamics Specialists Conference*, Paper 82-1434, San Diego, CA, 1982.

[32] Longman, R.W., and Horta, L.G., "Actuator Placement by Degree of Controllability Including the Effect of Actuator Mass," *Proceedings of the Seventh VPI&SU/AIAA Symposium on Dynamics and Control of Large Flexible Spacecraft*, Blacksburg, VA, 1989.

[33] Longman, R. W., and Alfriend, K. T., "Energy Optimal Degree of Controllability and Observability for Regulator and Maneuver Problems," *The Journal of the Astronautical Sciences*, Vol. 38, No. 1, Jan-Mar 1990, pp. 87-103.

[34] Maghami, P.G., and Joshi, S.M., "Sensor/Actuator Placement for Flexible Space Structures," *American Control Conference*, San Diego, CA, May 23-25, 1990.

[35] Maghami, P.G., and Joshi, S.M., "Sensor-Actuator Placement for Flexible Structures with Actuator Dynamics," *AIAA Guidance, Navigation, and Controls Conference*, New Orleans, LA, August 12-14, 1991.

[36] Meirovitch, L., *Analytical Methods in Vibrations*, Macmillan, New York, 1967, Chapter 4.

[37] Middleton, D., *An Introduction to Statistical Communication Theory*, McGraw-Hill, New York, 1960.

[38] Moore, B.C., "Principal Component Analysis in Linear Systems: Controllability, Observability, and Model Reduction," *IEEE Transactions on Automatic Control*, Vol. AC-26, No. 1, Feb. 1981, pp. 17-32.

[39] Noble, B., and Daniel, J.W., *Applied Linear Algebra*, Prentice-Hall, Inc., Englewood Cliffs, NJ, 1977, p.440.

[40] Rowell, L.F., "Techniques for Assessment of Flexible Structure Control Performance," *AIAA 26th Aerospace Sciences Meeting*, Jan 11-14, 1988, Reno, NV, AIAA 88-0677.

[41] Salama, M., Rose, T., and Garba, J., "Optimal Placement of Excitations and Sensors for Verification of Large Dynamic Systems," *28th Structures, Structural Dynamics, and Materials Conference*, Monterey, CA, April 1987.

[42] Skelton, R.E., *Dynamic System Control: Linear System Analysis and Synthesis*, Wiley, New York, 1988.

[43] Skelton, R.E., and DeLorenzo, M.L., "Selection of Noisy Actuators and Sensors in Linear Stochastic Systems," *Journal of Large Scale Systems, Theory and Applications*, Vol.4, April 1983, pp. 109-136.

[44] Viswanathan, C. N., Longman, R. W., and Likins, P. W., "A Definition of the Degree of Controllability - A Criterion for Actuator Placement," *Proceedings of the Second VPI&SU/AIAA Symposium on Dynamics and Control of Large Flexible Spacecraft*, Blacksburg, VA, June, 1979.

[45] Viswanathan, C. N., Longman, R. W., and Likins, P. W., "A Degree of Controllability Definition: Fundamental Concepts and Application to Modal Systems," *Journal of Guidance, Control and Dynamics*, Vol. 7, 1984, pp. 222-230.

[46] Williams, T., "Closed Form Grammians and Model Reduction for Flexible Space Structures," *IEEE Transactions on Automatic Control*, Vol.35, March 1990, pp. 379-382.

Structural Modification Techniques for Robust Actively Controlled Structures

S.S. Rao
School of Mechanical Engineering
Purdue University, West Lafayette, IN 47907

T.S. Pan
IBM, Poughkeepsie, NY 12602

and

V.B. Venkayya
Wright Laboratory
Wright-Patterson Air Force Base, OH 45433

I. INTRODUCTION:

In recent years, the problem of simultaneous design of space structures and their controls has received considerable attention due to the reductions realizable in the control effort and structural weight and due to the improvements achievable in the closed-loop performance of the system. The integrated structures and controls problem has been solved by several investigators as an optimization problem with constraints on several response characteristics such as transient response, actuator forces, magnitudes and locations of nonstructural masses, closed loop eigenvalues and damping

CONTROL AND DYNAMIC SYSTEMS, VOL. 57

parameters [1-4].

The mathematical model used to represent the physical system usually contains inaccuracies and does not represent the real system. In space structural design, a nominal model of the system is selected and a controller is designed for that model. The design process is expected to take care of the uncertainties of the mathematical model in order to guarantee the stability of the designed control system. The modelling errors in the control system are of two types - structered and unstructured errors. The structured uncertainties are due to the errors in the assumed passive damping coefficients and the natural frequencies of the structure. These are caused by the inaccuracies in the material properties and geometric variables. The unstructured uncertainties are due to the errors caused in the development of the equations of motion such as those introduced by the reduced order models. A robust control system should be stable under all possible uncertainties without deterioration of the performance.

This paper considers the robustness improvement of actively controlled structures through structural modifications. Several investigators have considered the problem of robustness in both frequency domain and time domain analyses [5]. Some of the models developed for use in time-domain state-space analysis are reviewed before deriving the equations of motion. Quantitative measures on time-varying nonlinear perturbation of an asymptotically stable linear system were developed by Patel et al [6] and Patel and Toda [7]. Yedavalli [8,9] obtained improved bounds on the linear perturbations of an asymptotically stable linear system. The methods presented in Refs. [6-9] are based on the solution of a certain Lyapunov equation for stability robustness. Thus there is certain conservativeness in all these methods in determining upper bounds on the parameter variations. Juang et. al [10] and Qiu and Davison [11] derived a robustness bound for asymptotically stable linear time invariant systems. This reference reduced the conservativeness of the bound by not solving the Lyapunov equation. Lim and Junkins [12] used the approach of Patel and Toda [7] for robust optimization; in addition, they minimized the eigenvalue sensitivity for improving the stability robustness. It was found that the maximization of the stability robustness is superior to the minimization of the eigenvalue sensitivity in producing robust designs. A pole placement strategy for obtaining a robust eigenstructure was considered by Rew et al [13] using state energy, control energy and stability robustness as objective functions. A similar study was presented by Juang et al [14]. Grandhi et al [15], Thomas and Schmit [16] and Khot and Veley [17] considered robustness in different ways in the design of actively controlled structures. This paper considers the problem of robustness improvement of actively controlled structures through structural modifications.

II. CONTROL DESIGN

The equations of motion governing the dynamic behavior of a structure are given by

$$\dot{\vec{q}} = [A]\,\vec{q} + [B]\,\vec{f} \tag{1}$$

$$\vec{y} = [C]\,\vec{q} \tag{2}$$

where $\vec{q}$ = state vector $(n \times 1)$, $\vec{f}$ = control vector $(p \times 1)$, $\vec{y}$ = output vector $(q \times 1)$, $[A]$ = plant matrix $(n \times n)$, $[B]$ = input matrix $(n \times p)$ and $[C]$ = output matrix $(q \times n)$. If the sensors and actuators are colocated, $[C] = [B]^T$. In terms of the modal solution of the structure, the matrices [A] and [B] can be expressed as

$$[A] = \begin{bmatrix} [0] & [I] \\ -[\omega^2] & -[2\,\zeta_i\omega_i] \end{bmatrix} \tag{3}$$

$$[B] = \begin{bmatrix} [0] \\ [\Phi]^T\,[D] \end{bmatrix} \tag{4}$$

where $[I]$ = identify matrix, $[\omega^2]$ = diagonal matrix of squares of natural frequencies of the structure, $[2\,\zeta_i\,\omega_i]$ = diagonal matrix with ζ_i denoting the ith modal damping ratio, $[\Phi]$ = modal matrix, and $[D]$ = controller force transformation matrix. The optimal control vector $\vec{f}$ is selected to minimize the following performance index:

$$J = \int_0^{\infty} \left(\vec{y}^T [Q] \vec{y} + \vec{f}^T [R] \vec{f} \right) dt \tag{5}$$

where $[Q]$ = a positive semidefinite state weighting matrix and $[R]$ = a positive definite control weighting matrix. The optimum feedback control law for a linear quadratic regulator is given by

$$\vec{f} = -[G]\,\vec{q} \tag{6}$$

with

$$[G] = [R]^{-1}\,[B]^T\,[P] \tag{7}$$

where $[G]$ = optimum gain matrix and $[P]$ is given by the solution of the algebraic Riccati equation:

$$[A]^T [P] + [P][A] + [Q] - [P][B][R]^{-1}[B]^T[P] = [0] \tag{8}$$

III. ROBUSTNESS INDICES

A good design of a controlled structure should be robust enough to maintain its stability as well as good performance even under the worst perturbations and errors in the system model or controller parameters. Hence the measures of stability and performance robustness are very critical in the design of a controlled structure. The stability robustness index, a measure of closed-loop system stability under uncertainty or modeling errors can be defined in different ways. A simple criterion to measure the stability robustness is the variation of the dominant eigenvalues between the nominal system and the perturbed system [5]. However, there is a possibility that the dominant eigenvalues may be at a sufficient distance from the origin but with totally different real and imaginary components. Hence a vector form is to be used as a criterion. To achieve the best description of stability robustness, more eigenvalues (instead of just the dominant ones) with normalized weights are included in defining the stability robustness index as:

$$\beta_{sr} = \sum_{i=1}^{2n} \omega_i \left| \lambda_i - \bar{\lambda}_i \right| \tag{9}$$

where β_{sr} is the stability robustness index, and λ_i and $\bar{\lambda}_i$ are the closed-loop eigenvalues of nominal system and perturbed system, respectively. The weighting coefficient w_i, denoting the importance of eigenvalue pair i, is determined by the horizontal distance of eigenvalue to imaginary axis in complex plane:

$$w_i = \frac{1}{S^*\text{real}(\lambda_i)} \tag{10}$$

$$S = \sum_{i=1}^{2n} \frac{1}{\text{real}(\lambda_i)} \tag{11}$$

This implies that a controlled system will be robust if its closed-loop eigenvalues are insensitive to changes in the environment. Hence, to maximize the system stability robustness, the stability robustness index should be minimized.

In addition to stability robustness, it is desirable to retain the performance unchanged when design variables are changed. Since the performance cannot remain the same, a performance robustness index is defined to preserve the performance robustness. From optimal linear quadratic regulator, if P is the steady state solution of the matrix Riccati equation, the performance index is given as a function of the initial state vector as

$$J = \vec{q}_o^{\,T} [P] \vec{q}_o \tag{12}$$

where $\vec{q}_o$ is the initial state vector. The performance index for perturbed system with the same controller can be written as

$$\bar{J} = \vec{q}_o^{\,T} \bar{P} \vec{q}_o \tag{13}$$

where $\bar{P}$ satisfies the Lyapunov equation

$$\bar{A}_{cl}^T \bar{P} + \bar{P}\bar{A}_{cl} + G^T R G + Q = 0 \tag{14}$$

in which, $\bar{A}_{cl}$, is the perturbed closed-loop system matrix:

$$\bar{A}_{cl} = \bar{A} - \bar{B}G \tag{15}$$

where the gain matrix for the controller, [G], is given by Eq. (7) and the matrix [P] satisfies Eq. (8). The performance robustness index is defined as the variation of the steady state performance index as:

$$\beta_{pr} = \left| \frac{J - \bar{J}}{J} \right| = \frac{\vec{q}_o^{\,T}(P - \bar{P})\vec{q}_o}{\vec{q}_o^{\,T} P \vec{q}_o} \tag{16}$$

It has been pointed out by Kleinman and Athans (1968) [18] that the optimal solution of Eq. (12) in general depends on the initial state $\vec{q}_o$ which may not be known always. The effect of $\vec{q}_o$ can be averaged by assuming $\vec{q}_o$ to be a uniformly distributed random vector whose covariance is given by the identity matrix. In such a case, the trace of P can be shown to be proportional to the expected value of J. Hence Eq. (16) can be written as

$$\beta_{pr} = \left| \frac{\mathrm{Tr}(P) - \mathrm{Tr}(\bar{P})}{\mathrm{Tr}(P)} \right| \tag{17}$$

Good dynamic response can be achieved for both nominal and perturbed systems if the real part of every eigenvalue is restricted to be smaller than a specified value as $\mathrm{Re}(\lambda) < -a$, with $a>0$. The LQ regulator can be modified [13] as follows with the requirement of (A,B) being controllable and (A,C) being observable.

$$\dot{x} = (A + aI)q + Bu = (A + aI - BG)q \tag{18}$$

$$u = -R^{-1} B^T \underset{\sim}{K} q = -Gq \tag{19}$$

where $[\underset{\sim}{K}]$ satisfies the relation:

$$(A+aI)^T\underset{\sim}{K} + \underset{\sim}{K}(A+aI) - \underset{\sim}{K}BR^{-1}B^T\underset{\sim}{K} + C^TQC = 0 \tag{20}$$

The characteristic equation of Eq. (18) is given by

$$\det [A + aI - BG) - \lambda I] = \det [A - BG - \lambda' I] = 0 \tag{21}$$

with $\lambda' = \lambda - a$. Since A is assumed to be stable, then $\text{Re}(\lambda') < 0$, and hence $\text{Re}(\lambda) < -a$. Thus, the system with controller, Eq. (19), provides a guaranteed stability margin.

IV. FORMULATION OF OPTIMIZATION PROBLEM

The structural design problem can be formulated with the cross-sectional areas of the truss members as design variables. Several objectives can be considered for optimization as: (1) minimization of weight, (2) minimization of stability robustness index, (3) minimization of performance robustness index, (4) minimization of the optimal performance index, (5) minimization of the Frobenious norm of the control gains, and (6) minimization of effective damping response time. The following constraints are to be considered in a general design problem: (1) minimum stability requirement, in terms of the negative values of the closed loop eigenvalues of the system, (2) bounds on the natural frequencies of the system (3) bounds on the closed loop damping ratios of the system, and (4) bounds on the cross-sectional areas of the members of the system. The design problem thus can be expressed as a multiple objective optimization problem which can be stated in a general form as follows [19]:

$$\text{Find} \qquad \vec{X} = \begin{Bmatrix} x_1 \\ x_2 \\ \cdot \\ \cdot \\ \cdot \\ x_n \end{Bmatrix}$$

$$\text{which minimizes} \qquad \vec{f}(\vec{X}) = \left\{ f_1(\vec{X}), f_2(\vec{X}), \ldots, f_k(\vec{X}) \right\}^T \tag{22}$$

subject to

$$g_j(\vec{X}) \leq 0 \; ; \quad j = 1,2,\ldots, m$$

where $\vec{X}$ is the design vector, $\vec{f}(\vec{X})$ is the vector of objective functions and $g_j(\vec{X})$ is the jth constraint function.

V. SOLUTION CONCEPT

For a single objective optimization problem, an optimum solution is defined as one that minimizes the objective function $f_i(\vec{X})$ subject to the constraints $g_j(\vec{X}) \leq 0$, $j = 1,2,\ldots,m$. Attempting to define a vector minimal point as one at which all components of the objective function vector $\vec{f}$ are simultaneously minimized is not an adequate generalization since such "utopia" points are seldom attainable. Thus a new optimality concept, different from that used in scalar optimization, is necessary to find a solution to the vector optimization problem. The concept of a Pareto-optimal solution has been found to be useful in this context [20].

Definition: A feasible solution X* is Pareto-optimal if there is no feasible $\overline{X}$ such that $f_i(\overline{X}) \leq f_i(X^*)$, $i=1,2,\ldots,k$, and $f_{i_0}(\overline{X}) < f_{i_0}(X^*)$ for at least one $i_0 \in [1,\ldots,k]$.

Alternately, a design vector X* is Pareto-optimal if there exists no feasible vector $\overline{X}$ which would decrease some objective function without causing a simultaneous increase in at least one other objective function. Unless a problem is convex, only a locally optimal solution can be guaranteed using standard mathematical programming techniques. The set of Pareto-optimal solutions usually consists of an infinite number of points and additional information is needed to order the Pareto-optimal set. This makes it possible to bring in additional considerations which are not included in the optimization model, thus making the multiobjective approach a flexible technique for most design problems. Several numerical techniques have been suggested for solving a vector optimization problem. Each method, in general, generates a different Pareto-optimal solution which reflects the decision maker's (DM's) preference structure.

V. SOLUTION METHODS

There are several methods developed for the solution of a multiple objective optimization problem. In order to have a common basis for comparison, and to avoid working with different objectives in different units, the objective functions f_i (X) are transformed into new objective functions (F_i) constructed as follows

$$F_i(X) = m_i \, f_i(X) \quad i = 1,2,...k. \tag{23}$$

where, the positive constant multipliers $m_1, m_2, ..., m_k$ are chosen so that

$$m_1 \, f_1(X_0) = m_2 \, f_2 \, (X_0) = ... = m_k \, f_k \, (X_0) = M \tag{24}$$

at any feasible starting vector X_0 and M is a specified constant. This scaling procedure ensures that all the objective functions are equal at a particular value of X_0. Hereafter, it will be assumed that he *k* objective functions correspond to the *k* scaled objective functions given by Eq. (23). Some of the techniques which are used to generate Pareto optimal solutions for the problem stated in Eq. (22) are given below [21].

A. GLOBAL CRITERION FORMULATION

This method belongs to a category of multiple objective optimization techniques which require no articulation of preferences on part of the decision maker once the problem objectives and constraints have been defined. This entails that the DM be willing to accept whatever solution is obtained by minimizing some global criterion F(X), for example, the sum of the squares of the relative deviations of the respective individual objective functions from the feasible ideal solutions. In other words, an optimum solution X* is found by minimizing

$$F(X) = \sum_{i=1}^{k} \left[\frac{F_i(X) - F_i(X^*_i)}{F_i(X^*_i)} \right]^p \tag{25}$$

subject o

$$g_j(X) \le 0 \quad j=1,2,...,m.$$

The value of p corresponds to the utility function of the DM and is usually taken as 2. The X_i^* is the feasible ideal solution corresponding to the ith objective function, and is obtained by minimizing $F_i(X)$ with respect to the constraint set $X \varepsilon S$. For $1 \le p < \infty$, each solution obtained by solving Eq. (25) is Pareto-optimal. Compromise solutions with $p = \infty$ correspond to min-max criterion for which Pareto-optimality is not guaranteed.

B. UTILITY FUNCTION FORUMATION

In this approach, the vector minimization problem (Eq. (22)) is converted to

$$\text{Maximize } U(\vec{f}) \tag{26}$$

subject to

$$g_j(X) \leq 0 \quad j=1,2,\ldots,m$$

where $U(\vec{f})$ is the utility function of multiple objective functions. The rationale for using $U(\vec{f})$ is that the DM has some utility associated with each of the *k* objective functions. A utility function U can have many forms. The most common form assumes that the DM's utility function is additively separable with respect to all the objective functions. Thus, if $U_i(F_i)$ is the utility function corresponding to the objective function F_i, an overall utility function U is defined as

$$U(\vec{F}) = \sum_{i=1}^{k} U_i(F_i). \tag{27}$$

An optimum solution vector X^* is found by maximizing the total utility $U(\vec{F})$ (Eq. (27)) subject to the constraint set $g_j(X) \leq 0$. A special form of Eq. (27) which has been extensively used in multiobjective problems is given by

$$U = -\sum_{i=1}^{k} w_i F_i(X) \tag{28}$$

where w_i is a scalar weighting factor associated with the ith objective function and indicates its relative importance. This additively separable form of the utility function (Eq. (28)) is also commonly referred to as the weighting method, and serves as a sufficient condition for the calculation of Pareto-optimal solutions.

The main advantage of the utility function formulation is its simplicity. It is easier to assess *k* unidimensional utility functions (U_i's) than to assess $U(\vec{F})$ directly. Similarly, it is easier to get w_i's from the decision maker. The disadvantages of this approach are that there are few cases where the utility function is really additively separable, and w_i depend not only on the achievement level of F_i but also the achievement level of F_i relative to F_j, for $i \neq j$. Further, if the problem is nonconvex, this approach may miss all but a finite number of Pareto-optimal solutions.

C. GOAL PROGRAMMING METHOD

In goal programming, there are two basic models: the Archimedian model and the Preemptive model. The Archimedian model deals with generation of candidate solutions whose criterion vectors are closest, in a weighted L_p metric sense to the utopian set in the criterion space. The preemptive model, on the other hand, generates solutions whose criterion vectors are most closely related in a lexicographic sense, to points in the utopian set. The Archimedian version of goal programming is considered in this work [22].

In the simplest version of Archimedian goal programming, a designer sets goals and relative weights for each of the objective functions that he/she wishes to attain. An optimum solution X* is then defined as the one that minimizes the weighted sum of the deviations from the set goals. Thus, the goal programming formulation of a multiobjective problem leads to

$$\text{Minimize} \quad \left[\sum_{j=1}^{k} w_j \left(d_j^+ + d_j^-\right)^p\right]^{1/p} \quad ; p \geq 1 \tag{29}$$

subject to

$$g_i(X) \leq 0 \quad i=1,2,\ldots,m \tag{30}$$

$$F_j(X) - d_j^+ + d_j^- = b_j \quad j=1,2,\ldots,k \tag{31}$$

$$d_j^+ \geq 0 \quad j=1,2,\ldots,k \tag{32}$$

$$d_j^- \geq 0 \quad j=1,2,...,k \tag{33}$$

$$d_j^+ \, d_j^- = 0 \quad j=1,2,...,k \tag{34}$$

where b_j are the goals sets by the designer for the *j*th objective function, and d_j^+ and d_j^- are the under- and over-achievement from the target goals for the *j*th objective function. The value of p is based on the utility function chosen by the designer. If the goals b_j are set equal to F_j^* obtained by minimizing the individual objective functions F_j, it is not possible to obtain an overachievement of the goals b_j's. Consequently, the d_j^- need not be defined. Thus the goal programming formulation given by Eqs. (29) - (34) reduces to

$$\text{Minimize} \quad \left[\sum_{j=1}^{k} w_j \left(d_j^+ \right)^p \right]^{1/p} \; ; \; p \geq 1 \tag{35}$$

subject to

$$\begin{aligned} g_i(X) &\leq 0 \quad i=1,2,...,m \\ d_j^+ &\geq 0 \quad j=1,2,...,k \\ d_j^+ &= F_j(X) - F_j^*(X) \quad j=1,2,...,k. \end{aligned} \tag{36}$$

The goal constraints in the above formulation are soft constraints in the sense that they do not restrict the original feasible region S. In effect, they augment the feasible region by casting S into higher dimensional space, thereby creating the augmented goal programming feasible region. However, if the goal vector is not chosen properly, there is no guarantee that the goal programming formulation will terminate at a Pareto-optimal solution.

D. LEXICOGRAPHIC METHOD

In the lexicographic method, the objectives are ranked in order of importance by the designer. An optimum solution X* is obtained by minimizing the objective functions, starting with the most important one and proceeding according to the order of importance of the objectives. The rationale for this method is that individuals tend to make decisions in this manner. Let the subscripts of the objectives denote not only the objective function number, but also the priority of the objective. The solution procedure is given as follows:

1) Starting with X_0, minimize $F_1(X)$ subject to the constraint set $g_j(X) \leq 0$. Let the resulting optimum solution be denoted as X^*_1 and F^*_1.

2) Starting from X^*_1, minimize $F_2(X)$ subject to the constraint set $g_j(X) \leq 0$, and an additional constraint of the form $0.95\ F^*_1 \leq F_1(X) \leq 1.05\ F^*_1$. Let the resulting solution be X^*_2, and $F^*_2 = F_2(X^*_2)$.

3) Proceeding as outlined in step (2), at the *i*th stage the resulting problem is given as: Starting from X^*_{i-1}, minimize $F_i(X)$ subject to the constraint set $g_j(X) \leq 0$, and *i*-1 additional constraints of the form $0.95\ F^*_j \leq F_j(X) \leq 1.05\ F^*_j$, $j=1,2,...,i-1$.

For a problem involving *k* criteria, there are a total of *k!* ways in which the objective functions can be ranked by the DM. Since the solution obtained using lexicographic method is fairly sensitive to the ranking of the objectives given by the DM, one should exercise caution in applying this method when some objective functions are of nearly equal importance.

VII. ROBUSTNESS ANALYSIS PROCEDURE

The effect of variation in the parameters of the structure on the robustness of the system can be studied using the following procedure. Start with an initial reference design of the structure and find the corresponding plant matrix A, input matrix B and the output matrix C. The optimal control gain G is found using the LQ regulator design technique. Next, change the design parameters by known percentage values and find the corresponding A, B and C matrices. The stability robustness index β_{sr} and the performance robustness index β_{pr} are found using Eqs. (9) and (17), respectively. The procedure is repeated with different parameter changes, (e.g. nominal design variables, damping ratio, density etc.). Then a graph between β_{sr} or β_{pr} and the change in the parameters can be plotted.

VIII. ROBUSTNESS DESIGN PROCEDURE

The structure can be designed by using a suitable multiobjective optimization technique as outlined earlier. The following procedure can be adopted for this purpose. A preliminary design, used as the nominal design, can be obtained from the requirements of stress and deformation. Corresponding to this nominal design, generate the plant input and output matrices. Formulate the multiobjective constrained optimization problem as outlined earlier. Minimize the individual objective functions and find the minima around the nominal design. Use a suitable multiobjective optimization approach to find a Pareto optimal (compromise) solution.

IX. ILLUSTRATIVE EXAMPLES

The robustness analysis and robustness-based design procedures presented earlier are illustrated with the help of two examples. The first example is a two bar truss, shown in Fig. 1, in which the actuators and sensors

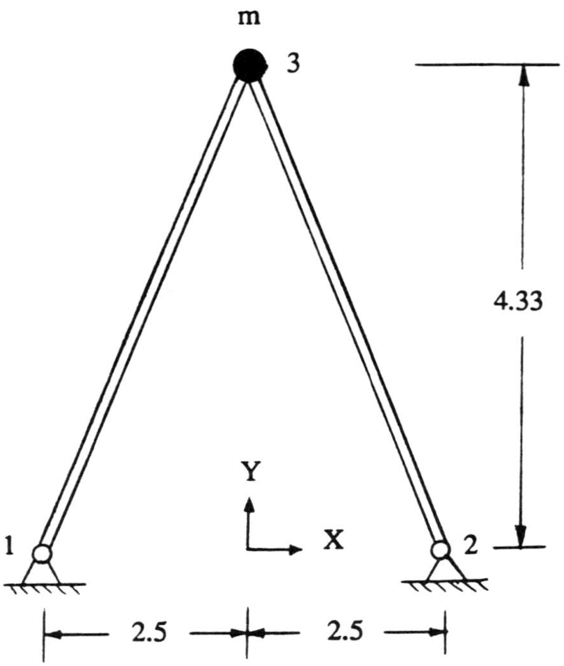

Figure 1 Two-Bar Truss

Table I

Data of two-bar truss

Nonstructural mass at node 3	1 unit
Lower bound on cross-sectional areas	0.01
Upper bound on cross-sectional areas	1.0
Nominal cross-sectional areas	0.1
Structural damping ratio	0.01
Young's modulus	10^7
Material density	4.6
Output weighting matrix [Q]	1000[I]
Input weighting matrix [R]	[I]
Lower bound on the closed-loop frequencies	20 rad/sec
Upper bound of the closed-loop frequencies	40 rad/sec
Upper bound on real part of perturbed closed-loop system	0

are colocated at node 3 in X and Y directions. The areas of cross-section of the two members are taken as design variables. The data of the structure and design specifications are given in Table I. The variation of the stability robustness index with changes in the two design variables is shown in Fig. 2. The influence of changes in the structural damping ratio on the stability robustness index is determined and the results are shown in Figs. 3 and 4. Here structural damping ratio is varied between 0 and 0.1 and the design variable changes are assumed to be -5% and -10%. The variations in the stability robustness index, the performance robustness index and the performance index with variation in the Young's modulus of the material are shown in Table II. The influence of changes in the output weighting matrix on β_{sr} and β_{pr} is indicated in Table III. All the figures and tables aid the designers in determining the approximate ranges of the parameters of the system in order to achieve any particular value of β_{sr} or β_{pr}. The optimum design problem is considered with three objectives, namely the minimization of weight, stability robustness index and performance robustness index with a permissible change of 5% in the design variables. When the individual objective functions are minimized under the stated constraints, the minimum values of weight, β_{sr} and β_{pr} are observed to be 23.598, 0.009502 and 0.0015899, respectively. The results given by the utility function, goal programming and lexicographic methods are shown in Table IV.

The second example is a two-bay truss, consisting of ten members and eight displacement degrees of freedom, as shown in Fig. 5. The data and the design specifications of the truss are given in Table V. The length of each bay (in the X-direction) is 50 inches and the height reduces from 36" at X = 0 to 24" at X = 100". The areas of cross-section of the ten members are chosen as the design variables. The actuators and sensors are colocated at nodes 1, 2, 3 and 4 and are assumed to act along the Y-direction. The variations of the stability and performance robustness indices when all the ten design variables are uniformly changed between the values of 0.1 and 0.5 are shown in Figs. 6-9. The optimum design problem is considered using the minimization of weight, β_{sr} and β_{pr} as

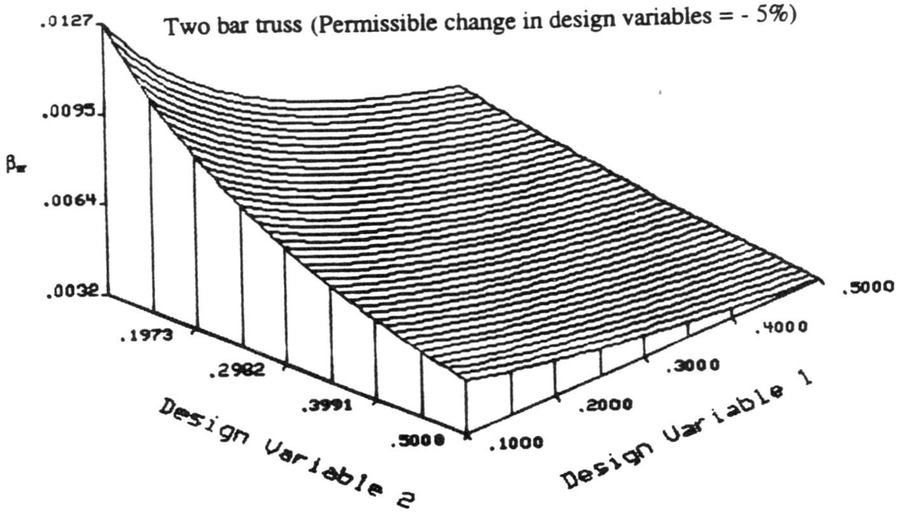

Figure 2 Stability Robustness Index versus Design Variables [3]

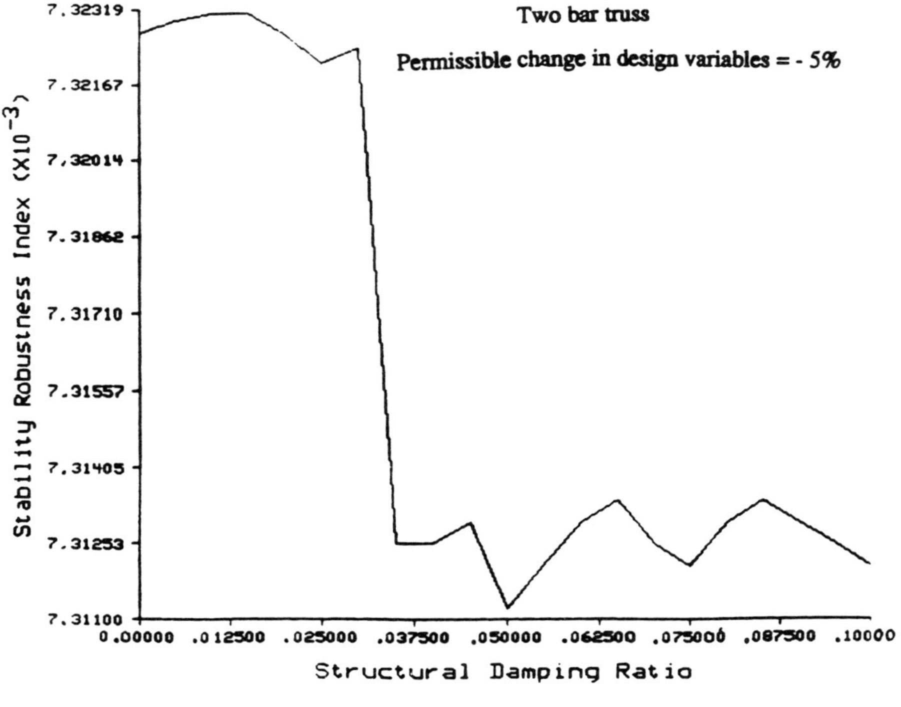

Figure 3 Variation of β_w with Structural Damping Ratio

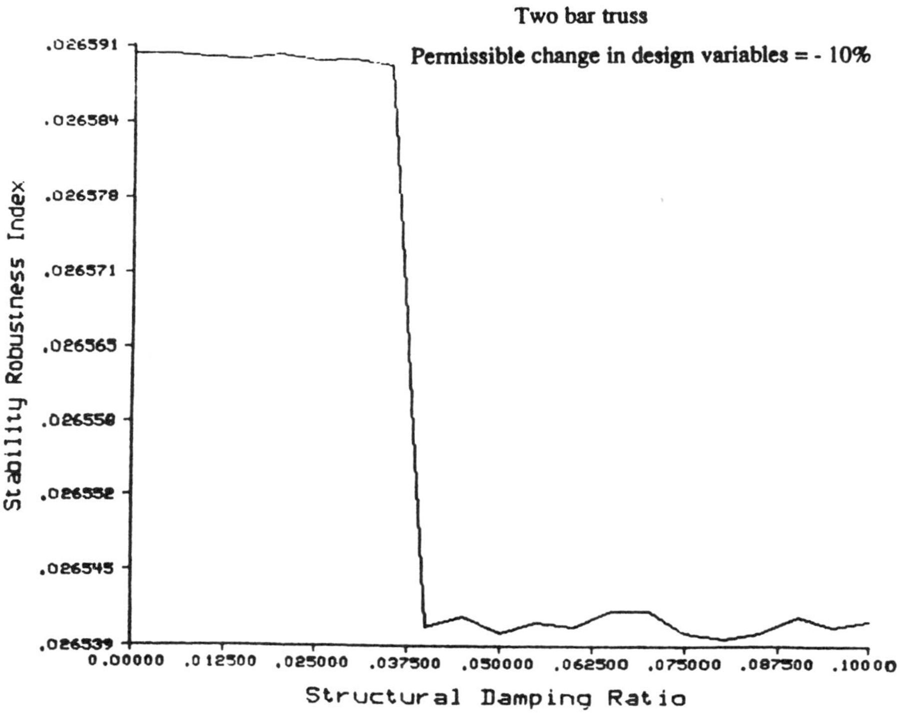

Figure 4 Variation of β_{sr} with Structural Damping Ratio

Table II
Effect of Variation in Young's Modulus

Young's Modulus	β_{sr}	β_{pr}	Performance Index
0.312500E+06	0.105365E-01	0.138324E-02	0.475665E+03
0.625000E+06	0.120861E-01	0.342239E-02	0.542370E+03
0.125000E+07	0.125931E-01	0.478706E-02	0.681617E+03
0.250000E+07	0.127294E-01	0.501379E-02	0.892266E+03
0.500000E+07	0.127650E-01	0.413170E-02	0.114687E+04
0.100000E+08	0.127736E-01	0.243972E-02	0.137735E+04
0.200000E+08	0.127762E-01	0.657325E-04	0.147823E+04
0.400000E+08	0.127750E-01	0.232127E-02	0.138979E+04
0.800000E+08	0.127758E-01	0.454950E-02	0.115054E+04
0.160000E+09	0.127767E-01	0.565636E-02	0.866480E+03
0.320000E+09	0.127767E-01	0.638549E-02	0.624165E+03

Table III

Effect of Coefficient of Output Matrix

Coefficient of Output Matrix	β_{sr}	β_{pr}	Performance Index
0.400000E+03	0.127755E-01	0.893862E-03	0.736905E+03
0.800000E+03	0.127747E-01	0.209998E-02	0.118867E+04
0.120000E+04	0.127740E-01	0.273048E-02	0.155007E+04
0.160000E+04	0.127723E-01	0.315626E-02	0.186061E+04
0.200000E+04	0.127715E-01	0.341708E-02	0.213712E+04
0.240000E+04	0.127700E-01	0.367024E-02	0.238916E+04
0.280000E+04	0.127683E-01	0.380359E-02	0.262216E+04
0.320000E+04	0.127675E-01	0.394992E-02	0.283994E+04
0.360000E+04	0.127668E-01	0.410320E-02	0.304527E+04
0.400000E+04	0.127653E-01	0.420803E-02	0.323989E+04
0.440000E+04	0.127636E-01	0.426687E-02	0.342557E+04
0.480000E+04	0.127630E-01	0.434907E-02	0.360339E+04
0.520000E+04	0.127621E-01	0.440142E-01	0.377431E+04
0.560000E+04	0.127603E-01	0.447826E-01	0.393906E+04
0.600000E+04	0.127594E-01	0.451510E-02	0.409822E+04
0.640000E+04	0.127588E-01	0.454517E-02	0.425245E+04
0.680000E+04	0.127571E-01	0.459293E-02	0.440215E+04
0.720000E+04	0.127554E-01	0.462976E-02	0.454768E+04
0.760000E+04	0.127546E-01	0.466137E-02	0.468947E+04
0.800000E+04	0.127538E-01	0.469942E-02	0.482772E+04
0.840000E+04	0.127524E-01	0.473370E-02	0.496276E+04
0.880000E+04	0.127516E-01	0.474985E-02	0.509473E+04
0.920000E+04	0.127506E-01	0.478344E-02	0.522391E+04
0.960000E+04	0.127489E-01	0.479688E-02	0.535038E+04
0.100000E+05	0.127474E-01	0.481362E-02	0.547438E+04

Table IV

Multiobjective optimum design of the two-bar truss

Quantity	Utility function method	Goal programming method (p = 2)	Lexicographic method (order: f_1, f_2, f_3)
$\vec{X} = \begin{Bmatrix} x_1 \\ x_2 \end{Bmatrix}$	$\begin{Bmatrix} 0.1309 \\ 0.1271 \end{Bmatrix}$	$\begin{Bmatrix} 0.1539 \\ 0.1348 \end{Bmatrix}$	$\begin{Bmatrix} 0.0518 \\ 0.0518 \end{Bmatrix}$
$\vec{f} = \begin{Bmatrix} f_1 = \text{weight} \\ f_2 = \beta_{sr} \\ f_3 = \beta_{pr} \end{Bmatrix}$	$\begin{Bmatrix} 59.3370 \\ 0.0105 \\ 0.0019 \end{Bmatrix}$	$\begin{Bmatrix} 66.4040 \\ 0.0096 \\ 0.0017 \end{Bmatrix}$	$\begin{Bmatrix} 23.8310 \\ 0.0199 \\ 0.0048 \end{Bmatrix}$

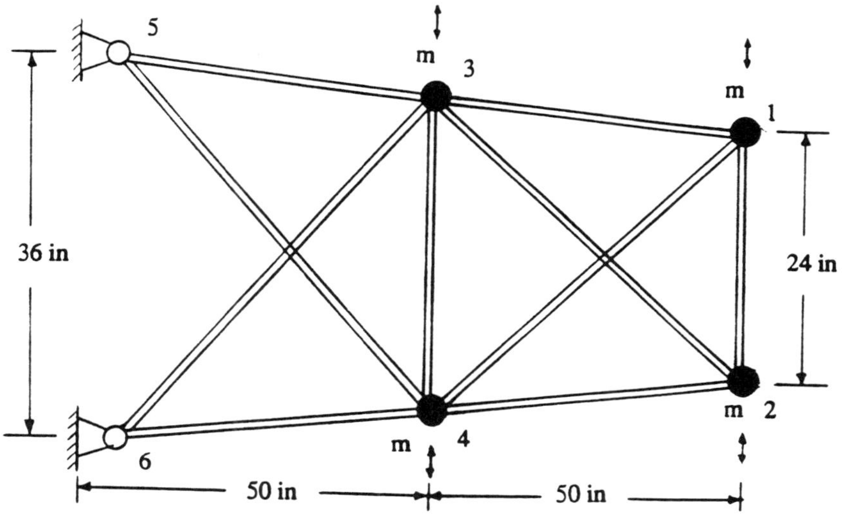

Figure 5 Two-Bay Truss

(m = 1.29, E = 10^7, Density = 0.1/32.2, Nominal areas = 0.1)

Table V

Data of ten-bar truss

Nonstructural mass at nodes 1,2,3 and 4	1.29
Lower bound on cross-sectional areas	0.001
Upper bound on cross-sectional areas	0.5
Nominal values of cross-sectional areas	0.1
Lower bound on frequencies of the closed-loop system	31.62 rad/sec
Young's modulus of the material	10^7
Density of the material	0.0031

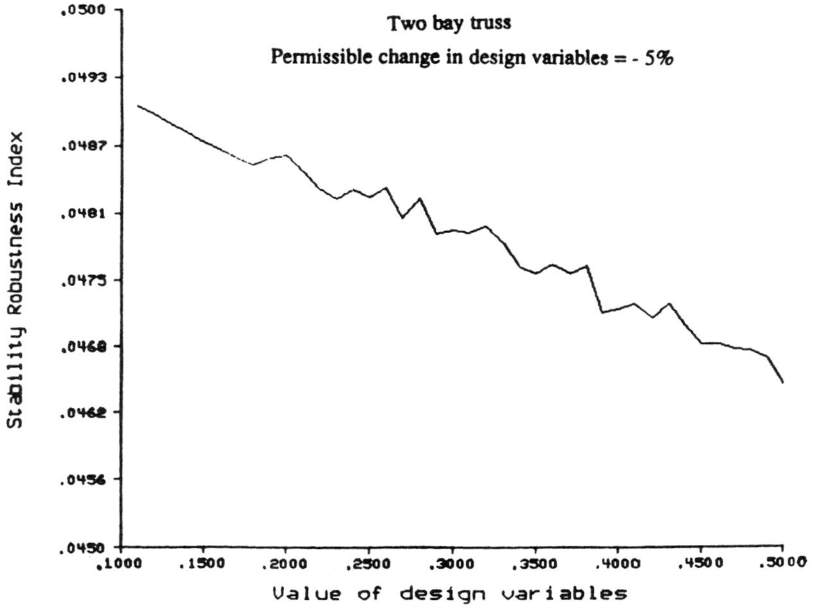

Figure 6 Stability Robustness Index versus Value of Design Variables

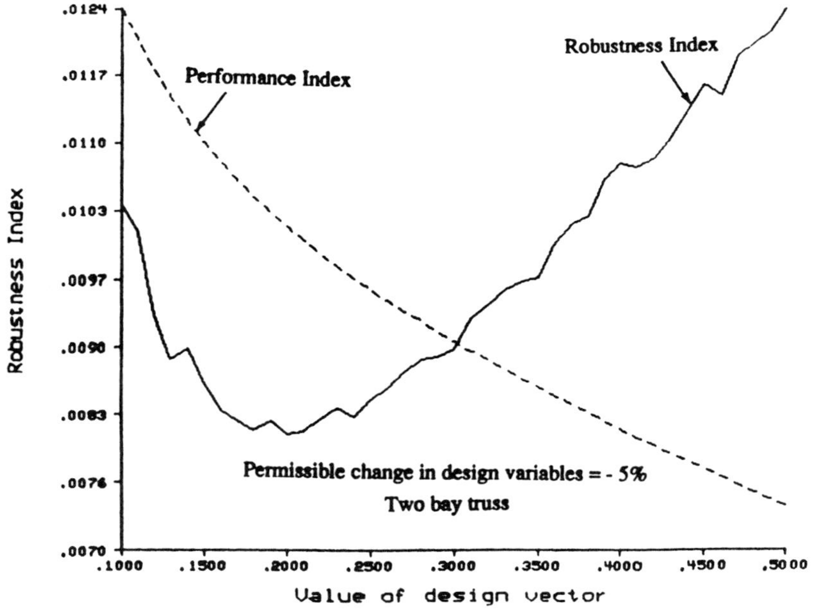

Figure 7 Robustness Index versus Value of Design Vector

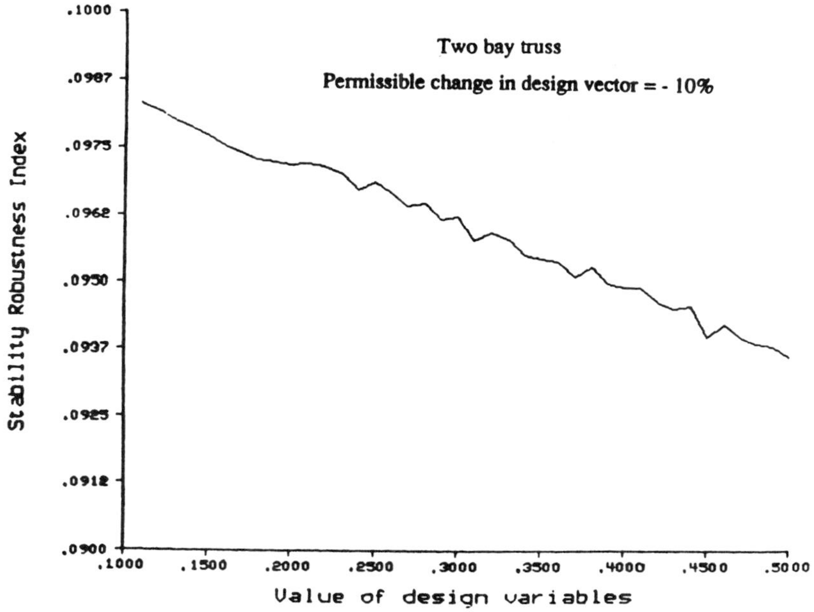

Figure 8 β_{sr} versus Value of Design Variables

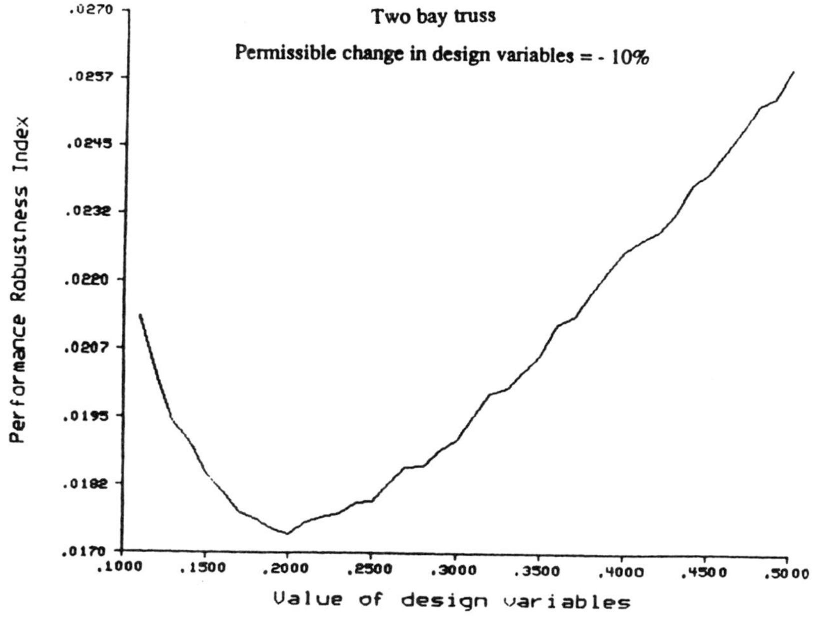

Figure 9 β_{pr} versus Value of Design Variables

objectives under the stated constraints. When individual objective functions are optimized, the minimum values of weight, β_{sr} and β_{pr} are found to be 0.2273, 0.0487 and 0.0098, respectively. The results obtained using the utility function, goal programming and lexicographic methods are shown in Table VI.

X. CONCLUDING REMARKS

The improvements of stability and performance robustness indices through structural modifications are considered in this work. The robustness indices defined have been found to be highly nonlinear with respect to the design variables and other structural parameters. The analysis procedure permits designers to find the ranges of the design or system parameters to achieve any specified values of stability or performance robustness index. The robustness improvement problem has also been formulated as a constrained multiobjective optimization problem. Several solution techniques are presented for numerical solution. The numerical results indicate the feasibility and effectiveness of the methods.

XI. ACKNOWLEDGEMENT

The first two authors acknowledge the support of this work through the Department of Air Force Contract No. F 33615-87-C-3256.

XII. REFERENCES

1. S.S. Rao, V.B. Venkayya and N.S. Khot, "Game Theory Approach for the Integrated Design of Structures and Controls," AIAA Journal, Vol. 26, 1988, pp. 463-469.

2. N.S. Khot, F.E. Eastep and V.B. Venkayya, "Simultaneous Optimal Structural/Control Modifications to Enhance the Vibration Control of

Table VI

Multiobjective design of ten-bar truss

Quantity	Utility function method	Goal programming method (p = 2)	Lexicographic method (order: f_1, f_2, f_3)
$\vec{f} = \begin{Bmatrix} f_1 \\ f_2 \\ f_3 \end{Bmatrix} = \begin{Bmatrix} \text{weight} \\ \beta_{sr} \\ \beta_{pr} \end{Bmatrix}$	$\begin{Bmatrix} 0.2290 \\ 0.0488 \\ 0.0120 \end{Bmatrix}$	$\begin{Bmatrix} 0.2274 \\ 0.0494 \\ 0.0085 \end{Bmatrix}$	$\begin{Bmatrix} 0.2484 \\ 0.0488 \\ 0.0103 \end{Bmatrix}$

Large Flexible Structure," AIAA Guidance and Control Conference, Snowmass, CA, April 1985.

3. S.S. Rao, T.S. Pan and V.B. Venkayya, "Robustness Improvement of Actively Controlled Structures through Structural Modifications," AIAA Journal, Vol. 28, 1990, pp. 353-361.

4. J. Junkins and D.W. Rew, "Unified Optimization of Structures and Controllers," Large Space Structures: Dynamics and Control, edited by S.N. Atluri and A.K. Amos, Springer-Verlag, Berlin, 1988, pp. 323-353.

5. D.B. Ridgely and S.S. Banda, "Introduction to Robust Multivariable Control," Flight Dynamics Laboratory, Air Force Wright Aeronautical Laboratory, Report AFWAL-TR-85-3102, February 1986.

6. R.V. Patel, M. Toda and B. Sridhar, "Robustness of Linear Quadratic State Feedback Designs in the Presence of System Uncertainity," IEEE Transactions on Automatic Control, Vol. 22, 1977, pp. 945-949.

7. R.V. Patel and M. Toda, "Quantitative Measures of Robustness for Multivariable Systems," Proceeding of Joint Automatic Control Conference, San Francisco, TP8-A, 1980.

8. R.K. Yedavalli, "Improved Measures of Stability Robustness for Linear State-space Models," IEEE Transactions of Automatic Control, Vol. 30, 1985, pp. 577-579.

9. R.K. Yedavalli, S.S. Banda, and D.B. Ridgely, "Time Domain Stability Robustness Measures for Linear Regulators," Journal of Guidance, Control and Dynamics, Vol. 8, No. 4, 1985, pp. 520-525.

10. Y. Juang, T. Kuo, and C. Hsu, "New Approach to Time-Domain Analysis for Stability Robustness of Dynamic Systems," Intl. J. of Systems Science, Vol. 18, No. 7, 1987, pp. 1363-1376.

11. L. Qiu and E.J. Davison, "New Perturbation Bounds for the Robust Stability of Linear State Space Models," Proceedings of the 25th Conference on Decision and Control, Athens, Greece, 1986, pp. 751-755.

12. K.B. Lim and J.L. Junkins, "Robustness Optimization of Structural and Control Parameters," 28th SDM Conference, Monterey, CA, April 1987, pp. 351-361, Paper No. AIAA-87-0791-CP.

13. D.W. Rew, J.L. Junkins and J.N. Juang, "Robust Eigenstructure Assignment by a Projection Method: Applications Using Multiple Optimization Criteria," AAS/AIAA Aerodynamics Conference, Kalispell, Montana, August 1987.

14. J. Juang, K.B. Lim and J.L. Jenkins, "Robust Eigensystem Assignment for Flexible Structures," AIAA Guidance, Navigation and Control Conference, Monterey, CA, 1987, pp. 117-123, Paper No. AIAA-87-2252-CP.

15. R.V. Grandhi, I. Haq and N.S. Khot, "Enhanced Robustness in Integrated Structural/Conrol Systems Design," 31st SDM Conference, Long Beach, CA, April 1990, pp. 268-276, Paper No. AIAA-90-1059-CP.

16. H.L. Thomas and L.A. Schmit, "Synthesis of Structural/Control Systems Robust with Respect to Design Variable Tolerances," 32nd SDM Conference, Baltimore, MD, April 1991, pp. 506-516, Paper No. AIAA-91-1156-CP.

17. N.S. Khot and D.E. Veley, "Use of Robustness Constraints in the Optimum Design of Space Structures," Journal of Intelligent Materials, Systems and Structures, Vol. 2, April 1991, pp. 161-176.

18. D.L. Kleinman and M. Athans, "The Design of Suboptimal Linear Time Varying Systems," IEEE Transactions on Automatic Control, Vol. AC-13,

April 1968, pp. 150-158.

19. S.S. Rao, "Optimization: Theory and Applications," Second Edition, John Wiley, New York, 1984.

20. W. Stadler, "Multicriteria Optimization in Mechanics," Applied Mechanics Reviews, Vol. 37, 1984, pp. 277-286.

21. S.S. Rao, "Multiobjective Optimizations in Structural Design in the Presence of Uncertain Parameters and Stochastic Process," AIAA Journal, Vol. 22, No. 11, 1984, pp. 1670-1678.

22. A. Charnes and W.W. Cooper, "Goal Programming and Multiple Objective Optimization - Part I," European Journal of Operations Research, Vol. 1, No. 1, 1977, pp. 39-54.

Active Control for Segmented Mirror Optical Systems

Jean-Noel Aubrun
Alain C. Carrier
Kenneth R. Lorell

Research & Development Division
Lockheed Missiles & Space Company, Inc.
3251 Hanover Street, Palo Alto, California 94304

I. INTRODUCTION

In 1608, a Dutch spectacles maker named Lippershey invented the telescope. A year later, Galileo built a telescope with an aperture of 40 mm and began a series of remarkable discoveries. Over the nearly 4 centuries since Lippershey's invention, instrumentation for astronomy has progressed in all aspects: collecting area, detector sensitivity, spatial resolution, spectral range, spectral resolution, and, most recently, temporal resolution.

The collecting area of Galileo's telescope was about 0.0013 m^2—roughly 60 times greater than that of the human eye. The premier astronomical telescope today is the Hale 200-in. (5-m) telescope, located on Mt. Palomar in southern California. The Hale has a collecting area of 20 m^2—four orders of magnitude larger than Galileo's telescope. It was constructed between 1929 and 1948, and 40 years of observations with it have dramatically changed our view of the cosmos [1]. The largest telescope in the world is the Soviet 6-m, completed in 1978. However, because of flaws in its design and fabrication, and problems associated with the site, the Soviet telescope does not have the performance of the Hale 200-in.

Because the speed of light is finite, looking at a distant object is equivalent to looking back in time. The answers to many of the most important questions in cosmology depend on the ability to look back in time to understand the physics of what astronomers call the "big bang"—the explosion of matter that created the universe. Thus, astronomers need instruments that can see the most distant, and hence the dimmest, objects in the universe. Because detector technology has taken a quantum leap in photon detection efficiency in recent years [with some charge-coupled device (CCD) detectors approaching 100% efficiency], there is relatively little room for improvement in sensitivity for focal-plane instruments. The key to studying dimmer and more distant objects, therefore, lies with increasing collecting area.

The 200-in. primary mirror of the Hale telescope is 2-ft thick and weighs nearly 35,000 lb. Nearly 8 years were needed to fabricate the mirror; 1-1/2 years were required just for the blank to cool after removal from the furnace. To build a 400-in. mirror (10-m aperture) with the same optical quality of the Hale mirror (i.e., the distortion due to the mass of the material is held constant), the blank for the primary mirror would have to be 8-ft thick and weigh 544,000 lb (heavier than a fully loaded Boeing 747). The structure required to support as well as provide precision pointing and tracking for such massive optics would be enormous. Clearly, it is impractical to try to scale up the design of the Hale telescope to sizes approaching 10 m.

A. NEW TECHNOLOGY

The Hale telescope was designed 60 years ago and went into operation 40 years ago. The fact that it is still astronomy's premier instrument indicates how difficult and expensive it is to solve the problems associated with the design and fabrication of large optical systems. One technique that employs new technology to obtain increased collecting area uses an array of smaller telescopes, coupled optically and mechanically, to provide a combined collecting area many times larger than that of the individual elements. The Multiple Mirror Telescope (MMT) on Mt. Hopkins in Arizona uses six 2-m primary mirrors on a single mount with special optics to combine the beams into a single focus. The proposed design for the National New Technology Telescope (NNTT) uses an array of four 8-m primary mirrors mounted in much the same way as the MMT primaries. The 8-m mirrors will be fabricated using a technique that is still under development. The mirrors will be specially lightweighted and prefigured during the casting process to render their manufacture practical in terms of weight, time, and cost. This technique is still highly experimental, however, and no operational mirror larger than 2 m has been built using it.

Although multiple mirror designs have many advantages, including the relative simplicity of the structure and mirror support system, a number of major difficulties are associated with this technique. In particular, the ability to provide for phasing of the separate beams (keeping the optical paths from the multiple reflectors constant to some fraction of the wavelength of light) is an especially difficult and as yet unsolved problem. Because multiple-mirror telescopes are independent devices, combining their images at one focal plane is also a source of difficulty. Not only must the images be precisely "stacked" (superposed) on each other, but the individual focal planes must be coplanar as well. This latter problem requires special consideration in the optical design so that the individual focal planes can be properly aligned.

Another approach to building a large-aperture telescope is to fabricate the optical components in sections, rather than attempting to polish a continuous surface from a single piece of glass. This is an especially attractive concept for spaceborne telescopes, because the problems of packaging a large optical system in the launch vehicle are much more tractable if the system can be launched in a folded configuration and then deployed on orbit.

B. SEGMENTED-MIRROR TECHNOLOGY

A monolithic reflector depends on the mechanical properties of its material to provide the dimensional stability required for good optical performance. A mirror built from segments relies on its support structure (mirror cell) for stiffness and rigidity, and on an active control system for precision alignment of the optical surface.

In the development of segmented mirrors used for astronomical instruments, three basic concepts have emerged. The simplest concept uses segments which move as pistons only (along an axis perpendicular to the reflecting surface). This concept is usually applied to a system which attempts to correct automatically for atmospheric disturbances [2]. The most complex type of segmented mirror has segments which are not only adjustable relative to one another, but are also deformable. Thus, automatic control must be applied to adjust the segment figure as well as intersegment alignment [3]. A third type of segmented mirror uses segments which are rigid (not deformable) but can be adjusted in two axes of tilt in addition to piston. This concept, which is intermediate in complexity between the piston-only mirror and the deformable mirror, was described by Robertson, Crane, and Hemstreet as part of a NASA project [4]. Robertson's mirror was 56 cm in diameter and used three segments. The segments were each supported on three actuators and could be controlled in three degrees of freedom.

The idea of using actively controlled rigid segments to build a very large reflector, as large as 30 m in diameter for an orbiting system, was investigated by Berggren and Lenertz for NASA [5]. NASA has also been involved in preliminary studies for a space-based infrared/radio astronomy telescope in the 10- to 20-m aperture range called the Large Deployable Reflector (LDR) [6]. An artist's concept of the LDR being assembled on-orbit is shown in Fig. 1. Probably the first large optical system to address all of the practical problems of design, fabrication, and operation of real hardware for segmented mirror optical systems is the W. M. Keck Ten Meter Telescope (TMT) [7]. The TMT, discussed later in this article, is shown in model form in Fig. 2. Unlike the segmented mirror developed by Robertson, the TMT primary mirror must also deal with the problems of a flexible support structure and disturbances of the segment-alignment system that typical laboratory devices do not experience.

II. TECHNICAL CHALLENGES OF LARGE SEGMENTED REFLECTORS

The performance of any optical system is a direct function of the accurate positioning of each of its components. For small and relatively rigid systems, initial alignment is usually achieved by careful manual adjustment. Thereafter, alignment is passively maintained by the mechanical rigidity of the optical bench. Typically, large segmented reflectors are characterized by the lack of temporal and spatial mechanical stability of the structure supporting the optical elements. As a result, an active control system is required to maintain the alignment of the segments. Designing and implementing such a control system

Fig. 1. Artist's concept of NASA's proposed LDR being assembled on-orbit.

presents several technical challenges in four main areas: sensing, actuation, modeling, and controls.

A. SENSING

Sensors must be used to measure the position of the segments with respect to some unique, stable reference rather than with respect to the underlying structure, which lacks dimensional stability. For the TMT [8,9] and other systems such as NASA's LDR, the relative position of each segment is measured with respect to adjacent segments. This is accomplished by position sensors mounted near the segment edges (edge sensors). A designated segment or a particular set of points is typically chosen as a reference for the whole system, and the positions of individual segments are obtained by a special algo-

Fig. 2. Scale model of the TMT.

rithm that applies the proper geometric transformation to the ensemble of sensor signals.

The edge sensors must have a number of special features. They must be accurate to a fraction of the wavelength of light, while their measurement range must accommodate the initially large misalignments caused by the support structure. The performance of the sensor system is also very sensitive to the geometry of the location of individual sensors on the panels and to the processing algorithm. In addition, edge sensing is not a complete answer to segment alignment. Because of electronic biases and drifts, as well as the need to provide an absolute basis to phase the segments with respect to each other, the overall figure must be optically measured for initial calibration and alignment of the system. Thus, the development of optical figure sensors is also a critical part of segmented telescope technology.

B. ACTUATION

The accuracy and dynamic range requirements for alignment actuators are the same as those for the sensors; however, they are more difficult to meet because of other requirements such as force output, bandwidth, and smoothness of operation. Friction and stiction, for example, can potentially result in uneven motions that can excite vibrations in the structure. Conventional designs have proven to be inadequate, and new design approaches have been necessary to achieve the required performance.

C. MODELING

Because segmented reflectors are so complex, the design of their mechanical system and of their control system depends heavily on mathematical models that accurately predict their structural behavior. Static and dynamic models of the segmented optics and its support structure are needed to ensure that the mechanical design is structurally sound and that it fulfills its functions prior to the construction of the telescope. These models are also needed to devise control strategies that will maintain the alignment of the segments to the desired accuracy.

D. CONTROL ALGORITHM

Whether edge sensors or optical figure sensors are used, the resulting control system is highly noncolocated, and dynamic coupling between the individual segments and the support structure may cause the segment alignment control system to become unstable at relatively low bandwidths. The action of one actuator influences the entire reflector and can be sensed by all the sensors. This behavior is the fundamental basis of the controls/structures interaction (CSI) phenomenon in large segmented reflectors and was demonstrated for the first time in the analytical study of the TMT summarized in the next section [10].

The large number of actuators and sensors involved in controlling a large segmented reflector poses an implementation problem. The TMT, for example, has 108 actuators and 168 sensors. Therefore, computation requirements and system integration are substantially more complex than for conventional control systems.

III. ANALYSIS OF THE W. M. KECK TEN METER TELESCOPE

The University of California/California Institute of Technology W. M. Keck Observatory Ten Meter Telescope (TMT) will be a high-performance astronomical instrument capable of unprecedented ground-based observations in both the visible and infrared portions of the spectrum. The TMT will be located on Mauna Kea in Hawaii at an elevation of 14,000 ft [11].

A unique feature of the TMT is its primary mirror, which is composed of 36 hexagonal segments. The segments will be individually oriented and positioned with a high degree of accuracy by an automatic control system. Each segment will be controlled by three linear actuators in three degrees of freedom: a translation along the mirror axis (piston mode) and two out-of-plane rotations (tilt modes). The other degrees of freedom will be constrained by a thin, flexible diaphragm [12].

Because of the size and flexibility of the TMT structure, the relatively large mass of the segmented primary mirror, and the low level of natural structural damping that is typical of such large systems, excitation of mechanical resonances by external, time-varying disturbance sources is considerably more severe than in a conventional telescope structure. In particular, wind loads will introduce pointing errors and create distortions in the mirror figure that will degrade optical performance. The frequency content of the disturbance as well as the distribution of resonant frequencies in the structure are critically important in evaluating the performance of the TMT. In addition, the spatial distribution of the disturbing forces is important because this distribution can directly affect the type of vibration modes most likely to be excited.

A second and perhaps more important difference between a conventional telescope and the TMT is the interaction between the structural dynamics and the alignment control system for the segments. The first vibration mode of the structure is at about 5 Hz. The first 50 modes are below 30 Hz. The overlap of control-system bandwidth with the modal frequency spectrum, a major problem in the control of large, flexible structures, seriously limits the effectiveness of the TMT segment-alignment system.

A. SYSTEM DESCRIPTION OF THE TMT

1. Telescope Structure

The structure of the TMT is comprised of three main sections: the yoke, the telescope tube, and the primary mirror support cell. Figure 3 illustrates the major components of the telescope and the locations of the primary and secondary mirrors within the structure. The complex geometry of the mirror support cell can be seen in the area beneath the primary mirror segments. Because the TMT is an altitude-azimuth design, the entire yoke is designed to

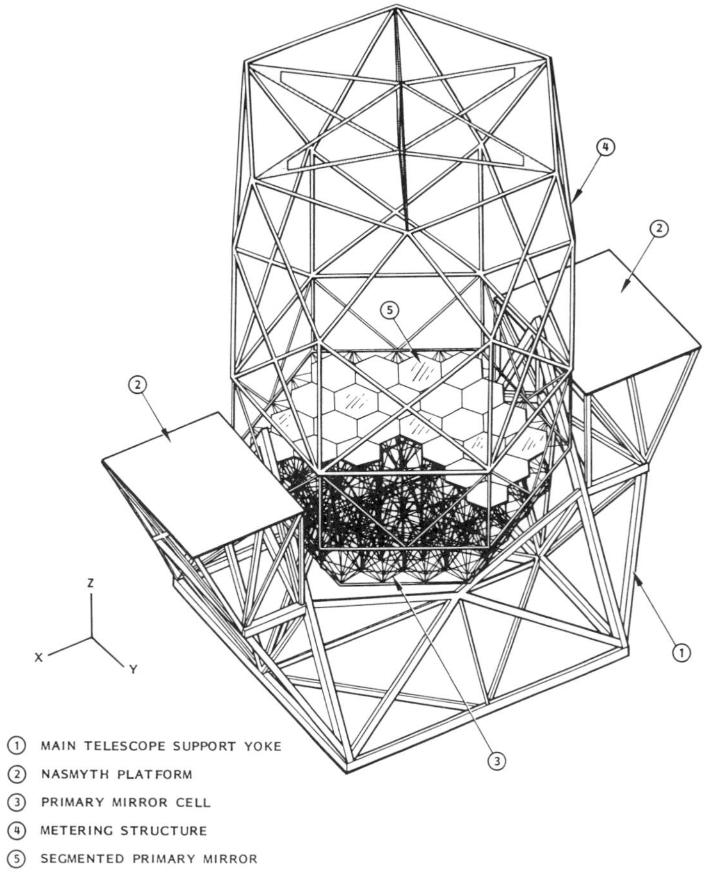

Fig. 3. Detailed structural model of the TMT with primary-mirror segments removed to show complexity of primary-mirror cell.

rotate about the Z (or vertical) axis, while the telescope tube can pivot on yoke supports located just above the Nasmyth platforms [13].

2. Optical System

The TMT's optical system is conventional in the sense that it is designed as a Cassegrain telescope with the standard provisions for instrumentation at the prime focus, Cassegrain focus, and two Nasmyth foci. In addition, there are provisions for "bent Cassegrain" foci which require the use of the same tertiary mirror used for the Nasmyth instruments.

Figure 4 is a plan view of the primary mirror. It shows the arrangement of the segments, each of which is figured as an off-axis hyperbola. The locations of the 108 actuators used to position the segments are also illustrated in Fig. 4, along with the three actuators which are independently controlled to provide a reference plane that effectively ties the mirror to the telescope structure.

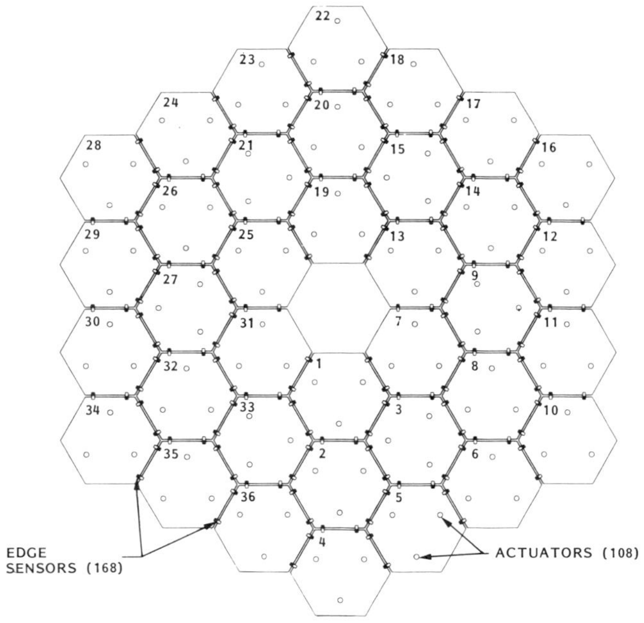

Fig. 4. Plan view of the TMT primary mirror showing locations of sensors and actuators.

The small tabs at the segment-to-segment interfaces indicate the locations of the 168 position sensors (only the segment edges that adjoin other segments contain sensors). These sensors are the differential capacitance type and are permanently affixed to the back surface of each segment. Misalignments between segments are measured by these sensors with a resolution of about 4 nm. The relative misalignments between adjacent segments are the only measurements used during the operation of the telescope to determine the quality of the mirror figure. The desired values of each sensor reading are, of course, determined by precisely calibrating the orientation and phasing of each segment in a special calibration mode prior to making observations.

3. Segment Alignment Control System

The segment alignment control system operates by measuring the displacements of the edges of each hexagon relative to the edges of its nearest neighbors using the capacitive sensors described in the preceding section. By processing the data from the sensors with the proper algorithm, it is possible to calculate the rotational and axial displacements of each of the 36 segments. Once these misalignments are known, commands are sent to the 108 actuators to exactly reorient the segments to bring the optical surface into the desired alignment. The process of measuring the edge misalignments, computing a new set of desired positions, and commanding the actuators is repeated several times a second. The frequency of the control process directly influences the degree to which the system can respond to external disturbances. A schematic block diagram showing the major components and signal flow paths of the segment alignment control system is shown in Fig. 5. The control law itself is equivalent to integral control, i.e., the actuator command $u = K_I \int e$ where e is the alignment error and K_I is a fixed gain.

B. TMT DYNAMIC MODEL

1. Finite-Element Model

A detailed finite-element model of the TMT was developed which contains all structural elements, including the support yoke, individual mirror segments with central post-and-diaphragm supports and whiffletrees, subcells and actuators, telescope metering structure, spider, and secondary mirror. The model contains about 1800 nodes corresponding to 7000 degrees of free-

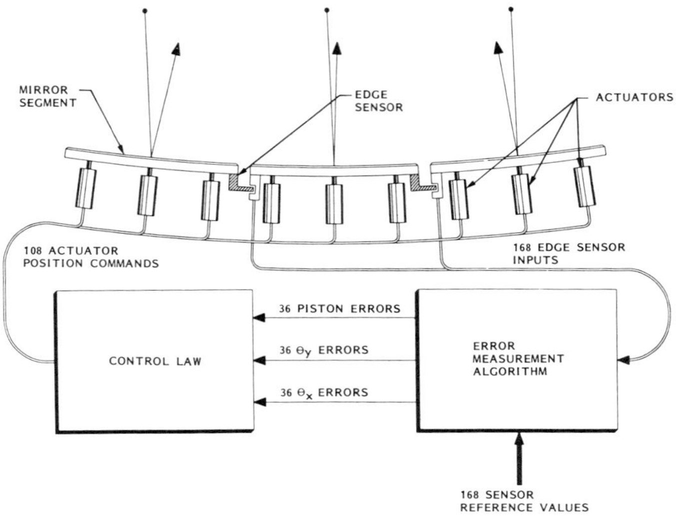

Fig. 5. TMT segment alignment control system.

dom and is shown graphically in Fig. 3 with some structural members and mirror segments removed for clarity.

2. Modeling Methodology

In the development of the finite-element model for the TMT, the inherent symmetry of order six was used to reduce the number of substructures. Figure 6 shows how the various substructures were assembled to generate the final model.

The number of degrees of freedom was reduced by grouping them together in related clusters. The resulting model contained a much smaller number of nodes and modal frequencies, and was mathematically much more tractable. Of course, some resolution was lost in the process, but the primary concern was with the global modes affecting the mirror cell, the mirror segments attached to it, and the motion of the secondary mirror. A total of 213 degrees of freedom were retained in the reduced model.

The model was developed using Lockheed Missiles & Space Company, Inc.'s (LMSC's) ASTRO computer program, which has both substructuring

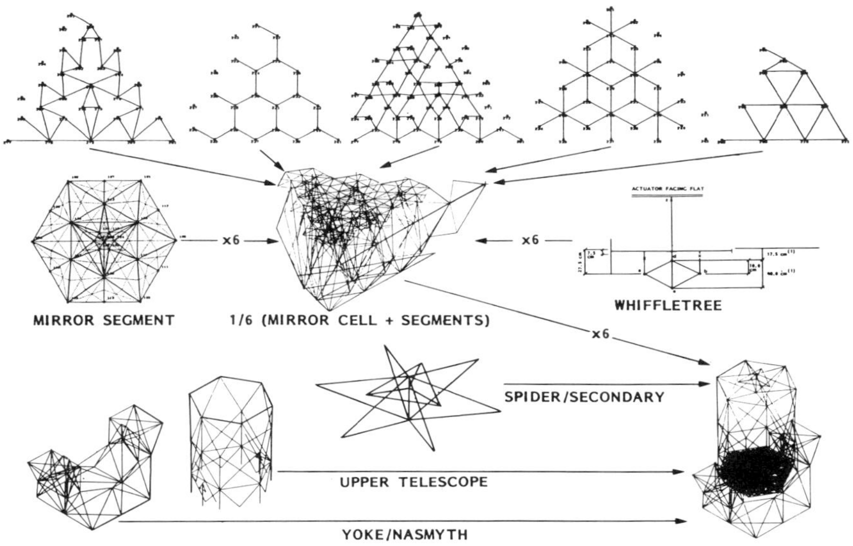

Fig. 6. Substructuring technique used to build the TMT structural model.

and reduction capabilities. Details of the program and the methods and procedures outlined above were described by [14,15,16,17].

The model used for control and simulation studies had an even smaller set of modes and a special correction term to ensure static fidelity (i.e., behavior at 0 frequency). This is essential when dealing with segmented optical systems.

3. Structural Dynamics Characteristics

The eigenvalue analysis of the system revealed that the lowest modal frequency is at 5.39 Hz, corresponding to global motions of the telescope about the elevation axis. The next set of vibration modes involves global deformations of the mirror cell with a gradual increase in the motion of individual segments. Each of the 36-segment/whiffletree systems has a fundamental resonance at 25 Hz (piston motion). When mounted on the mirror cell, elastic coupling with the cell spreads the modal frequencies around this value [18]. This can be seen in the histogram of Fig. 7, which shows the interesting grouping around 25 Hz. After mode 54, a departure is observed which corresponds to global modes of the telescope structure. In addition, after mode 80, a new

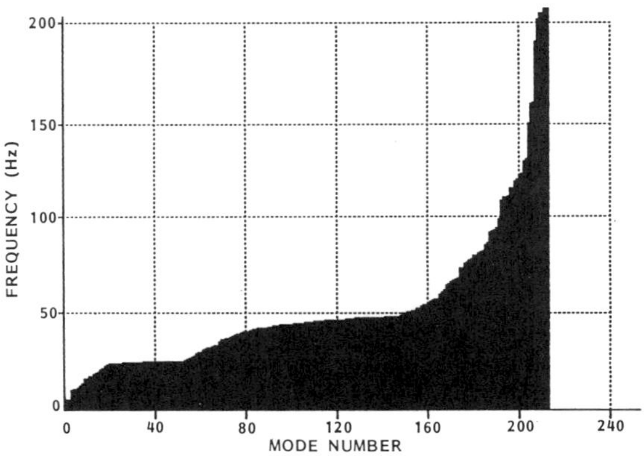

Fig. 7. Modal histogram derived from the TMT structural model showing frequency clustering in the 25-Hz mirror/whiffletree resonance region.

grouping is observed which is related to the transverse motions of the segments.

C. SIMULATION RESULTS

1. Open-Loop Analysis

A 20-s simulated time history was generated using a state-space model developed from the structural dynamic model described in the preceding section, with the integrator gain K_I set to zero. The noise input was obtained from a random-number generator, so that aerodynamic forces from the wind acting on the mirror segments could be simulated. The main output of the simulation consists of the relative position errors of each of the 108 mirror segment attachment points. This relative position error gives an important measure of the surface figure error. It varies from one segment to another because of the wind force distribution and the differences in structural response over the mirror cell. A global picture of a 5-s segment of these 108 time histories is shown in Fig. 8. Although this type of plot is not very accurate, it gives a visual overview of the general response of the system. The traces are organized by groups of three, each group corresponding to a mirror segment. Each row of six segments represents one sixth of the primary mirror. The vertical scale in the upper left corner is in nanometers.

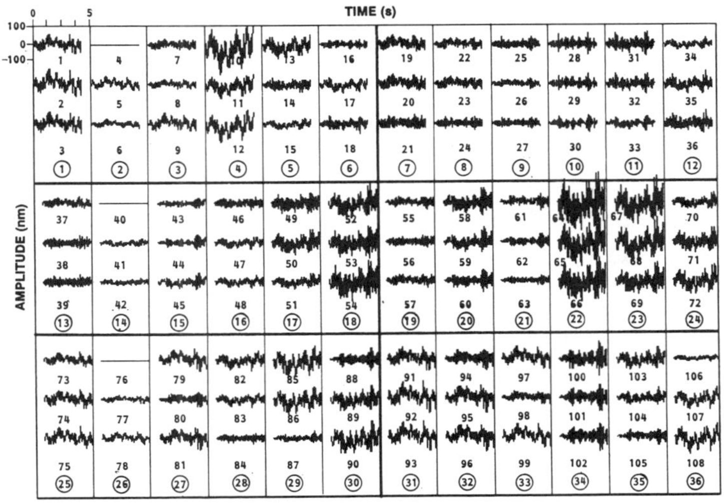

Fig. 8. 108 open-loop time histories of the segment actuator motions of the TMT during a wind-induced disturbance.

2. Closed-Loop Stability Analysis

The roots of the system were computed for values of K_I between 0 and 4π, which translates into a control bandwidth varying between 0 and 2 Hz. Because of the adverse interaction between the control system and the structural dynamics, the damping ratios of some of the structural modes decrease as the control bandwidth increases. When one of these damping ratios decreases to zero or goes negative, the system becomes unstable.

The major result of the closed-loop stability analysis is that the control system design is limited to a 0.5-Hz bandwidth due to structural interactions around 22 Hz. Mode 18 (the first mode to go unstable) is a collective mode. It involves simultaneous motions of the mirror segments whose dynamics are coupled to each other by the mirror cell. This phenomenon could not be seen on the single segment experiment [19], although the natural frequency of that system was similar (25 Hz). This result highlights the importance of the coupling effects in large structures which use multiple-input, multiple-output control systems, and demonstrates the necessity to take extreme care in analyzing large-scale dynamical systems.

D. OPTICAL SYSTEM PERFORMANCE

The operation and performance of the control system have been described in the preceding sections and illustrated in Fig. 8. For a telescope, however, the improvement in optical system performance is the true measure of how successfully a disturbance rejection system is operating. To quantify this improvement, the point spread functions for the primary mirror were calculated with and without the control system operating. Using purely geometrical optics, the position of the image of a star at the telescope prime focus was computed as a function of the time-varying motion of the telescope structure and the orientations of the 36 primary mirror segments. The calculation required sorting the 34,500 points that were generated in the time-domain simulations.

Figures 9, 10, and 11 represent the way in which the total optical energy from all of the segments is distributed at the prime focus with and without the use of the control system. Figure 9 shows how, with the control system turned off, the energy is spread out, because structural vibrations cause the segments to move and consequently blur the resulting image. This figure was made by simply plotting the x-y coordinates of each point generated in the time-domain simulation.

Figures 10 and 11 are pseudo three-dimensional plots of the energy at the prime focus that illustrate the optical performance with and without the

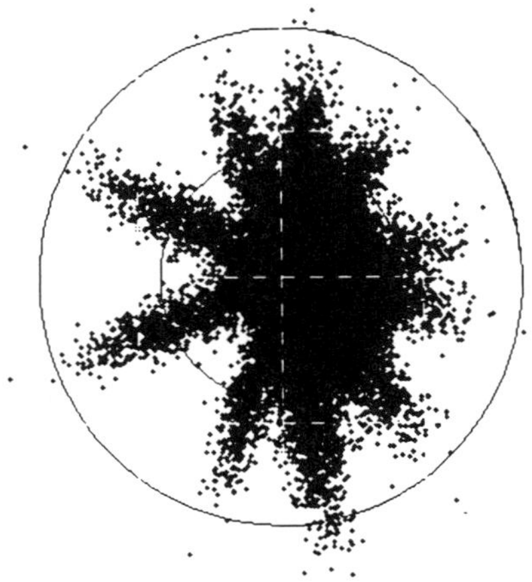

Fig. 9. X-Y plot of optical energy at the TMT prime focus with the segment-control system off.

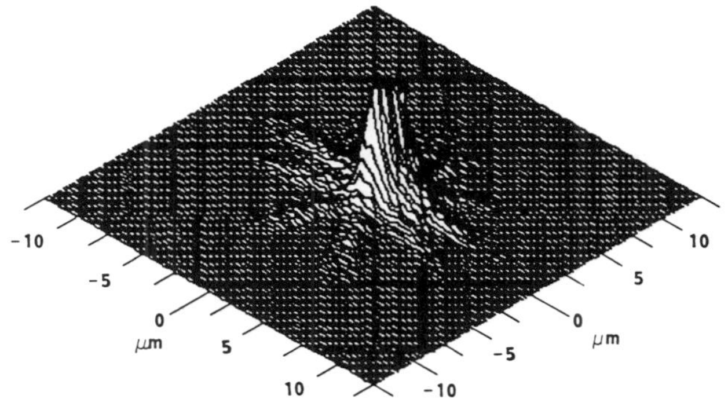

Fig. 10. Three-dimensional plot of prime-focus optical energy with the segment-control system off.

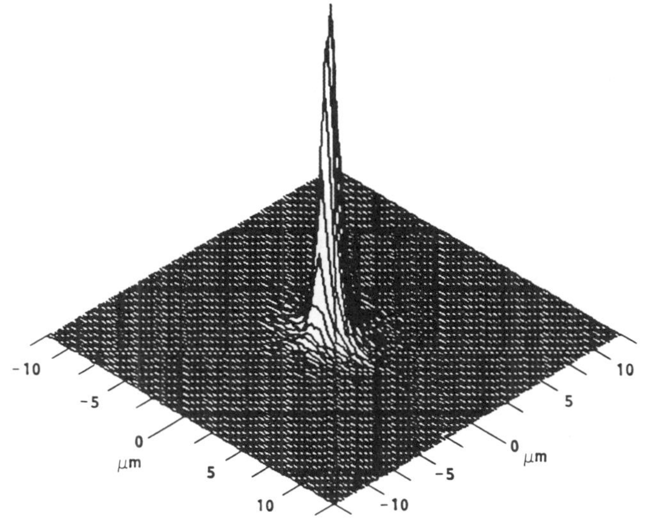

Fig. 11. Three-dimensional plot of prime-focus optical energy with the segment-control system engaged.

use of the segment control system. They were made by mapping the prime focus onto a grid of 100 × 100 square bins, each bin 0.25 μm on a side. The number of points falling in each bin was then summed, and the resulting total for each bin was plotted against the coordinates of the bin [20]. The fact that the spike in Fig. 11 is about a factor of three higher than that of Fig. 10 and that the dispersion at the base is substantially less, indicates that the segment alignment control system has an important impact on the TMT's optical performance.

IV. A LABORATORY TESTBED FOR THE DEVELOPMENT OF SEGMENTED OPTICS CONTROL SYSTEM TECHNOLOGY

The analytical study of the TMT showed that the dynamic behavior of a large segmented optical system is far more complex than that of the simple beam and plate-like structures used in a number of laboratory experiments to test control design methodologies.

This section describes an experimental apparatus developed at the LMSC Research & Development Division called the Advanced Structures/ Controls Integrated Experiment (ASCIE). ASCIE is designed to perform experiments that simulate the complex dynamic behavior of a large segmented optical system. It is an experimental means of addressing in an integrated way the problems associated with structural dynamics, distributed control of multiple-input multiple-output systems, optics, electronics, actuator and sensor design, and digital implementation. This section presents the control system architecture and details of the actuation and sensing system hardware.

A. REQUIREMENTS FOR A LABORATORY TESTBED

Few laboratory experiments have been performed to demonstrate the CSI phenomenon or develop methods to alleviate its adverse effects. These experiments generally use relatively simple test specimens (such as beams and plates) and do not address the technical challenges presented by complex, three-dimensional structures such as large segmented optical systems.

To validate both control strategy and implementation for realistic systems, a laboratory testbed must satisfy four major criteria:

- Dynamics: The structure must approximate the fundamental dynamic characteristics of a three-dimensional large structure, i.e., low-frequency modes, high modal density, and global mode shapes that properly reflect the coupling of the subelements of the structure.
- Performance: Performance of the testbed must be of comparable quality to that of an actual system, so that the range of sensing and actuation will be similar.

- Functionality: The testbed must be able to perform the essential functions needed for the various system missions, e.g., static and dynamic segment alignment, vibration suppression, fine pointing, tracking, and slewing.
- Technology Integration: Experiments conducted with the testbed must address the interdisciplinary nature of the research. Control algorithms, structural dynamics, optics, electronics, as well as actuator, sensor, and distributed multiprocessor technologies are closely related in most advanced systems of interest.

B. ASCIE SYSTEM DESCRIPTION

The design of ASCIE used the above criteria to develop an appropriate testbed for studying the interaction between actively controlled optical elements and structural dynamics. A typical large optical system can be represented by a generic Cassegrain telescope with a segmented primary mirror. The active optical elements are the primary-mirror segments and the secondary mirror. The distributed systems of actuators and sensors dynamically interact with each other and with the supporting structure.

The ASCIE consists of a 2-m, seven-segment, actively controlled primary mirror supported by a lightweight, flexible truss structure. The optical system emulates an f/1.25 Cassegrain telescope. The six peripheral segments are controlled in three degrees of freedom using specially developed actuators [21]. The schematic drawing of the ASCIE shown in Fig. 12 illustrates the features discussed below.

1. Structure

One of the most fundamental design goals, as well as one of the most difficult to implement accurately, is a strong, lightweight truss structure whose structural-dynamic characteristics are representative of a large, flexible spaceborne system. A design of this type requires a careful tradeoff between the need for the structure to support itself in the 1-g laboratory environment versus the need to keep the frequency of the first bending mode as low as possible.

The ASCIE utilizes thin-wall stainless steel tubing in a unique geometric configuration to achieve the correct balance between strength and flexibility. The structure consists of a central hexagonal box made from a series of struts which are attached to each other with specially designed fittings. The segment-positioning actuators and the segments themselves are supported by

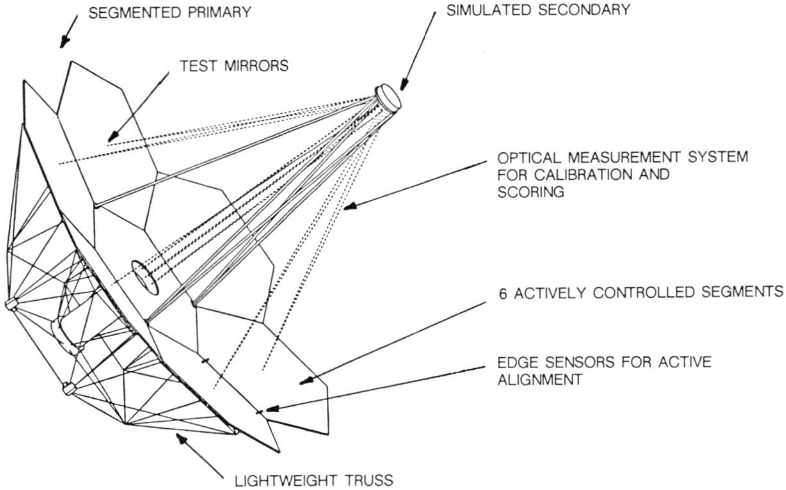

Fig. 12. Schematic drawing of the ASCIE testbed illustrating main components and features.

subcell structures cantilevered from each of the central hexbox's six faces. This structure exhibits the desired structural flexibility, dynamic coupling, and low-resonance frequencies that are characteristic of much larger structures.

2. Segmented Primary Mirror

The ASCIE primary mirror was designed to emulate the critical properties of a real segmented mirror. These properties include segmentation geometry, intersegment spacing, segment mass, inertia and stiffness, and optical focal ratio. Because the ASCIE is control-system oriented, and because of the difficulty and added expense of fabricating actual optical-quality segments from glass, the segments were fabricated from flat plates of aluminum.

The seven-segment primary consists of a ring of six actively controlled hexagonal segments surrounding a fixed center segment which acts as a reference. The active segments are attached to segment-positioning actuators with special three-degree-of-freedom flexures. These flexures permit individual segments to have two rotational degrees of freedom (tilt) and one translational degree of freedom (piston). Each segment is controlled by three actuators, and the entire active primary has a total of 18 actuators.

The relative displacement between the edges of adjacent segments is measured by an ensemble of 24 edge sensors. The edge sensors provide information about segment-to-segment relative displacements as well as the absolute displacements from the fixed center reference segment.

3. Segment-Positioning Actuators

The key to the precision control of the ASCIE primary mirror is the use of high-performance segment-positioning actuators. These actuators must have extremely low noise levels, be able to generate substantial forces over a wide mechanical range, and support a segment in a 1-g field. They must also have a bandwidth sufficient to accommodate the spectra of expected disturbances. In addition, thermal energy dissipation must be minimized.

Because conventional actuators were unable to meet one or more of the performance requirements for the ASCIE, a new actuator design was specially developed at LMSC [22]. Figure 13 is a schematic diagram of this actuator and the electronic control system that drives it. The actuator utilizes flex pivots, thus avoiding the problems of stiction and friction, and operates under

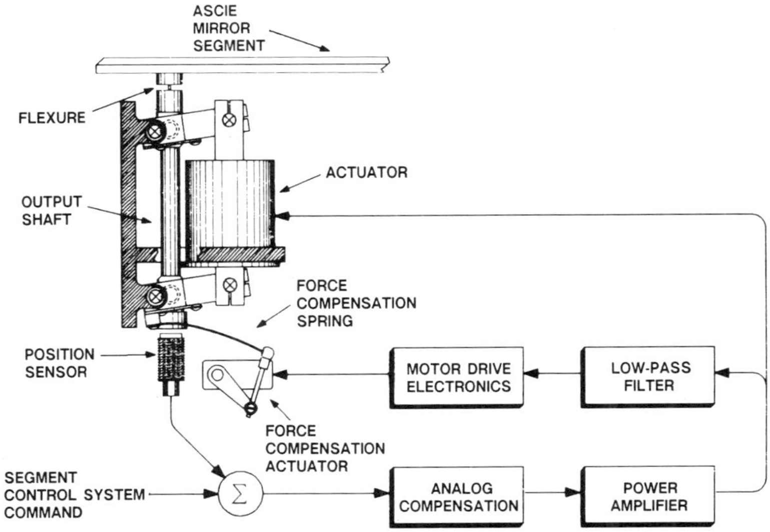

Fig. 13. Schematic drawing of the ASCIE segment-positioning actuator illustrating the use of closed-loop control, force offloading, and four-bar flexural linkage.

closed-loop control by measuring the location of the output shaft with an inductive position sensor.

A separately controlled automatic system is used to provide force offloading. This system uses a special control loop to measure the current in the main actuator. A small servomotor, shown as the force compensation actuator in Fig. 13, is commanded to move a leaf spring which applies a force to the linkage mechanism until the direct-current portion of the current in the main actuator is essentially zero.

Table I presents the performance characteristics of the ASCIE segment-positioning actuator. Figure 14 shows an operational prototype of the actuator with part of its outer case cut away to expose the internal mechanism.

4. Optical Scoring System

The optical calibration and scoring system (OCSS) developed for ASCIE provides independent measurements of segment tilt errors. It uses a laser mounted inside the central tube that forms the hub of the ASCIE structure. Surrounding the laser is an array of two-axis proportional photodetectors which act as the ASCIE "focal plane." The laser beam is directed at the facetted secondary mirror, which splits the beam into six parts, each subbeam directed at one segment.

Table I. Operational characteristics and features of the ASCIE segment-positioning actuator.

Parameter	Value
Dynamic Range	100,000:1
Total Mechanical Range	±1 mm
Noise-Equivalent Position	20 nm (rms, measured using a 100-Hz filter)
Friction/Stiction	None
Typical Statis Power Required	10 mW
Maximum Available Bandwidth	140 Hz
Maximum Available Force	±45 N
Weight	700 g
Operational Features	• Soft startup and shutdown • Automatic force unloading • All-analog electronics

Fig. 14. ASCIE segment-positioning actuator.

To emulate a real optical surface, small reflecting flats mounted on each segment reflect the beams back to the secondary. The six subbeams are then returned to the focal-plane detectors surrounding the laser after a second reflection off the secondary mirror. The motion of the focused laser spots on each of the six detectors provides a direct measurement of the angular motion of the corresponding segment. Figure 15 shows the laser path on the laboratory hardware. The bright spots on the panels are the locations of the six reference flats.

5. Active Secondary Mirror

The ASCIE secondary mirror is equipped to provide two-axis, active beam-steering control. This feature is important because the optical scoring system relies on the fact that the secondary mirror will not introduce unwanted motion into the measurement beams. Thus, an active closed-loop control system has been implemented which aligns the secondary to the focal plane and removes all relative angular motion between the secondary and the reference.

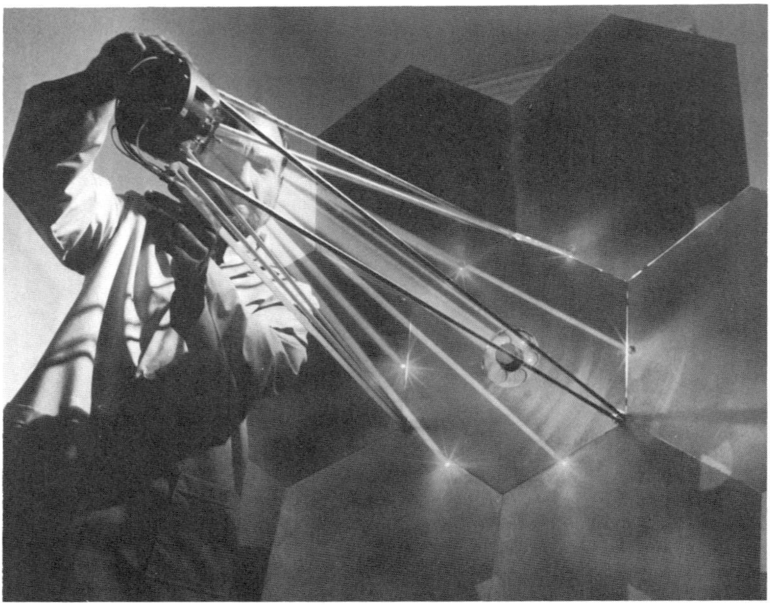

Fig. 15. Laser path for the ASCIE optical calibration and scoring system.

C. LABORATORY HARDWARE

Figures 16 and 17 show the ASCIE optomechanical hardware in the laboratory. Figure 16 shows the segmented primary mirror and the secondary mirror mounted on its hexapod support structure. The focal-plane optics and sensors are also visible. The truss support structure, segment-position actuators, edge sensors, and central hexbox/support tube are shown in Fig. 17.

Figure 18 depicts the arrangement of the ASCIE control hardware. The ASCIE optomechanical hardware communicates directly with a set of custom interface electronics that processes the signals from the edge sensors (inputs) and the control computer (outputs). An array processor coupled to a high-performance superminicomputer is used to compute the control outputs. This array-processor/host-computer configuration can directly access and store all the states of the system, an important feature for controller performance analysis. In addition, the gain matrix can be changed between processor cycles, thus making gain and scaling adjustments as easy as with conventional analog hardware.

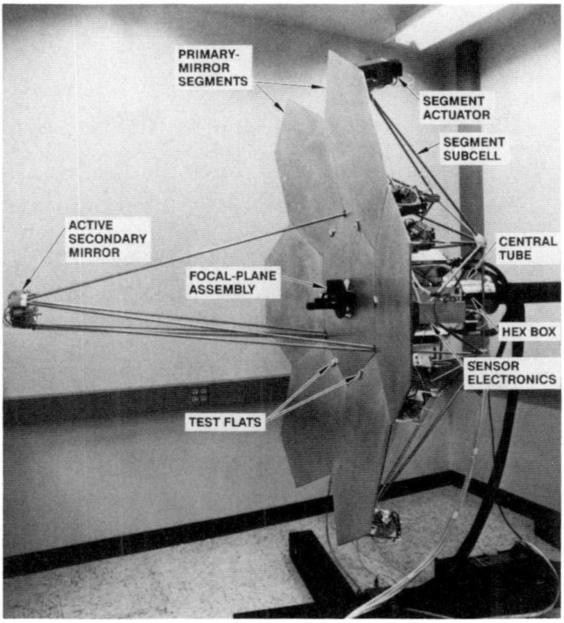

Fig. 16. ASCIE optomechanical hardware in the laboratory showing secondary mirror, focal plane, and actively controlled segments.

Control of the experiment, including a display showing the levels of all the sensor outputs and actuator commands in addition to the computed piston and tilt errors for each segment, is provided by a personal computer (PC) coupled to both the main control computer and the interface electronics.

V. MODELING AND CONTROL OF THE ASCIE LABORATORY TESTBED

The high dimensionality and unusual complexity of segmented optical systems pose formidable challenges to structural analysts, test engineers, and control engineers. Often, the analysts and engineers involved with the control and modeling of these systems must rely less on usual (sometimes empirical) practices and more on the power of the computers and the reliability of mathematical algorithms.

The structural dynamicist must accurately predict the dynamics of a system with several thousand degrees of freedom in 100 or more natural modes. The test engineer must extract, from hundreds of measured responses, the dy-

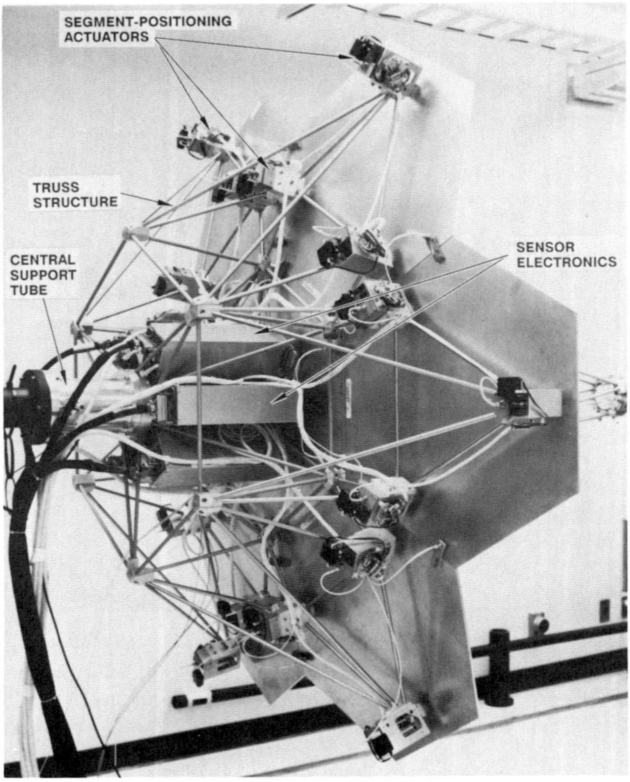

Fig. 17. ASCIE optomechanical hardware in the laboratory showing truss structure, actuators, and sensors.

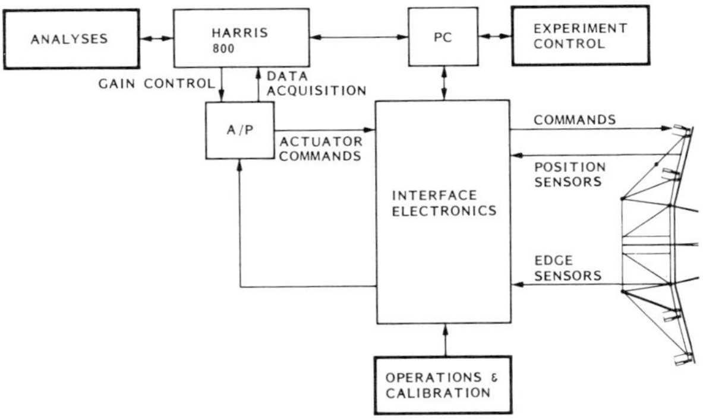

Fig. 18. Schematic of the ASCIE control hardware showing the interconnection between the experiment and control processors.

namic characteristics of a system with tens of degrees of freedom and tens of natural modes grouped in clusters. The control engineer must devise, from the partial and erroneous knowledge of the dynamics of the system, control strategies that will guarantee optical-quality performance and will damp the structural vibrations in tens of natural modes.

These challenges have been overcome on the ASCIE laboratory testbed in the three critical areas of structural analysis, dynamic testing, and control of optical performance. This represents a major step in developing and proving the technology needed to build future segmented optical systems.

A. DYNAMICS CHARACTERISTICS

A preliminary design phase determined the correct geometry, mass, and stiffness properties for the various truss elements. The design goal was to tune the dynamic response of the system so that it would be traceable to large-scale space or ground systems, while keeping the system compact enough to fit in a laboratory and be physically realizable in a 1-g environment. A critical goal in this type of design is to ensure that the structural deformations due to gravity (gravity sag) remain smaller than the available stroke of the segment position actuators.

The finite-element analysis was conducted with the same techniques used for the TMT study described earlier. The initial ASCIE structural model contained 798 degrees of freedom. The actuator mechanisms were then included in the finite-element model, so that a complete control analysis model could be developed later.

Table II compares the characteristics of the ASCIE and TMT structures. Despite the large differences in mass and size, the structural frequencies and the expected performance are similar.

A comparison of modal frequency distributions illustrates an important aspect of the ASCIE design. Figure 19 shows modal frequency as a function of mode number for the first 100 modes of the TMT (upper right), 60 modes of ASCIE (upper left), and 30 modes of a simple cantilever beam (lower left). For ASCIE, modes were calculated both with the actuator rotors locked to the stator (locked case) and with the rotors left completely free to move (unlocked case). These two cases represent the two extremes of infinite-control band-

Table II. ASCIE compared with TMT.

Parameters	ASCIE	TMT
Primary-Mirror Diameter (m)	2	10
Number of Segments	7	36
Segment Diameter (m)	0.6	1.9
Number of Actuators	18	108
Number of Edge Sensors	24	168
Total Mass (kg)	75	200,000
First Mode (Hz)	7.2	5
First Critical Mode Frequency (Hz)	12	18
Expected Phasing Error (nm)	30	30
Expected Tilt Error (arcsec)	0.1	0.03

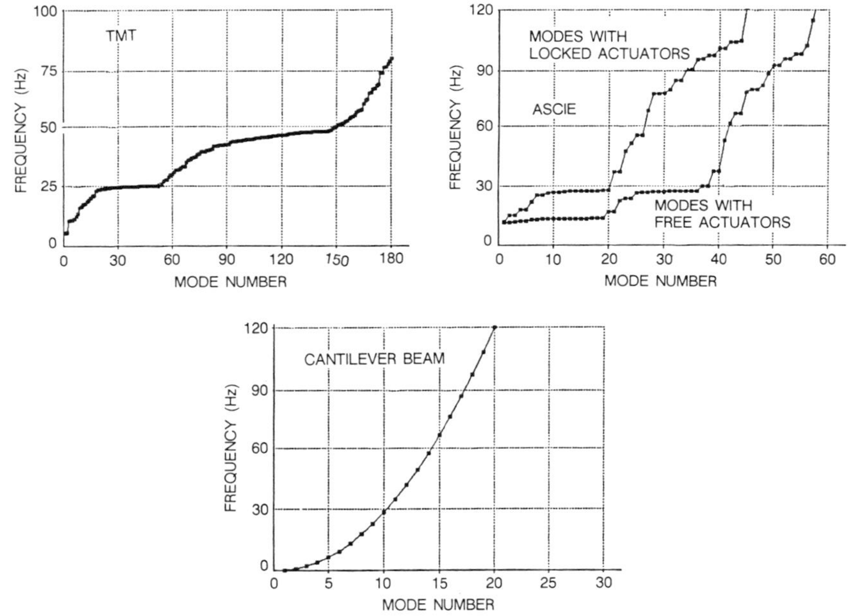

Fig. 19. Comparative modal histograms of the TMT, ASCIE, and a cantilever beam showing the close similarity between ASCIE and TMT.

width and open position servo loops, respectively. In both cases, there is a remarkable similarity between the TMT and ASCIE modal spectra: both exhibit modal grouping, a phenomenon typical of large structures with a high degree of symmetry. In comparison, the cantilever beam has modes that increase rapidly with mode number and whose frequencies are well separated.

B. BASELINE CONTROL IMPLEMENTATION

The baseline-design ASCIE segment-control system is a direct emulation of the controller used for the TMT and was intended to provide baseline performance data for comparison with more sophisticated control laws. The relative position of each segment with respect to its three adjacent neighbors is provided by a set of 24 edge sensors (4 per segment). These inputs are sent to the control algorithm, which uses them to compute commands to be sent to each of the 18 segment control actuators. Figure 20 shows the physical arrangement of the edge sensors and segment position actuators.

The baseline control law is of the form

$$P_{n+1} = P_n - KG(S - S_{ref})_n \tag{1}$$

where P is the vector of actuator position commands, S the vector of edge sensor measurements, K a scalar gain, and G a rectangular matrix that maps the 24 sensor signals into 18 actuator positions.

The new values for the actuator positions (P_{n+1}) are calculated from the current positions (P_n) by adding the error $(S - S_{ref})_n$. The integral control na-

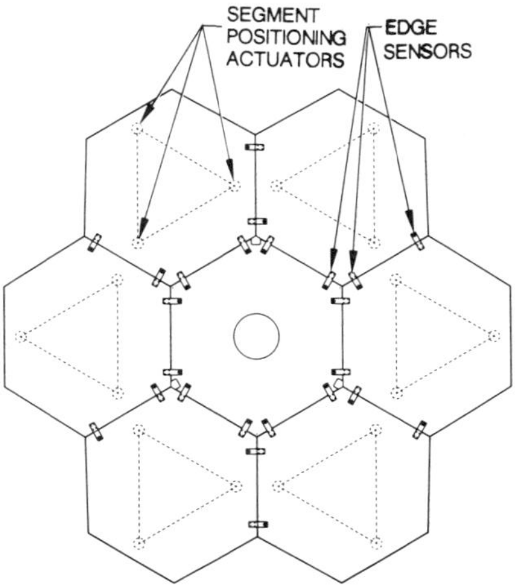

Fig. 20. Plan view of the ASCIE primary mirror showing the arrangement of the sensors and actuators.

ture of the control law is evident from Eq. (1), which is that of a simple digital integrator.

The matrix form of Eq. (1) has been implemented in the array processor that controls the ASCIE and runs at 200 updates per second. The gain matrix, KG, can be changed between array-processor cycles; thus, instantaneous and noninterfering control strategy changes can be made while the system is operating with the loop closed.

C. EXPERIMENTAL TEST RESULTS: SIX-SEGMENT CONTROL

ASCIE has been successfully run with all six segments actively controlled. Control-system performance has been independently determined using both the calculated tilt and piston errors based on edge-sensor measurements, and tilt-only errors based on OCSS measurements.

Figure 21 is a photograph of the display of the control computer used by the ASCIE operator. The display shows (clockwise from the upper left corner) the outputs of the 24 edge sensors (full scale is 100 μm), the control commands

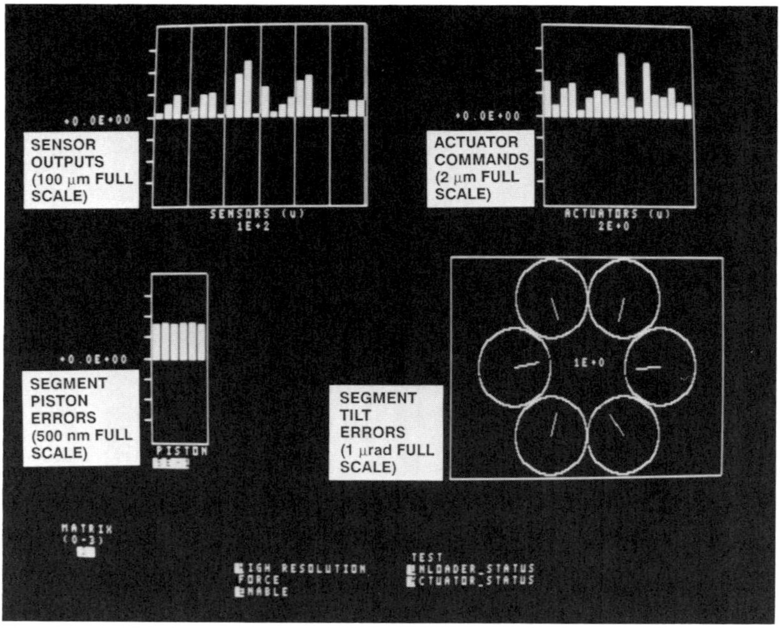

Fig. 21. Screen of the ASCIE control computer showing the sensor readings, actuator commands, and segment position errors.

sent to the 18 segment actuators (full scale is 2 μm), the computed piston error of each segment (full scale is 500 nm), and the angular tilt error of each segment, in two axes (the radius of each circle is 1 μrad).

The errors displayed in Fig. 21 are computed in the rms mode. Computed tilt and piston errors are on the order of 300 nm and 800 nrad, respectively. While these errors are somewhat worse than the desired performance of the TMT segment-control system, they indicate the level of performance that is possible using a relatively straightforward control algorithm. Improved performance requires higher gains to provide a higher bandwidth and better disturbance rejection. The limitations on control-system bandwidth based on CSI are discussed below.

D. EXPERIMENTAL DEMONSTRATION OF CSI

Figure 22 shows the output of the OCSS computer, which displays a cumulative sampled ray trace from each of the six photodetectors located at the ASCIE focal plane. Each sample is displayed as a dot on the screen, and the accumulation of dots over a 60-s sample time is displayed superposed on a set of calibration circles. All six photodetector outputs have been adjusted so that their origins are coincident with the origin of the calibration circles. If the

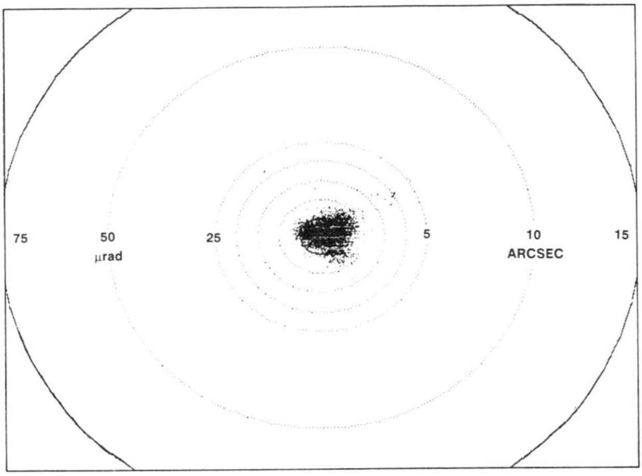

Fig. 22. OCSS computer display showing distribution of optical energy at the ASCIE focal plane when the baseline segment-control system is operating normally.

segment-control system is operating correctly, the pattern of dots from the six segments will be virtually a point at the origin. Because noise, vibration, and control-system errors distort the image and degrade the optical system performance, the combined samples tend to generate a distributed image at the center of the calibration circles.

When the gain for the segment control system is increased beyond the limit of stability, oscillations are induced by the control system into the backup structure which supports the segments. This oscillation is the CSI phenomenon.

Figure 23 is a plot of the OCSS computer output similar to that shown in Fig. 22, except that the gain has been increased by a factor of 1.4. The star pattern is generated when structural resonance, caused by interaction with the control system, moves the segments in an uncontrolled fashion. These motions are extremely small. The angles range from 25 to 50 μrad. It is interesting to observe the similarity between the plot of Fig. 9, generated from the TMT time-domain simulation, and the experimentally derived plot of Fig. 23. The close match between the forms of the structural dynamics in the TMT and in ASCIE (as shown in the histograms of Fig. 20) is verified by the virtually identi-

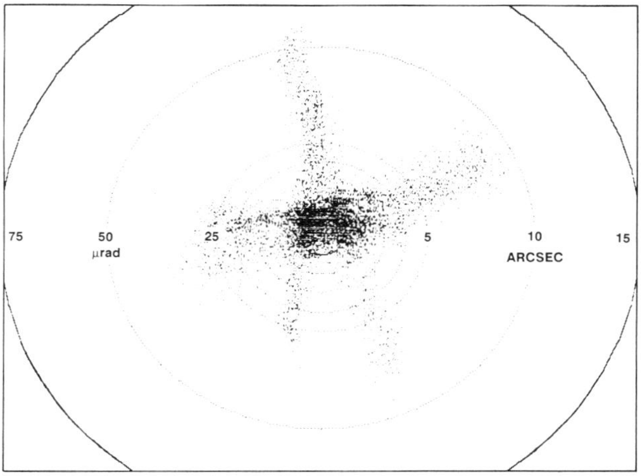

Fig. 23. OCSS computer display showing distribution of optical energy at the ASCIE focal plane when the baseline segment-control system has induced CSI.

cal way in which the two systems spread optical energy when their respective structures vibrate.

The plot in Fig. 24 was generated by connecting a spectrum analyzer to the sensor signal from one of the segment actuators. The plot shows the magnitude of the frequency spectrum of the motion of the actuator. In plot (a), the gain is below the stability margin, and the actuator response corresponds to the focal-plane display of Fig. 22. Plot (b) shows the spectrum when the gain is increased beyond the stability margin (corresponding to Fig. 23). The peak occurs at a frequency of 16.2 Hz and represents the first mode to be driven unstable by CSI. (In the TMT, the first mode to be driven unstable is at 22.8 Hz.)

E. ROBUST MODERN CONTROL DESIGN

The baseline control system is quasistatic—it ignores the flexibility of the structure that supports the mirror segments. The structure of such a control system limits the achievable segment-alignment performance, not the capabil-

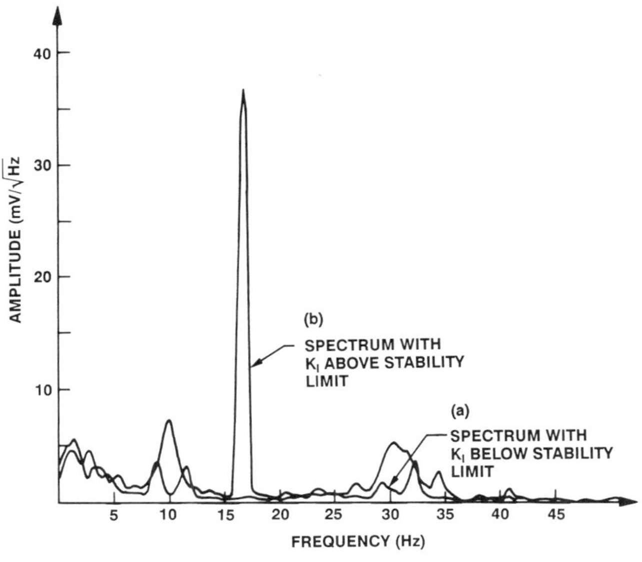

Fig. 24. Energy spectra for a single ASCIE segment-positioning actuator when (a) the baseline segment-control system is operating normally, and (b) the gain has been increased to induce CSI.

ity of the sensor and the actuator hardware. The way to achieve better optical performance is to design a control system that takes advantage of knowledge of the dynamics of the mirror-segment support structure. The achievable performance is then limited by the completeness and accuracy of the available dynamic information, and capabilities of the control hardware.

This section describes the process of building a dynamic model, evaluating the accuracy of this model, and designing a control system based on the available dynamic information and its accuracy.

1. Control Design Model

A 586-mode finite-element model is used to build the control design model. For control design purposes, the finite-element model is unnecessarily complex and numerically intractable; it is therefore replaced by a model with fewer modes. This reduced model should be as simple as possible to reduce the computation and the complexity of the control laws, and to relax the compute power requirements on the real-time digital computer that will run these laws. However, it must accurately describe the static and dynamic characteristics of ASCIE in the frequency range of interest (below 30 Hz).

Three criteria are used to select modes from the finite-element model to form the control design model: controllability/observability, disturbability, and static fidelity. Controllability/observability is an important concept used in control theory to quantify the regulation capability of the control system hardware. Figure 25 is a graphical representation of the relative controllability/observability of the first 50 modes of the ASCIE testbed predicted by the finite-element model. The 28 most controllable/observable modes give an accurate static and dynamic description of the system below 30 Hz, and they are kept to form the control design model.

Like the mathematical models of most systems with multiple symmetries, the ASCIE control design model can be dynamically and input/output decoupled into several smaller models—two in this case. The larger submodel has 22 modes, and the smaller one has 6 modes. As a result of the open-loop decoupling, the control design problem is broken down into two smaller problems that can be numerically solved faster and more reliably. The global and decoupled control designs give the same overall control performance if the decoupling is exact. The structure of the controller, which reflects the de-

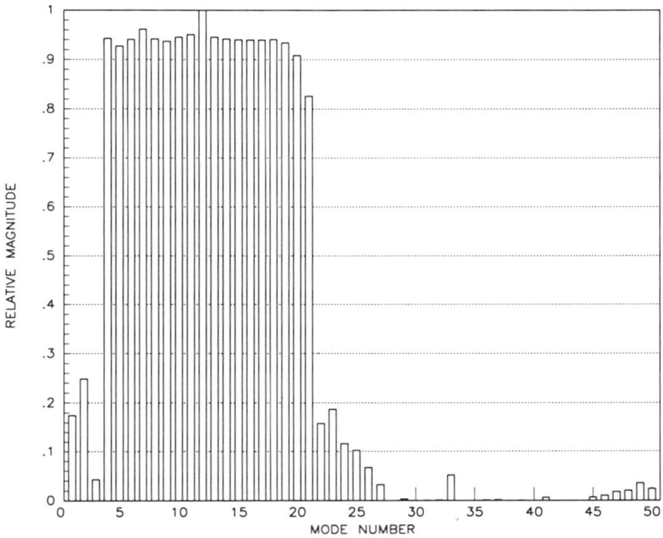

Fig. 25. ASCIE modal controllability/observability products.

coupled structure of the control design model (Fig. 26), can be exploited in a real-time distributed processing implementation. Such implementation will likely be needed in future segmented optical systems.

2. Modeling Error Assessment

Some knowledge of the modeling errors is needed in control synthesis to define the stability robustness requirements that a controller must meet. The lack of accuracy of the finite-element model, the model reduction, and the input/output decoupling are three potential sources of modeling errors. A quick evaluation reveals several significant differences between the

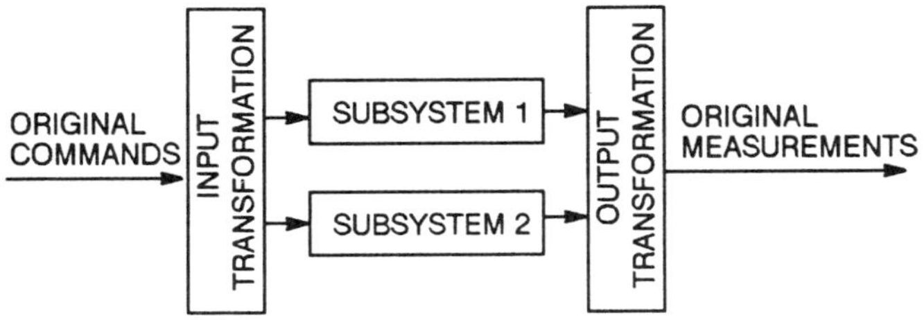

Fig. 26. Open-loop system decoupling: SUBSYSTEM 1 and SUBSYSTEM 2 can be controlled independently of one another.

finite-element model and real system structural dynamics. This is illustrated in Figs. 27 and 28, which compare the piston response of one of the mirror segments submitted to a piston command predicted by the finite-element model to the actual response measured by a transfer function analyzer. The modal frequencies of the lowest-frequency modes are known to no better than 10%. Beyond 35 Hz, the finite-element model is totally unreliable, and the predicted and actual responses are 180° or more out of phase.

Further analysis reveals that all the modeling errors are dominated by the lack of accuracy of the finite-element model.

3. Control Objectives and Constraints

The segment alignment control system must correct the mirror segment misalignments caused by the gravity loads, the thermal loads, and the quasistatic disturbances, so that the primary mirror will retain the optical quality of a single mirror.

For the purpose of the control design, a fictitious disturbance composed of 18 independent forces applied at 18 evenly distributed locations on the primary mirror is defined. The spectrum and the magnitude of this disturbance are sized to produce a 1-mm mirror static displacement and a 10-μm-amplitude jitter at the resonant frequencies of the first structural vibration modes. These values, according to current experimental and theoretical data,

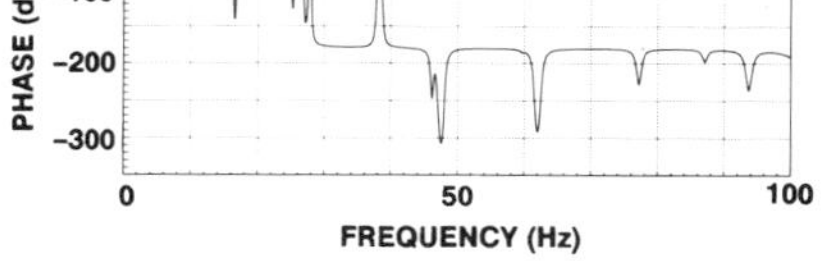

Fig. 27. Theoretical mirror segment piston response to a piston command.

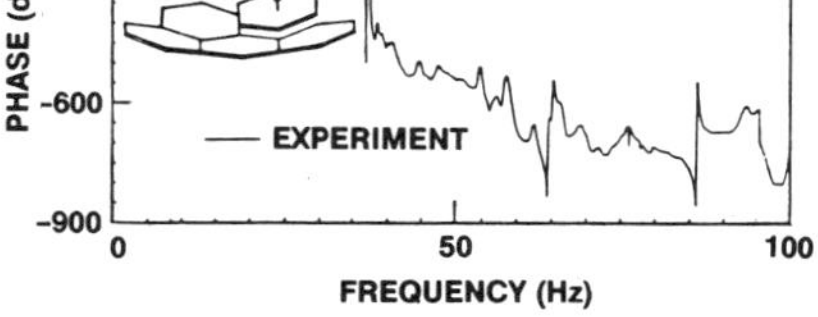

Fig. 28. Experimental mirror segment piston response to a piston command.

are typical ground telescope primary mirror surface deformations caused by the gravity loads and the winds coming through the dome. The nominal performance objective is to reduce the amplitude of the mirror displacements due to the fictitious disturbance just defined to below 100 nm at all frequencies.

To perform well, the control system must account for the limited actuator authority, measurement imperfections, and inaccuracy of the dynamic information available. The last constraint is particularly severe here, because the control design model is derived from a finite-element model which has not been tuned to the real hardware. It translates into strict stability robustness requirements: 10% stability margin with respect to modal frequency errors, 100% stability margin with respect to modal damping errors, and 60 dB/decade of rolloff in the loop gain starting at 16 Hz to gain stabilize the structural modes of the system beyond 35 Hz.

Figure 29 is a schematic of the various control-design objectives and constraints showing the external disturbances and measurement errors as inputs, the actuator authority and segment-alignment errors as penalized outputs, and the modeling errors as feedback perturbations on the nominal control-design model.

4. H_∞ Control-Design Methodology [23]

A control-design method that trades performance, parameter stability margins, and gain margins is needed to synthesize the control laws which must

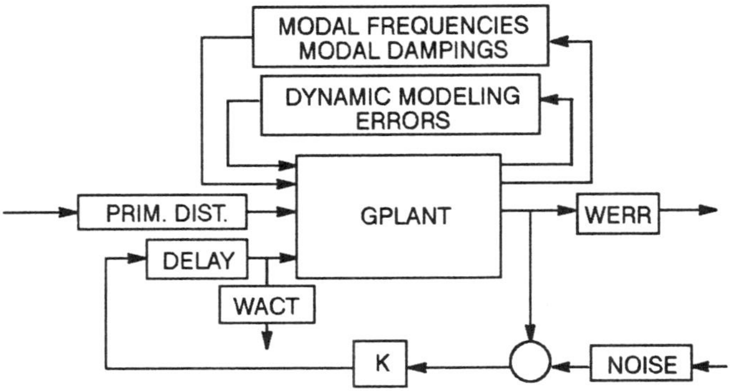

Fig. 29. Summary of control design objectives and constraints.

provide both for vibration suppression and for integral control action to counter the effects of the gravity loads and low-frequency disturbances.

When the dynamics of the system and the spectrum of the disturbances are accurately known, the Linear Quadratic Gaussian (LQG) synthesis technique is preferred, because it produces the best performance control law. When the dynamics of the system is incompletely or not accurately modeled, however, the LQG paradigm performance predictions are unreliable, and the closed-loop system may not be stable.

The High-Authority Control/Low-Authority Control [24] is an LQG-based two-step procedure that addresses the instability problem in the presence of neglected system dynamics. The procedure combines an LQG controller (HAC) for performance with colocated rate feedback (LAC) to add damping into the neglected structural vibration modes destabilized by the HAC controller. Colocated rate feedback is an inherently stabilizing control strategy, so that the dynamics of the neglected modes does not need to be known accurately for the LAC to be effective. The HAC/LAC methodology has been applied successfully to several structural control problems [25], but it does not address the robust stability problem directly, and its implementation still requires a somewhat accurate control-design model.

The H_∞ control-design method is used here to derive the segment alignment control laws because of its ability to trade performance and stability requirements, and its ability to determine whether given performance and stability margin specifications can be met simultaneously. Given bounds on the parametric and the dynamic modeling errors, and bounds on the spectra of the disturbances, H_∞ control-design methods define a combination of modeling errors and disturbance spectra within these bounds that will result in the worst possible closed-loop performance. They also design a linear controller that minimizes this worst-case performance.

5. Analytical Control Synthesis Results

The following results show some of the analytical performance and stability margin characteristics of the H_∞ controller for the largest control-design subsystem defined in section V.E.1.

Figure 30 shows the open- and closed-loop responses of the subsystem to the fictitious disturbance used in the design (top two curves), and the actuator authority needed to achieve the above disturbance attenuation (bottom curve). Up to 10 dB of vibration attenuation is achieved around 16 Hz and 25 Hz.

Figure 31 shows the stability margin of the closed-loop system to dynamic output plant multiplicative uncertainties. The loop gain rolls off at 60 dB per decade starting at 30 Hz. The minimum stability margin, 0.8 around 14 Hz, shows that any loop gain variation less than 80% in any one or multiple channels at the output of the plant would not cause instability.

Figure 32 shows the stability margin of the closed-loop system to individual modal frequency errors. The effect of each frequency uncertainty on the closed-loop system is modeled as a scalar feedback perturbation whose gain depends on the magnitude of the error. The stability of each loop can be tested using a Nyquist plot. The errors are normalized so that whenever the Nyquist plot crosses the real axis between +1 and −1, the stability margin is at least 10%. The figure shows that the control system has more that 10% stability margin in each modal frequency error. The worst margin, obtained for the first mode, is 20%.

6. Experimental Test Results: H_∞ Versus Baseline Six-Segment Control

This section compares experimental performance of the H_∞ controller and the baseline controller. The baseline controller ignores the flexibility of the truss structure that supports the mirror segments, and the H_∞ controller

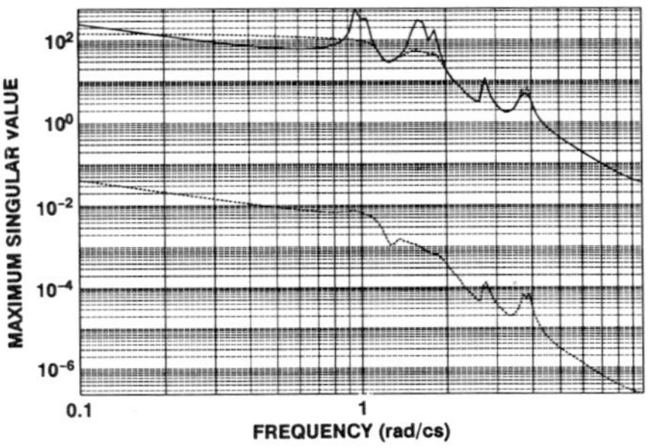

Fig. 30. Frequency responses showing predicted disturbance rejection (unit: 100 nm) and control effort (lbf) for the H_∞ segment-control system.

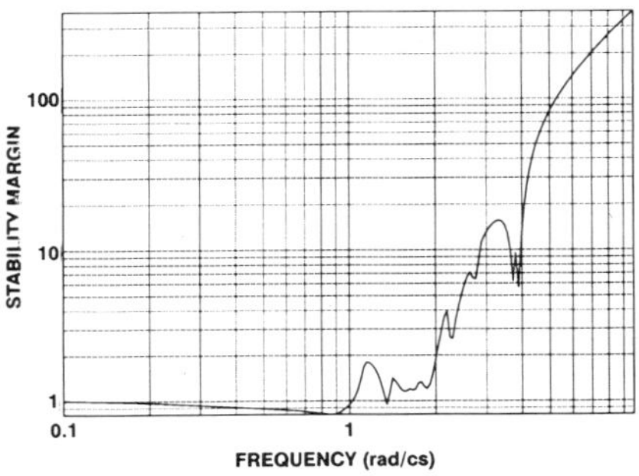

Fig. 31. H_∞ segment-control system stability margin to output plant multiplicative uncertainties.

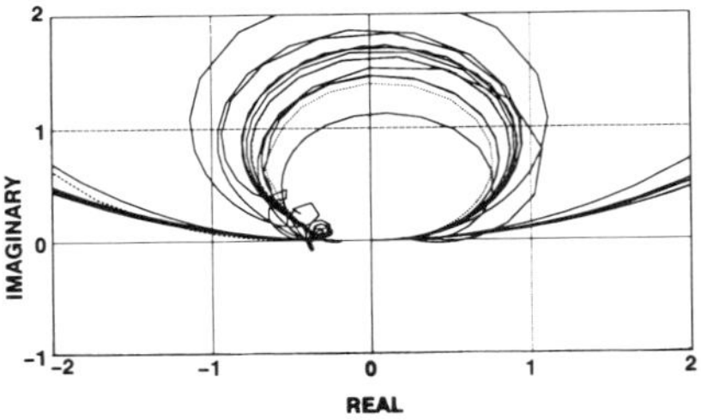

Fig. 32. H_∞ segment-control system stability margin to individual modal frequency errors.

uses some inaccurate dynamic information on the segmented primary mirror and its support structure. Two series of tests are devised: one to compare segment-alignment performance, the other to compare structural-vibration suppression.

a. Segment-Alignment Test Results. In the segment-alignment tests, square waves are applied at the input of the system. The evaluation is performed by comparing the closed-loop time responses of one of the ASCIE actuator input error. The tests show how well the control system rejects input step disturbances. The tests also give valuable information on the segment-alignment bandwidth. Figures 33 and 34 compare the closed-loop time responses to a

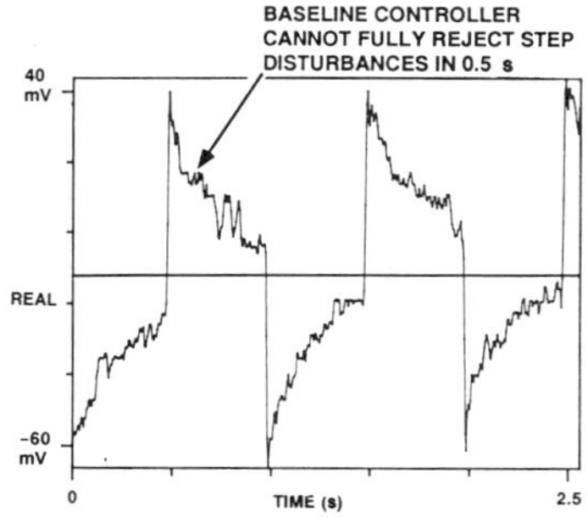

Fig. 33. Closed-loop input error to 1-Hz square wave (baseline controller).

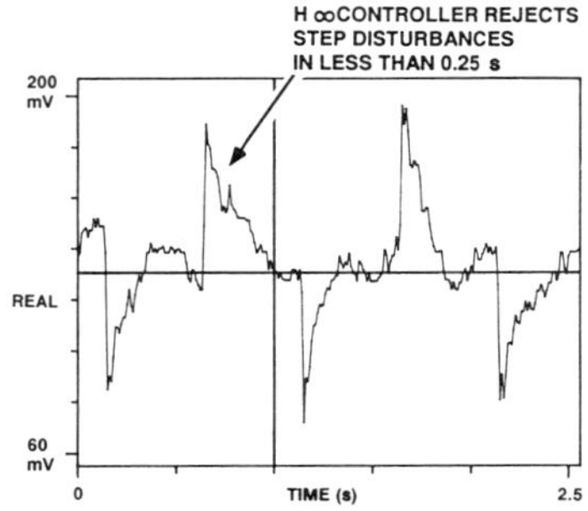

Fig. 34. Closed-loop input error to 1-Hz square wave (H_∞ controller).

0.1-Hz square wave. The baseline control law cannot fully reject the steps in 0.5 s, while the H_∞ control law needs less than half this time to perform the rejection. The H_∞ control segment-alignment bandwidth is 3 to 5 times larger than the baseline control segment-alignment bandwidth.

b. Vibration-Attenuation Test Results. In the vibration-attenuation tests, white-noise signals are applied at the input of the system. The evaluation is performed by comparing the closed-loop power spectra of one of the ASCIE input errors. The results show the attenuation or amplification of dynamic input disturbances and demonstrate how well the CSI instability is solved.

In a first test, the white-noise signal is distributed to the various actuator inputs to excite the second natural vibration mode of the system. Figures 35 and 36 show that the H_∞ control law achieves 10 dB of disturbance attenuation, while the baseline control law amplifies the disturbances by 10 dB.

In a second test, the white-noise signal is distributed to the various actuator inputs to excite the 17th natural vibration mode of the system. Figures 37 and 38 show that the H_∞ control law achieves 10 dB of disturbance attenuation, while the baseline control law amplifies the disturbances by a few dB.

The whole series of 28 tests revealed that the H_∞ control law achieved disturbance attenuation in 23 out of 28 controlled modes, the attenuation going from a few dB to 15 dB.

c. Conclusion. The experimental results demonstrate the ability of H_∞ control-design methods to deal with very inaccurate models of complex systems and to give better performance than classical control methods. H_∞ con-

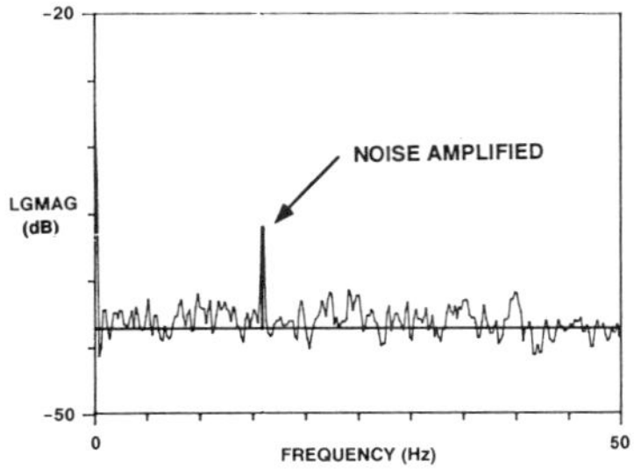

Fig. 35. Closed-loop input error to random excitation in mode 2 (baseline controller).

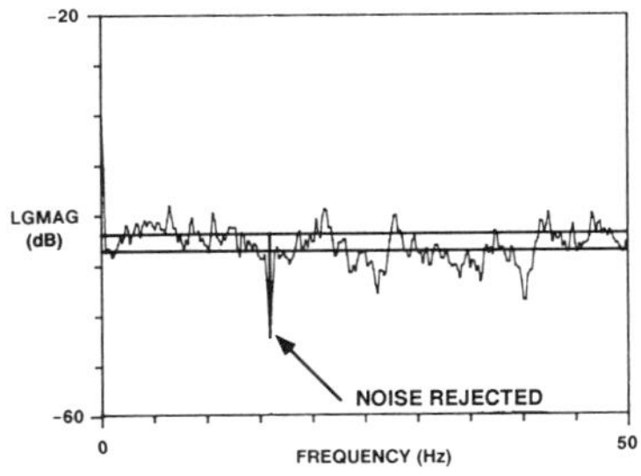

Fig. 36. Closed-loop input error to random excitation in mode 2 (H_∞ controller).

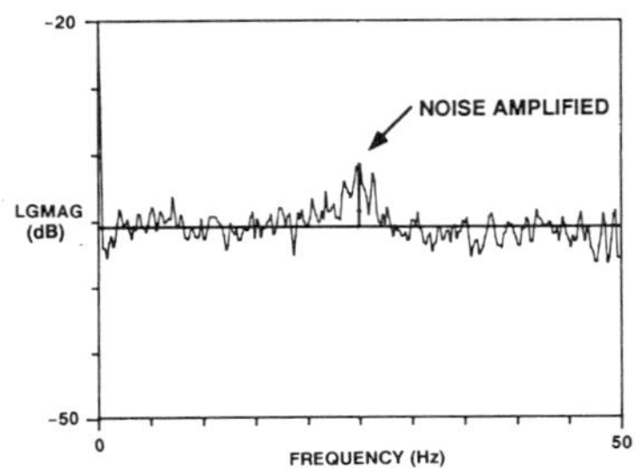

Fig. 37. Closed-loop input error to random excitation in mode 17 (baseline controller).

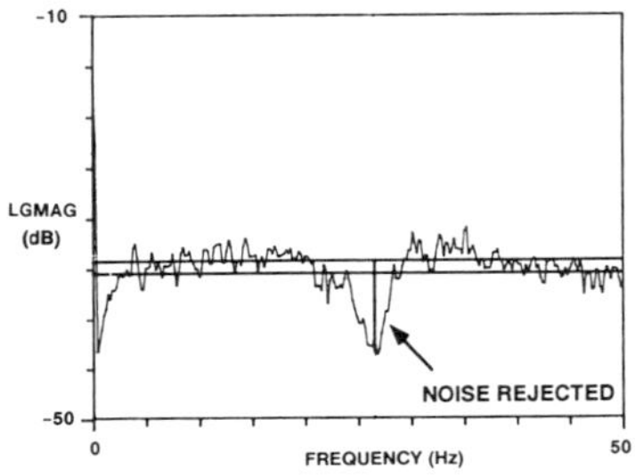

Fig. 38. Closed-loop input error to random excitation in mode 17 (H_∞ controller).

trollers do not merely avoid CSI by not controlling the system at the resonant frequencies (as do notch filters for example). They also attenuate structural vibrations.

F. DYNAMIC CHARACTERIZATION FROM MEASURED INPUT-OUTPUT DATA

When lack of accuracy of the mathematical model prevents the control designer from achieving the desired performance objectives, the theoretical dynamics of the system to control must be tuned to the dynamics of the system hardware, or a new dynamic model must be constructed from observed responses.

This section illustrates techniques used to develop a dynamic model of the ASCIE segmented optics and support structure from measured responses only. The dynamic characterization of ASCIE is a challenging task because hundreds of measured responses covering tens of degrees of freedom must be processed, and because tens of natural modes grouped in clusters must be extracted. On actual hardware, model identification techniques have been demonstrated only on small systems. Also, no technique has addressed the computational complexity associated with structures with high modal density.

1. Data Acquisition

Steady-state responses to sinusoidal excitations are measured and analyzed to determine the system admittance matrix at one frequency, and frequency sweep is used to determine the system admittance matrix over a grid of frequencies. The frequency response measurement procedure uses all the actuators to simultaneously excite the structure, and all the sensors to simultaneously measure the structural responses. The transfer functions in all degrees of freedom are measured in a single path. The spectral responses are analyzed as the spectral data become available, and the results are used to excite the system preferentially according to the system's principal gains and corresponding principal actuator force vectors. The mean-squares force input is adjusted to maintain a given mean-squares displacement response. This excitation technique maintains high-quality measurements and avoids the problems associated with system ill-conditioning.

2. On-Line Modal Testing

On-line modal testing consists of experimentally isolating the modes of a system to measure their characteristics. It was first proposed by Lewis and Wrisley [26]. It requires tuning the polarities and the relative magnitudes at one frequency of several shakers located at discrete points on the structure to produce a pure mode response. Classical modal tuning techniques require a lot of instrumentation and expensive data-acquisition systems, and they can be difficult to implement. Experimental factors such as measurement noncolocated with the excitations, delays, nonproportional damping, lack of orthogonality of the mode shapes in high modal density regions, and ill-conditioning affect these techniques or make them fail. This section presents an effective on-line modal testing technique that is relatively insensitive to the above experimental factors. The technique was developed to perform limited modal surveys on ASCIE.

The approximate modal tuning objective is to maximize the root-mean-squares displacements for a given root-mean-squares force excitation. It is a generalization of the gain concept used for single-input single-output systems. It requires tuning the relative phases and the relative magnitudes of the various sine forces to maximize the response. Only a preliminary measurement of the transfer function matrix and a singular value analysis of this matrix are needed to find the force vector that maximizes the response at a given frequency. Gain tuning can therefore easily and reliably be achieved experimentally.

For lightly damped systems, at a resonant frequency, the maximum admittance gain is obtained when the actuator forces are combined so as to excite a pure mode. Gain and modal tuning are therefore nearly equivalent. This is the theoretical basis for our modal tuning procedure.

The tuned force and response vectors at a resonant frequency yield the mode shapes at the actuator and sensor locations respectively. To isolate the target mode, the polarities and relative magnitudes of the actuator forces must be adjusted according to the mode shape at the actuator locations, and a linear combination of all the responses must be taken according to the mode shape at the sensor locations. Once the target mode is isolated, a frequency sweep can be performed to measure its modal characteristics.

The effectiveness of this procedure is illustrated in Figs. 39 and 40. The ASCIE admittance matrix from the 18 mirror segment actuators to the 24 edge position sensors (Fig. 39) is measured between 22 and 33 Hz. The singular value decomposition of the admittance matrix at 27 Hz is then computed to extract the mode shape information of the dominant mode around this frequency. This mode shape information in turn is used to isolate the corresponding mode as explained above, and a single-input single-output sine sweep is performed. Figure 40 shows the resulting measured modal response. The phase curve shows that the phase drops by 180° as the sweep passes through the resonant frequency as it should for a pure mode response. The

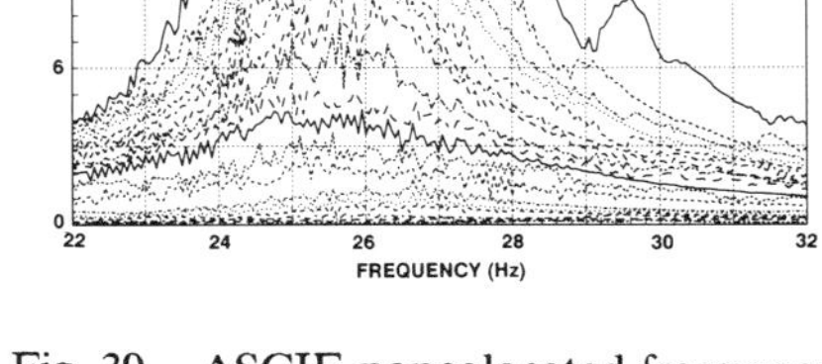

Fig. 39. ASCIE noncolocated frequency response in high modal density region.

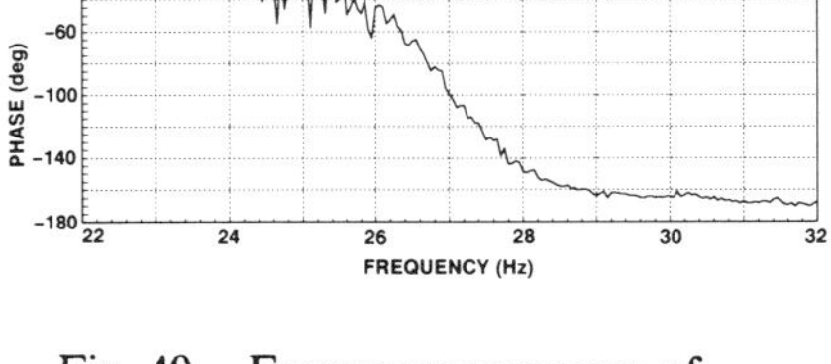

Fig. 40. Frequency response of dominant mode around 27 Hz extracted by on-line modal tuning.

modal frequency and the damping ratio of the isolated mode, obtained by curve fitting the modal response, are 26.9 Hz and 3.3%, respectively.

3. Dynamic Characterization of ASCIE

With a small number of sensors and actuators, it is experimentally impossible to isolate all the modes of the complex ASCIE structure. Analytical techniques that can determine the modal properties of the structure without isolating the modes must be used. This section illustrates the application of analytical techniques specially developed to construct a modal representation of the ASCIE segmented optics and support structure from its measured colocated response. Figure 41 shows the principal gains of this response between 6 and 50 Hz.

a. Approach. The characterization is treated as a curve-fitting problem in the frequency domain. The error criterion is the least-squares difference between the theoretical and the measured frequency responses. Its minimization is treated as a classical two-step initialization/least-squares optimization problem. However, because of the complexity of the present problem, a lot of effort is devoted to the initialization procedure. The main idea consists of isolating the modes according to their spectral and spatial separation to estimate their dynamic characteristics.

b. Spectral Partitioning. The spectral partitioning is performed by visual inspection of the frequency response. The dominant modes with similar frequencies are grouped together. Accordingly, the ASCIE colocated frequency

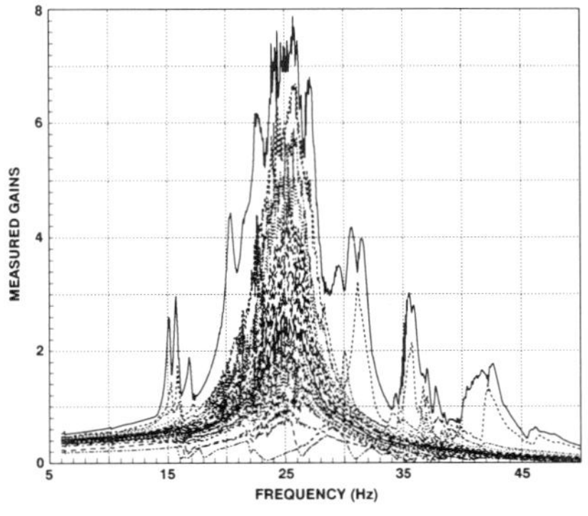

Fig. 41. Measured colocated frequency response of ASCIE.

response is partitioned over six frequency ranges: 1 to 13 Hz, 13 to 18 Hz, 20 to 29 Hz, 28 to 33 Hz, 33 to 39 Hz, and 39 to 50 Hz.

c. Analytical Modal Separation. On each frequency range, an analytical modal separation algorithm [27] is used iteratively to extract the nonparametric frequency response of one or several modes at a time from the measured frequency response. This algorithm operates like the gain-based modal tuning technique of section V.E.2. It extracts modes on the basis of the dominance of the modes' response in some frequency range around their resonant frequencies. Extracting a single-mode response involves determining the proper distribution of excitation forces and the proper linear combination of the measured responses (i.e., the mode shapes at the actuator and the sensor locations, respectively) as in on-line modal tuning. Similarly, extracting a multiple-mode response involves determining multiple distributions of excitation forces and multiple linear combinations of the measured responses. Figure 43 illustrates a single application of the analytical modal-separation algorithm to extract the modal response of the four dominant natural modes of the ASCIE segmented optics and support structure between 13 and 18 Hz from its colocated frequency response (Fig. 42). The frequency response of the residual system (Fig. 44) clearly illustrates that the modal response of the dominant modes is well extracted.

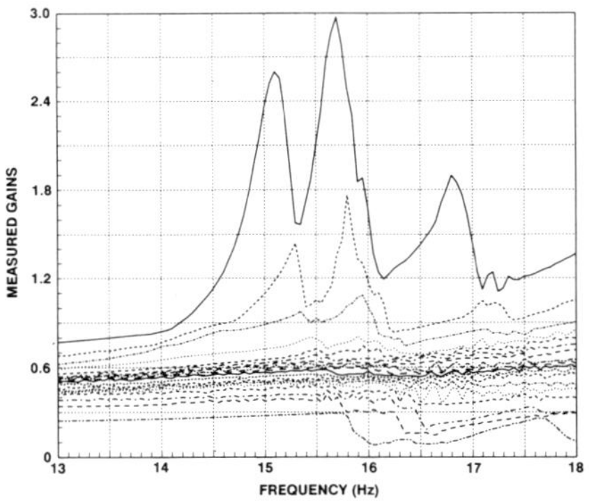

Fig. 42. Measured colocated frequency response of ASCIE between 13 and 18 Hz.

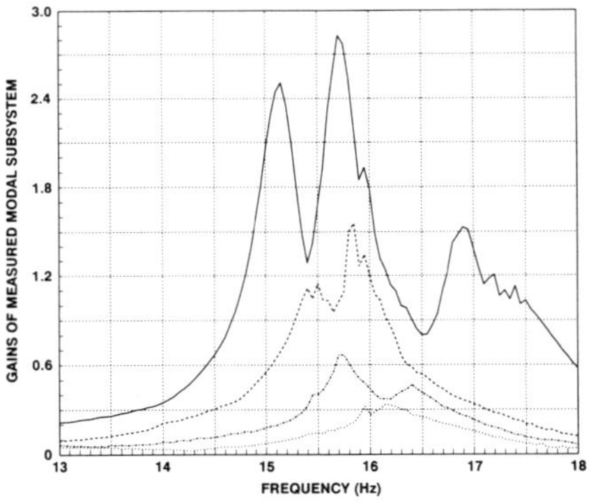

Fig. 43. Nonparametric frequency response of ASCIE four-mode model between 13 and 18 Hz.

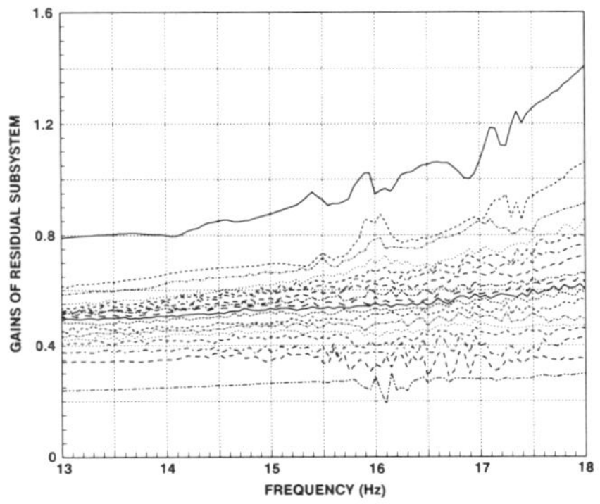

Fig. 44. Residual ASCIE frequency response between 13 and 18 Hz after the estimated modal response of a four-mode model is removed.

d. Extracting Modal Properties From Nonparametric Modal Responses. Each nonparametric modal response is analyzed to estimate the dynamic characteristics of the modes it contains. Nonparametric modal responses are simple, involving only a few modes, so that their analyses can be carried out effectively. Special numerical algorithms [27] are used for this purpose. Figure 45 is the result of applying these algorithms to the nonparametric modal response of Fig. 43. It shows the calculated frequency response of a theoretical four-mode model.

e. Optimization of Theoretical Model Constructed From Modal Subsystems. Once all the nonparametric modal responses have been analyzed, the charac-

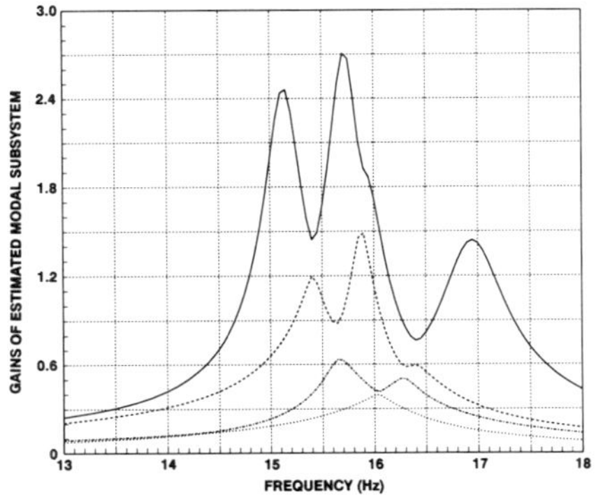

Fig. 45. Calculated ASCIE four-mode model frequency response between 13 and 18 Hz.

teristics of all the modes in each frequency range are combined to build a linear state-space modal representation of ASCIE. This modal representation is then tuned to best predict the measured frequency response in the least-squares sense. A Newton-Raphson optimization algorithm is used to carry out this task. However, because the optimization involves several thousands modal parameters, it is not carried out in the usual vector form. Instead, the structure of the identification problem is exploited to break down the optimization problem into two linear least-squares problems which are solved explicitly, and a nonlinear least-squares problem of relatively small size involving only the modal frequencies and the damping ratios of the predicted modes. All three optimization problems are solved successively and iteratively.

Figure 46 shows the principal gains of the optimal model obtained by globally tuning a 50-mode model to the measured ASCIE colocated frequency response. In this case, 4024 modal parameters were optimized, and 324 transfer functions were fitted simultaneously. Figure 47 compares some local results, namely the measured and the calculated frequency responses for one colocated actuator sensor pair. The only significant difference between the two responses is around the 50-Hz boundary and corresponds to the contributions of the unmodeled modes outside of the frequency range being analyzed.

Figure 48 gives the modal frequency and damping ratio histograms corresponding to the identified 50-mode model. It shows that modal clusters with as many as five modes/Hz have been resolved.

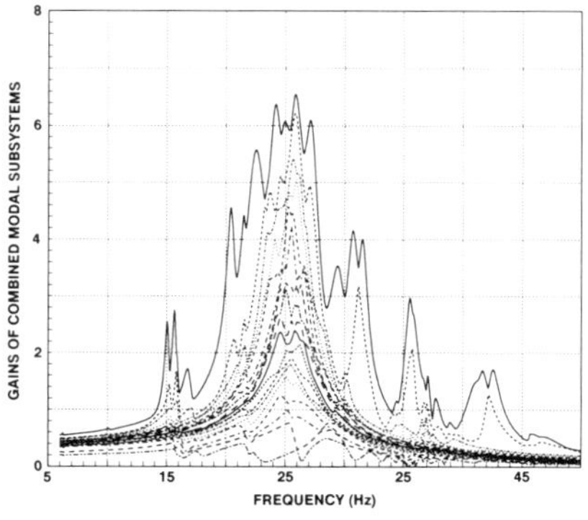

Fig. 46. Calculated colocated frequency response of ASCIE.

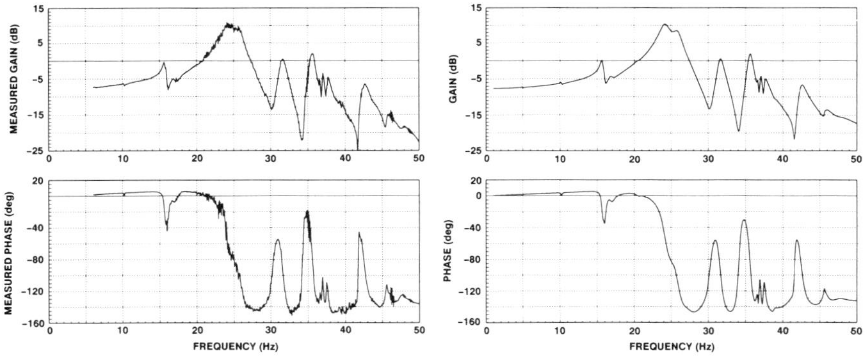

Fig. 47. Example of colocated transfer functions showing excellent agreement between predicted and measured responses.

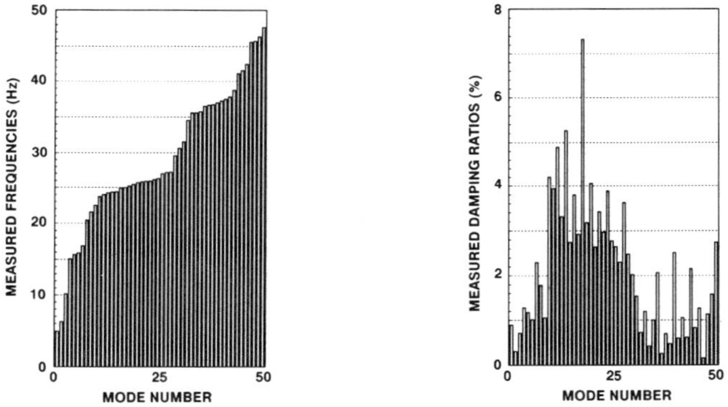

Fig. 48. Modal frequencies and damping ratios of ASCIE 50-mode model.

f. Conclusion. This section describes the application, to ASCIE, of the system identification techniques developed to characterize the dynamics of complex structures from frequency response data. Multiple excitations at a single frequency are used to acquire the frequency responses. At each frequency, the system is excited preferentially according to its principal gains to avoid the problems associated with system ill-conditioning. Spectral separation and spatial separation are used concurrently in on-line modal tuning techniques and off-line modal separation techniques. These techniques can isolate pure modal responses in high modal density clusters for estimating their characteristics. These estimated modal characteristics can be combined to construct a linear model and to initialize a frequency-domain multilevel least-squares optimization procedure. The optimized model matches accurately the measured responses and meets the requirements for high-performance control design.

VI. SUMMARY

This article describes the analysis and design of control systems that position optical elements in large segmented reflectors. Designing and implementing a high-performance segment alignment control system are extremely difficult, because the positioning and accuracy requirements for this type of control system are so difficult to meet, and because the control system interacts with the vibration modes of the structure supporting the segmented reflector. Analysis of the TMT shows how the interaction between the segment-control system and structural dynamics affects the optical performance of the telescope and limits the bandwidth of the control system. A laboratory test bed, called ASCIE, has been developed to investigate the technological issues related to control of segmented mirrors. Details of the ASCIE hardware and results of experiments with ASCIE are described. ASCIE made it possible to conduct pioneering experiments in the control of structures with significant modeling uncertainties, and to investigate the dynamic characterization of structures with high modal density.

VII. REFERENCES

1. R. Rhodes, "Reflected Glory: How They Built Palomar," *American Heritage of Invention and Technology* 1, 1, 1985.
2. A. Buffington, F. S. Crawford, R. A. Muller, A. J. Schwernin, and R. G. Smits, "Active Image Restoration With a Flexible Mirror," *Proc. SPIE* 75, pp. 90–96, 1976.
3. E. R. Washwell, D. L. Edwards, A. H. Gelders, R. J. Serviss, and D. H. Wilson, "Active Mirror Segment Figure Control," *Proc. SPIE* 365, pp. 42–52, 1982.
4. H. J. Robertson, R. Crane, and H. S. Hemstreet, *Active Optical System for Spaceborne Telescopes*, NASA Contractor Report CR-66297, 1966.
5. R. R. Berggren and G. E. Lenertz, "*Feasibility of a 30-meter Space-Based Laser Transmitter*, NASA Contractor Report CR-134903, 1975.

6. P. N. Swanson, S. Gulkis, T. B. Kuiper, and M. Kiya, "Large Deployable Reflector (LDR): A Concept for an Orbiting Submillimeter-Infrared Telescope for the 1990s," *Opt. Eng.* 22, pp. 725–731, 1983.
7. J. E. Nelson, *Mirror Segmentation of the 10 Meter Telescope, Design Considerations*, University of California Ten Meter Telescope Report No. 3, 1977.
8. T. S. Mast and J. E. Nelson, "Figure Control for a Fully Segmented Telescope Mirror," *Appl. Opt.* 21, pp. 2631–2641, 1982.
9. J-N. Aubrun, K. R. Lorell, T. S. Mast, and J. E. Nelson, "Dynamic Analysis of the Actively Controlled Segmented Mirror of the W. M. Keck Ten-Meter Telescope," *IEEE Control Systems* 7, p. 6, 1987.
10. J-N. Aubrun, K. R. Lorell, T. W. Havas, and W. C. Henninger, *An Analysis of the Segment Alignment Control System for the W. M. Keck Observatory Ten Meter Telescope*, Keck Observatory Report No. 143, 1985.
11. J. E. Nelson, T. S. Mast, and S. M. Faber, *The Design of the Keck Observatory and Telescope (Ten Meter Telescope)*, Keck Observatory Report No. 90, 1985.
12. T. S. Mast and J. E. Nelson, *Figure Control for a Fully Segmented Telescope Mirror*, Keck Observatory Report No. 77, 1981.
13. S. J. Medwadowski, *Conceptual Design for the Structure of the Ten Meter Telescope*, Keck Observatory Report No. 59, 1981.
14. T. W. Havas, *ASTRO Program Manual, Version I*. Lockheed Missiles & Space Company, Inc., LMSC-D051785, 1978.
15. C. Lanczos, "An Iteration Method for the Solution of the Eigenvalue Problem of Linear Differential and Integral Operators," *J. Res. Nat. Bur. Stand.* 45, pp. 255–282, 1950.
16. J. H. Wilkinson, *The Algebraic Eigenvalue Problem*, Clarendon Press, Oxford, 1965.
17. R. J. Guyan, "Reduction of the Stiffness and Mass Matrices," *AIAA J.* 3, 2, 1965.
18. B. Sridhar, J-N. Aubrun, and K. R. Lorell, "Analytical Approximations of the Dynamics of a Segmented Optical System," *Proc. 4th IFAC Symp. Control of Distributed Parameter Systems*, Los Angeles, CA, 1986.

19. G. Gabor, *The Mirror Control System for the U.C. Technical Demonstration Prototype*, Keck Observatory Report No. 77, 1982.

20. J-N. Aubrun and K. R. Lorell, *The Effect on Image Quality of the Ten Meter Telescope Segment Control System Dynamics*, Keck Observatory Report No. 162, 1987.

21. K. R. Lorell, J-N. Aubrun, D. F. Zacharie, and E. O. Perez, "A Control Technology Testbed for Large Segmented Reflectors," *IEEE Control Systems* 9, 6, 1989.

22. K. R. Lorell, J-N. Aubrun, D. F. Zacharie, and E. O. Perez, "Development of a Precision Wide Dynamic Range Actuator for Use in Active Optical Systems," *Proc. 23rd Aerospace Mechanisms Symp.*, NASA Marshall Space Flight Center, AL, 1989.

23. A. Carrier, *Modeling and Shape Control of a Segmented-Mirror-Telescope*, Ph.D. thesis, Department of Aeronautics and Astronautics, Stanford University, Mar 1990.

24. J-N. Aubrun, "Theory of the Control of Structures by Low-Authority Controllers," *AIAA J. Guidance and Control* 3, 5, p. 444, 1980.

25. *ACOSS 12 (Active Control of Space Structures)*, Final Report for DARPA, Lockheed Missiles & Space Company, Inc., LMSC-D833023, Jul 1983.

26. R. Lewis and D. Wrisley, "A System for Excitation of Pure Natural Modes of Complex Structures," *J. Aerospace Sciences* 17, 11, pp. 705–722, Nov 1950.

27. A. Carrier and J-N. Aubrun, "Modal Characterization of the ASCIE Segmented Optics Testbed: New Algorithms and Experimental Results," *Proc. 5th NASA/DoD CSI Technology Conf.*, Lake Tahoe, NV, Mar 1992.

2284_ACRR

Robust Control Experiments for Two Highly Elastic Space Structures

Emmanuel G. Collins, Jr., Douglas J. Phillips,
David C. Hyland, and James A. King

Harris Corporation
Government Aerospace Systems Division
Melbourne, Florida 32902

I. INTRODUCTION

Many large space system concepts will require active vibration control systems to satisfy critical performance requirements such as line-of-sight (LOS) accuracy and constraints on rms surface roughness. In order for these concepts to become operational it is imperative that the benefits of active vibration control be practically demonstrated in ground-based experiments. Here, we discuss two experiments in flexible structure control conducted as part of the NASA Control Structures Interaction (CSI) Guest Investigator Program. These experiments involved control design and implementation for the Mini-MAST Structure at NASA Langley Research Center [1,2] and the ACES Structure at NASA Marshall Space Flight Center [3,4] and successfully demonstrate the substantial performance benefits achievable via active control.

One of the unifying features in the two experiments was the use of the Maximum Entropy/Optimal Projection approach to design robust, reduced-order controllers [5–15]. In particular, Maximum Entropy design was used to achieve robustness with respect to parametric uncertainty while the Opti-

CONTROL AND DYNAMIC SYSTEMS, VOL. 57

mal Projection representation of the necessary conditions for reduced-order optimal control was used to design reduced-order controllers. The initial step in the reduced-order design process was balanced controller reduction [16] of a full-order Maximum Entropy or LQG compensator. It was found in our control design efforts, especially those for the Mini-MAST structure that this initial step in the reduced-order design process often gave controllers that were effectively optimal controllers. Hence, many of the reduced-order controllers implemented were actually obtained via a straight-forward application of balanced controller reduction. However, our experience also discovered cases where it was actually necessary to use more complex algorithms to obtain the optimal reduced-order controllers.

Another unifying feature of the two experiments was the design and implementation of decentralized controllers. It was seen in both experiments that decentralized controllers were able to provide significant attenuation of the modes that were most important to performance. Centralized controllers were also implemented in the Mini-MAST experiment and as would be expected the increased complexity enabled the achievement of even greater performance improvement.

This chapter is organized as follows. In section II we describe the Maximum Entropy/Optimal Projection approach to robust reduced-order control design. In section III we describe the Mini-MAST experiment. Section IV then provides a similar exposition of the ACES experiment. Section V presents some final conclusions.

II. THE MAXIMUM ENTROPY/OPTIMAL PROJECTION APPROACH TO CONTROL DESIGN

Maximum Entropy/Optimal Projection design [5–15] assumes an uncertain plant of the form

$$\dot{x}(t) = (A + \Delta A)x(t) + Bu(t) + D_1 w_1(t) \tag{1}$$

$$y(t) = Cx(t) + w_2(t) \tag{2}$$

$$z(t) = E_1 x(t) \tag{3}$$

where $x \in \mathbb{R}^{n_x}$, $u \in \mathbb{R}^{n_u}$, $y \in \mathbb{R}^{n_y}$, $z \in \mathbb{R}^{n_z}$, $w_1 \in \mathbb{R}^{n_x}$ is white disturbance noise with nonnegative definite intensity $\bar{V}_1$, $w_2 \in \mathbb{R}^{n_y}$ is white observation noise with positive definite intensity V_2, and w_1 and w_2 are uncorrelated. The state matrix A is assumed to be in real normal form such that A is block diagonal with scalar blocks corresponding to the real poles of the system and 2×2 blocks of the form $\begin{bmatrix} -\nu_i & \omega_i \\ -\omega_i & -\nu_i \end{bmatrix}$ corresponding to the flexible modes. It is assumed that the uncertainty ΔA is of the form

$$\Delta A = \sum_{i=1}^{n_\alpha} \delta_i A_i, \quad |\delta_i| \le \alpha_i, \quad \alpha_i \ge 0 \tag{4}$$

where n_α denotes the number of uncertain modes and each A_i is of the form

$$A_i = \text{block-diag}\{0, \ldots, 0, \begin{bmatrix} 0 & 1 \\ -1 & 0 \end{bmatrix}, 0, \ldots, 0\} \tag{5}$$

corresponding to uncertainty in ω_i, the frequency of the ith mode.

Maximum Entropy/Optimal Projection design allows the synthesis of a fixed order dynamic compensator,

$$\dot{x}_c(t) = A_c x_c(t) + B_c y(t) \tag{6}$$
$$u(t) = -C_c x_c(t) \tag{7}$$

where $x_c \in \mathbb{R}^{n_c}$ and $n_c \le n_x$, that minimizes the steady-state performance criterion

$$\begin{aligned} J((A_c, B_c, C_c)) &\triangleq \lim_{t\to\infty} E[x^{\mathrm{T}}(t) R_1 x(t) + u^{\mathrm{T}} R_2 u(t)], \\ R_1 &= E_1^{\mathrm{T}} E_1, \quad R_2 > 0 \end{aligned} \tag{8}$$

subject to a constraint on a modified Lyapunov equation which is a function of δ_i and A_i. If we define

$$V_1 = D_1 \bar{V}_1 D_1^{\mathrm{T}}, \quad \Sigma \triangleq B R_2^{-1} B^{\mathrm{T}}, \quad \bar{\Sigma} \triangleq C^{\mathrm{T}} V_2^{-1} C \tag{9}$$

then the controller gains are given by

$$A_c = \Gamma(A_s - \Sigma P - Q\bar{\Sigma}) G^{\mathrm{T}} \tag{10}$$
$$B_c = \Gamma Q C^{\mathrm{T}} V_2^{-1} \tag{11}$$
$$C_c = R_2^{-1} B^{\mathrm{T}} P G^{\mathrm{T}} \tag{12}$$

where $Q, P, \hat{Q}$ and $\hat{P}$ are nonnegative definite solutions of the design equations

$$\begin{aligned} 0 = A_s Q + Q A_s^{\mathrm{T}} + \bar{V}_1 - Q\bar{\Sigma}Q + \sum_{i=1}^{n_\alpha} \alpha_i^2 A_i Q A_i^{\mathrm{T}} \\ + \sum_{i=1}^{n_\alpha} \alpha_i^2 A_i \hat{Q} A_i^{\mathrm{T}} + \tau_\perp Q\bar{\Sigma}Q\tau_\perp^{\mathrm{T}} \end{aligned} \tag{13}$$

$$0 = A_s^{\mathrm{T}} P + P A_s^{\mathrm{T}} + R_1 - P\Sigma P + \sum_{i=1}^{n_\alpha} \alpha_i^2 A_i^{\mathrm{T}} P A_i + \sum_{i=1}^{n_\alpha} \alpha_i^2 A_i^{\mathrm{T}} \hat{P} A_i + \tau_\perp^{\mathrm{T}} P\Sigma P Q \tau_\perp \tag{14}$$

$$0 = (A_s - \Sigma P)\hat{Q} + \hat{Q}(A_s - \Sigma P) + Q\bar{\Sigma}Q - \tau_\perp Q\bar{\Sigma}Q\tau_\perp^{\mathrm{T}} \tag{15}$$

$$0 = (A_s - Q\bar{\Sigma})^{\mathrm{T}}\hat{P} + \hat{P}(A_s - Q\bar{\Sigma}) + P\Sigma P - \tau_\perp^{\mathrm{T}} P\Sigma P\tau_\perp \tag{16}$$

$$\text{rank } \hat{Q} = \text{rank } \hat{P} = \text{rank } \hat{Q}\hat{P} = n_c \tag{17}$$

$$\tau = \hat{Q}\hat{P}(\hat{Q}\hat{P})^{\#}, \quad (\cdot)^{\#} \text{denotes the group generalized inverse} \tag{18}$$

and τ (the optimal projection matrix) has the factorization

$$\tau = G^{\mathrm{T}}\Gamma, \quad G, \Gamma \in \mathbb{R}^{n_c \times n_x}. \tag{19}$$

The Maximum Entropy design equations for full-order control law design are simply (13)–(16) with $\tau_\perp = 0$. In this case $\Gamma = G = I_{n_x}$ in (10)–(12). If we additionally assume perfect knowledge of the system (i.e., $\alpha_i = 0$ for $i = 1, \ldots, n_\alpha$), then (13) and (14) become the standard LQG Riccati equations.

Maximum Entropy design is closely related to recent results on parameter-dependent Lyapunov functions and variations of the Popov criteria [17–19]. In [20] it is shown that the Maximum Entropy Covariance equation describes the average covariance for a Cauchy uncertainty probability distribution.

The design equations (13)-(18) can be solved by using a homotopy algorithm [13,14]. As illustrated in Figure 1, this homotopy algorithm allows the deformation of an LQG controller into a full order Maximum Entropy controller. The Maximum Entropy controller is then reduced to an appropriate order by using an indirect controller reduction method. It is important that this initial reduced-order controller approximately solves the Maximum Entropy/Optimal Projection design equations to within a small error, although it is not even required to be a stabilizing controller. This can be achieved by beginning with a low authority LQG design and/or incorporating a sufficiently high level of uncertainty in the Maximum Entropy design since both of these tactics tend to make the controller dynamics less complex which in turn makes reduction easier. In practice a slight modification of the balanced controller reduction algorithm of Youssuff and Skelton [16] is currently used as the indirect controller reduction method. Once this initial reduced-order controller is obtained, the algorithm is used to deform this controller into a Maximum Entropy/Optimal Projection controller. Then if a higher authority controller is desired, the final step of the algorithm is to increase the controller authority to a desirable level.

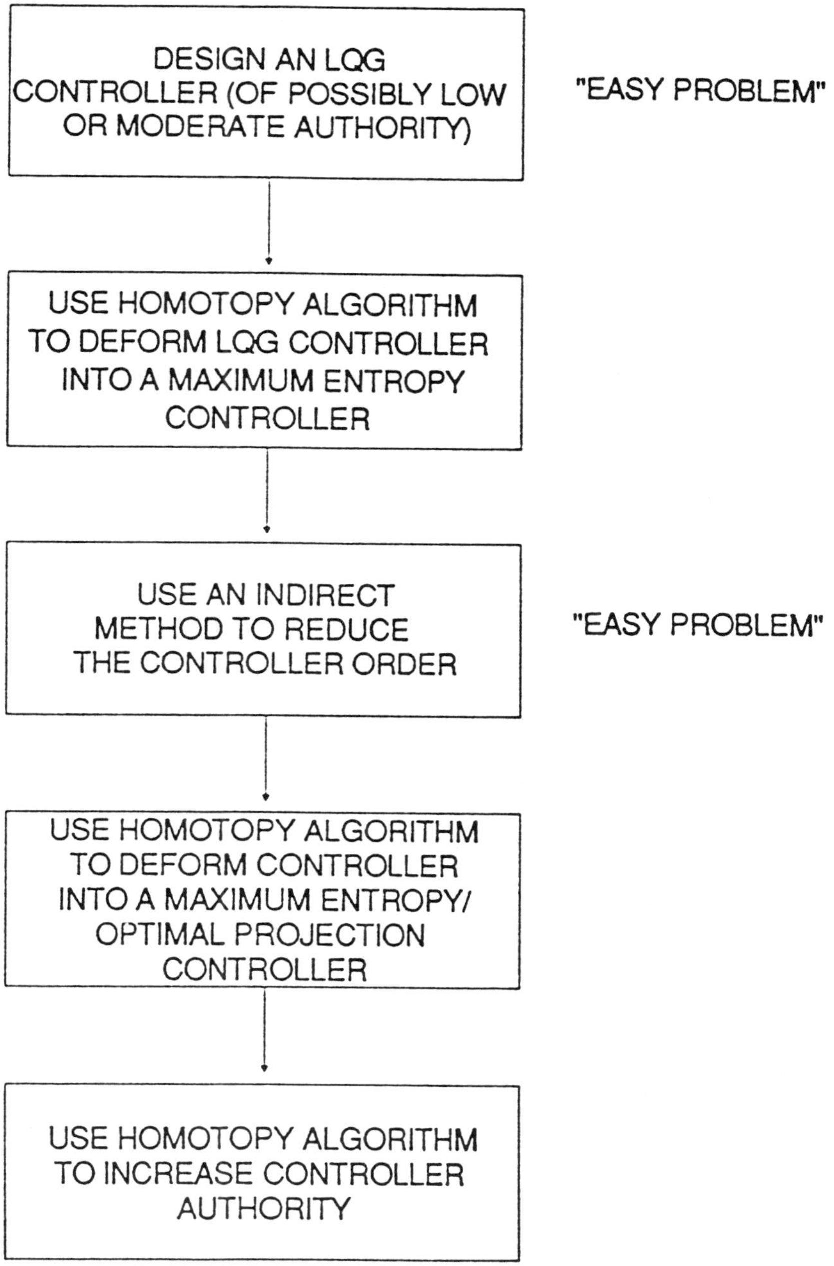

Figure 1. The practical use of the Maximum Entropy/Optimal Projection Control Design Algorithm.

In the Mini-MAST experiment it was found that the final two steps were actually unnecessary. That is, for each design the weights of the initial LQG controller could always be chosen such that the initial reduced-order controllers, which were obtained by balancing, had performance that could not be significantly improved by better satisfying the Maximum Entropy/ Optimal Projection equations and these initial reduced-order controllers were also of sufficiently high authority. This desirable feature of the design process was a nice by-product of the precompensation methodology used in the design of the Mini-MAST controllers. This precompensation will be discussed in the next section.

III. THE MINI-MAST TEST EXPERIMENT

A. DESCRIPTION OF THE MINI-MAST TESTBED AND MODELS

Here, we provide a brief description of the Mini-MAST experimental testbed, whose schematic is shown in Figure 2. This testbed is located at NASA Langley Research Center, Hampton, VA. A more detailed description is provided in [21].

The basic Mini-MAST test article is a generic space truss designed and manufactured by Astro Aerospace Corporation. The tubing members of the truss are made of graphite epoxy. The truss beam is deployable and retractable and has a triangular cross section. The total height of the truss is 20.16 meters and the truss consists of 18 bays, each of which is 1.12 meters in height.

The actuators and sensors available for control design implementation, disturbance generation, and performance evaluation are shown in Figure 2. The only actuators available for control are three torque wheel actuators that are mounted on the tip plate (top of Bay 18) parallel to the global x, y and z reference axes.

The torque wheels provide both torsional and bending torque loads to the Mini-MAST. These DC permanent-magnet motors have a rated peak output of 50 ft-lbs at 50 volts and 9.6 amps.

The available control sensors are six Sundstrad QA-1400 servo accelerometers and three Watson angular rate gyros. Four accelerometers are located at the beam tip (Bay 18) and two are located on the mid platform (Bay 10). These sensors measure linear acceleration in the global x and y directions. The three rate sensors are located at the beam tip (Bay 18) and measure pitch (about the x-axis), roll (about the y-axis), and yaw (about the z-axis).

Fifty-one Kaman KD-2300 proximity probes (i.e, displacement sensors) are installed on the support structure along the Mini-MAST. These devices can be used for control but were primarily intended for structural dynamic testing and performance evaluation. Our experimentation used the three Kaman sensors at Bay 18 for performance evaluation. These sensors are oriented normal to the faces of the truss at each of the three vertices.

Three Unholtz-Dickie 50-lb shakers are attached at Bay 9 for disturbance generation. These shakers have the same orientation as the Kaman sensors at Bay 18.

Two models of the Mini-MAST were provided by NASA Langley Research Center. The first model, Model 1, was used to generate the reduced order models that were used to design the decentralized controllers and had good correspondence to experimental data below 10 Hz. A second model, Model 2, that had even better correspondence to experimental data under 10 Hz was provided later in the program. This close correspondence is illustrated by Figure 3. Model 2 was used to generate the reduced order model that was used to design the centralized controllers. The final evaluation model for each of the control designs was the full-order representation of Model 2, discretized at 80 Hz, the sample frequency chosen for control law implementation. Delay states were included in the evaluation model to account for the one sample period computational delay.

Both of the models consisted of essentially three parts: the torque wheel dynamics, the finite element generated structural dynamics, and the 20 Hz analog Bessel filters used to filter the sensor outputs. The dynamics of the three torque wheels were quite similar to one another and significantly contributed to the plant dynamics over a wide bandwidth. For example, the dynamics of Torque Wheel X provided 75° of phase lead at 0.8 Hz, the frequency of the first bending modes, and provided a magnitude variation of over 50 dB from 1 Hz to 100 Hz. The 20 Hz Bessel filters were chosen instead of lower frequency Bessel filters to allow better control of the second bending modes which had frequencies of around 6 Hz. Only one shaker at Bay 9, Shaker-A, was used to disturb the structure. The three Kaman sensor measurements (Displacements-A, B and C) at Bay 18 were used for performance evaluation. The inputs used for control law implementation were the three torque wheels at Bay 18, Torque-X, Torque-Y and Torque-Z, while the outputs used were two Bay 18 accelerometers, Acceleration-1X and Acceleration-2Y, the Bay 10 accelerometers, Acceleration-X and Acceleration-Y, and the torsional rate gyro at Bay 18, Rate Gyro-Z. As opposed to Model 1, which contained 18 structural modes, the structural dynamics for Model 2 were composed of 28 modes. These modes are described in Table I. Most of the additional modes are a set between 15–20 Hz, involving the diagonal members and the tip plate.

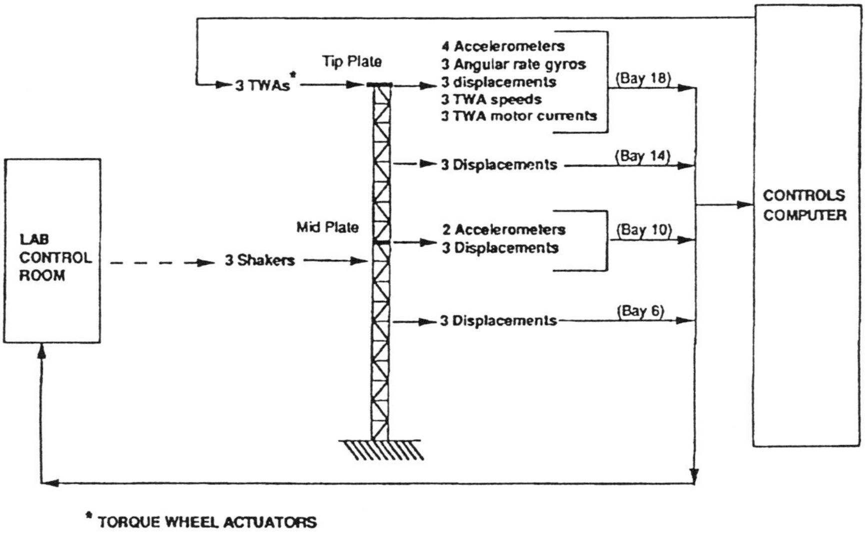

Figure 2. The Mini-MAST test article.

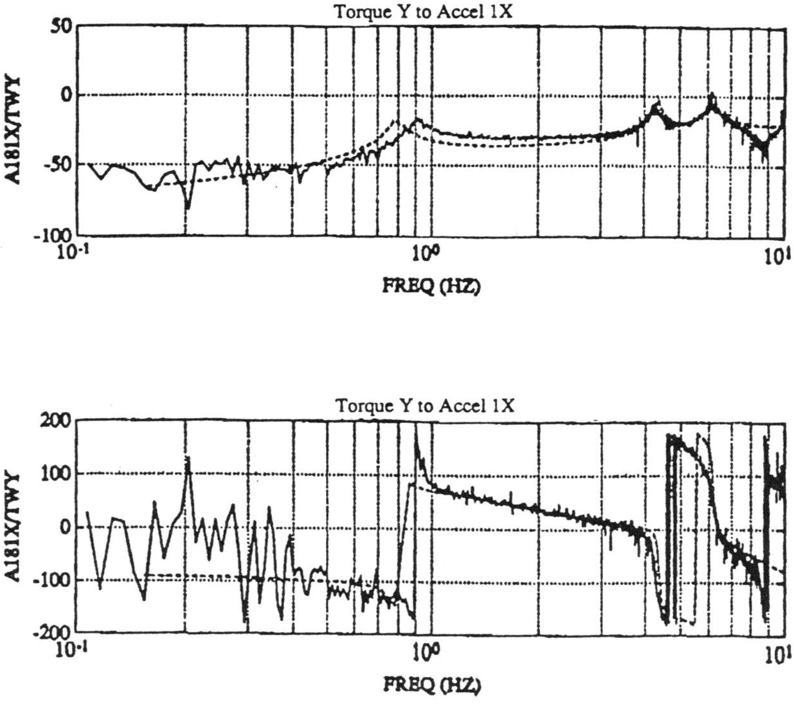

Figure 3. Torque-Y to Acceleration-1X: Model 2 bode plots vs. experimental frequency response function (Experimental = Solid, Model 2 = Dashed).

Table I. The structural modes of Model 2.

Mode Number	Predicted Frequency (Hz)	Predicted Damping	Description
1	0.83	.018	1st Y Bending
2	0.83	.018	1st X Bending
3	4.37	.012	1st Torsion
4	6.38	.010	2nd Y Bending
5	6.44	.010	2nd X Bending
6	14.72	.005	Tip Plate
7	14.83	.005	Diagonal
8	15.42	.005	1st Axial
9	15.57	.005	Tip Plate
10	15.60	.005	Tip Plate
33	17.17	.005	Diagonal
117	20.29	.005	Tip Plate
118	21.80	.005	Second Torsion
119	23.75	.005	Both Plates
120	25.83	.005	Both Plates
121	31.62	.005	3rd Y Bending
122	33.10	.005	3rd X Bending
123	39.14	.005	3rd Torsion
126	42.46	.005	4th Y Bending
127	45.95	.005	4th X Bending
128	54.14	.005	Tip Plate
129	57.25	.005	4th Torsion
132	60.85	.005	Mid Plate
133	62.68	.005	Tip Plate
134	66.04	.005	Tip Plate
135	72.18	.005	5th Y Bending
136	72.37	.005	5th X Bending
137	73.34	.005	5th Torsion

B. PROBLEM DEFINITION AND THE INTRODUCTION OF A PRE-COMPENSATION METHODOLOGY FOR CONTROL DESIGN

The Mini-Mast structure can be viewed as the secondary support tower of a precision optical structure. To achieve high accuracy line-of-sight pointing in such a structure, it is important to minimize the relative displacement of the tip of the beam with respect to the base. Hence, the primary objective of this experiment was to design controllers that provide substantial reduction of the displacement of the tip of the Mini-MAST structure.

It was assumed that the Mini-MAST would be subjected to a .1 sec pulse of 50 N amplitude from Shaker-A. It was evident from the magnitude Bode plots from Shaker-A to Displacements A, B and C of Bay 18 (see e.g. Figure 4) that if Shaker-A is used to excite the Mini-MAST with a pulse of infinitesimal duration, then the open-loop displacements at the tip are primarily due to the excitation of the first bending modes of the X and Y axes (at .8 Hz) with minor contributions from the first torsional mode (at 4.4 Hz) and the second bending modes of the X and Y axes (at 6.1 Hz and 6.2 Hz). Since the disturbance pulse was actually of finite duration, the contributions of the latter three modes were even further reduced.

From the above analysis it was apparent that the performance objectives required the control of no more than 5 modes with the first bending mode pair being the two most important modes to attenuate. The control design thus required the control of these 5 modes without destabilizing the remaining higher frequency modes, all of which are above 14 Hz. Recognize that the experimental data discussed in subsection A that was used to verify the accuracy of the two models was only valid below 10 Hz. Thus, we did not have confidence in our models above 10 Hz and wished to ensure stability of these modes by having the compensators roll off these modes to achieve gain stabilization.

An additional motivation for using a roll-off strategy for achieving the desired high frequency stabilization was that we used continuous-time models and design theory for the control designs. The controllers were then discretized for implementation at 80 Hz. This discretization tends to distort the higher frequency dynamics of the continuous-time compensators which can cause instability when the discretized controllers are implemented. The use of compensator roll-off for high frequency stabilization was used to reduce the importance of high frequency compensator dynamics.

It is important to note that, for this problem, the desired roll-off could not be achieved by a straight-forward application of LQG design theory. This fact is illustrated by Figure 5 which shows an LQG compensator designed using a reduced-order model of the Torque-X to Accel-2Y transfer function, whose full-order magnitude Bode plot is shown in Figure 9. The reduced model included only one mode above 10 Hz, a 20.3 Hz mode. The LQG compensator was designed to attenuate the first bending mode at .8 Hz. The compensator shown in Figure 5 has sufficient gain at the first bending mode to achieve significant performance improvement but gain stabilizes the 20 Hz plant mode by notching it and has enough gain above 20 Hz to potentially destabilize the unmodeled modes. This type of compensator was clearly undesirable.

The desired roll-off could have been ensured by assuming norm-bounded uncertainty and using either H_∞ [22-27] or μ-synthesis [28,29] control design. Alternatively, one could have used frequency weighted LQG [27,30,31] to attempt to achieve the desired roll-off. In this experiment, a "precompensation" strategy was used, which, like frequency weighted LQG, is heuristically based. However, the precompensation methodology does have some attractive features. In particular, it avoids the great increase in dimensionality of the design model that is sometimes resultant from using H_∞, μ-synthesis, or frequency weighted LQG methods. Also, this methodology is very easy to implement.

The basic idea behind the precompensation methodology is very simple and is illustrated in Figures 6 and 7. As shown in Figure 6, one simply embeds the precompensation filters (in this case roll-off filters) $C_u(s)$ and $C_{y'}(s)$ in the plant a priori and designs the Maximum Entropy/Optimal Projection controller $\hat{H}(s)$ for this modified design plant. Then, as illustrated in Figure 7, the precompensation dynamics are included in the implemented compensator $H(s)$. It is not difficult to show that the closed loop transfer function $G_{c\ell}(s)$ satisfying

$$\begin{bmatrix} z(s) \\ y'(s) \end{bmatrix} = \begin{bmatrix} G_{c\ell}(s) \end{bmatrix} \begin{bmatrix} w_1(s) \\ u(s) \end{bmatrix} \tag{20}$$

is identical in Figures 6 and 7. Hence, this methodology ensures that if $\hat{H}(s)$ is stabilizing in the feedback loop of Figure 6, then $H(s)$ is stabilizing in the feedback loop of Figure 7. In addition, (20) also guarantees that the transfer function between $z(s)$ and $w_1(s)$ is preserved, thus ensuring the preservation of attenuation from w_1 to z. However, it is necessary to check after the design process as to whether the compensator achieved the desired roll-off (or robustness to high frequency unstructured uncertainty). The actual roll-off filters used are described in the next section.

C. CONTROL DESIGN STRATEGY AND MODELS

A very practical control design and implementation strategy was used in this experiment. First, controllers with simple architectures (i.e., controllers

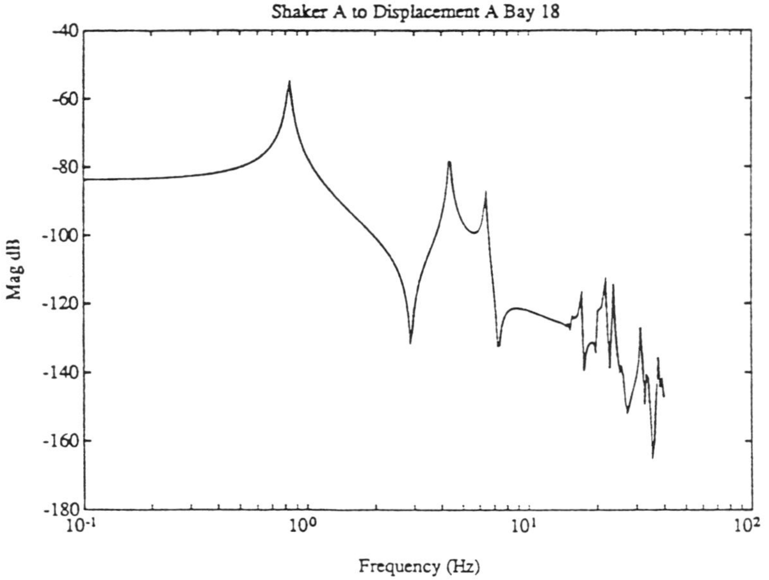

Figure 4. Bode magnitude plot of Shaker-A to Displacement-A of Bay 18 for Model 2.

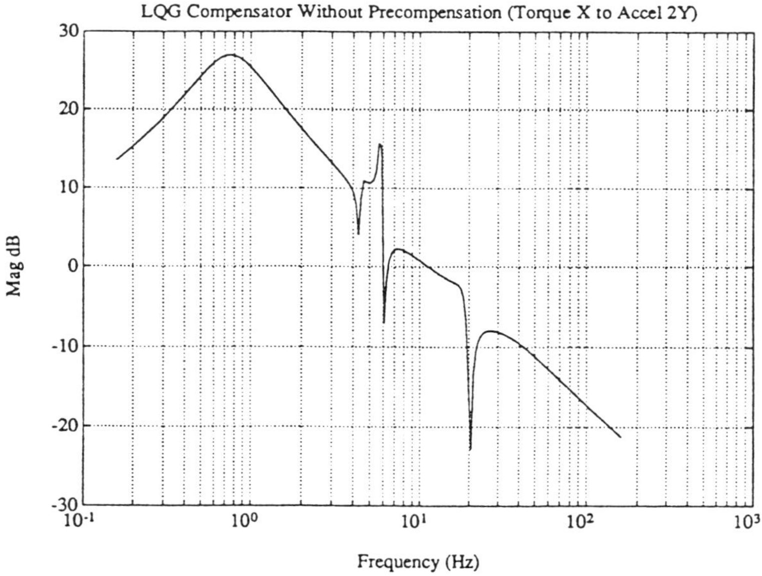

Figure 5. Frequency response magnitude of an LQG Compensator Designed without using precompensation.

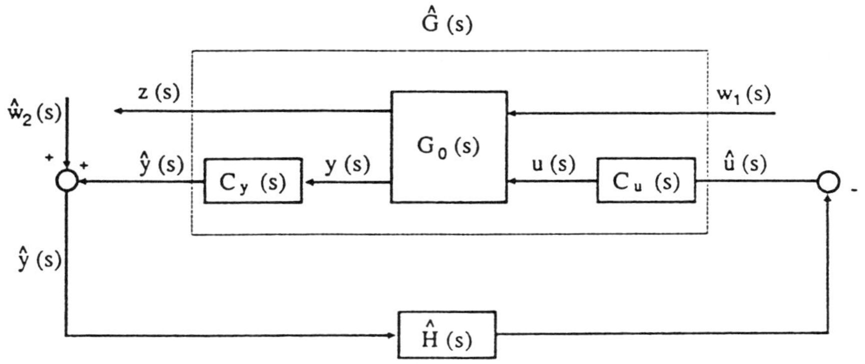

Figure 6. Design configuration for the Precompensation Methodology.

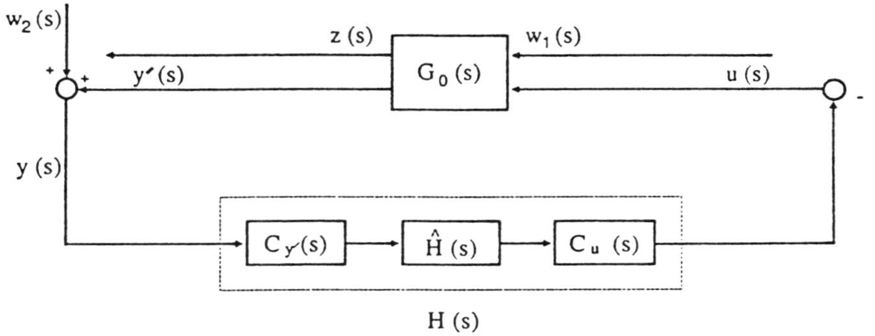

Figure 7. Implementation configuration for the Precompensation Methodology.

which are decentralized, reduced order and use few sensors) were developed. Decentralized designs are more desirable for actual flight implementation because they ease the processor requirements and are more fault tolerant, but can be more difficult to design because decentralization requires a careful analysis of the physics of the structure. Subsequently, centralized control laws were developed in an attempt to improve the performance. The centralized design with the best performance did significantly improve the performance of the "best" decentralized design.

The subcontrollers of the decentralized design were each single-input, single-output (SISO). Although the advantages usually stated for modern control usually center around multi-input, multi-ouput (MIMO) control problems, there are also advantages for SISO systems. In particular, whereas classical control design relies heavily on the designer's intuition and experience to design the feedback filter and is very iterative, modern control helps to automate the design of the feedback filter, generating a filter which for many problems (including the problem considered here) is in a meaningful sense optimal. In addition, modern control methods do generalize naturally to MIMO control problems. Hence, in a strategy which employs first decentralized control (involving possibly SISO control design) and then centralized (MIMO) control it is natural to make primary use of one design tool.

From Figure 8 the transfer function from Torque-Z to Rate Gyro-Z was dominated by the first torsional mode at 4.4 Hz. Hence, it appeared feasible to use simple decentralized constant gain feedback from Rate Gyro-Z to Torque-Z to achieve high attenuation of the torsional mode. A full-order discrete-time model of the system was developed using Model 1 to represent the system at the 80 Hz sampling frequency. This model was used to perform a root locus and it was determined from this root locus that the "optimal" gain was K=10. *All* of the implemented controllers were designed assuming that this feedback loop was closed. These designs added feedback loops that used only the accelerometers and the X and Y torques. The use of the Rate Gyro-Z to Torque-Z feedback essentially eliminated the influence of the torsional mode on the remaining loops as illustrated by the dotted line in Figure 9. The resultant performance improvement in the torsional motion is shown in Figure 10.

Next, we considered attenuating the first bending modes in each axis by feeding back accelerometer outputs to the X and Y torque wheels. Since Acceleration-1X and Acceleration-2X provided essentially the same information about the acceleration of the Mini-MAST tip in the X-direction, we decided to use only one of these accelerometers and chose Acceleration-1X. Likewise we chose to use Acceleration-2Y instead of both Acceleration-2Y and Acceleration-1Y. The remaining accelerometers, Acceleration-X and

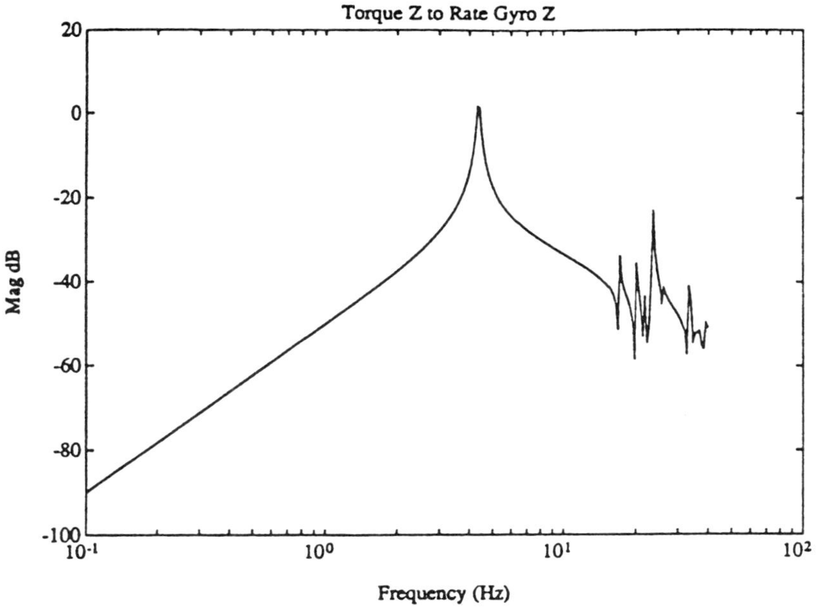

Figure 8. Bode plots of Torque-Z to Rate Gyro-Z for Model 2.

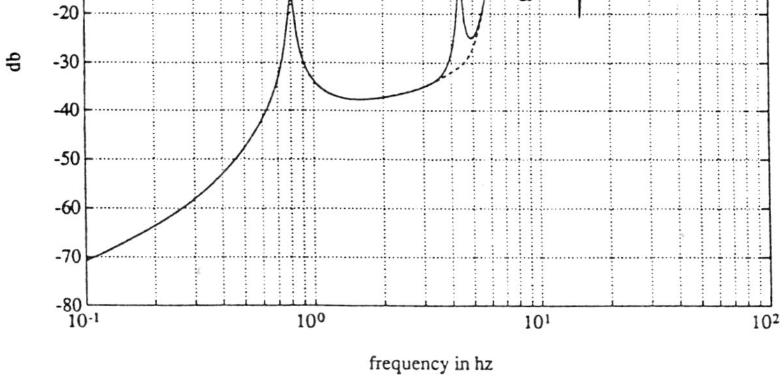

Figure 9. The effectiveness of connstant gain rate feedback in eliminating the influence of the Torsional Mode (without feedback = solid, with feedback = dashed).

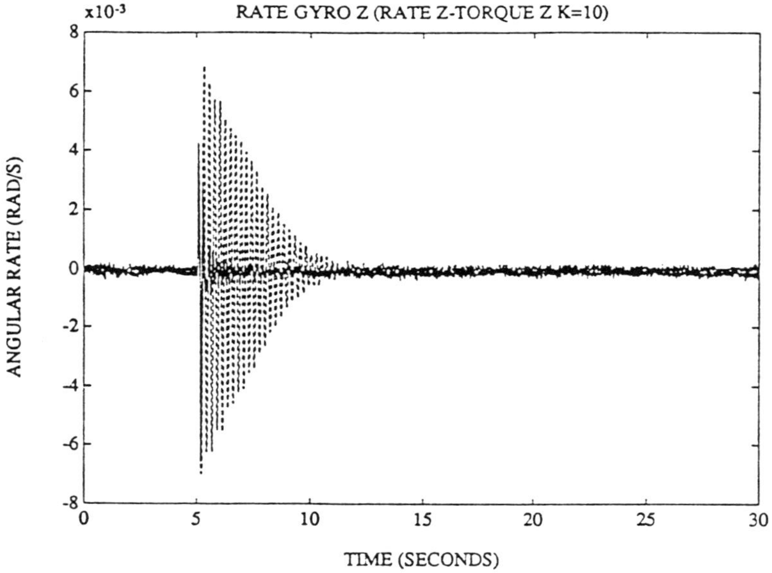

Figure 10. Open loop vs. closed loop rate Gyro-Z response for constant gain feedback from rate Gyro-Z to Torque-Z.

Acceleration-Y, were located near the center of the Mini-MAST at Bay 10.

It was apparent from the frequency responses of the models that the first bending modes were not very prominent in the four transfer functions from Torque-X and Torque-Y to Acceleration-X and Acceleration-Y. However, the first bending modes were prominent in the four transfer functions from Torque-X and Torque-Y to Acceleration-1X and Acceleration-2Y. Hence, we initially chose to feed back only the latter two accelerometers to attenuate the first bending mode. Furthermore, it was evident that the dominant transfer functions were from Torque-X to Acceleration-2Y and Torque-Y to Acceleration-1X. We thus chose to design feedback laws for the two loops corresponding to these transfer functions.

Reduced-order continuous-time models were developed for the transfer functions Torque-X to Acceleration-2Y and Torque-Y to Acceleration-1X. First order all-pass filters were included in each model to account for the phase lag due to computational delay. Precompensation dynamics were added to each reduced-order model to yield the models actually used to design the Maximum Entropy/Optimal Projection controllers. The Torque-X to Acceleration-2Y design model and the Torque-Y to Acceleration-1X design model each contained 20 states.

The precompensation for each of the two control problems was chosen so that the modified design plant appeared to have a rate measurement and a true torque input within the control bandwidth. The precompensation thus consisted of dynamics to integrate the accelerometer output and dynamics to cancel the effects of the actuator dynamics below 10 Hz. This choice of precompensation added the desired roll-off to the original design plant and also allowed us to develop a baseline rate-feedback design. Because the transfer functions for each of the two control problems were so similar, we decided to use the same precompensation for both. Referring to Figure 6, the precompensation used is given by

$$C_u(s) = \frac{s + 62.8}{s^2 + .628s + .3944}, \quad C_y(s) = 1. \tag{21}$$

Due to the requirements of Maximum Entropy Design, each of the modified design models was placed in real normal form. Thus the A matrix of each of the modified design models was block-diagonal with one of the blocks of the form $\begin{bmatrix} -\nu_1 & \omega_1 \\ -\omega_1 & -\nu_1 \end{bmatrix}$ corresponding to the first bending mode. The disturbance weighting matrix, V_1, and the performance state weighting matrix, R_1, were each chosen to be block diagonal with the only nonzero block corresponding to the position of the first bending mode in the A matrix. The nonzero block of V_1 is given by $\begin{bmatrix} 0 & 0 \\ 0 & \alpha \end{bmatrix}$ and the nonzero block of R_1 is given by $\begin{bmatrix} \alpha & 0 \\ 0 & 0 \end{bmatrix}$ where α is a scalar parameter which was used to determine the controller authority. This choice of weighting matrices was equivalent to assuming that the system was disturbed by a force actuator and the performance output was a displacement sensor that sees only the first bending mode when an impulse is placed in the system by the force actuator. Also, we chose $R_2 = V_2 = 1$.

The centralized designs used a reduced-order, continuous-time model which contained 12 structural modes. This model also included first-order all-pass filters to account for the phase lag due to computational delay. Precompensation dynamics were added to the reduced order model which yielded a design model with 54 states.

It should be noted that each of the models used for (centralized or decentralized) control design included modes with frequencies above the first five modes in the bandwidth of interest. These modes were included to increase the accuracy of the first five modes in the reduced-order design model. That is, if none of the higher frequency modes were included, the phases and magnitude of the first five modes were significantly distorted.

The precompensation dynamics that we used for the centralized controllers differed from that used for the decentralized controllers. The reason for this change was primarily an attempt to eliminate some of the low frequency oscillation that appeared in the output of the displacement sensors and ultimately limited the achievable performance of the decentralized designs. We initially attributed this low frequency oscillation to the migration of the low frequency poles of the precompensation dynamics towards the right half plane, although subsequent control designs and analyses revealed that this was not actually the case. Referring again to Figure 6, the precompensation used is given by

$$C_u(s) = \text{diag}\{C_{u,1}(s), C_{u,2}(s)\}, \quad C_y(s) = I \tag{22}$$

where

$$C_{u,1}(s) = C_{u,2}(s) = \frac{39.0625 \times 10^4}{(s^2 + 18.8755 + 62.5)(s^2 + 45.969 + 625)}. \tag{23}$$

$C_{u,1}(s)$ and $C_{u,2}(s)$ are 4 Hz Butterworth filters and hence provide roll-off between the first and second bending modes.

The weighting scheme used for the centralized controllers is very straight-forward. Let the state space representation corresponding to the modified design plant $\hat{G}(s)$ in Figure 3 be given by

$$\dot{x}(t) = Ax(t) + B\hat{u}(t) + D_1 w_1(t) \tag{24}$$

$$\hat{y}(t) = Cx(t) + \hat{w}_2(t) \tag{25}$$

$$z(t) = E_1 x(t) \tag{26}$$

where the basis of (24)–(26) is chosen such that A is in real normal form. Here, it is assumed that $w_1 \in \mathbb{R}$ denotes the input corresponding to Shaker-A and

$$z = \begin{bmatrix} z_a \\ z_b \end{bmatrix}, \quad E_1 = \begin{bmatrix} E_{1,a} \\ E_{1,b} \end{bmatrix} \tag{27}$$

where $z_A \in \mathbb{R}^3$ denotes a performance vector whose elements correspond to Displacements A, B and C of Bay 18, $z_b \in \mathbb{R}^3$ denotes the performance vector whose elements correspond to Displacements A, B and C of Bay 10. The weights R_1, R_2, V_1 and V_2 in the control problem of section II were chosen as

$$R_1 = \alpha E_{1,a}^{\mathrm{T}} E_{1,a} + \beta E_{1,b}^{\mathrm{T}} E_{1,b}, \quad V_1 = \alpha D_1 D_1^{\mathrm{T}}, \quad R_2 = V_2 = 1 \tag{28}$$

where the scalar parameter α was used to determine the controller authority. The scalar parameter β was chosen to be zero for the initial designs but was given a finite value in latter designs in order to penalize the Bay 10 displacements. The use of the parameter β was motivated by the fact that the dominant behavior at Bay 10 was due to the second bending mode pair. Thus, β was essentially used to reduce the influence of the second bending modes on the displacement responses at Bay 18.

D. DECENTRALIZED CONTROL DESIGNS

Now, we describe in more detail the decentralized controllers discussed in the previous subsection and present the resultant performance improvement. Four controllers are described, each of which was designed using the precompensation methodology of subsection III.B. Each controller was also designed assuming that a decentralized constant gain feedback loop from Torque-Z to Rate Gyro-Z (with gain K=10) was closed. This feedback loop essentially eliminated the effects of the first torsional mode. Hence, although the controllers described used only Accelerometers-1X, 2Y, X and Y for sensing and Torques X and Y for actuation, it is important to keep in mind that they were implemented in conjunction with the constant gain feedback law from Torque-Z to Rate Gyro-Z.

Each of the four decentralized controllers included at least two subcontrollers; one fed back Accelerometer-2Y to Torque-X while the other fed back Accelerometer-1X to Torque-Y. These two subcontrollers were designed using the precompensation methodology of subsection III.B with precompensation dynamics given by (21). Controller 1 consisted of two subcontrollers. Referring to Figure 6, these subcontrollers were designed by simply choosing $\hat{H}(s)$ equal to a constant. Since the precompensation was chosen to make the design plants $\hat{G}(s) \triangleq C_{y'}(s)G(s)C_u(s)$ for each subcontroller appear to have a rate measurement and a true torque input within the control bandwidth, Controller 1 approximated a rate feedback controller and was used as a baseline design. Controller 2 also included only two subcontrollers and was designed without assuming uncertainty in any of the first five modes. Controller 3 was similar to Controller 2 except that it was designed assuming uncertainty in the second bending mode pair. Controller 4 consisted of Controller 3 plus two additional subcontrollers designed using classical control concepts. One of these subcontrollers fed back Accelerometer-Y to Torque-X while the other subcontroller fed back Accelerometer-X to Torque-Y. The subcontrollers were discretized using the bilinear transformation with frequency prewarping to match the continuous-time and discrete-time representations of the subcontroller at the frequency of the second modes (about 6 Hz). At implementation the gain of each of the five decentralized controllers was increased as much as possible to maximize the performance.

Controller 1: A Pseudo Rate Feedback Controller

This controller consisted of two subcontrollers, one from Accelerometer-2Y to Torque-X and one from Accelerometer-1X to Torque-Y. This controller was our least complex and, as previously mentioned, was used as a baseline design. Although this controller did provide substantial increase in the performance, it will subsequently be seen that more complex controllers achieved much better performance.

Controller 2: A Reduced Order LQG Controller Using Bay 18 Accelerometers

Like Controller 1, this controller consisted of two subcontrollers, one from Accelerometer-2Y to Torque-X and one from Accelerometer-1X to Torque-Y. However, the feedback laws in this case were designed using LQG. The decentralized feedback laws for each loop were each of order 9 (including the precompensation) so that the implemented controller was of order 18.

Controller 3: A Reduced Order Maximum Entropy Controller Using Bay 18 Accelerometers

This controller was a robustified version of Controller 2 and was of order 20. It was designed assuming uncertainty in the second bending mode pair and, as illustrated by Figure 11, the Maximum Entropy design robustified the corresponding notches for these two modes. This uncertainty description was originally motivated by an analysis of Controller 2 which indicated that the controllers were sensitive to uncertainties in the second bending modes. However, due to the high fidelity of the models provided by

NASA, subsequent analysis revealed that robustness was not really needed. Nevertheless, the Maximum Entropy design yielded useful gain margin at the frequency of the second bending modes and allowed us to scale the subcontrollers to further attenuate the influence of the second bending modes on the performance. Also, note that the small 20 Hz notch appearing in Figure 11 was *not* needed to gain stabilize the corresponding 20 Hz plant mode. This notch appeared solely as a consequence of optimality and contrasts with the large 20 Hz notch of Figure 5 which was needed to gain stabilize the corresponding plant mode.

Controller 4: Controller 3 plus a Classical Controller Using Bay 10 Accelerometers

This controller consisted of Controller 3 plus two subcontrollers, $H_1(z)$ and $H_2(z)$ which were designed using classical control concepts. $H_1(z)$ was used in the feedback path from Accelerometer-Y to Torque-X while $H_2(z)$ was used in the feedback path from Accelerometer-X to Torque-Y. These two subcontrollers were designed to further attenuate the second bending modes and are given respectively by

$$H_1(z) = K_1 z/(z - .9245)$$
$$H_2(z) = K_2 z/z(z - .9245)^2$$

where K_1 and K_2 are constant gains. The implemented compensator was of order 24. It should be noted here that the only reason that all other controllers were designed in the s-domain was the greater maturity of the theory and design tools available for continuous-time robust control. Classical control was used here because of the simplicity of the design problem.

Figure 12 shows the performance for each of the controllers as evidenced from Displacement-A of Bay 18. The results were very close to the predicted results. As desired, each successive controller improved on the performance of the prior controllers.

E. CENTRALIZED CONTROLLERS

Next, we describe three of the centralized controllers discussed in subsection C and present the resultant performance improvement for each controller. As was the case for the decentralized controllers, Controllers 1 thru 4, each of these centralized controllers was designed assuming that a decentralized constant gain feedback loop from Torque-Z to Rate Gyro-Z (with gain K=10) was closed, which effectively eliminated the effects of the first torsional mode. These controllers were designed after the decentralized controller experiments had been performed. It was during those experiments that we discovered we could rely more on the provided models (below 10

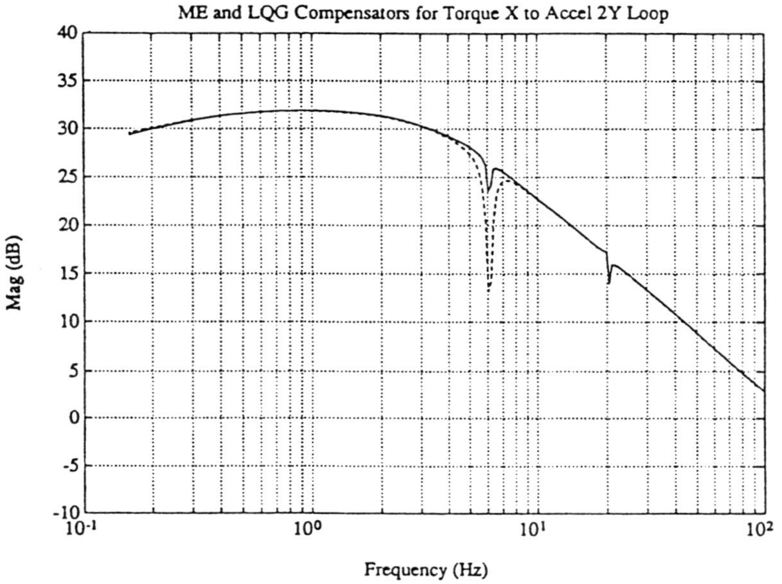

Figure 11. Comparison of the magnitude bode plots for the Acceleration-2Y to Torque-X $\hat{H}(s)$ subcontrollers of controllers 2 and 3 (controller 2 = solid, controller 3 = dashed).

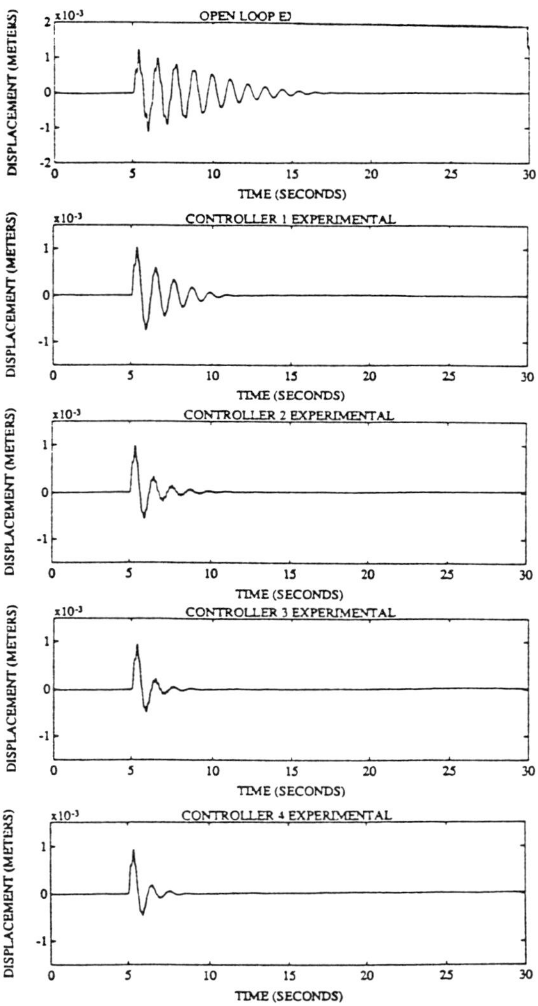

Figure 12. Open loop vs. closed loop responses of Displacement A of Bay 18 for controllers 1 thru 4.

Hz) and that robustness in the control bandwidth was not a significant issue.

Each of the five controllers was designed using the precompensation methodology of subsection B with precompensation dynamics given by (22) and (23). The design weights are given by (28) with α and β being the design parameters. The first centralized controller, Controllers 5, fed back Bay 18 Accelerometers-2Y and 1X to Torques-X and Y. Controllers 6 and 7 also fed back Bay 10 Accelerometers-Y and X to Torques-X and Y. The difference between these two controllers was due to the amount of performance penalty that was placed on the displacements at Bay 10. None of these latter three controllers was designed assuming uncertainty in the first five modes. The controllers were discretized using the bilinear transformation with frequency prewarping to match the continuous-time and discrete-time representations of the controller at the frequency of the second bending modes (about 6 Hz). As with the decentralized controllers, at implementation the gain of each of these controllers was increased as much as possible to maximize the performance.

It is of interest to note that although the centralized controllers were designed assuming 20 Hz Bessel filters to process the sensor outputs, each of the controllers was originally inadvertently implemented with 10 Hz Bessel filters which provided substantial phase delay in the controller bandwidth. Nevertheless, each controller was able to stabilize the system and obtain substantial performance improvement after the controller gain was reduced by a factor of 2. In addition, one of the controllers, Controller 6, was stabilizing without reducing its gain.

Controller 5: A Reduced-Order LQG Controller Using Bay 18 Accelerometers

This controller was the least complex centralized controller that was implemented and was designed to feed back Accelerometers-2Y and 1X to Torques-X and Y. The displacements at Bay 10 were not penalized in the design process (i.e., $\beta = 0$ in (28)). No uncertainty was assumed in the first five modes. The controller is of order 26. The performance of this compensator, which uses two sensors, was comparable to the performance of Controller 4, a decentralized controller which used four sensors, and had better performance than the decentralized controllers using only the two sensors employed by Controller 5.

Controller 6: A Reduced-Order LQG Controller Using Bay 10 and Bay 18 Accelerometers

This controller was designed to feed back Accelerometers-2Y, 1X, X and Y to Torques-X and Y. The displacements at Bay 10 were not penalized in the design process (i.e., $\beta = 0$ in (28)) and no uncertainty was assumed in the first five modes. The controller was of order 25. This centralized controller yielded significantly better performance than the highest performance obtained by a decentralized controller.

Controller 7: A Final Reduced-Order LQG Controller Using Bay 10 and Bay 18 Accelerometers

This controller used the same four sensors and two actuators as Controller 6. The design process was essentially identical to that of Controller 6 except that α and β in (28) were chosen so that $\beta/\alpha = 5$. The controller was of order 33. This controller yielded the best performance of any of the decentralized or centralized controllers that were implemented.

The increasing performance improvement of Controller 4 (the highest performance decentralized controller) and Controllers 5 thru 7 is shown in Figure 13 by viewing the response of Displacement-A of Bay 18. These responses are very close to the predicted performance.

F. DISCUSSION

This experiment successfully demonstrated high performance control law design and implementation for the Mini-MAST testbed, a flexible structure which has features representative of future spacecraft. The controllers were designed using the LQG and full-order Maximum Entropy design, and balanced controller reduction in conjunction with a precompensation methodology.

The precompensation methodology used in this experiment allows the designer to use classical insights to precompensate the original plant model at the inputs and outputs to develop a "nice" design plant for modern control design. The precompensation is appended to the modern controller to obtain the controller that is to be implemented. In this experiment the precompensation was chosen to provide high frequency roll-off (i.e., robustness with respect to unstructured uncertainty) and was also used to make the design plant appear to have rate sensors and torque inputs in the control bandwidth. This precompensation methodology allows the designer to obtain controller configurations that might be difficult or impossible to obtain using straight-forward applications of modern control theory. This strategy appears to be a useful means of integrating classical control concepts with modern control. In fact, it has already proved useful in designing controllers for an important benchmark problem [32,33].

The control design and implementation approach used was to start with simple controllers (i.e., reduced-order controllers, decentralized controllers and/or controllers using relatively few sensors and actuators) and increase controller complexity to increase performance. This strategy was effective.

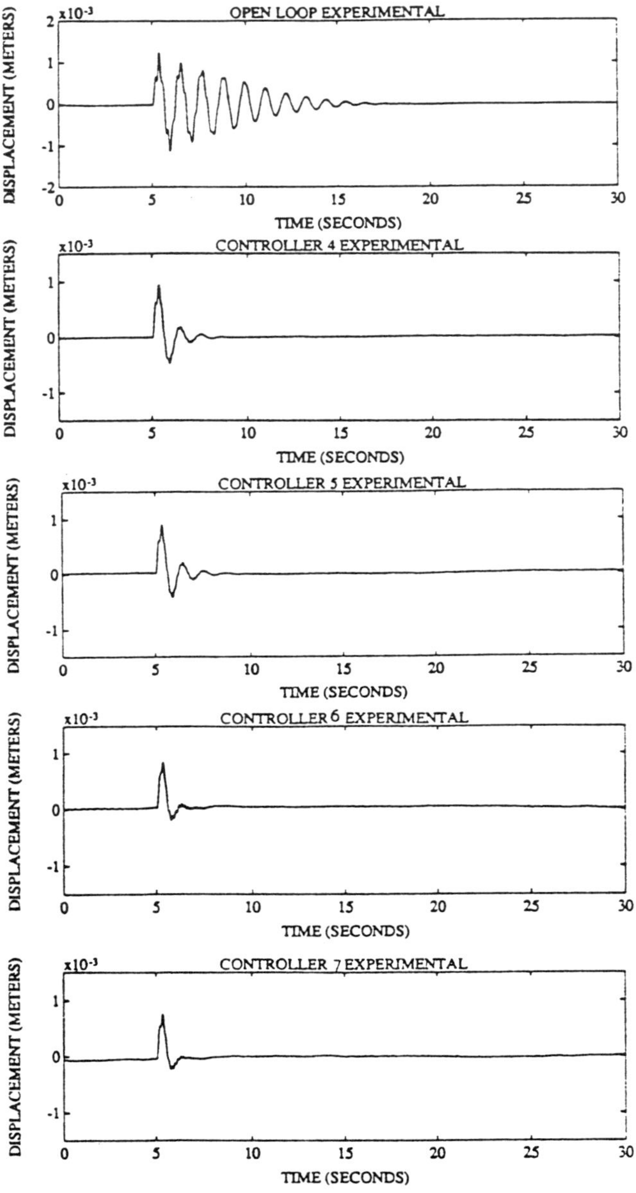

Figure 13. Open loop vs. closed loop responses of Displacement A of Bay 18 for controllers 4 thru 7.

The first four controllers each had a decentralized structure. The first controller was our simplest and did achieve some performance improvement. As the complexity of the controllers was increased by first adding order and then using additional sensors and controller order, the performance progressively improved. The performance was further improved by allowing the controllers to have a centralized structure.

IV. THE ACES EXPERIMENT

A. DESCRIPTION OF THE ACES STRUCTURE

The ACES experimental testbed is located at NASA MSFC. The basic test article, a spare Voyager Astromast, is a deployable, lightweight (about 5 pounds), lightly damped beam, which is approximately 45 feet in length. The ACES configuration, shown in Figure 14, consists of an antenna and counterweight legs appended to the Astromast tip and the pointing gimbal arms at the Astromast base. Overall, the structure is very flexible and lightly damped. It contains many closely spaced, low frequency modes (more than 40 modes under 10 Hz). A detailed description of the ACES testbed is presented in [34].

The goal of the control design was to position the laser beam in the center of the detector. The sample rate for digital implementation was fixed at 50 Hz. The actuators and sensors available for controller implementation are listed in Figure 14. In our control design and implementation we used 8 control inputs and 8 measurement outputs. The inputs were the X and Y torques of the Image Motion Compensation gimbals, the X and Y torques of the Advanced Gimbal System, and the X and Y forces of the two Linear Momentum Exchange Device packages. The measurements consisted of the X and Y Detector position outputs, the X and Y Base Gyro rate outputs and the X and Y outputs of the Linear Momentum Exchange Device accelerometers. The disturbances were chosen to be position commands to the Base Excitation Table. The table motion is regulated by an analog controller which allows any type of table movement within the frequency limitations of the hydraulic actuation system.

B. MODELING AND CONTROL STRATEGIES

A finite element model was supplied by MSFC. We generated frequency response functions (FRF's) using test data from the structure and used these FRF's to evaluate the finite element model.

In general, the finite element approach predicted significantly different frequency responses than those obtained by actually testing the structure. An illustration of this is given in Figure 15 which shows that for the Advanced Gimbal System-Y to Base Gyro-Y loop the finite element model

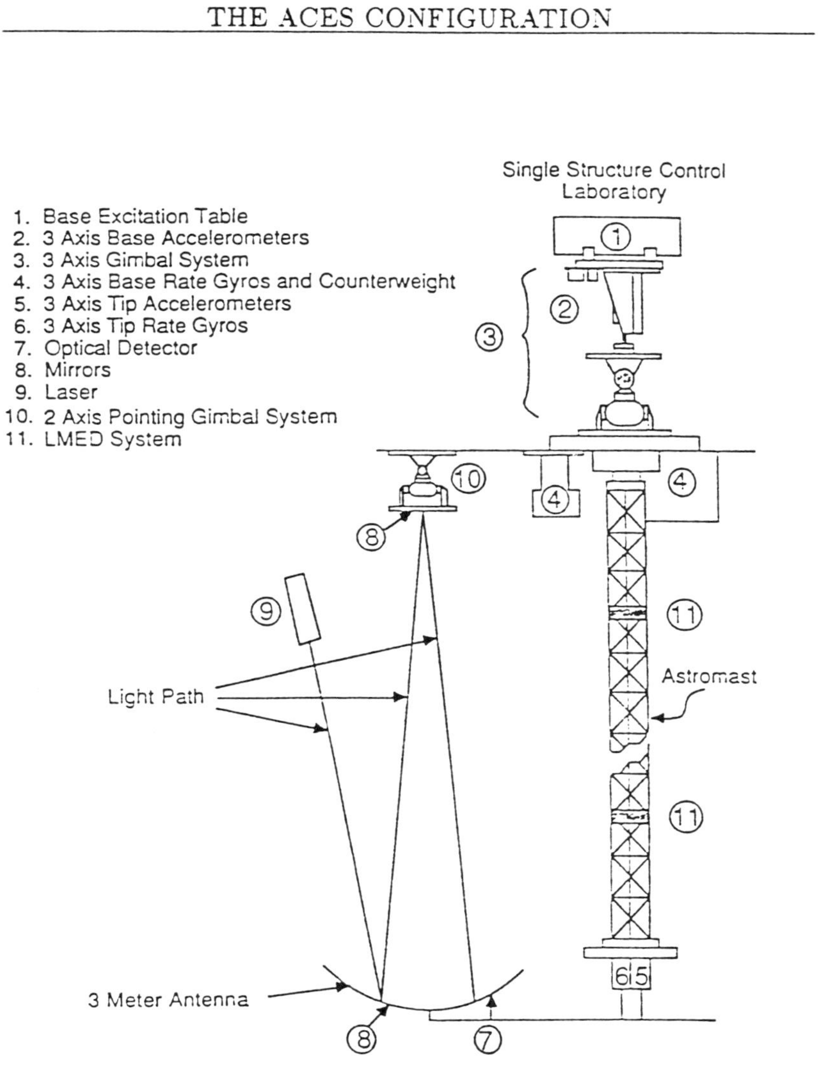

Figure 14. The ACES test structure.

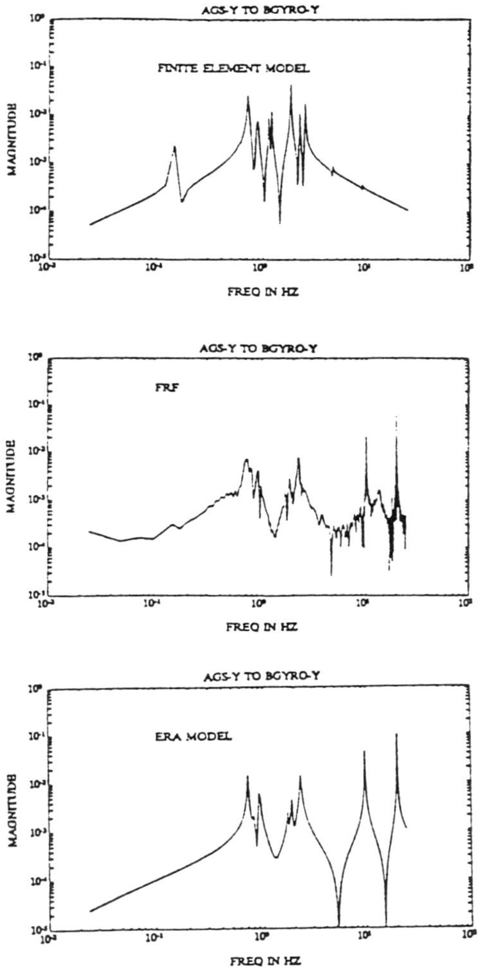

Figure 15. A comparison of the Finite Element Model (top), Frequency Response Function (center) and Eigensystem Realization Algorithm Model (bottom) for the Advanced Gimbal System-Y to Base Gyro-Y transfer function.

Bode plot does not contain the 1.7 Hz mode which most influenced the open loop line-of-sight performance.

The finite element model also does not show the large magnitude peaks corresponding to some of the modes past 8 Hz. In addition, it can be shown that the finite element model predicts positive real behavior for the Advanced Gimbal System-Y to Base Gyro-Y loop, while the FRF reveals that between 2 Hz and 4 Hz the phase lags $-90°$ by as much as $25°$ (even when the computational delay due to the 50 Hz sample rate is not taken into account). This phase lag is probably due to actuator and sensor dynamics. Similar magnitude and phase results also hold for the Advanced Gimbal System-X to Base Gyro-X loop.

The FRF's of the Image Motion Compensation-Y to Detector-X and Image Motion Compensation-X to Detector-Y loops revealed that those loops are influenced very little by the flexible modes of the structure. It follows that the Image Motion Compensation gimbals are not capable of controlling flexible modes to improve line-of-sight performance. Thus, if one considers the four actuator inputs (Image Motion Compensation-X and Y, Advanced Gimbal System-X and Y) and the four sensor outputs a (Detector-X and Y, Base Gyro-X and Y), it is not necessary to feed back the Base Gyro outputs to the Image Motion Compensation gimbals since the Base Gyro outputs primarily contain information about the behavior of the flexible modes which the Image Motion Compensation gimbals cannot control. In addition, the Detector outputs do not contain much (if any) useful information for the Advanced Gimbal System gimbals that is not already provided by the Base Gyro outputs. Thus, it is expected that the achievable performance cannot be significantly improved by feeding back the Detector output to Advanced Gimbal System gimbals or the Base Gyro outputs to the Image Motion Compensation gimbals. As illustrated by Figure 16, analysis of test data also revealed that within the decentralized structure described above there were four dominant loops: Advanced Gimbal System-X to Base Gyro-X, Advanced Gimbal System-Y to Base Gyro-Y, Image Motion Compensation-X to Detector-Y, and Image Motion Compensation-Y to Detector-X.

Due to time and budgetary constraints we decided to develop control design models by using the Eigensystem Realization Algorithm (ERA) [35–36] to develop state space models of the four dominant loops instead of modifying the finite element model. It is important to note that since all test data was collected by using the control sensors and actuators, models generated from this data should account for the sensor and actuator dynamics.

Due to a change during the project in the internal configuration of the Linear Momentum Exchange Devices we delayed control design activities

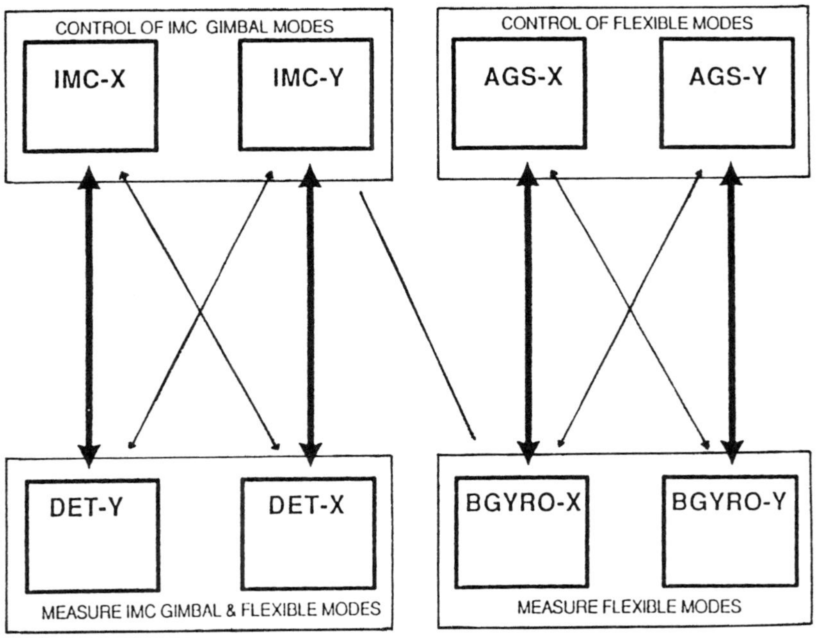

Figure 16. The pattern of strong and weak interactions in the system dynamics of the ACES Structure.

involving these devices. Even when the internal configuration was finalized it was difficult to obtain good fidelity ERA models due to device nonlinearities (such as stiction and saturation) and the resultant poor quality test data. Ultimately, it was decided to simply feed back the colocated Linear Momentum Exchange Device accelerometers to the corresponding Linear Momentum Exchange Device force axis and to determine simple dynamics for these controllers by using crude models of the loops, developed by information provided to us by MSFC and our knowledge of proof mass devices. As will be discussed in the next subsection, these colocated Linear Momentum Exchange Device loops did provide significant performance improvement.

The control design models for the Advanced Gimbal System-X to Base Gyro-X and Advanced Gimbal System-Y to Base Gyro-Y loops were respectively 17th and 19th order continuous-time realizations. First-order all-pass filters were incorporated into these models to account for the computational delay. The control design models for the loops Image Motion Compensation-X to Detector-Y and Image Motion Compensation-Y to Detector-X were both 4th order discrete-time design models. We chose to design the Advanced Gimbal System to Base Gyro feedback loops using continuous-time models instead of discrete-time models because of a greater maturity in the theory and software development for the continuous-time setting.

Figure 15 shows a comparison of the ERA model of the Advanced Gimbal System-Y to Base Gyro-Y loop with the FRF generated from the test data. It is seen that the ERA model closely resembles the FRF.

C. CONTROL DESIGN FOR THE ACES STRUCTURE

Once we had settled on a decentralized controller architecture, control design for the ACES structure was essentially a three step process:

(1) design for the Image Motion Compensation gimbal to Detector loops,
(2) design for the Advanced Gimbal System gimbal to Base Gyro loops, and
(3) design for the Linear Momentum Exchange Device force to colocated accelerometer loops.

Below, we give details of the design procedure for each step along with experimental data describing the resultant performance improvement. Experimental data describing the performance improvement of the integrated controller (i.e., the controller with all feedback loops closed) is delayed until the next section.

1. Design Process for the Image Motion Compensation to Detector Loops

The design processes for the Image Motion Compensation-X to Detector-Y loop and the Image Motion Compensation-Y to Detector-X loops were essentially identical. As illustrated by Figure 17, for each loop the control problem was formulated as a disturbance tracking problem. Optimal Projection control laws [10,11] $H(z)$ were synthesized by minimizing the quadratic cost function of Figure 17. The controllers implemented were the limiting controllers as $\varepsilon \rightarrow 0$ and thus, the implemented controllers contained integrators which were able to effectively eliminate the line-of-sight biases. For both loops the design models were 4th order while the controllers implemented were 3rd order.

The resultant performance improvement is illustrated by Figure 18, which shows open-loop and closed-loop responses for the x axis of the Detector to a pulse command to the x axis of the Base Excitation Table. Notice that the feedback loops were able to eliminate the system biases and also improved the line-of-sight performance by providing low frequency tracking.

2. Design Process for the Advanced Gimbal System to Base Gyro Loops

The design processes for the Advanced Gimbal System-X to Base Gyro-X loop and the Advanced Gimbal System-Y to Base Gyro-Y loop (subsequently to be referred to as the X and Y loops) were very similar. For the X-loop the modes that most influenced line-of-sight performance were 1.4 Hz and 2.4 Hz bending modes. Similarly, for the Y-loop the modes that most influenced line-of-sight performance were 1.7 Hz and 2.3 Hz bending modes.

Maximum Entropy/Optimal Projection synthesis was used to synthesize continuous-time controllers of varying orders, control authorities and robustness. These controllers were then discretized by using the bilinear transformation with frequency prewarping.

The Maximum Entropy/Optimal Projection design equations are given by (13)–(18). In our control designs we chose R_1 and V_1 to penalize all of the modes less than 3 Hz since these modes dominate line-of-sight performance. We also chose $R_2 = V_2 = \rho$. The scalar parameter ρ was then used to determine the controller authority. The Maximum Entropy robustness design was used to incorporate uncertainty in each of the modal frequencies and proved to be crucial in developing robust controllers which yielded significant performance improvement when implemented. The effectiveness of the Maximum Entropy approach is illustrated by the compensator Bode plots of Figure 19–21.

Figure 19 describes the influence of Maximum Entropy uncertainty

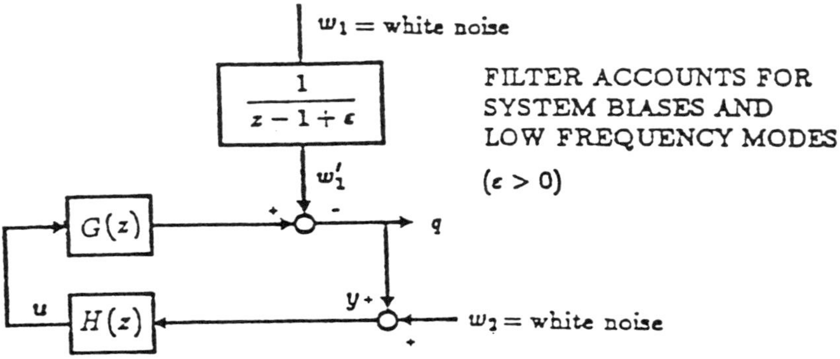

- CONTROLLERS WERE DESIGNED BY MINIMIZING

$$J(\varepsilon) = \lim_{k \to \infty} E[q^{\mathrm{T}}(\varepsilon)q(\varepsilon) + \rho u^{\mathrm{T}}(\varepsilon)u(\varepsilon)], \quad \rho > 0.$$

Figure 17. The tracking problem for the Image Motion Compensation to Detector transfer function.

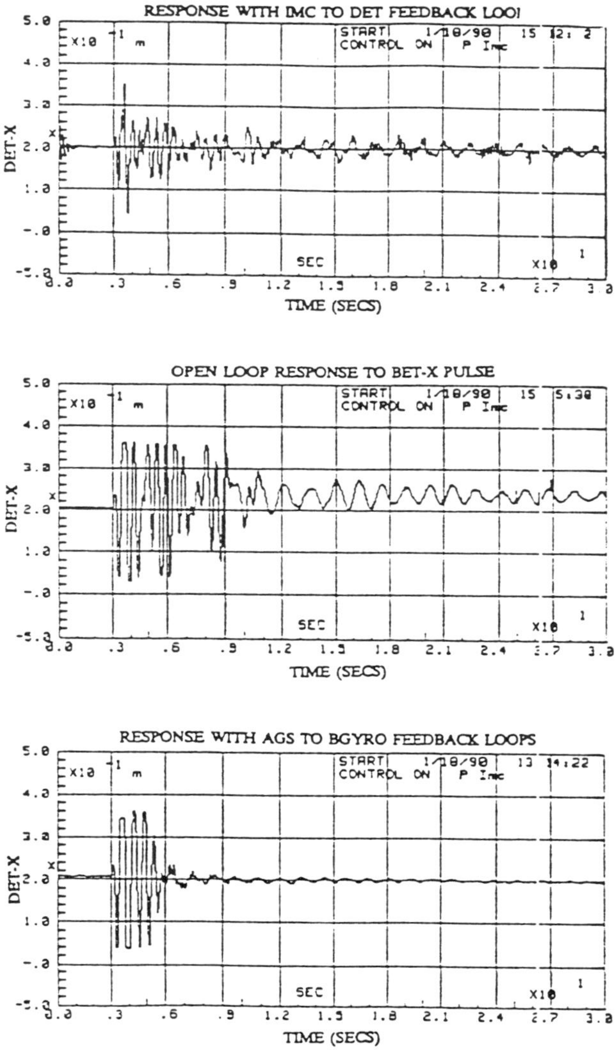

Figure 18. The Detector-X response to a Base Excitation Table-X pulse with the Image Motion Compensation to Detector Controller (top), in the Open Loop (center) and with the Advanced Gimbal System to Base Gyro Controller (bottom).

design on the phase of a full-order compensator in the performance region (i.e., less than 3 Hz). The phase of the LQG compensator varies widely over this frequency interval and was thus sensitive to phase uncertainty in the design model. As one might expect, these designs were nonrobust and were unstable when implemented. However, by assuming frequency uncertainty in all the modes below 3 Hz, the Maximum Entropy design became positive real in the performance region tending toward rate feedback as the uncertainty level was increased. Thus, the Maximum Entropy design provided the needed stability robustness in the performance region.

Figure 20 describes the influence of Maximum Entropy uncertainty design on the magnitude shape of a full-order compensator in the performance region. Notice that the Maximum Entropy compensator magnitudes are smoother than those of the LQG compensators, thus providing performance robustness. Another implication is that the Maximum Entropy designs yield robust controllers that are effectively reduced order controllers. In practice, the full-order Maximum Entropy designs actually provide insight into the choice of the order of the compensator and are a numerical aid in synthesizing reduced order controllers.

The higher authority controllers notched the high frequency modes that had high gain, which for the Advanced Gimbal System-Y to Base Gyro-Y loop are the two highest frequency modes shown in the ERA model of Figure 15. As illustrated by Figure 21, Maximum Entropy design was able to robustify the controller notches. That is, the controller notches were increased in both width and depth. The transfer function of the notch of the LQG controller shown in Figure 21 is given by

$$\frac{1+2(.0043)(s/61.39)+(s/61.39)^2}{1+2(.0303)(s/61.24)+(s/61.24)^2}$$

while the transfer function of the notch of the Maximum Entropy controller shown in Figure 21 is given by

$$\frac{1+2(.0086)(s/61.04)+(s/61.04)^2}{1+2(.2259)(s/63.09)+(s/63.09)^2}.$$

Notice that the Maximum Entropy controller notch obtained increased width by separating the frequencies of the complex pole and zero pairs and increased depth by decreasing the ratio of the damping ratio of the complex zero pair to the damping ratio of the complex pole pair.

The controllers which yielded the best performance when implemented were a 4th order controller for the X-loop and a 6th order controller for the Y-loop. The resultant performance improvement is illustrated in Figure 18 which shows open and closed loop responses of the x axis of the Detector

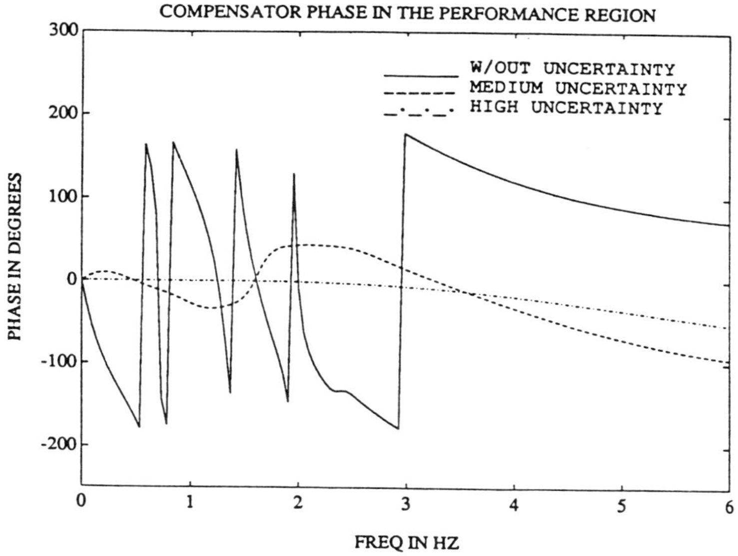

Figure 19. Phase of Maximum Entropy Compensators for the Advanced Gimbal System to Base Gyro Transfer Functions for various levels of uncertainty.

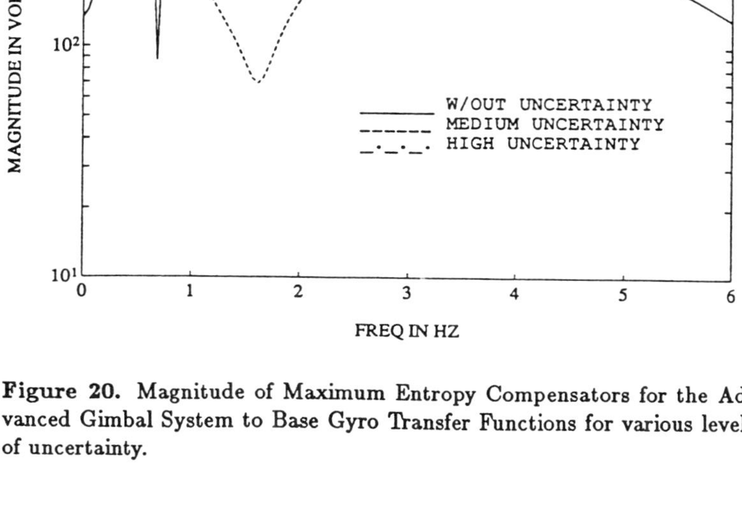

Figure 20. Magnitude of Maximum Entropy Compensators for the Advanced Gimbal System to Base Gyro Transfer Functions for various levels of uncertainty.

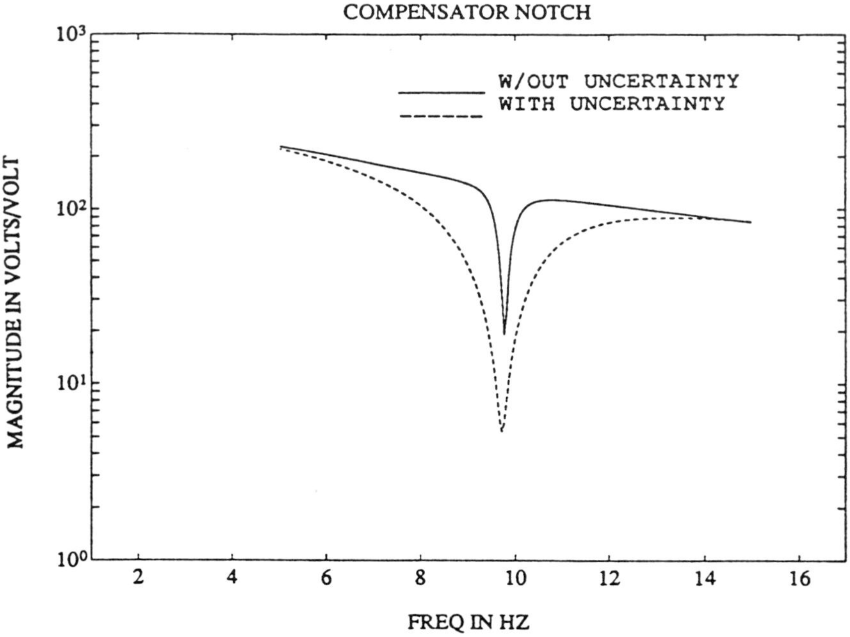

Figure 21. Illustration of Maximum Entropy notch robustification for the high frequency modes.

to pulse commands to the x axis of the Base Excitation Table. Significant performance improvement was achieved.

3. Design Process for the Linear Momentum Exchange Device Force to Accelerometer Loops

In this subsection and hereafter the two-axis Linear Momentum Exchange Device closest to the base will be referred to as Linear Momentum Exchange Device-1 while the two-axis Linear Momentum Exchange Device closest to the tip will be referred to as Linear Momentum Exchange Device-2. The control design was based on feeding back each of the four colocated accelerometer outputs to the corresponding Linear Momentum Exchange Device proof-mass axis. Thus, the four loops utilized for control design were (i) Linear Momentum Exchange Device-1X to Acceleration-1X, (ii) Linear Momentum Exchange Device-1Y to Acceleration-1Y, (iii) Linear Momentum Exchange Device-2X to Acceleration-2X, and (iv) Linear Momentum Exchange Device-2Y to Acceleration-2Y. It was assumed that the open loop dynamics of each of the four loops was identical so that the same controller $H(s)$ can be utilized in each loop. The transfer function from the beam velocity (at the given Linear Momentum Exchange Device location and along a given axis) to the force applied (by the Linear Momentum Exchange Device along the same axis) was assumed to be as shown below where D, k_m, k_d, and m are motor constants.

$$\frac{f_c}{\dot{x}_p} = \frac{H(s)s^2 + Ds + k_m k_d}{s^2 + (D/m)s + k_m k_d/m}.$$

(Note that the controller $H(s)$ multiplies only s^2 in the numerator.)

The design goal was to choose $H(s)$ such that the above transfer function is positive real at low frequency (say around 1 Hz) and remains positive real to some significantly higher frequency (say 10 Hz) in order to provide damping to the beam modes in this frequency band. To satisfy this objective while limiting the low frequency stroke, $H(s)$ was chosen to be the cascade of a second-order low-pass filter and first order high pass filter and was thus of the form

$$H(s) = \frac{ks}{(s+\alpha)(s^2 + 2\varsigma w_n s + w_n^2)}.$$

The low and high pass portions of the controller were discretized separately by using the bilinear transformation with frequency prewarping and the resultant discrete-time controller was then implemented in each of the loops. When integrated with the feedback controllers involving the Image Motion Compensation and Advanced Gimbal System gimbals, the Linear

Momentum Exchange Device loops improved the line-of-sight performance especially by reducing the influence of the higher frequency vibration in the Detector responses.

D. PERFORMANCE OF THE INTEGRATED CONTROLLER

The integrated controller combined the three controllers described above. Since the order of the controllers for the Image Motion Compensation gimbal to Detector loops, the Advanced Gimbal System gimbal to Base Gyro loops, and the colocated Linear Momentum Exchange Device loops were respectively 6th, 10th and 12th order, the integrated controller consisted of 28 states.

The two measures of performance that were computed to compare the open-loop and closed-loop performance are the mean ($\bar{q}$) and the standard deviation (σ), defined respectively [34, p. 2–24] as follows where for a given sensor output N is the number of recorded samples over a specified time interval and $\{q_i\}_{i=1}^{N}$ is the sequence of sampled values.

$$\bar{q} \triangleq \frac{1}{N}\sum_{i=1}^{N} q_i \quad \sigma \triangleq (\sum_{i=1}^{N} q_i^2 - N\bar{q}^2)^{1/2}/(N-1)^{1/2}.$$

The performance improvement in dB for both the mean and standard deviation is defined as 20 times the log to the base 10 of the open loop performance divided by the closed loop performance.

For a pulse command to Base Excitation Table-X, Table II shows the resultant performance improvement in the Detector-X response as the controllers were combined. The relatively large improvements in mean are large open-loop Detector biases which were effectively eliminated using feedback. In general, the mean values were dominated by the size of the initial biases which varied with each test. Figure 22 shows a direct comparison of the open loop response of the x axis of the Detector to the closed loop response with the integrated controller.

The results of this experiment illustrate that simple controllers (i.e., reduced order and decentralized controllers) can provide very significant performance improvement for some flexible structure control problems. The total design model contained 45 states. However, the integrated controller contained only 28 states and had a decentralized architecture. This reduction in controller complexity is very important for the development of practical controllers due to substantial limits on throughput capability of even the most advanced space-qualified processors.

V. CONCLUSIONS

The Mini-MAST and ACES experiments described in this chapter both successfully demonstrated the very significant performance benefits achievable via active control technology. The controllers were designed using the Maximum Entropy/Optimal Projection approach to robust, reduced-order control design and hence the two experiments validated the utility of this approach for flexible structure control. In addition, the Mini-MAST experiment demonstrated the effectiveness of a precompensation method for obtaining controller roll-off in order to achieve robustness to high frequency, unmodeled dynamics. Likewise, the ACES experiment illustrated the ability of the Eigensystem Realization Algorithm to identify high fidelity models of flexible structures.

The control design and implementation approach used was to start with simple controllers (i.e., reduced-order controllers, decentralized controllers and/or controllers using relatively few sensors and actuators) and increase controller complexity to increase performance. This strategy proved to be very effective. In both experiments, as the complexity of the controllers was increased, the performance progressively improved.

Due to the implementation of reduced-order, decentralized controllers in the two experiments (in addition to the reduced-order, centralized controllers implemented in the Mini-MAST experiment), the results of both experiments illustrate that this class of controllers can provide very significant performance improvement for some flexible structure control problems.

Our experience in the NASA CSI Guest Investigator program and in implementing controllers on Harris testbeds [37–38] leads to the following recommendations for future flight experiments.

1. Design the structure to allow decentralized control design (even if centralized controllers can yield better performance).
2. Using the best model available without in-flight system identification, design highly robust decentralized controllers as the first controllers to be implemented.
3. Design the mission to allow in-flight system identification using a proven identification tool such as the Eigensystem Realization Algorithm [35,36].
4. Using the system identification model, design a sequence of controllers of increasing complexity and performance improvement.
5. Successively implement the sequence of controllers. If instability is sensed, then revert back to a controller that is known to be stabilizing.

These steps should only be viewed as a rough outline. Many variations of this scheme are of course possible.

Table II. The Detector-X response to a Base Excitation Table-X as the controllers were combined.

Controller (Controller order)	Mean Error & Standard Deviation	dB Improvement	Test Result Time Histories for X LOS
Open Loop	mean = 0.0329	———	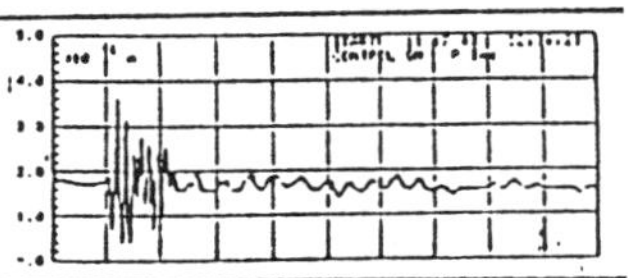
	S.D. = 0.0256	———	
IMC (6 states)	mean = 8.5×10^{-5}	51.8	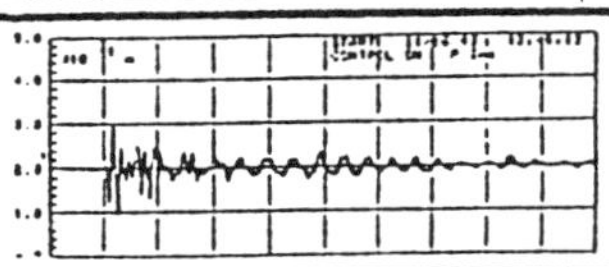
	S.D. = 0.01742	3.34	
IMC + AGS (16 states total)	mean = 8.4×10^{-5}	51.9	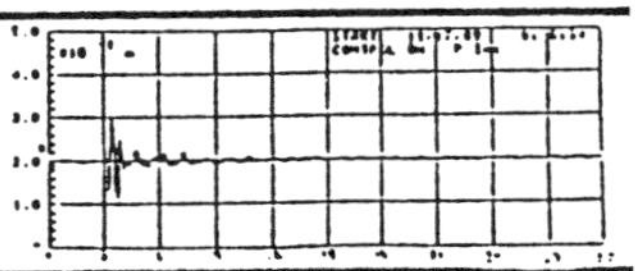
	S.D. = 0.0107	7.58	
IMC + AGS + LMED (28 states total)	mean= 4.2×10^{-5}	57.9	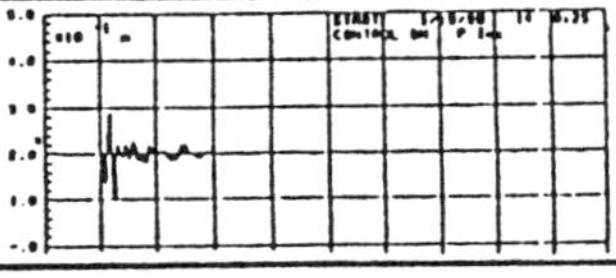
	S.D. = 9.0×10^{-3}	9.0	

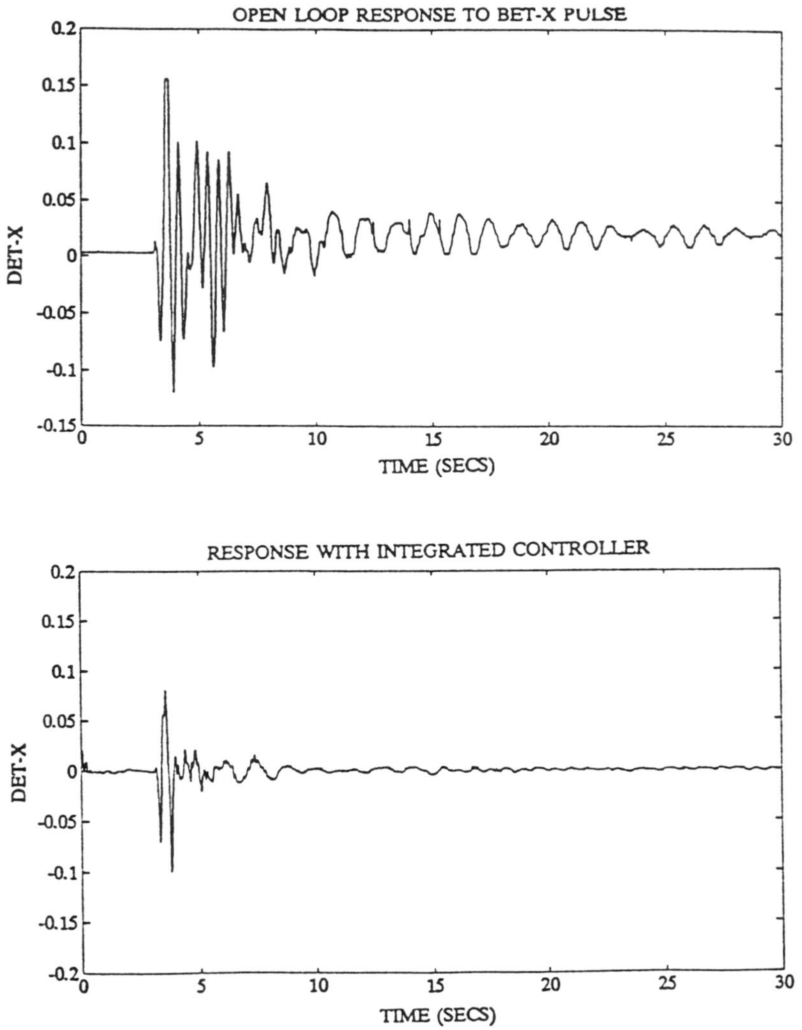

Figure 22. The Detector-X response to a Base Excitation Table-X pulse in the open loop (top) and with the Integrated Controller (bottom).

REFERENCES

1. E. G. Collins, Jr., J. A. King, D. J. Phillips, and D. C. Hyland, "High Performance, Accelerometer-Based Control of the Mini-Mast Structure at Langley Research Center," NASA Contractor Report 4377, May 1991.
2. E. G. Collins, Jr., J. A. King and D. J. Phillips, "High Performance, Accelerometer-Based Control of the Minimast Structure," *AIAA J. Guid. Contr. Dyn.*, to appear.
3. E. G. Collins, Jr., D. J. Phillips, and D. C. Hyland, "Design and Implementation of Robust Decentralized Control Laws for the ACES Structure at Marshall Space Flight Center," NASA Contractor Report 4310, Langley Research Center, July 1990.
4. E. G. Collins, Jr., D. J. Phillips, and D. C. Hyland, "Robust Decentralized Control Laws for the ACES Structure," *Contr. Sys. Mag.*, 62–70, April 1991.
5. D. C. Hyland, "Minimum Information Stochastic Modelling of Linear Systems with a Class of Parameter Uncertainties," *Proc. Amer. Contr. Conf.*, 620–627, Arlington, VA, June 1982.
6. D. C. Hyland, "Maximum Entropy Stochastic Approach to Controller Design for Uncertain Structural Systems," *Proc. Amer. Contr. Conf.*, 680–688, Arlington, VA, June 1982.
7. D. C. Hyland and D. S. Bernstein, "The Optimal Projection Equations for Fixed-Order Dynamic Compensation," *IEEE Trans. Autom. Contr.* **29**, 1034–1037 (1984).
8. D. S. Bernstein and D. C. Hyland, "The Optimal Projection/Maximum Entropy Approach to Designing Low-Order, Robust Controllers for Flexible Structures," *Proc. IEEE Conf. Dec. Contr.*, 745–752, Fort Lauderdale, FL. December 1985.
9. D. S. Bernstein and S. W. Greeley, "Robust Controller Synthesis Using the Maximum Entropy Design Equations," *IEEE Trans. Autom. Contr.* **31**, 362–364 (1986).
10. D. S. Bernstein, L. D. Davis, and S. W. Greeley, "The Optimal Projection Equations for Fixed-Order, Sampled-Data Dynamic Compensation with Computation Delay," *IEEE Trans. Autom. Contr.***31**, 859–862 (1986).
11. D. S. Bernstein, L. D. Davis, and D. C. Hyland, "The Optimal Projection Equations for Reduced-Order, Discrete-Time Modelling, Estimation and Control," *AIAA J. Guid. Contr. Dyn.***9**, 288–293 (1986).
12. D. S. Bernstein and D. C. Hyland, "Optimal Projection for Uncertain System (OPUS): A Unified Theory of Reduced-Order, Robust Control

Design," in *Large Space Structures: Dynamics and Control*, S. N. Atluri and A. K. Amos, eds., Springer-Verlag, New York, 1988.

13. S. Richter and E. G. Collins, Jr., "A Homotopy Algorithm for Reduced Order Compensator Design Using the Optimal Projection Equations," *Proc. 28th IEEE Conf. Decis. Contr.*, 506–511, Tampa, FL, Dec. 1989.
14. E. G. Collins, Jr., and S. Richter, "A Homotopy Algorithm for Synthesizing Robust Controllers for Flexible Structures Via the Maximum Entropy Design Equations," *Third Air force/NASA Symposium on Recent Advances in Multidisciplinary Analysis and Optimization,* 324–333, San Francisco, CA, September 1990.
15. D. S. Bernstein and D. C. Hyland, "The Optimal Projection Approach to Robust, Fixed-Structure Control Design," in *Mechanics and Control of Space Structures,* 237–293, J. L. Junkins, Ed., AIAA, 1990.
16. A. Yousuff and R. E. Skelton, "A Note on Balanced Controller REduction," *IEEE Trans. Autom. Contr.***29** 254–256 (1984).
17. W. M. Haddad and D. S. Bernstein, "Robust Stability and Performance Analysis for State Space Systems Via Parameter-Dependent Lyapunov Functions: The Popov Criterion Revisisted," *Proc. IEEE Conf. Dec. Contr.*, Brighton, U.K., December 1991.
18. W. M. Haddad and D. S. Bernstein, "Parameter-Dependent Lyapunov Functions, Constant Real Parameter Uncertainty, and the Popov Criterion in Robust Analysis and Synthesis," submitted to *IEEE Trans. Autom. Contr.*
19. D. S. Bernstein, W. M. Haddad, D. C. Hyland, and F. Tyan, "A Parameter Dependent Lyapunov Function for the Maximum Entropy Design Approach," submitted to *Syst. Contr. Letters.*
20. S. R. Hall, D. G. MacMartin, and D. S. Bernstein, "Incoherence, Equipartition, and Covariance Averaging in the Analysis of Uncertain Systems," submitted for publication.
21. R. Pappa, et al., *Mini-MAST CSI Testbed User's Guide,* NASA Langley Research Center, March 1989.
22. B. A. Francis, "A Course in H^∞ Control Theory," in "Lecture Notes in Control and Information Sciences," Vol. 88, Springer-Verlag, 1987.
23. B. A. Francis, and J. C. Doyle, "Linear Control Theory with an H_∞ Optimality Criterion," *SIAM J. Contr. Opt.* **25**, 815–844 (1987).
24. J. C. Doyle, K. Glover, P. P. Khargonekar and B. A. Francis, "State-Space Solutions to Standard H_2 and H_∞ Control Problems," *IEEE Trans. Autom. Contr.***34** (1989).

25. D. S. Bernstein, "LQG Control with an H_∞ Performance Bound: A Riccati Equation Approach," *IEEE Trans. Autom. Contr.* **34**, 293–305 (1989).
26. J. A. King and R. D. Irwin, "Issues in the Application of H^∞ Control to Large Space Structures," Southeastern Symposium on System Theory, Cookeville, TN, March 1990.
27. P. G. Clar, A. Carrier, and A. E. Bryson, Jr., "Comparison of Two Methods for Causing Roll-Off in Control System Design," *AIAA Guid. Nav. Contr. Conf.*, 1734–1741, August 1990.
28. J. C. Doyle, "Structured Uncertainty in Control System Design," *Proc. IEEE Conf. Decis. Contr.* 260–265, December 1985.
29. G. J. Balas, C.-C. Chu, and J. C. Doyle, "Vibration Damping and Robust Control of the JPL/AFAL Experiment Using μ-Synthesis," *Proc. IEEE Conf. Decis. Contr.*, 2689–2694, December 1989.
30. N. K. Gupta, "Frequency-Shaped Cost Functionals: Extension of Linear-Quadratic-Gaussian Design Methods," *J. Guid. Contr.* **3** (1980).
31. R. D. Hefner and D. L. Mingori, "Suboptimal Controller Design Using Frequency Domain Constraints," *Proc. AIAA Guid. Contr. Conf.*, August 1982.
32. E. G. Collins, Jr., J. A. King, and D. S. Bernstein, "Robust Control Design for the Benchmark Problem Using the Maximum Entropy Approach," *Proc. Amer. Contr. Conf.*, 1935–1936, Boston, MA, June 1991.
33. E. G. Collins, Jr., J. A. King, and D. S. Bernstein, "Application of Maximum Entropy/Optimal Projection Design Synthesis to the Benchmark Problem," submitted to *AIAA J. Guid. Contr. Dyn.*
34. R. D. Irwin, V. L. Jones, S. A. Rice, D. J. Tollison and S. M. Seltzer, *Active Control Technique Evaluation for Spacecraft (ACES),* Final Report to Flight Dynamics Lab of AF Wright Aeronautical Labs, Report No. AFWAL-TR-88-3038, June 1988.
35. J. N. Juang and R. S. Pappa, "An Eigensystem Realization Algorithm for Modal Parameter Identification and Model Reduction," *J. Guid. Contr. Dyn.* **8**, 620–627 (1985).
36. J. N. Juang and R. S. Pappa, "Effects of Noise on Modal Parameters Identified by the Eigensystem Realization Algorithm," *J. Guid. Contr. Dyn.* **8**, 620–627 (1985).
37. D. J. Phillips, E. G. Collins, Jr., and D. C. Hyland, "Experimental Demonstrations of Active Vibration Control of Flexible Structures," *Proc. IEEE Conf. Dec. Contr.*, 2024–2029, Honolulu, HI, December 1990.

38. D. J. Phillips, D. C. Hyland, E. G. Collins, Jr., and J. A. King, "The Multi-Hex Prototype Experiment," *Proc. IEEE Conf. Dec. Contr.*, Brighton, U.K., December 1991.

Applications of an Optimization-Based Integrated Controls–Structures Design Methodology For Flexible Space Structures

Peiman G. Maghami
Suresh M. Joshi
Ernest S. Armstrong

NASA Langley Research Center
Hampton, Virginia 23665

I. INTRODUCTION

Current spacecraft design customarily involves an iterative sequence of designs performed separately within the structural and control disciplines. Structural design is first performed based on loading considerations that would occur during launch, reboost, or component operational maneuvers. Sizes and masses of mission-related components are estimated and a structure is designed which maintains the desired component relationships during operations. A controller is next designed for the fixed structure to orient, guide, and/or move the spacecraft to obtain the required performance. The control design must also provide satisfactory closed-loop stability and robustness properties. If the nominal structure does not admit an adequate control design, it is returned to the structural discipline for modification. After modification, the structure is returned to the control discipline for redesign. This iterative process continues until a satisfactory compromise is found between the mission and control requirements. This separate-discipline approach has been successfully used in the past and works well in cases where a relatively high-stiffness structure is attainable and where nonstructural components are concentrated masses and inertias, or where performance requirements are not stringent. However, future space structures and performance requirements do not fit this category.

A number of future space missions will utilize large flexible structures in low-Earth and geostationary orbits. Example missions include space science platforms, space processing facilities, and Earth observation systems. Such missions typically require large distributed-mass components such as booms, solar arrays, and antennas whose dimensions range from a few meters to possibly hundreds of meters. In order to minimize the costs of construction, launching, and operations, it is necessary to make the structure as light as possible. However, the combination of large size and low structural mass leads to increased flexibility and makes more difficult the control of the structure and its components to a specified precision in attitude and shape.

The combination of flexibility and low mass creates a special dynamical interaction problem which is characteristic of large flexible space structures. In the design of controllers for physical systems, there almost always has to be some trade-off performed between design-model accuracy and mathematical complexity. The more accurate analysis models often require computational time that is too excessive to qualify them as design models for control purposes. Also, many of the most widely used multivariable controller design techniques work best for moderate-order, linear, time-invariant design models. Typically, in practice, high-order nonlinear models are linearized about some operating condition and have their model order reduced to produce design models which conform to computational limitations or compensator implementation constraints. These practicalities introduce modeling errors in the form of unmodeled dynamics which must be accounted for in the controller design process. Space structure controller design models are generally found through some order-reduction procedure applied to a high-order analysis model obtained from finite element techniques. The order-reduction process essentially deletes a portion of the finite element model to produce a lower-order controller design model. Although no longer contained in the design model, the unmodeled-dynamics, represented by the deleted portion, still appear in the measurement vector and, therefore, can be affected by model-based control inputs. Care must be taken in the design of controllers for lightly damped systems that the control inputs do not "spillover" [1] into the unmodeled dynamics with disastrous destabilizing effects.

Controls-structures interaction (CSI) in the form of destabilizing spillover has been verified in simple Earth-based laboratory experiments as well as in the design, analysis, ground development, and test and/or flight operation of space systems in industry [2]. The current approach to solving CSI problems is to design the spacecraft to avoid undesired dynamical interaction. This generally requires either stiffening the structure or slowing down the control system response. Stiffening the structure simplifies the control design problem in that the predominant dynamics tend toward rigid body but is costly in terms of mass as well as launch packaging and weight, leading to increased fuel consumption. Slowing down the control response produces control inputs which have less of a chance of producing destabilizing effects but is costly in terms of reduced performance capability. Neither approach is completely satisfactory. What is needed is a new design approach which has the capability of avoiding the damaging aspects of controls-structures interaction while at the same time identifying and taking advantage of any beneficial aspects.

The development and experimental validation of such an approach is a primary purpose of NASA's Controls-Structures Interaction (CSI) Program [3]. Fundamentally, the CSI program recognizes the high degree of coupling that exists between the control and structural disciplines when dealing with flexible structures. For example, controllers designed to be robust to unmodeled dynamics may need very low gain (and, therefore, produce low performance)[4]. In order to maintain high performance, it would then be necessary to redesign the structure to increase the frequencies of the higher modes likely to be affected by spillover. If a structural member thickness is changed, the dynamics will change, which would then change the control law and the required actuator sizes (and masses). This would, in turn, again change the structural model. One of the goals of the CSI program is to employ this coupling property to integrate the structures and controls design disciplines. Rather than design control and structural aspects of space structures separately in the sequential manner described earlier, approaches are being developed for a unified controls-structures modeling, analysis, and design method which would allow a complete iteration on all critical (control and structure) design variables in a single integrated computational framework.

This paper considers integrated design for missions which include flexible spacecraft with no articulated appendages and require fine attitude pointing and vibration suppression (e.g., large space antennas). Missions which involve flexible spacecraft with articulated multiple payloads, missions requiring large-angle slewing of spacecraft without articulated appendages, and missions which include general nonlinear motion of a flexible spacecraft with articulated appendages and robot arms are not considered. Two controller strategies are considered, namely, the static and dynamic dissipative controllers, which are well known for their stability robustness in the presence of unmodeled dynamics and parametric uncertainties [4-5]. The approach selected for the development of integrated controls-structures design methodology is an optimization-based procedure employing mathematical programming techniques. The optimization approach allows freedom and variety in selecting the potentially large number of design variables. The optimization approach is also the one commonly used in the field of structural design. In fact, many papers describing optimization-based integrated design have appeared in the literature for the last twelve years. An excellent literature survey can be found in [6]. However, it has only recently been that applications have gone beyond analytical studies with fictitious models, relatively simple laboratory apparatus, or unrealistic control designs. Among some of the recent efforts, the works by Balakrishnan [7] using infinite-dimensional models, and by Bossi, et al. [8] using Q-parameterization theory along with shape optimization, the studies by Maghami, et al. [9-11] on the integrated design of large structures (with thousands of degrees of freedom) utilizing dissipative compensators, and a paper by Woodard, et al. [12] on integrated optimization with varying actuator mass, may be mentioned. One of the unique features of the CSI program is the fact that integrated designs will be fabricated and the methodology laboratory validated. This places a constraint of realism on the design process that is not normally found elsewhere.

The purpose of this paper is to describe a portion of the integrated design methodology and results from the CSI program. This work differs from past studies in several aspects. The design treats two large realistic applications - a derivative of a geostationary platform designed in support of Earth observation sciences [13] and the initial phase of an evolutionary laboratory structure in the CSI program [14]. A design similar to the integrated design reported herein using the CSI model is actually being constructed at NASA's Langley Research Center for the purpose of validating the integrated design methodology. Also we recognize that the effect of the control system on the structural mass and constraints can be significant when the magnitude of the control effort determines the mass of the control system and its power supply [15]. This realistic effect is accounted for by the inclusion of a mathematical expression coupling the actuator masses and the controller gains. Finally, the controller formulation utilizes recently developed theory for dissipative controllers not previously employed in applications of this size.

We begin with a description of the mathematical model of the structure and controller design methods. Next, an overview of the optimization-based integrated design procedure employed in this paper is provided. Afterwards, descriptions of the structure, design variables and constraints, and integrated design problem statements are given for each of the applications. Finally, results from the integrated design process are discussed. An appendix is provided giving the fundamentals of the dissipative controller theory.

II. MATHEMATICAL MODEL

The linear, time-invariant, mathematical model of a flexible space structure is given by

$$M\ddot{x} + D\dot{x} + Kx = \tilde{B}u \tag{1}$$

$$y_p = \tilde{C}_p x \quad ; \quad y_r = \tilde{C}_r \dot{x} \tag{2}$$

where x is an $n \times 1$ displacement vector; M is the positive-definite inertia matrix; D is the open-loop damping matrix; K is the nonnegative-definite stiffness matrix; $\tilde{B}$ is an $n \times m$ control influence matrix; u is an $m \times 1$ control input vector; y_p and y_r are, respectively, $p \times 1$ position and rate measurements; and $\tilde{C}_p$ and $\tilde{C}_r$ are the corresponding output influence matrices. The second-order representation of the structure given in Eq. (1) is obtained through finite element modeling. The order of a large flexible space structure (LFSS) can be quite large. For design and analysis purposes the order of the system is reduced to a design size using a modal truncation approach, wherein every mode in the controller bandwidth is retained and the remaining modes are truncated. The system equations in modal coordinates for the retained modes are written as

$$M_r\ddot{q}_r + D_r\dot{q}_r + K_r q_r = \Phi^T \tilde{B}u \equiv \Gamma^T u \tag{3}$$

where q_r is an $r \times 1$ vector of modal amplitudes; M_r, D_r, K_r are, respectively, the generalized inertia, damping and stiffness matrices; and Φ is an $r \times n$ matrix whose columns are the r open-loop eigenvectors associated with the included modes. If the mode shapes are normalized with respect to the inertia matrix, and modal damping is assumed, then $M_r = I_{r\times r}$, $D_r = Diag\{2\zeta_1\omega_1, \ldots, 2\zeta_r\omega_r\}$, and $K_r = Diag\{\omega_1^2, \ldots, \omega_r^2\}$ where ω_i and ζ_i, $i = 1, \ldots, r$ are the open-loop frequencies and damping ratios. Here, the sensors are collocated and compatible with the actuators, whereby

$$y_p = \Gamma q_r \quad ; \quad y_r = \Gamma \dot{q}_r \tag{4}$$

Note that the collocation of the sensors and actuators is necessary for the implementation of the dissipative controllers (see the Appendix). This collocation guarantees the system to be minimum phase; i.e, all transmission zeros are in the left-half plane and will enhance the stability robustness of the overall system.

Defining the state vector

$$z = \{\, q_r \quad \dot{q}_r \,\}^T \tag{5}$$

the dynamics of the system, assuming no appreciable sensor and/or actuator dynamics, can be written in a first-order form

$$\dot{z} = Az + Bu \tag{6}$$

where

$$A = \begin{bmatrix} 0 & I_{n \times n} \\ -K_r & -D_r \end{bmatrix} \quad ; \quad B = \begin{bmatrix} 0 \\ \Gamma^T \end{bmatrix}$$

$$y_p = \begin{bmatrix} \Gamma & 0 \end{bmatrix} z \equiv C_p z \quad ; \quad y_r = \begin{bmatrix} 0 & \Gamma \end{bmatrix} z \equiv C_r z$$

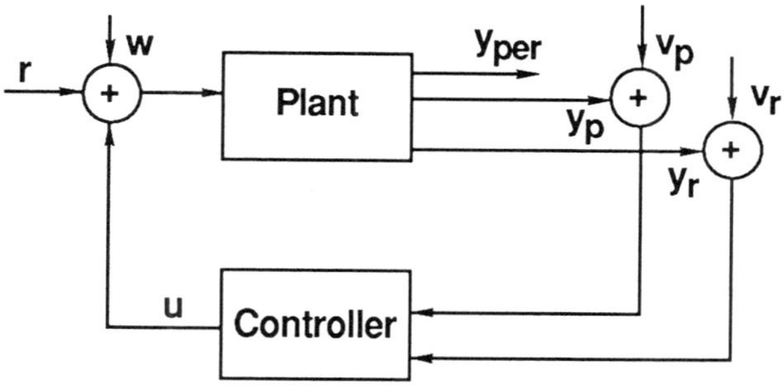

Figure 1. Block Diagram for a general system configuration

The general configuration of the controlled structure is illustrated in the block diagram shown in Figure 1. In this figure, w, v_p, and v_r represent disturbances at the input, position output, and the rate output, respectively; r denotes the reference signal vector; and y_{per} refers to those measurements which are used for performance evaluations and not in the feedback loop. In this configuration, the system equations are written as

$$\dot{z} = Az + Bu + Hw \tag{7}$$

$$y_p = C_p z + S_p v_p \quad ; \quad y_r = C_r z + S_r v_r \tag{8}$$

$$y_{per} = C_{per} z$$

where H, S_p, and S_r are influence matrices for the disturbances w, v_p, and v_r, respectively; and C_{per} is the output influence matrix associated with y_{per}. It is assumed that w, v_p, and v_r are uncorrelated, zero-mean stationary white-noise processes with unit intensities.

III. CONTROLLER DESIGN METHODS

Control system design for LFSS is a challenging problem because of their special dynamic characteristics, which include: a large number of structural modes within the controller bandwidth, low, closely-spaced structural mode frequencies, very small inherent damping, and lack of accurate knowledge of the parameters. In order to be implementable, the controller must be of a reasonably low order and must also satisfy the performance specifications (i.e., root-mean-square (RMS) pointing error, closed-loop bandwidth). It must also have robustness to "nonparametric" uncertainties (i.e., unmodeled structural modes), and to "parametric" uncertainties (i.e., errors in the knowledge of the design model). Two major categories of controller design methods for LFSS are "Model-Based" controllers (MBC) and "Dissipative" controllers. An MBC generally consists of a state estimator (a Kalman-Bucy filter or an observer) followed by a linear-quadratic regulator. The state estimator utilizes the knowledge of the "design" model (consisting of the rotational rigid-body modes and a few elastic modes) in its "prediction" part. Using multivariable frequency-domain design methods, such controllers can be made robust to unmodeled structural dynamics, that is, the "spillover" effect can be overcome [4]. However, such controllers generally tend to be very sensitive to uncertainties in the design model, in particular, to uncertainty in the structural mode frequencies [4-5]. An analytical explanation of this instability mechanism may be found in [5]. Achieving robustness to real parametric uncertainties is, as yet, an unsolved problem, although considerable research activity is in progress in that area.

A. STATIC DISSIPATIVE CONTROLLER

In view of the sensitivity problem of MBC's, dissipative controllers, which utilize collocated and compatible actuators and sensors, offer an attractive alternative. Details of dissipative controller theory may be found in the Appendix. Dissipative controllers utilize special passivity-type input/output properties of the plant and offer robust stability in the presence of both nonparametric and parametric uncertainties [4-5]. The simplest controller of this type is the constant-gain dissipative controller. Using collocated torque actuators and attitude and rate sensors, the constant-gain dissipative control law is given by:

$$u = -G_p y_p - G_r y_r \tag{9}$$

where y_p and y_r are the measured ($3m \times 1$, where m is the number of 3-axis sensors) attitude and rate; G_p and G_r are $3m \times 3m$ symmetric, positive-definite gain matrices. This control law has been proven to give guaranteed closed-loop stability despite unmodeled elastic modes, parameter errors, certain types of actuator and sensor nonlinearities (such as saturation and dead zone; see the Appendix for details), and first-order actuator dynamics [4]. Robust stability is generally not guaranteed if the actuator dynamics are second or higher order, or in the presence of digitization, time delays, and control hysterisis. However, if acceleration feedback is permissible, it is possible to obtain guaranteed stability for second-order actuator dynamics

[5,16]. For practical space implementation, rate measurements may be obtained from rate gyros and attitude measurements may be provided by star sensors. However, if linear velocity is required; e.g., in the case of a ground-based experiment, velocity signals can be obtained by integrating accelerometer signals with the aid of washout filters to asymptotically remove the accelerometer bias. The inclusion of washout filters would, however, hinder the dissipativity of this control design. Finally, the drawback of this controller is that the achievable performance is inherently limited because of its simple mathematical structure.

B. DYNAMIC DISSIPATIVE CONTROLLER

In order to obtain higher performance while still retaining the highly desirable robust stability, dynamic dissipative compensators can be used. The main characteristic of all dissipative controllers is that they do not rely on the knowledge of the design model to ensure stability, although they utilize it to obtain the best possible performance. An n_c-order (two-level) dynamic dissipative controller is given by:

$$\dot{x}_c = A_c x_c + B_c y_r \tag{10}$$

$$u_c = -G x_c - G_r y_r - G_p y_p \tag{11}$$

where A_c, B_c, and G are the compensator system, input, and output matrices, respectively. G_r and G_p are symmetric, positive-definite rate and position gain matrices. Furthermore, A_c is strictly Hurwitz, and the Kalman-Yacubovich relations [17] hold:

$$A_c^T P + P A_c = -Q \tag{12}$$

$$G = B_c^T P \tag{13}$$

where $P = P^T > 0$, and $Q = Q^T > 0$.

Equations (10)-(13) represent a two-level controller, wherein the inner loop consists of static position-plus-rate feedback and the outer loop consists of a dynamic compensator. This controller assures robust asymptotic stability regardless of unmodeled structural dynamics or parametric uncertainties [5,16]. In the absence of zero-frequency rigid-body modes (e.g., for a ground-based experiment), G_p and G_r can be null matrices without destroying the robust asymptotic stability; that is, the inner loop is not required. Further results for systems with zero-frequency modes are given in the Appendix. Similar to the static dissipative controller, this control law has been proven to give guaranteed closed-loop stability despite unmodeled elastic modes and parameter errors. However, robust stability is generally not guaranteed if the actuator dynamics, nonlinearities, and time delays are present or if digital implementation is used. Similar to the static dissipative controller, for practical space implementation, rate measurements may be obtained from rate gyros and attitude measurements may be provided by star sensors. However, if linear velocity is required; e.g., in the case of a ground-based experiment, velocity signals can be obtained by integrating accelerometer signals with the aid of washout filters to asymptotically remove the accelerometer bias. The inclusion of washout filters would, however, hinder the dissipativity of this control design.

IV. OPTIMIZATION-BASED INTEGRATED CONTROLS-STRUCTURES DESIGN

The methodology employed in this study uses optimization techniques to integrate the separate disciplines of control and structural design. Fundamentally, the integrated design goals and requirements are formulated as a nonlinear programming problem wherein selected controls-structures design variables are chosen, subject to structural and performance constraints, to optimize an objective function whose construction captures the essence of the integrated design problem. It is felt that this approach takes advantage of the highly developed optimization technology in both the controls and structures fields and provides the designer an opportunity to obtain nonintuitive results.

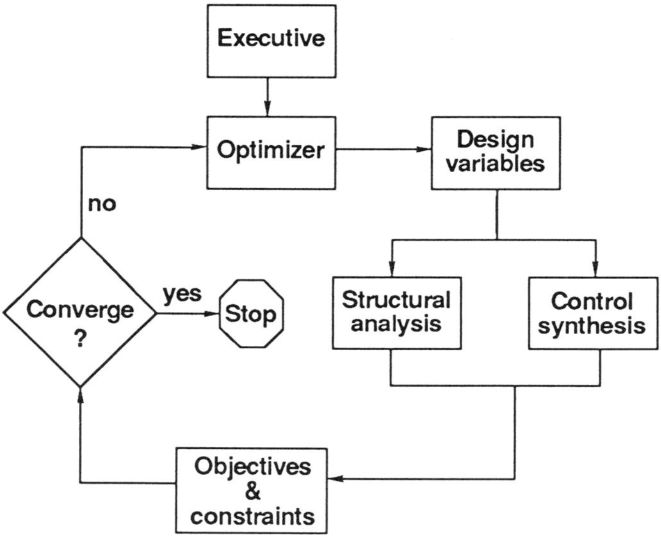

Figure 2. Flow diagram for the integrated Controls-Structures Optimization

Nominally, an initial control-structure configuration defines a set of design variables which are available for modification. Typical controller design variables are sizes and location of sensors and actuators, and elements of dynamic compensator system matrices. Structural design parameters could be the dimensions of structural members along with mass and densities of materials. An example of a realistic integrated design problem would be, for some given input or maneuver, to design a spacecraft which provides the best possible dynamic response amplitude reductions (below the amplitudes of a nominal spacecraft) with no increase in structural mass or control system energy requirements. As indicated by the preceding examples, most integrated design objectives and constraints can be expected to be dependent functions of the design variables which are obtained as the outputs of some separately developed computer control synthesis [18] or structural analysis packages [19]. Integrated design is then a computationally intensive process requiring super-computer technology.

A flow chart summarizing the optimization-based integrated design process is given in Figure 2. A user-provided executive program drives an optimization module which, at each stage, (1) uses the current values of the design variables to compute the objective function and constraints, (2) evaluates the objective function and constraints for optimality and feasibility and, (3) if necessary, generates a new set of design variables defining a control-structure configuration with better integrated design characteristics. Aspects of the computer implementation of this process are discussed in [20].

A version of the integrated design software tool CSI-DESIGN, which is under development at the NASA-Langley Research Center, was used to perform the numerical studies reported herein. CSI-DESIGN is intended for research purposes and is composed of public domain software. It uses in-core data transfer and is ultimately limited in the size of problems it can successfully treat (The package has control, structural and optimization modules linked as shown in Figure 2). The control compensator models used in this study were static and dynamic dissipative. The benefits and motivation for using this control strategy for the control of flexible structures is discussed in the Appendix. A description of the contents of the CSI-DESIGN structural module may be found in [20]. Employing a four-processor Alliant FX-80 digital computer, integrated optimization was performed using the Automated Design Synthesis (ADS) software [21]. The interior penalty function method of ADS was used to solve the nonlinear programming problems. In this method, the constrained optimization problem is transformed into an unconstrained problem through creation of a pseudo-objective function which is the sum of the original objective function and an imposed penalty function (which is a function of the constraints) [22]. The Reverse-Cuthill-McKee algorithm [23] for minimizing the bandwidth of the banded stiffness and mass matrices was used to reduce computational requirements. Additionally, analytical expressions for eigenvalue/eigenvector sensitivity (with respect to the structural design variables) [24-25] were used in the integrated design process to approximate the eigenvalues and eigenvectors at design points which are in the neighborhood of the nominal design point. This approximation was in the form of a first-order Taylor series approximation and resulted in substantial computational savings since it removes the need for costly computation of structural eigenvalues and eigenvectors at many of the optimization moves.

V. INTEGRATED DESIGN OF THE EPS MODEL

The Earth Pointing System (EPS), shown in Figure 3, is a multi-user platform concept, and consists of a 10-bay, 30-meter long truss structure with two radial rib antennas (7.5m and 15m diameter) at the ends. All the members (i.e., constituting the truss, the antennas, and the antenna supports) are assumed to be hollow tubes with circular cross-section and 1.59 mm thickness. The antennas are fixed with respect to the truss, so that the problem is that of controlling the pointing and vibration of the entire structure. It is assumed that a three-axis control moment gyro (CMG) and collocated attitude and rate sensors, located at the center-of-mass of the structure, are used for accomplishing the control. No sensor and actuator dynamics are considered. The modal frequencies for the first ten flexible modes are given in Table I. The first flexible mode had a frequency of 0.58 Hz, corresponding to a mode of the large antenna support structure, while the first flexible mode of the truss structure was at about 6.6 Hz. A design model consisting of three rigid-body modes and the first ten flexible modes of the structure was used in the design process. A 0.5 percent open-loop modal damping ratio was assumed for the flexible modes.

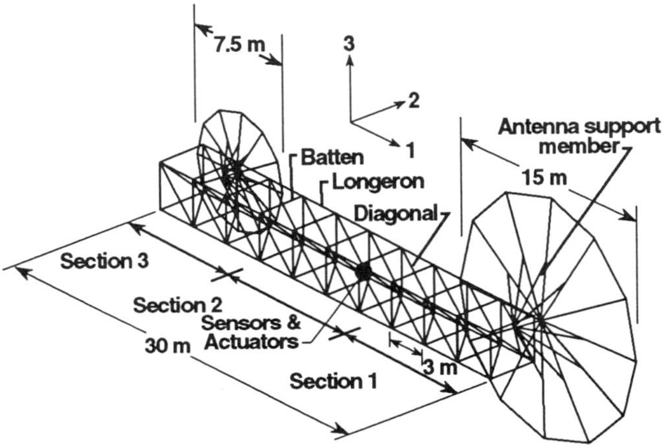

Figure 3. Schematic of the Earth Pointing System

The integrated design problem was formulated as a single-objective optimization problem. The structural design variables used were outer diameters of the truss and antenna support members with the thicknesses fixed. In particular, the truss was divided into three sections shown in Figure 3, and the outer diameters of the longerons, battens, and diagonals within each section constituted nine structural design variables. Two additional structural design variables were the outer diameters of the support members for the two antennas, thus constituting a total of 11 structural design variables. The number of control design variables varies depending on the type of controller used. For the static dissipative design, the elements of the Cholesky factorization matrices of the positive-definite position and rate gain matrices, G_p and G_r, are chosen as design variables.

$$G_p = L_p L_p^T \quad ; \quad G_r = L_r L_r^T \tag{14}$$

In dynamic dissipative designs, the elements of the compensator matrices, A_c and B_c, the elements of weighting matrix Q, and the elements of the Cholesky factorization matrices, L_p and L_r, are used as design variables.

Within this configuration, the following two design problems are considered.

Table I. Open-loop Modal Frequencies for the Nominal EPS Model

Mode Number	Frequency (Hz)
1	0.58
2	0.73
3	0.91
4	2.40
5	2.99
6	3.18
7	6.65
8	7.36
9	16.31
10	17.73

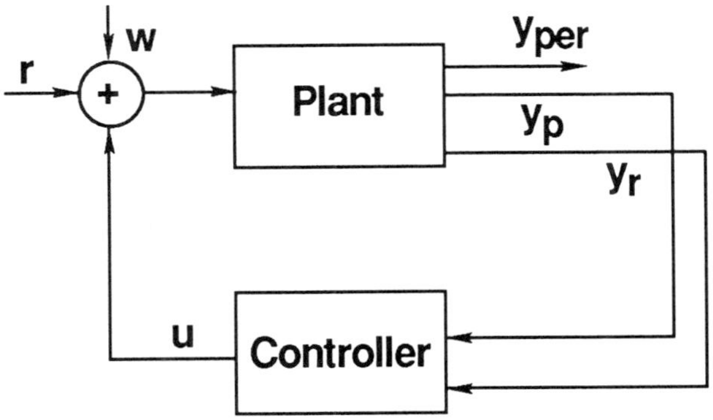

Figure 4. Block diagram of the system configuration for EPS design problem I

A. DESIGN PROBLEM I

The system configuration for the first design problem is illustrated in block diagram shown in Figure 4. Here, white-noise disturbances of unit intensity are applied to the structure at the same locations as the control inputs, and no disturbances at the outputs are considered. The system equations are given as

$$\dot{z} = Az + Bu + Bw \tag{15}$$

$$y_p = C_p z \quad ; \quad y_r = C_r z \tag{16}$$

$$y_{per} \equiv e = C_{per} z$$

In this design problem, the closed-loop performance measure is the steady-state RMS pointing error, e, at the large antenna due to white-noise disturbance of unit intensity at the inputs. In order to achieve a realistic design, constraints are placed on the steady-state average control power and the total structural mass. Additional side-constraints are placed on the structural design variables for safety and practicality reasons. Lower bounds are placed on these variables to satisfy structural integrity requirements against buckling and stress failures. On the other hand, upper bounds are placed on these variables to accommodate manufacturing limitations. Thus, the first design problem is:

Minimize the steady-state RMS pointing error at the large antenna:

$$J = \lim_{t \to \infty} \left[Tr(E\{e(t)e^T(t)\}) \right]^{\frac{1}{2}} \tag{17}$$

with respect to the tube outer diameters $d_i, (i = 1, 2, ..., 11)$, and the elements of L_p, L_r, (where L_p and L_r are 3×3 lower-triangular Cholesky factors of G_p and G_r given in Eq. (9)), subject to the constraints:

$$\lim_{t \to \infty} \left[\mathrm{Tr}(E\{u(t)u^T(t)\}) \right] \leq E_{max} \tag{18}$$

$$M_{str} \leq M_{max} \tag{19}$$

$$d_{min} \leq d_i \leq d_{max} \tag{20}$$

where e is the 3×1 attitude vector at the large antenna; $Tr(.)$ denotes trace of (.); $E(.)$ represents expected value of (.); and M_{str} denotes the structural mass. The subscripts min and max define the range of allowable values.

The steady-state covariance of the pointing error, P_e, is computed from the steady-state covariance of the state vector, i.e.,

$$P_e \equiv \lim_{t \to \infty} Tr\{E(ee^T)\} = Tr\{C_{per} P_z C_{per}^T\} \tag{21}$$

where $P_z = \lim_{t \to \infty} E[zz^T]$ denotes the steady-state covariance of the state which is determined from the solution of the following Liapunov equation [26]

$$A_{cl} P_z + P_z A_{cl}^T = -BB^T \tag{22}$$

in which A_{cl} is the closed-loop system matrix and B is the control (input) influence matrix.

Table II. Integrated Design of the EPS Model-Design Problem I (Static Dissipative Controller)

	RMS Pointing μrad	Structural Mass kg	Actuator Mass kg	Control Power N^2m^2
Initial Design	73.6	442.06	150	2.98
Control-Optimized Design	26.78	442.06	150	3.00
Integrated Design (without act. mass)	16.78	437.34	150	3.00
Integrated Design (with act. mass)	17.01	400.32	298.73	3.00

The results for this design problem are summarized in Table II. An initial control design was first obtained for the nominal structure. A decoupled controller was designed, using a pole placement technique, to provide adequate performance for each of the three rigid body modes, with no consideration given to the flexible modes. That is, the static-dissipative control gain matrices were diagonal with elements chosen to give satisfactory closed-loop frequency and damping for the rigid-body dynamics, and to maintain the RMS pointing error within the required tolerance. The nominal structural mass is 442.04 kg, while the actuator mass was assumed constant at 150 kg.

With the average control power E_{max} constrained at $3N^2m^2$, the initial design gave an RMS pointing error of 73.6 microradians. The conventional design approach was next followed, wherein the control gains (12 elements of the two Cholesky factors L_p and L_r) were optimized for the fixed nominal structure. This "control-optimized" design yielded an RMS pointing error of 26.8 microradians. Next, an integrated design was performed, wherein both the structural and control design variables were allowed to change simultaneously. This resulted in an RMS error of 16.8 microradians, which represents a 37-percent reduction over the conventional design. The structural mass was slightly lower than the nominal design. The upper and lower bound values, initial values, and the optimal values of the structural

Table III. Optimization Results for the Structural Design Variables - EPS Model (Design Problem I)

Design Var	Section No.	Initial Values, m	Final Values, m
1 (longeron)	1	0.051	0.107
2 (batten)	1	0.051	0.030
3 (diagonal)	1	0.051	0.025
4 (longeron)	2	0.051	0.066
5 (batten)	2	0.051	0.010
6 (diagonal)	2	0.051	0.010
7 (longeron)	3	0.051	0.066
8 (batten)	3	0.051	0.041
9 (diagonal)	3	0.051	0.058
10 (support)	large antenna	0.051	0.149
11 (support)	small antenna	0.051	0.010

design variables are summarized in Table III and Figure 5. The integrated design redistributed the structural mass from the battens and diagonals of the last two sections of the main bus (closest to the small antenna) and small antenna support members to the large antenna support members and the section of the main bus closest to the large antenna, thus increasing the stiffness of these sections. This behavioral trend may be attributed to a trade off between structural controllability and observability and its excitability by disturbances. In other words, the stiffness (or flexibility) of the structure is redistributed to establish a balance between the ability of the control system to fine point the structure efficiently and the ability of the structure to reject disturbances. The modal frequencies of the first ten flexible modes of the redesigned EPS structure (redesigned through the integrated design process) are presented in Table IV. These frequencies indicate that the first three frequencies associated with the small antenna support structure have been reduced substantially, mainly because the pointing performance of the small antenna was not included in the performance metric. On the other hand, the frequency of the modes associated with the large antenna support structure has increased considerably (for example, the first modal frequency increased from 0.58 HZ to 1.79 Hz), thus making the large antenna/support structure less sensitive to disturbances at the inputs or the outputs. At the same time, the first flexible frequency of the main truss decreased from about 6.6 Hz to about 6.0 Hz, making the main truss more sensitive to disturbances but more controllable (observable). The elements of the 3 × 3 lower-triangular Cholesky factorization matrices L_p and L_r are given in Table V for the nominal controller design, the control-optimized design, and the integrated controls-structures design. The attitude and rate gain matrices may be computed from the Cholesky matrices by using Eq. (14).

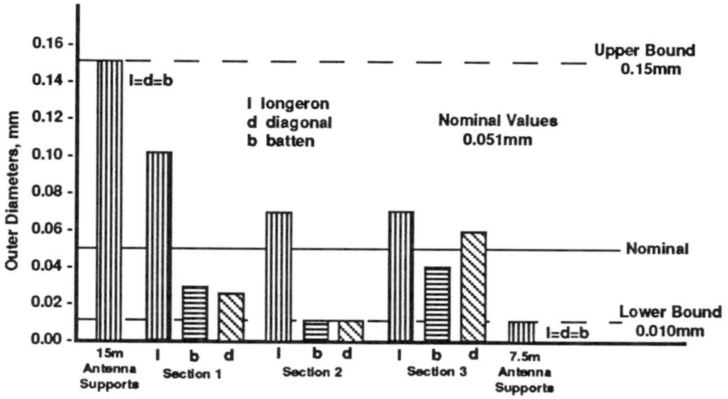

Figure 5. Optimized structural parameters for EPS design problem I

In order to evaluate the effect of varying the actuator mass in the integrated design process, the actuator mass was allowed to vary by relating it to the infinity norms of the gain matrices (a worst-case scenario), i.e.,

$$M_{act} = \alpha|u|_\infty \equiv \alpha\{|G_r|_\infty|y_r|_\infty + |G_p|_\infty|y_p|_\infty\} \tag{23}$$

in which α is a constant scalar taken to be 1.005 and $|.|_\infty$ denotes the infinity norm of $|.|$. It is noted that $|y_p|_\infty$ and $|y_r|_\infty$ were assumed fixed a priori to be $0.1deg$ and $0.1deg/s$, respectively. For this case, the actuator mass increased from 150 kg to 298.7 kg, while the RMS pointing error and the structural mass were essentially unaffected. This is attributed to the fact that the structure is rather stiff and is not affected by small masses.

B. DESIGN PROBLEM II

The system configuration for the second design problem is illustrated in the block diagram shown in Figure 6. Here, the attitude and rate sensors are assumed to be polluted with noise modeled as zero-mean white-noise processes. Furthermore, no disturbances at the inputs are considered. With this configuration, the system equations become

Table IV. Modal Frequencies for the Redesigned EPS Model

Mode Number	Frequency (Hz)
1	0.17
2	0.22
3	0.23
4	2.75
5	3.11
6	3.30
7	5.98
8	6.62
9	8.58
10	14.13

Table V. Elements of Cholesky Matrices for the Attitude and Rate Gain Matrices - EPS Model (Design Problem I)

Design Var.	Nominal Design	Control-optimized Design	Integrated Design
$L_p\,(1,1)$	120.0	188.7	254.3
$L_p\,(2,1)$	0.0	7.1	1.2
$L_p\,(3,1)$	0.0	3.9	-1.7
$L_p\,(2,2)$	180.0	179.2	245.0
$L_p\,(3,2)$	0.0	7.4	0.6
$L_p\,(3,3)$	240.0	189.9	255.7
$L_r\,(1,1)$	90.0	212.5	223.2
$L_r\,(2,1)$	0.0	-17.4	-2.9
$L_r\,(3,1)$	0.0	15.3	8.3
$L_r\,(2,2)$	150.0	272.1	327.9
$L_r\,(3,2)$	0.0	-33.0	-1.0
$L_r\,(3,3)$	200.0	293.8	403.3

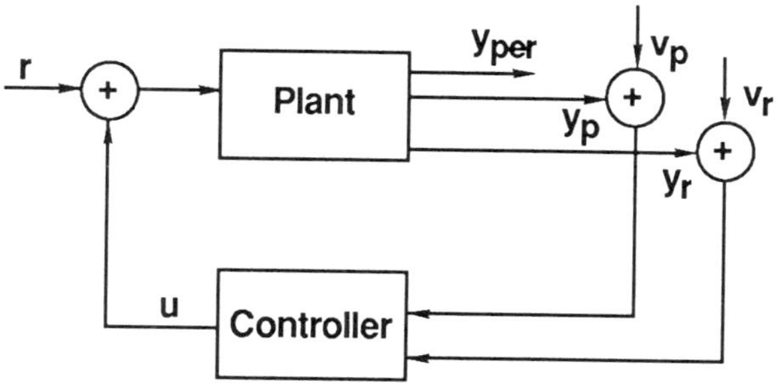

Figure 6. Block diagram of the system configuration for EPS design problem II

$$\dot{z} = Az + Bu \tag{24}$$

$$y_p = C_p z + S_p v_p \quad ; \quad y_r = C_r z + S_r v_r \tag{25}$$

$$y_{per} \equiv e = C_{per} z$$

where v_p and v_r denote the attitude and rate sensors noise vectors, and e represents the RMS pointing error at the large antenna. In this design problem, a dual-objective formulation is pursued, wherein both the total mass (structure plus actuator) and a measure of the closed-loop performance are optimized. The closed-loop performance measure is chosen as the sum of the time constants of the closed-loop eigenvalues (the reciprocals of the absolute value of the real part of the closed-loop eigenvalues) which represents a measure of the transient performance of the system, i.e., the smaller the time constants are the faster the system dissipates transient disturbances. Define the closed-loop or controlled performance, P_c, as the inverse of the sum of the time constants of the system, such that

$$P_c = \frac{1}{\sum[\frac{1}{|\mathrm{Re}(\lambda_i)|}]} \tag{26}$$

Now, in order to minimize the total mass and maximize the closed-loop performance, an objective function consisting of a convex combination of the total mass and the inverse of the controlled performance P_c is minimized. The design problem is to minimize

$$J = \beta \frac{(M_{str} + M_{act})}{(M_{str}^o + M_{act}^o)} + (1 - \beta) \frac{\frac{1}{P_c}}{\frac{1}{P_c^o}} \tag{27}$$

with respect to structural and/or control design variables, subject to a constraint on the RMS pointing error at the large antenna

$$\lim_{t \to \infty} \left[\text{Tr}(E\{e(t)e^T(t)\}) \right] \leq e_{max} \tag{28}$$

as well as side constraints on structural member sizes

$$d_{min} \leq d_i \leq d_{max} \tag{29}$$

Note that both the mass term and the term for the controlled performance in Eq. (27) are normalized with respect to their corresponding initial values. In Eq. (27), M_{str}, M_{act} denote the truss mass and the actuator mass, and the superscript "o" denotes the initial value of the corresponding variable. The steady-state RMS pointing error is computed from the steady-state covariance of the state vector, i.e.,

$$P_e \equiv \lim_{t \to \infty} Tr\{E(ee^T)\} = Tr\{C_{per} P_z C_{per}^T\} \tag{30}$$

where $P_z = \lim_{t \to \infty} E(zz^T)$ denotes the closed-loop steady-state covariance of the state which is determined from the solution of the following Liapunov equation [26]

$$A_{cl} P_z + P_z A_{cl}^T = -BG_p W_p G_p^T B^T - BG_r W_r G_r^T B^T \tag{31}$$

in which the attitude measurement noise intensity matrix, $W_p = E(v_p v_p^T)$, and the rate measurement noise intensity matrix, $W_r = E(v_r v_r^T)$, are assumed to be diagonal, i.e., $W_p = I_{3\times3}(arc-s)^2$ and $W_r = 2 \times I_{3\times3}(arc-s/s)^2$. The pointing tolerance e_{max} in Eq. (27) is taken as $11 \mu rad$.

The parameter β allows for a trade-off between the two objectives. As β is varied from 1 to 0, more emphasis is placed on the closed-loop performance and less on the total mass or the cost. Furthermore, in order to satisfy certain fine-pointing requirements, an upper bound constraint is placed on the RMS error at the large antenna due to white measurement noise in the attitude and rate sensors. Additional side constraints are also placed on the structural design variables for safety and practicality concerns. Figures 7(a)-7(d) show, respectively, the behavior (normalized relative to the nominal design) of the objective function J, controlled performance (a measure of transient performance), structural mass, and actuator mass as a function of the trade-off parameter β. As β approaches zero, the design becomes "performance" or "control" dominated, as evidenced by the sharp increase in the controlled performance (see Figure 7(b)) and the substantial increases in the structural and actuator masses (see Figures 7(c) and 7(d)). On the other hand, as β approaches one, the controlled performance diminishes considerably, while

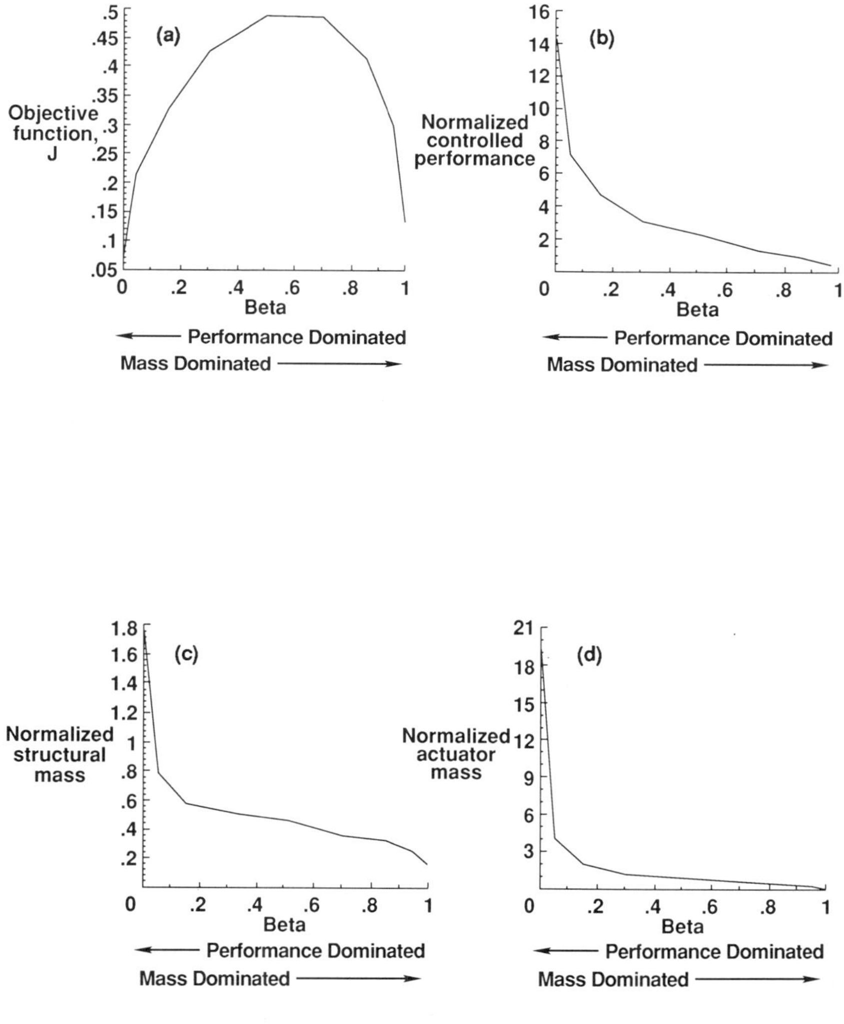

Figure 7. Controls-Structures trade study for the EPS Model

Table VI. Integrated Design of the EPS Model-Design Problem II (Static Dissipative Controller) (RMS $< 11\mu$rad)

	Normalized Controlled Performance	Normalized Structural Mass	Normalized Actuator Mass	Normalized Total Mass
Initial Design	1.0	1.0	1.0	1.0
Control-Optimized Design $\beta = 0.50$	1.44	1.0	1.33	1.01
Integrated Design $\beta = 0.50$	2.34	0.47	0.76	0.95

Table VII. Optimization Results for the Structural Design Variables - EPS Model (Design Problem II)

Design Var	Section No.	Initial Values, m	Final Values, m
1 (longeron)	1	0.051	0.010
2 (batten)	1	0.051	0.010
3 (diagonal)	1	0.051	0.010
4 (longeron)	2	0.051	0.018
5 (batten)	2	0.051	0.010
6 (diagonal)	2	0.051	0.010
7 (longeron)	3	0.051	0.010
8 (batten)	3	0.051	0.010
9 (diagonal)	3	0.051	0.010
10 (support)	large antenna	0.051	0.114
11 (support)	small antenna	0.051	0.083

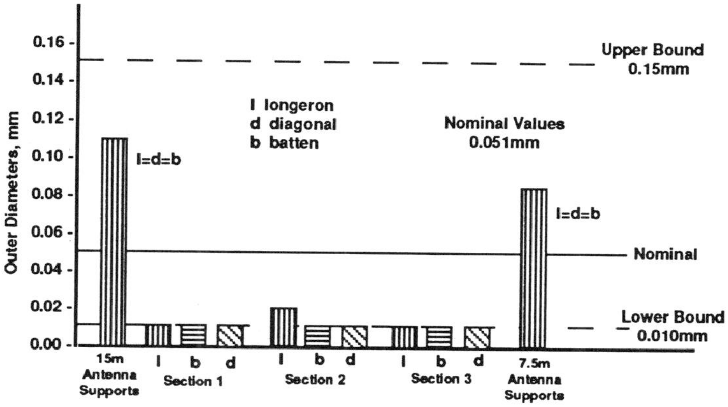

Figure 8. Optimized structural parameters for EPS design problem II

Table VIII. Modal Frequencies for the Redesigned EPS Model (Design Problem II)

Mode Number	Frequency (Hz)
1	1.79
2	2.04
3	2.16
4	3.48
5	3.76
6	4.63
7	5.52
8	5.74
9	9.26
10	9.71

the structural mass and actuator mass decrease, giving a "mass" or "cost" dominated design. One may use these trade-off results to obtain a design that gives a good controlled performance with acceptable cost levels by choosing a right value for β. For example, with $\beta = 0.50$, Table VI compares the corresponding integrated design with a control-optimized (conventional) design. An initial design to achieve good rigid-body performance (same as the first design problem) was first obtained. This design required an actuator mass of about 171 kg (computed from Eq. (23)), which along with the nominal structural mass of 442.04 kg and payload mass of 5168 kg, resulted in a total spacecraft mass of 5781.04 kg. The controlled performance measure P_c was about 0.003. Then, a control-optimized design was performed by optimizing the control variables only. This resulted in a 44-percent increase in the control performance, but also increased the total mass by 1 percent (57.17 kg). An integrated design was next carried out, which resulted in a 234-percent increase in the control performance along with lower total mass over the initial design (6 percent less than the total mass of the control-optimized design, or about 353 kg). Integrated design reduced the structural mass by 53 percent and decreased the actuator mass by 24 percent. The optimization data and results for the structural design variables are given in Table VII and Figure 8. It is observed that the structural mass was reduced and redistributed, i.e., mass was taken from the main bus structure and was added to the antenna support members. In other words, most of the longeron, batten, and diagonal dimensions of the main bus structure were reduced to lower bound values, while the antenna support members increased drastically in size. The open-loop modal frequencies of the redesigned EPS structure (obtained through integrated design process) are presented in Table VIII, where it is seen that the first flexible frequency corresponding to the large antenna support structure increased to near 1.8 Hz, with a reduction of truss flexible frequencies by as much as 50 percent at the higher end of the spectrum. In this configuration, the antennas are less susceptible to disturbances at the inputs or the outputs, while the overall structure has become more controllable (observable). The closed-loop eigenvalues are presented in Figures 9, 10, and 11 for the nominal design, the control-optimized design, and the integrated controls-structures design, respectively. These figures indicate that integrated design is more effective than the conventional design in moving the closed-loop eigenvalues of the system farther into the left-half plane (such that the system can dissipate transient disturbances faster), thus clearly demonstrating the advantage of the integrated design. Also, the Cholesky factorization matrices L_p and L_r (associated with the attitude and rate gain matrices) are given in Table IX for the nominal design, the control-optimized design, and the integrated design.

Summarizing the results obtained from the first and the second design problems, both clearly show the advantage of integrated design over the conventional approach in that the integrated design produces a better overall design. The main advantage of integrated design is its ability to obtain a better overall design and not necessarily just a reduction of the total mass. Moreover, comparison the behavioral trends of the structural design variables for the first and second design problems indicate that such trends could vary considerably depending on the design objectives and constraints of the integrated design problem.

Table IX. Cholesky Matrices for the Attitude and Rate Gain Matrices - EPS Model (Design Problem II)

Design Var.	Nominal Design	Control-optimized Design	Integrated Design
$L_p(1,1)$	120.0	201.9	178.4
$L_p(2,1)$	0.0	1.2	-0.005
$L_p(3,1)$	0.0	1.5	-0.001
$L_p(2,2)$	180.0	254.3	178.4
$L_p(3,2)$	0.0	1.8	0.012
$L_p(3,3)$	240.0	254.6	178.3
$L_r(1,1)$	90.0	184.0	206.0
$L_r(2,1)$	0.0	4.3	0.001
$L_r(3,1)$	0.0	-1.8	0.002
$L_r(2,2)$	150.0	251.5	205.5
$L_r(3,2)$	0.0	-2.4	-0.006
$L_r(3,3)$	200.0	252.0	200.9

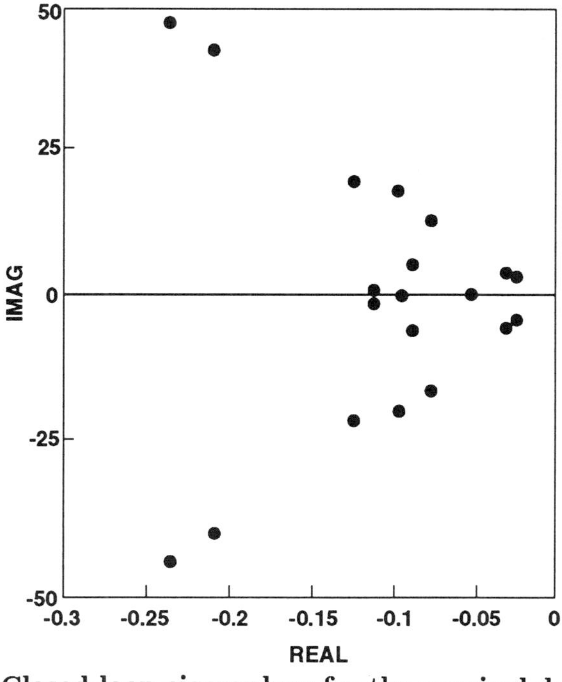

Figure 9. Closed-loop eigenvalues for the nominal design - EPS design problem II

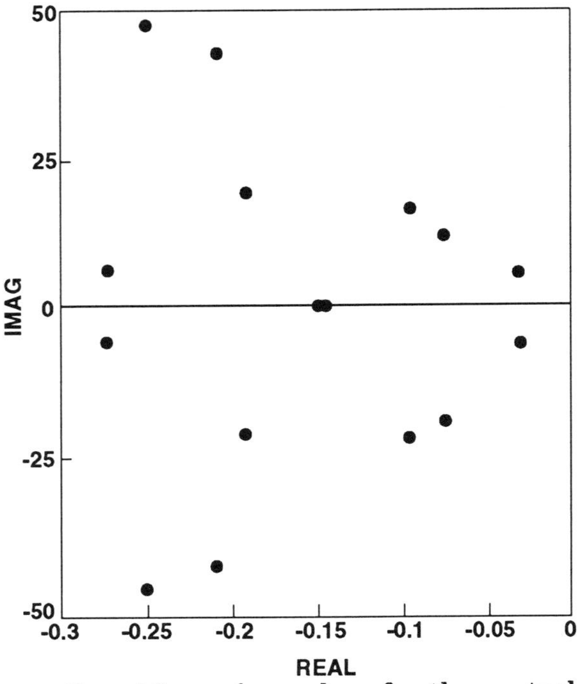

Figure 10. Closed-loop eigenvalues for the control-optimized design - EPS design problem II

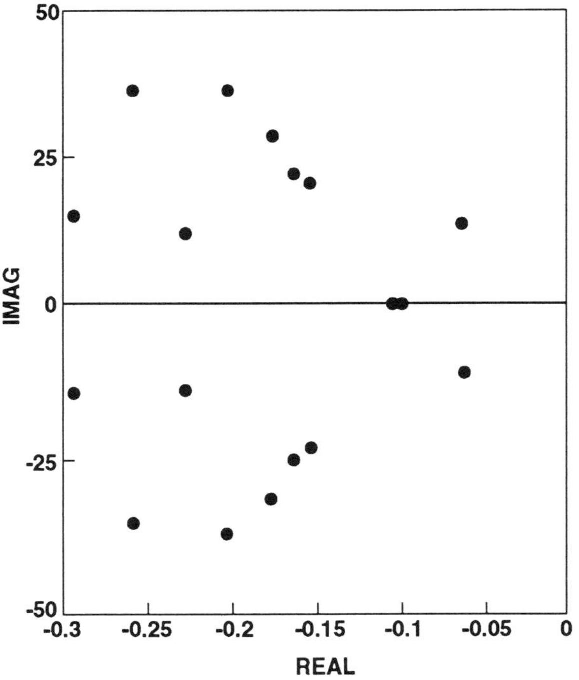

Figure 11. Closed-loop eigenvalues for the integrated controls-structures design - EPS design problem II

VI. INTEGRATED DESIGN OF THE CSI EVOLUTIONARY MODEL

This section considers the application of the integrated design methodology to the CSI Evolutionary Model. The Phase-Zero Evolutionary Model, shown in Figure 12, basically consists of a 62-bay central truss (each bay 10 inches long), two vertical towers, and two horizontal booms. The structure is suspended from the ceiling (about 840 inches above the main truss) using two cables as shown. A laser source is mounted at the top of one of the towers, and a reflector with a mirrored surface is mounted on the other tower. The laser beam is reflected by the mirrored surface onto a detector surface 660 inches above the reflector. Eight proportional bi-directional gas thrusters, with maximum output force of 4.4 lbs each, provide the input actuation, while collocated servo accelerometers provide output measurements. The finite element model of the system has 3216

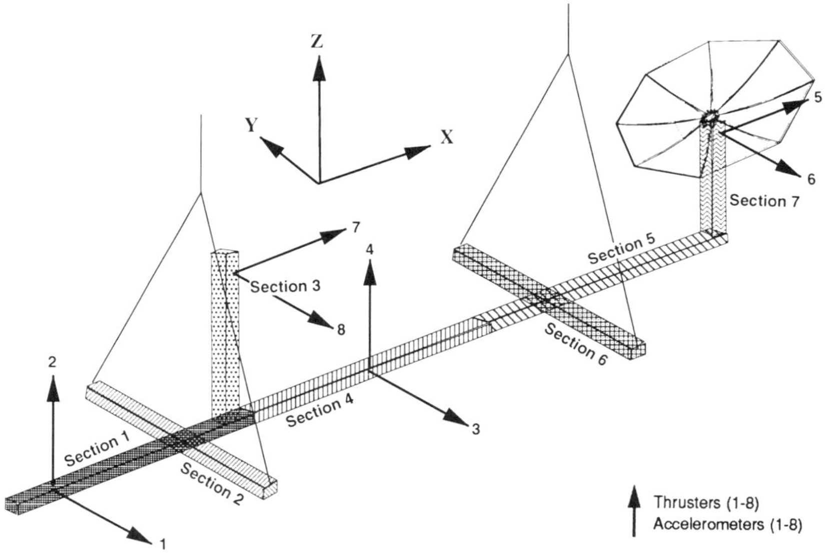

Figure 12. Schematic of the CSI Phase-Zero Evolutionary Model

degrees of freedom; therefore, the bulk of the computational effort is required for the solution of the structural eigenvalue problem of that size. The design model consisted of the first 20 modes of the structure, including six suspension modes (modes due to the suspension of the structure and gravity) and 14 flexible modes. A modal damping ratio of 0.1 percent was assumed. The first 20 modes of the nominal Phase-0 model are presented in Table X. The first six modes which range from 0.12 Hz to 1.16 Hz are the suspension modes (or nonzero frequency rigid-body modes). The first two bending modes (lateral and vertical) and the first torsional mode of the structure are modes 7, 8 and 9. The system configuration for this design problem is illustrated in the block diagram shown in Figure 13. Here, white-noise disturbances of unit intensity are applied to the structure at the same locations as the control inputs, and no disturbances at the outputs are considered. With no appreciable sensor and actuator dynamics considered, the system equations are written as

$$\dot{z} = Az + Bu + Bw \tag{32}$$

$$\begin{aligned} y_r &= C_r z \\ y_{per} \equiv e &= C_p z \end{aligned} \tag{33}$$

Table X. Modal Frequencies for the CSI Phase-0 Evolutionary Model

Mode Number	Frequency (Hz)
1	0.122
2	0.126
3	0.173
4	0.680
5	0.704
6	1.159
7	1.572
8	1.676
9	2.085
10	3.867
11	3.972
12	4.127
13	4.172
14	5.788
15	6.456
16	6.568
17	6.777
18	7.806
19	8.732
20	9.396

where the performance vector $e(t)$ is an 8×1 vector corresponding to the displacements at the eight accelerometer locations. Here, only velocity measurements are used for feedback, and they are obtained by integrating the accelerometer signals with the aid of wash-out filters. It is noted that position feedback is not necessary for asymptotic stability since the structure is open-loop stable.

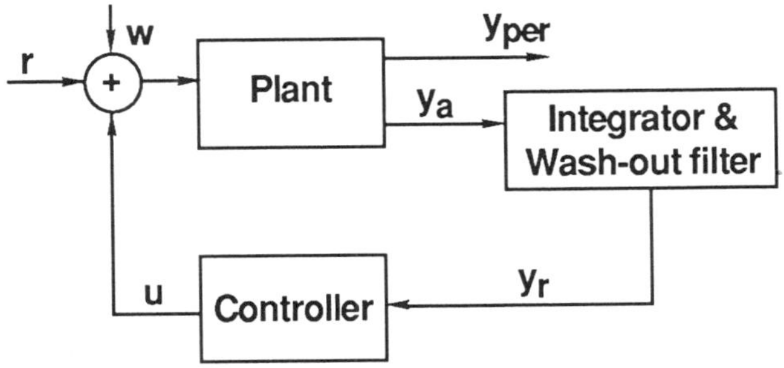

Figure 13. Block diagram of the system configuration for the CSI Evolutionary Model

To perform the integrated design, the structure was divided into seven sections (see Figure 12), three sections in the main bus, and one section each for the two horizontal booms and two vertical towers. Three structural design variables were used in each section, namely, outer diameters of the longerons, the battens, and the diagonals, making a total of 21 structural design variables. The design was posed in the form of a nonlinear optimization problem, wherein the steady-state average control power in the presence of a white noise input disturbance with unit intensity (i.e., standard deviation intensity = 1 lbf) is minimized, with the steady-state RMS position error at the eight accelerometer locations and total mass constrained to required values. That is, the problem solved was:

Minimize

$$\lim_{t \to \infty} Tr\{E\{u(t)u^T(t)\}\} \tag{34}$$

subject to

$$\lim_{t \to \infty} \left[Tr(E\{e(t)e^T(t)\})\right]^{\frac{1}{2}} \leq e_{max} \tag{35}$$

and a constraint on the total mass

$$M_{tot} \leq M_{max} \tag{36}$$

Here, e_{max} was taken to be 1.7 inches and M_{max} was chosen equal to the total mass of the nominal Phase-0 structure at 2.0 $lb - s^2/in$.

A. STATIC DISSIPATIVE CONTROLLER

As mentioned earlier, only velocity measurements are used for feedback, which are obtained by processing the accelerometer outputs. Thus, the static dissipative controller is given by

$$u = -G_r y_r \tag{37}$$

Here, the rate gain matrix was an 8×8 diagonal matrix, i.e., there were eight control design variables which along with the 21 structural design variables resulted in a total of 29 design variables for the design optimization.

The results of the design optimization are summarized in Table XI. A conventional or "control-optimized" design was first performed (with the structural design fixed at the nominal values), which required an average steady-state control power of 19.34 lb^2.

Table XI. Integrated Design of the CSI Evolutionary Model (Static Dissipative Controller)

	RMS Displacement, in	Control Power, lb^2	Transient Performance, P_c
Control-Optimized Design	1.70	19.34	0.0048
Integrated Design	1.70	11.41	0.0038

Next, an integrated design was performed , wherein the average control power was minimized with respect to both control and structural design variables. The results (see Table XI) indicate an average control power of 11.41 lb^2, a reduction of about 40 percent in the average control power over the conventional design. The initial and final values of the structural design variables, along with the corresponding lower bound and upper bound values, are presented in Table XII and Figure 14. Keeping in mind that the tube diameters of the nominal CSI Evolutionary structure are 0.367 inches for the longerons and battens, and 0.349 inches for the diagonals, it is observed from Table XII and Figure 14 that sections 1 and 2 of the main bus (furthest away from the reflector) and the laser tower are

Table XII. Optimization Results for the Structural Design Variables - CSI Evolutionary Model (Static Dissipative Controller)

Design Var.	Section No.	Initial Values, in	Final Values, in
1 (longeron)	1	0.3125	0.999
2 (batten)	1	0.3125	0.633
3 (diagonal)	1	0.3125	0.958
4 (longeron)	2	0.3125	0.124
5 (batten)	2	0.3125	0.644
6 (diagonal)	2	0.3125	0.144
7 (longeron)	3	0.3125	0.999
8 (batten)	3	0.3125	0.453
9 (diagonal)	3	0.3125	0.658
10 (longeron)	4	0.3125	1.000
11 (batten)	4	0.3125	0.181
12 (diagonal)	4	0.3125	0.407
13 (longeron)	5	0.3125	0.425
14 (batten)	5	0.3125	0.143
15 (diagonal)	5	0.3125	0.138
16 (longeron)	6	0.3125	0.122
17 (batten)	6	0.3125	0.392
18 (diagonal)	6	0.3125	0.145
19 (longeron)	7	0.3125	0.142
20 (batten)	7	0.3125	0.183
21 (diagonal)	7	0.3125	0.427

considerably stiffened, while the horizontal booms and the reflector tower became more flexible, partly to satisfy the mass constraint. Generally, in those sections that showed an increase in stiffness, the longerons increased in size more than the diagonals and the battens, since they are most effective in changing the stiffness of a section. The trends in Table XII may be attributed to a trade off between structural controllability, observability, and excitibility. The areas near the disturbance sources (actuator locations) were stiffened in order to reduce the sensitivity of the structure to external disturbances at those locations, while ensuring that no appreciable loss of controllability and/or observalibility occured (as observed from the transient controlled performance values in Table XI). It should be noted that the small reduction in the transient controlled performance of the integrated design over the conventional design in Table XI is mainly because no measure of the transient performance was included in the optimization process. Moreover, although the transient controlled performance is lower in the integrated design case, it is still quite acceptable. The modal frequencies of the first twenty modes of the redesigned Phase-0 structure (redesigned through the integrated design process) are presented in Table XIII. These frequencies indicate that the first six frequencies associated with the suspended structure have not been changed significantly, mainly because the changes in the structure can effect these frequencies only through changing the location of the center of mass of the structure, and not directly as for the flexible modes. On the other hand, the frequency of the flexible modes, particularly the second and third flexible modes, have increased considerably (as much as 40-percent). The second flexible mode frequency increased from 1.68 Hz to 2.33 Hz and the third flexible mode frequency from 2.08 Hz to 2.65 Hz, making these modes and the structure less sensitive to disturbances at the inputs. The diagonal elements of the rate gain matrix are given in Table XIV for the control-optimized design and the integrated controls-structures design.

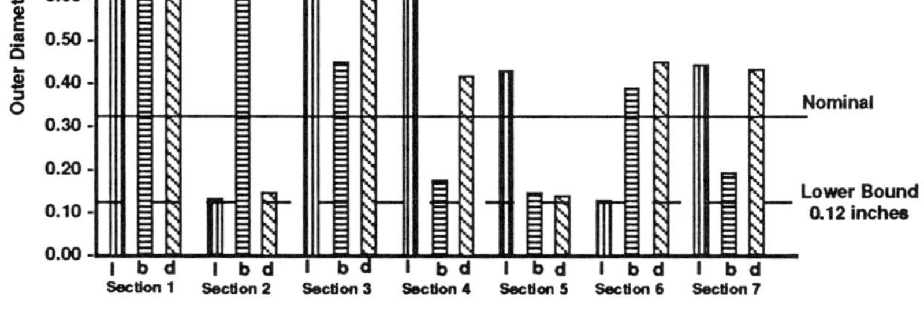

Figure 14. Optimized structural parameters for the CSI Evolutionary Model - static dissipative controller

Table XIII. Modal Frequencies for the Redesigned CSI Evolutionary Model (Static Dissipative Controller)

Mode Number	Frequency (Hz)
1	0.122
2	0.127
3	0.173
4	0.625
5	0.684
6	1.191
7	1.642
8	2.334
9	2.651
10	3.829
11	3.995
12	4.226
13	4.598
14	5.379
15	7.092
16	7.639
17	7.910
18	8.599
19	8.994
20	9.020

Table XIV. Elements of Rate Gain Matrix for the CSI Evolutionary Model (Static Dissipative Controller)

Design Var.	Control-optimized Design	Integrated Design
$G_r\,(1,1)$	1.525	1.263
$G_r\,(2,2)$	0.409	0.477
$G_r\,(3,3)$	1.234	1.203
$G_r\,(4,4)$	0.448	0.440
$G_r\,(5,5)$	2.126	1.674
$G_r\,(6,6)$	2.122	1.671
$G_r\,(7,7)$	0.352	0.869
$G_r\,(8,8)$	0.680	0.626

B. DYNAMIC DISSIPATIVE CONTROLLER

The next two evolutionary model designs performed were conventional (control-optimized) and integrated designs with a dynamic dissipative controller. Since the system has no zero-frequency modes, G_p and G_r were taken to be null matrices. The dynamic dissipative controller represented by Eqs. (10) and (11) was used with block-diagonal compensator state matrix A_c (consisting of eight second-order blocks), and compensator input influence matrix B_c, i.e.,

$$A_c = \begin{bmatrix} A_{c1} & 0 & \dots & 0 \\ 0 & A_{c2} & \dots & 0 \\ \vdots & \vdots & \ddots & \vdots \\ 0 & 0 & \dots & A_{c8} \end{bmatrix} \tag{38}$$

$$B_c = \begin{bmatrix} B_{c1} & 0 & \dots & 0 \\ 0 & B_{c1} & \dots & 0 \\ \vdots & \vdots & \ddots & \vdots \\ 0 & 0 & \dots & B_{c8} \end{bmatrix} \tag{39}$$

where A_{ci} and B_{ci}, $i = 1, 2, \dots, 8$ are, respectively, 2×2 matrices and 2×1 vectors, defined as

$$A_{ci} = \begin{bmatrix} 0 & 1 \\ -\alpha_i & -\beta_i \end{bmatrix} \quad ; \quad B_{ci} = \begin{bmatrix} 0 \\ 1 \end{bmatrix} \tag{40}$$

Furthermore, the weighting matrix Q in Eq. (12) is assumed to be diagonal, i.e.,

$$Q = Diag(q_1, q_2, \ldots, q_{16}) \tag{41}$$

Here, the scalar variables $\alpha_i, \beta_i, i = 1, 2, \ldots, 8$ and $q_j, j = 1, 2, \ldots, 16$ were chosen for the control design variables. Thus, the number of control design variables was 32, making the total number of design variables 53. Table XV shows the results of the designs with dynamic dissipative controller. The conventional design reduces the control power by over 50-percent (8.31 lb^2) compared to the static dissipative case. The integrated design reduces the average control power by another 30-percent (over the conventional design) to 5.91 lb^2. Table XV also indicates that the transient controlled performance is slightly lower for the integrated design case. Similar to the static dissipative design, this reduction in performance may be attributed to the fact that the optimization process did not include any measure of transient performance. Moreover, although comparison of Tables XI and XV implies that the transient controlled performance of the dynamic dissipative design was quite lower than the static dissipative design, one should keep in mind that in the dynamic dissipative design, this performance measure includes contributions from the compensator poles, and thus it is not readily comparable with the static dissipative values.

Table XV. Integrated Design of the CSI Evolutionary Model (Dynamic Dissipative Controller)

	RMS Displacement, *in*	Control Power, lb^2	Transient Performance, P_c
Control-Optimized Design	1.70	8.31	0.0026
Integrated Design	1.70	5.91	0.0024

The optimization data and results for the structural design variables are summarized in Table XVI and Figure 15. The trends are quite similar to the static dissipative design given in Table XII. As a matter of fact, the two optimal structures are quite close to each other (within 20-percent of each other), i.e., the structure that is optimal for the static dissipative design is also optimal for the dynamic dissipative design. The modal frequecies of the optimal structure for the dynamic dissipative design are summarized in Table XVII. These frequencies are also close (within 5-percent) to their corresponding values for the static dissipative design. Finally, the singular value plots for the input-to-output transfer functions of the dynamic dissipative compensator are presented in Figures 16(a)-16(h) for the control-optimized design and the integrated controls-structures design. The singular value plots for the control-optimized design and

Table XVI. Optimization Results for the Structural Design Variables CSI Evolutionary Model (Dynamic Dissipative Controller)

Design Var.	Section No.	Initial Values, in	Final Values, in
1 (longeron)	1	0.3125	0.999
2 (batten)	1	0.3125	0.547
3 (diagonal)	1	0.3125	1.000
4 (longeron)	2	0.3125	0.120
5 (batten)	2	0.3125	0.540
6 (diagonal)	2	0.3125	0.120
7 (longeron)	3	0.3125	1.000
8 (batten)	3	0.3125	0.354
9 (diagonal)	3	0.3125	0.579
10 (longeron)	4	0.3125	1.000
11 (batten)	4	0.3125	0.138
12 (diagonal)	4	0.3125	0.367
13 (longeron)	5	0.3125	0.357
14 (batten)	5	0.3125	0.120
15 (diagonal)	5	0.3125	0.120
16 (longeron)	6	0.3125	0.120
17 (batten)	6	0.3125	0.308
18 (diagonal)	6	0.3125	0.120
19 (longeron)	7	0.3125	0.120
20 (batten)	7	0.3125	0.135
21 (diagonal)	7	0.3125	0.332

integrated design are somewhat similar except for input/output channel No. 1 and 7 (see Figures 16(a) and 16g)), where considerable reduction in the singular values, mainly in the lower end of the frequency spectrum, and increase in the bandwidth is observed in the integrated design case over the conventional design. The elements of the compensator state matrix A_c are given in Table XVIII for the control-optimized design and the integrated controls-structures design. Also, the elements of the weighting matrix Q are presented in Table XIX for both designs.

The results obtained for both the static and dynamic dissipative controllers clearly show that integrated controls-structures design methodology can yield a substantially superior overall design over the conventional sequential design scenario.

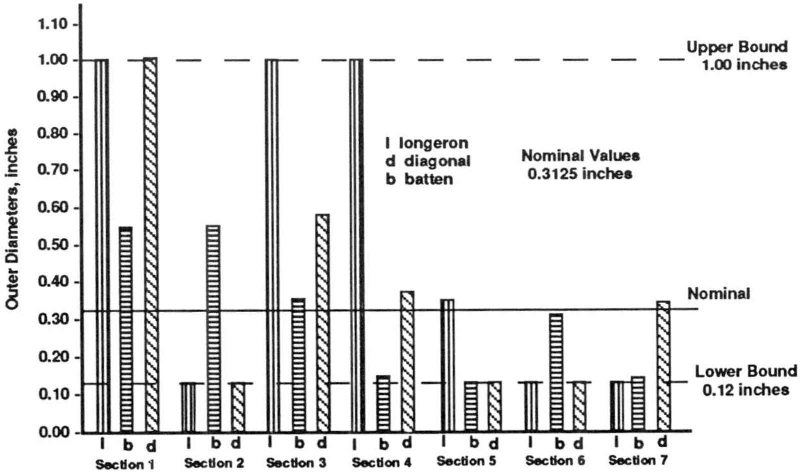

Figure 15. Optimized structural parameters for the CSI Evolutionary Model - dynamic dissipative controller

Table XVII. Modal Frequencies for the Redesigned CSI Evolutionary Model (Dynamic Dissipative Controller)

Mode Number	Frequency (Hz)
1	0.122
2	0.127
3	0.173
4	0.629
5	0.690
6	1.186
7	1.578
8	2.238
9	2.576
10	3.761
11	3.782
12	4.174
13	4.320
14	5.196
15	6.443
16	7.351
17	7.876
18	8.549
19	8.922
20	8.982

Table XVIII. Control Design Variables for the CSI Evolutionary Model (Dynamic Dissipative Controller)

Design Var.	Control-optimized Design	Integrated Design
$A_c(2,1)$	11.400	45.204
$A_c(2,2)$	66.351	107.918
$A_c(4,3)$	82.667	72.342
$A_c(4,4)$	92.846	80.691
$A_c(6,5)$	20.989	35.454
$A_c(6,6)$	70.983	94.716
$A_c(8,7)$	89.514	74.961
$A_c(8,8)$	92.700	79.441
$A_c(10,9)$	27.749	43.161
$A_c(10,10)$	42.645	88.555
$A_c(12,11)$	21.109	44.916
$A_c(12,12)$	50.887	91.610
$A_c(14,13)$	21.495	38.261
$A_c(14,14)$	100.329	86.221
$A_c(16,15)$	22.814	29.081
$A_c(16,16)$	81.224	90.669

Table XIX. Control Design Variables for the CSI Evolutionary Model (Dynamic Dissipative Controller)

Design Var.	Control-optimized Design	Integrated Design
$Q(1,1)$	1903.7	13117.1
$Q(2,2)$	3585.4	10156.0
$Q(3,3)$	3126.9	2632.3
$Q(4,4)$	3155.5	2472.4
$Q(5,5)$	3704.8	8693.4
$Q(6,6)$	3054.8	5610.7
$Q(7,7)$	3008.7	2292.9
$Q(8,8)$	3047.9	2254.9
$Q(9,9)$	4088.5	12123.5
$Q(10,10)$	2279.1	8354.5
$Q(11,11)$	3018.3	11099.9
$Q(12,12)$	3302.3	9466.5
$Q(13,13)$	4205.1	6252.3
$Q(14,14)$	2455.0	3939.0
$Q(15,15)$	3711.6	4858.3
$Q(16,16)$	2917.6	3167.5

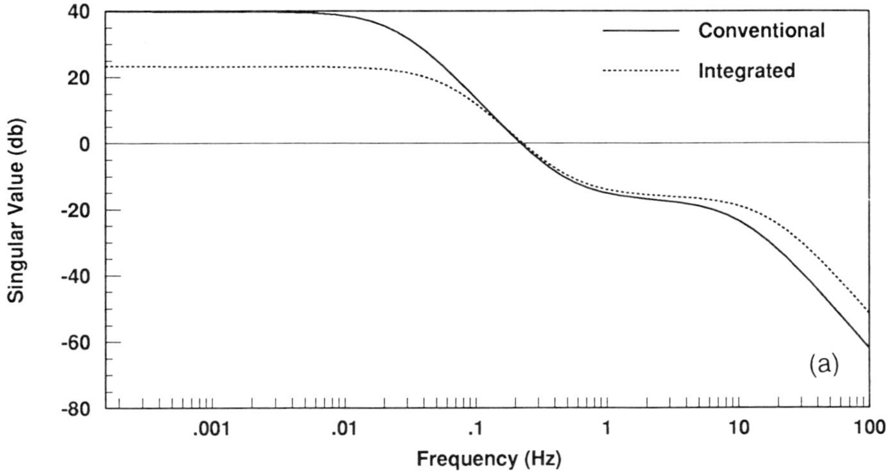

Figure 16(a). Singular value plots for input/output channel No. 1 of the dynamic dissipative controller - CSI Evolutionary Model

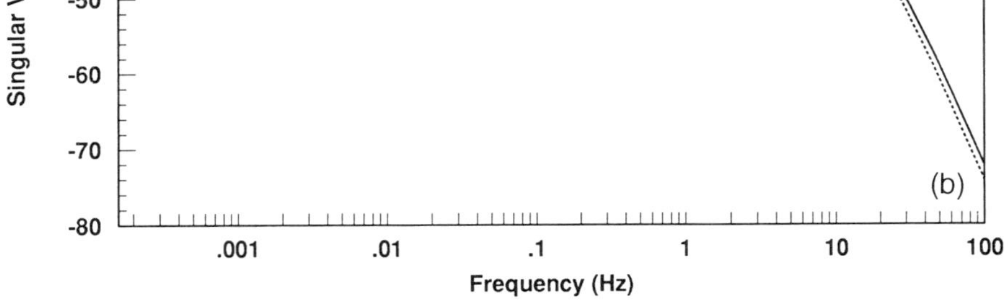

Figure 16(b). Singular value plots for input/output channel No. 2 of the dynamic dissipative controller - CSI Evolutionary Model

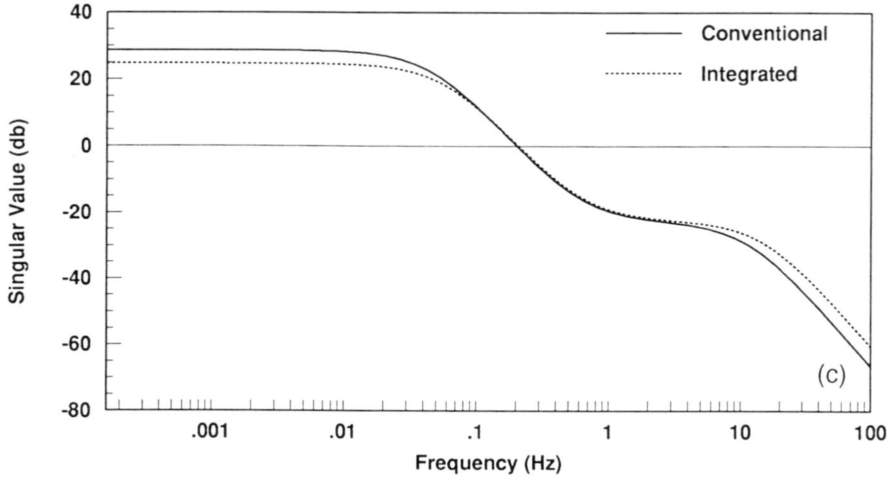

Figure 16(c). **Singular value plots for input/output channel No. 3 of the dynamic dissipative controller - CSI Evolutionary Model**

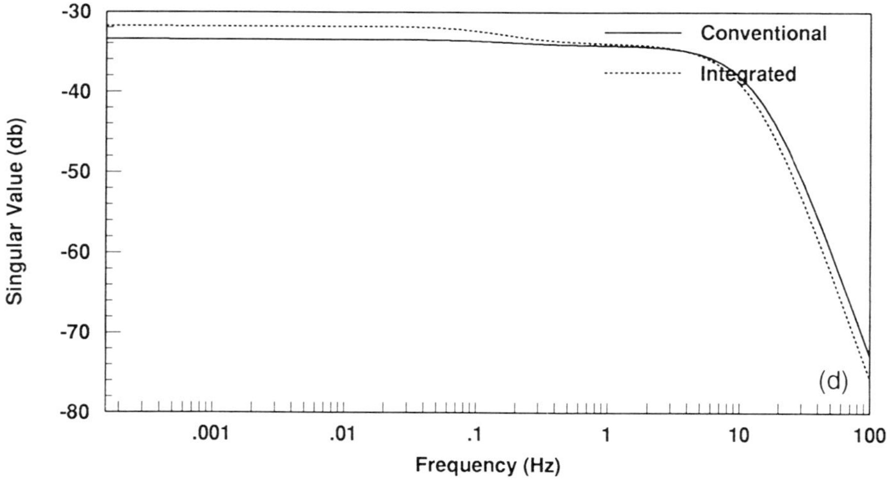

Figure 16(d). **Singular value plots for input/output channel No. 4 of the dynamic dissipative controller - CSI Evolutionary Model**

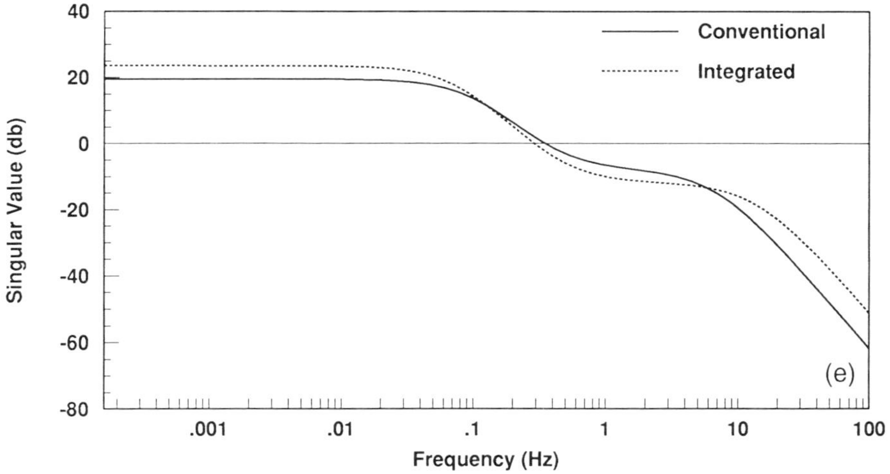

Figure 16(e). Singular value plots for input/output channel No. 5 of the dynamic dissipative controller - CSI Evolutionary Model

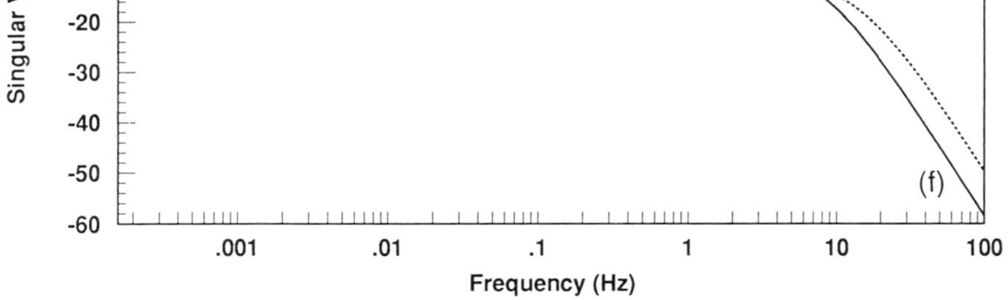

Figure 16(f). Singular value plots for input/output channel No. 6 of the dynamic dissipative controller - CSI Evolutionary Model

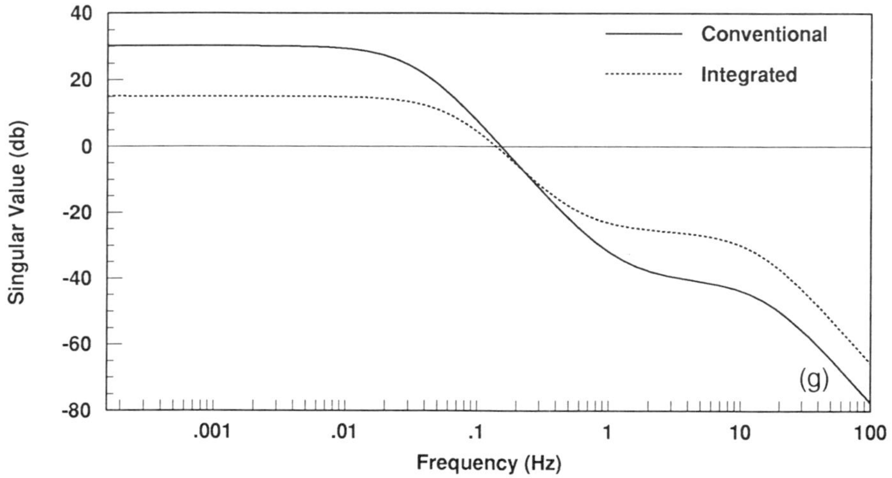

Figure 16(g). Singular value plots for input/output channel No. 7 of the dynamic dissipative controller - CSI Evolutionary Model

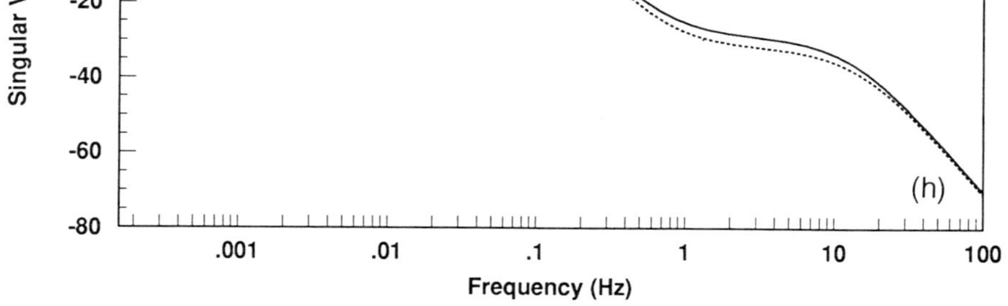

Figure 16(h). Singular value plots for input/output channel No. 8 of the dynamic dissipative controller - CSI Evolutionary Model

VII. CONCLUDING REMARKS

An optimization-based approach has been developed for performing integrated controls-structures design of a class of flexible spacecraft. The approach formulates the problem as a constrained optimization problem, wherein the set of design variables consists of both control and structural design variables. The approach uses static and dynamic dissipative control laws, which provide robust stability in the presence of parametric and nonparametric uncertainties in the model. The approach was demonstrated by application to integrated designs of a generic model of a space platform concept, as well as to a model of the Controls-Structures Interaction Program's Evolutionary Model, which is a ground-based experimental testbed developed and constructed at NASA-Langley Research Center. The numerical results obtained indicate that the integrated design approach can yield substantially superior spacecraft design as compared to the traditional sequential design approach. For example, a 40-percent reduction in the pointing error is observed along with a slight reduction in mass, or an almost two-folds increase in the controlled performance is indicated with more than 5-percent reduction in the overall mass of the spacecraft (a reduction of hundreds of kilograms). Furthermore, the automated nature of the integrated design approach can accommodate a wide variety of design specifications and requirements. A practical software tool (CSI-DESIGN) is being developed for performing integrated designs. Research is presently in progress for incorporating other types of control laws, (such as H_2 or H_∞) in the integrated design process. Future plans also include hardware validation of the integrated design approach by constructing a laboratory test article based on the redesigned CSI Evolutionary Model, and comparing the performance of the redesigned structure with that of the original CSI Evolutionary Model.

APPENDIX
I. DISSIPATIVE CONTROLLERS

This appendix briefly presents the main results concerning dissipative controllers for flexible structures. Dissipative controllers utilize the passivity properties of the system which result from the use of collocated and compatible actuators and sensors, such as torque actuators and attitude and rate sensors, or force actuators and position and velocity sensors. In general, dissipative controllers are designed to control both rigid and elastic modes. In certain cases, the zero-frequency rigid modes may not be present. This may occur, for example, when paired torquers are used in a balanced configuration to control only the elastic motion [4], or in the case of ground-based test articles. In such cases, no position feedback is required; only velocity feedback is sufficient to accomplish robust stability, and the dissipative compensator degenerates to a positivity controller. In the following both the static and dynamic dissipative controllers are discussed. The concepts and the underlying theory is based largely on the original work of Popov [27].

A. STATIC DISSIPATIVE CONTROLLERS

The constant-gain or static dissipative control law is given by:

$$u = -G_p y_p - G_r y_r \tag{A.1}$$

where G_p and G_r are $m \times m$ symmetric, positive-definite, proportional and rate gain matrices, and y_p and y_r represent the $m \times 1$ position and rate measurement vectors. This control law is known to give guaranteed asymptotic stability (of the entire system consisting of both rigid and flexible modes) regardless of unmodeled elastic modes or parameter uncertainties. (It was also shown in [4] that stability is maintained even if small imprecision exists in the collocation of the actuators and sensors). Furthermore, if G_p and G_r are diagonal, this robust stability property is carried over in the presence of:
1) monotonically increasing actuator nonlinearities, rate sensor nonlinearities belonging to the $[0, \infty)$-sector, and position sensor nonlinearities belonging to the $(0, \infty)$-sector. [Note: A single-valued function $\psi(\nu)$ is said to belong to the $(0, \infty)$ sector if $\psi(0) = 0$ and $\nu\psi(\nu) > 0$ for $\nu \neq 0; \psi$ is said to belong to the $[0, \infty)$ sector if $\nu\psi(\nu) \geq 0$], and
2) stable actuator dynamics $g_a(s) = k/(s + a)$, provided that $g_p/g_r < a$, where g_p and g_r denote the appropriate diagonal elements of G_p and G_r. A possible drawback of these controllers is that the performance can be limited because of the structure of the controller. G_p and G_r can be designed to minimize a quadratic performance function or to obtain closed-loop eigenvalues close to the desired locations in the least-square sense [4].

When the zero-frequency modes are absent, G_p can be zero, and $G_r = G_r^T > 0$ is sufficient for asymptotic stability. Furthermore, the closed-loop system is robust to $[0, \infty)$ sector actuator and sensor nonlinearities and first-order actuator dynamics [4].

B. DYNAMIC DISSIPATIVE CONTROLLERSs

In order to obtain better performance while still retaining the guaranteed robustness to unmodeled dynamics and parameter uncertainties, we consider a class of dynamic dissipative controllers. We shall first consider the case where both zero-frequency rigid modes and flexible modes are present.

Assuming for simplicity that the plant has three (one per axis) torque actuators and collocated attitude and rate sensors, the 3×3 transfer function from the torque input to the attitude rate output is given by:

$$G'(s) = \frac{J^{-1}}{s} + \sum \frac{\Phi_i \Phi_i^T s}{s^2 + 2\rho_i \omega_i s + \omega_i^2} \tag{A.2}$$

where J is the moment-of-inertia matrix, and Φ_i, ρ_i, and ω_i denote the rotational mode shape matrix, damping ratio, and natural frequency of the ith structural mode. The transfer funtion $G'(s)$ is positive-real, as defined below [28]:

Definition- A rational matrix-valued function $T(s)$ of the complex variable s is said to be positive-real if $T(s)$ is real when s is real, and $T(s) + T^T(s^*) \geq 0$ for $Re[s] \geq 0$, where * denotes the complex conjugate.

Scalar positive-real (PR) functions have a relative degree (i.e., the difference between the degrees of the denominator and numerator polynomials) of -1, 0, or 1 [29]. It can also be shown that PR matrices have no transmission zeros or poles in the open right-half of the complex plane, and that the poles on the imaginary axis are simple and have nonnegative-definite residues. By applying the maximum modulus theorem, it can be shown that it is sufficient to check the positive semi-definiteness of $T(s)$ only on the imaginary axis ($s = j\omega, 0 \leq \omega < \infty$). Suppose $(\mathcal{A}, \mathcal{B}, \mathcal{C}, \mathcal{D})$ is an nth-order minimal realization of $T(s)$. From [28], a necessary and sufficient condition for $T(s)$ to be positive real is that there exists an $n \times n$ symmetric positive-definite matrix P, and matrices W and L such that

$$\begin{aligned} \mathcal{A}^T P + P\mathcal{A} &= -LL^T \\ \mathcal{C} &= \mathcal{B}^T P + W^T L \\ W^T W &= \mathcal{D} + \mathcal{D}^T \end{aligned} \tag{A.3}$$

This result is generally known in the literature as the Kalman-Yacubovich lemma. A stronger concept along these lines is "strictly positive-real (SPR) systems". However, there are several nonequivalent definitions of SPR, all of which require the system to have all poles in the open left-half plane [30]. For the purpose of this paper, we define a less restrictive class of "strongly PR" systems as follows:

Definition- A rational matrix-valued function $T(s)$ of the complex variable s is said to be strongly positive-real if $T(s)$ is real when s is real, and $T(s) + T^T(s^*) > 0$ for $Re[s] \geq 0$.

The obvious difference between the above definition and the definition of PR systems is that the "$\geq$ " has been replaced by strict inequality. The difference between the "strongly PR" and "SPR" systems is that the latter have poles only in the open left half-plane, whereas the former can have poles on the $j\omega$-axis.

The transfer function from u to y_p is given by:

$$G(s) = \frac{G'(s)}{s} \tag{A.4}$$

It can be seen that $G(s)$ is not PR. However, $G'(s)$, the transfer function from u to y_r, is PR.

Definition- The compensator $K(s)$ is said to stabilize a plant $G(s)$ if the closed-loop system consisting of stabilizable and detectable realizations of K and G in the standard feedback configuration is asymptotically stable (a.s.).

Suppose a controller $K(s)$ is represented by the $n_k th$-order minimal realization:

$$\dot{x}_k = A_k x_k + B_k y_p \tag{A.5}$$

$$y_k = C_k x_k + D_k y_p \tag{A.6}$$

The input to the plant is given by:

$$u = -y_k \tag{A.7}$$

Define

$$A_z = \begin{bmatrix} A_k & 0 \\ C_k & 0 \end{bmatrix} \quad B_z = \begin{bmatrix} B_k \\ D_k \end{bmatrix} \quad C_z = [0 \quad I_3] \tag{A.8}$$

We present the following stability result which was proved in [16].
Theorem 1.- Suppose
i) A_k is strictly Hurwitz
ii) There exists an $(n_k + 3) \times (n_k + 3)$ matrix $P_z = P_z^T > 0$ such that

$$A_z^T P_z + P_z A_z = -Q_z = -diag(L_k^T L_k, 0_3) \tag{A.9}$$

where L_k is a $3 \times n$ matrix such that (L_k, A_k) is observable, and $L_k(sI - A_k)^{-1} B_k$ has no transmission zeros in $Re[s] \geq 0$
iii)

$$C_z = B_z^T P_z \tag{A.10}$$

iv)$K(s) = C_k(sI - A_k)^{-1}B_k + D_k$ has no transmission zeros at the origin. Under these conditions, the controller $K(s)$ stabilizes $G(s)$.

Remark 1.- In Theorem 1, if L_k is an $n_k \times n_k$ nonsingular matrix, then the observability and minimum-phase property in (ii) are satisfied and the closed-loop system is a.s.

Remark 2.- The controller K stabilizes the full plant; i.e., the system consisting of the *rigid modes*, the *elastic modes*, and the *compensator state vector* (x_k) is a.s. The asymptotic stability is guaranteed *regardless of the number of modes in the model, or parameter uncertainties*. The order of K can be chosen to be any number ≥ 3. In other words, this results enables the design of a controller of any desired order, which robustly stabilizes G. A procedure for designing K is to choose $Q_z = diag(Q_k, 0_3)$ where $Q_k = Q_k^T > 0$, and to choose a stable A_k and matrices B_k and C_k so that (9) and (10) are satisfied. Using (8) and denoting

$$P_z = \begin{bmatrix} P_{z1} & P_{z2} \\ P_{z2}^T & P_{z3} \end{bmatrix}$$

where P_{z1} is $n_k \times n_k$ and P_{z3} is 3×3, conditions (ii) and (iii) of Theorem 1 can be expanded as:

$$\begin{aligned} & P_{z1}A_k + A_k^T P_{z1} + P_{z2}C_k + C_k^T P_{z2}^T = -L_k^T L_k \\ & P_{z2}^T A_k + P_{z3}C_k = 0 \\ & B_k^T P_{z1} + D_k^T P_{z2}^T = 0 \\ & B_k^T P_{z2} + D_k^T P_{z3} = I \end{aligned} \tag{A.11}$$

In addition, P_z must be positive definite. Because of the large number of free parameters $(A_k, B_k, C_k, D_k, L_k)$ it is generally not straightforward to use (11) to obtain the compensator, and this remains an area of continuing research. Another method is to use the following s-domain equivalent of Theorem 1 (See [16]).

Theorem 2.- $K(s)$ stabilizes $G(s)$ if $K(s)$ has no transmission zeros at $s = 0$, and $K(s)/s$ is strongly positive real.

The condition that $K(s)/s$ be strongly PR is sometimes much easier to check than the conditions of Theorem 1. For example, let $K(s) = diag[K_1(s), K_2(s), K_3(s)]$, where

$$K_i(s) = k_i \frac{s^2 + \beta_{1i}s + \beta_{0i}}{s^2 + \alpha_{1i}s + \alpha_{0i}} \tag{A.12}$$

It is straightforward to show that $K(s)/s$ is strongly PR iff (for i = 1, 2, 3), $k_i, \alpha_{0i}, \alpha_{1i}, \beta_{0i}, \beta_{1i}$ are positive, and

$$\alpha_{1i} - \beta_{1i} > 0 \tag{A.13}$$

$$\alpha_{1i}\beta_{0i} - \alpha_{0i}\beta_{1i} > 0 \tag{A.14}$$

For higher order K_i's, the conditions on the polynomial coefficients are harder to obtain. One systematic procedure for obtaining such conditions for higher order controllers is the application of Sturm's theorem [29]. Symbolic manipulation codes can then be used to derive explicit inequalities. The controller design problem can be subsequently posed as a constrained optimization problem which minimizes a given performance function. For the case of fully populated $K(s)$, however, there appear to be no straightforward methods.

The controller $K(s)$ (Eqs. A.5, A.6) is not strictly proper because of the direct transmission term D_k. From a practical viewpoint, it is sometimes desirable to have a strictly proper controller because it attenuates sensor noise as well as high-frequency disturbances. Furthermore, the most common types of controllers, which include the LQG as well as the observer/pole placement controllers, are strictly proper (they have a first-order rolloff). In addition, the realization in (A.5) and (A.6) does not utilize the rate measurement y_r. The following result from [16] states that K can be realized as a strictly proper controller, wherein both y_p and y_r are utilized.

Theorem 3.- The plant $G(s)$ is stabilized by the controller K' given by:

$$\dot{x}_k = A_k x_k + [\,B_k - A_k L \quad L\,] \begin{bmatrix} y_p \\ y_r \end{bmatrix} \tag{A.15}$$

$$u_k = C_k x_k \tag{A.16}$$

where L is a solution of:

$$D_k - C_k L = 0 \tag{A.17}$$

There are many possible solutions for L. The solution which minimizes the Frobenius norm of L is:

$$L = C_k^T (C_k C_k^T)^{-1} D_k \tag{A.18}$$

For the case with no zero-frequency modes, the dynamic dissipative controller degenerates to a "positivity" controller which uses the feedback of only the velocity. For this case, the robust stability is guaranteed with an "SPR" dynamic compensator [31-32]. In terms of the state equations, it is sufficient for robust stability that the compensator satisfies the Kalman-Yacubovich conditions (3) with $(\mathcal{A}, L)$ observable. A simpler sufficient condition, which also involves fewer parameters, is that $\mathcal{D} = 0$, and L is a (lower or upper) triangular matrix.

The problem of designing dissipative controllers which also provide optimal performance has been a subject of active research. It has been shown [4] that the static dissipative controller minimizes a certain quadratic performance index. Methods based on eigenstructure assignment with dissipativity constraints have also been recently proposed [33]. For the case with no zero-frequency modes, it was shown in [34] that the selection of LQ regulator and estimator weighting matrices in a certain manner results in an LQG controller that is also SPR, and, therefore, robustly stabilizes the plant. For the case where zero frequency modes are present, design methods based on numerical minimization of the "distance" between a nominal LQG controller and dynamic dissipative controllers was suggested in [16].

Another approach for the case with zero-frequency modes is a two-level controller, wherein an inner loop static dissipative controller with position and velocity feedback is used, and a dynamic dissipative (SPR) velocity feedback controller is used in the outer loop. This approach was discussed in [5].

In spite of some progress, the problem of designing *optimal* dissipative controllers remains an area of continuing research.

REFERENCES

1. M. Balas, "Trends in Large Space Structures: Fondest Hopes, Wildest Dreams", *IEEE Trans. Auto Control* **AC–27**, pp. 522–535 (1982).
2. G. Ketner, "Survey of Historical Incidences With Controls-Structures Interaction and Recommended Technology Improvements Needed to Put Hardware in Space", Battelle, Pacific Northwest Laboratory, Richland WA, Report no. PNL–6846 (1989).
3. J. Newsom, W. Layman, H. Waites and R. Hayduk, "The NASA Controls-Structures Interaction Technology Program", NASA TM–102752 (1990).
4. S. Joshi, "*Control of Large Flexible Space Structures*", Vol. 131, Lecture Notes in Control and Information Sciences, Berlin Springer-Verlag (1989).
5. S. Joshi and P. Maghami, "Dissipative Compensators for Flexible Space-craft Control", *American Control Conference*, San Diego CA (1990).
6. S. Rao and T. Pan, "Robustness Improvement of Actively Controlled Structures Through Integrated Structural/Control Design", Flight Dynamics Directorate, Wright Laboratory, Report no. WL–TR–91–3058 (1991).
7. A. Balakrishnan, "Combined Structures-Controls Optimization of Lattice Trusses", *Computer Methods in Applied Mechanics and Engr.* **94** (1992).
8. J. Bossi, K. Hunziker, and R. Kraft, "Integrated Control/Structure Design", NASA CR–182020 (1990).

9. P. Maghami, S. Joshi, E. Armstrong and J. Walz, "Integrated Controls-Structures Design Methodology Development for a Class of Flexible Spacecraft", *Third Air Force/NASA Symposium on Recent Advances in Multidisciplinary Analysis and Optimization*, San Francisco CA (1990).
10. P. Maghami, S. Joshi and K. Lim, "Integrated Controls-Structures Design: A Practical Tool for Modern Spacecraft", *American Control Conf.*, Boston MA (1991).
11. P. Maghami, S. Joshi, K. Elliot and J. Walz, "Integrated Design of the CSI Evolutionary Structure: A Verification of the Design Methodology", *Fifth NASA/DoD Controls/Structures Interaction Technology Conf.*, Lake Tahoe NV (1992).
12. S. Woodard, S. Padula, P. Graves and B. James, "An Optimization Method for Controlled Space Structures with Variable Actuator Mass", *Fourth NASA/DoD Controls/Structures Interaction Technology Conf.*, Orlando FL (1990).
13. J. Ramler and R. Durrett, " NASA's Geostationary Communications Platform Program", *AIAA 10th Communications Satellite Systems Conference*, Orlando FL (1984).
14. W. Belvin, L. Horta and K. Elliot, " The LaRC CSI Phase-0 Evolutionary Model Testbed: Design and Experimental Results", *Fourth NASA/DoD Controls/Structures Interaction Technology Conf.*, Orlando FL (1990).
15. J. Oneda and R. Haftka, "An Approach to Simultaneous Structure-Control Optimization of Large Flexible Spacecraft", *AIAA Journal* **25**, pp.1133–1138 (1987).
16. S. Joshi, P. Maghami and A. Kelkar, "Dynamic Dissipative Compensator Design for Flexible Space Structures", *AIAA Guidance, Navigation, and Control Conference*, New Orleans LA (1991).
17. S. Lefschetz, *Stability of Nonlinear Systems*, Academic Press, New York NY (1965).
18. M. Jamshidi and C. Herget, eds., *Advances in Computer-Aided Control Systems Engineering*, Amsterdam, North Holland (1985).
19. S. Fenves, N. Perrone, A. Robinson and W. Schnobrich, eds., *Numerical and Computer Methods in Structural Mechanics*, Academic Press, New York NY (1973).
20. W. Belvin and K. Park, "Computer Implementation of Analysis and Optimization Procedures for Control-Structure Interaction Problems", *AIAA Dynamics Specialist Conference*, Long Beach CA (1990).
21. G. Vanderplaats, "ADS-A Fortran Program for Automated Design Synthesis-Version 1.10", NASA CR–177985 (1985).
22. G. Vanderplaats, *Numerical Optimization Techniques for Engineering Design: With Applications*, McGraw-Hill New York NY (1984).
23. J. George and J. Liu, *Computer Solution of Large Sparse Positive-Definite Systems*, Prentice Hall, Englewood Cliffs NJ (1981).
24. S. Kenny, *Eigensensitivity Analysis for Space Structures with Applications in Design Optimization*, Master's Thesis, Old Dominion University, Norfolk VA (1990).
25. S. Kenny, G. Hou and W. Belvin, "Eigensensitivity in Integrated Design", *Fourth NASA/DoD Controls/Structures Interaction Technology Conf.*, Orlando FL (1990).

26. R. Stengel, *Stochastic Optimal Control: Theory and Aplications*, John Wiley & Sons, Inc., New York NY (1986).
27. V. Popov, *Hyperstability of Control Systems*, Springer-Verlag, Berlin (1973).
28. B. Anderson, "A System Theory Criteron for Positive-Real Matrices", *SIAM J. Control* **5** (1967).
29. M. Valkenberg, *Introduction to Modern Network Analysis*, John Wiley & Sons, New York NY (1965).
30. R. Lozano-Leal and S. Joshi, "Strictly Positive Real Functions Revisited", *IEEE Trans. Auto. Control* **35**, pp. 1243–1245 (1990).
31. R. Benhabib, R. Iwens and R. Jackson, "Stability of Large Space Structure Control Systems Using Positivity Concepts", *AIAA J. Guidance and Control* **4**, pp. 487–494 (1981).
32. M. McLaren and G. Slater, "Robust Multivariable Control of Large Space Structures Using Positivity", *AIAA J. Guidance, Control and Dynamics* **10**, pp. 393–400 (1987).
33. P. Maghami, S. Gupta and S. Joshi, "Design of Dissipative Low-Authority Controllers Using an Eigensystem Assignment Technique", *American Control Conference*, Chicago IL (1992).
34. R. Lozano-Leal and S. Joshi, "On the Design of Dissipative LQG-Type Controllers", *in* "Recent Advances in Robust Control" (P. Dorato and R. Yedavalli, eds.), IEEE Press, New York NY, pp. 251–252 (1990).

Adaptive Linear System Identification and State Estimation

Chung-Wen Chen

Mars Mission Research Center
North Carolina State University
Raleigh, North Carolina 27695-7921

Jer-Nan Juang

Spacecraft Dynamics Branch
NASA Langley Research Center
Hampton, Virginia 23665-5225

Jen-Kuang Huang

Department of Mechanical
Engineering and Mechanics
Old Dominion University
Norfolk, Virginia 23529-0247

I. INTRODUCTION

For active control of dynamic systems using a state feedback strategy, having accurate models is very important. In the control system the controller issues control commands based on control laws and the current states of the system; however, the control laws are formulated and programmed in a computer based on the available model, and the states are also estimated based on the model by a state estimator. To fulfill a control task, this model must be able to describe the dynamic behavior of the system adequately. Without a good model a control system cannot work satisfactorily; in some cases, the control system might even become unstable.

Usually there are two different ways to obtain a model for a dynamic system. The first is an analytical approach; it starts from deriving governing equations based on physical laws, and then simplifies and/or solves the equations and transforms them into a desired form. One example is the finite element method. The second is an experimental approach; it processes on the measured input and output data obtained from experiments or operations to find a model directly, such as some modal testing techniques. In general, the models obtained through analytical ways need further confirmation and refinement using experimental results before its use for control design.

System identification is the technique which deals with building mathematical models for dynamic systems based on their input and output data [1]. Besides controls, this technique is also important in many other disciplines, such as economics, communication, and system dynamics [1]. Mathematical models allow researchers understand more about the properties of the system, so that they can explain, predict or control the behaviors of the system.

However, for large flexible space structures, accurate models are not easy to obtain. Because of the great flexibility of the structures, the gravity effects usually hinder obtaining accurate system models from ground testing. In some cases ground testing is even not feasible because the structures have to be deployed or assembled on-orbit. In addition, after putting into service in space, the system characteristics of flexible space structures still can vary due to such factors as temperature gradient, reorientation, or material deterioration. Therefore, for better control performance the system model needs to be updated frequently. Consequently, on-line adaptive system identification is required for a satisfactory control of flexible space structures. Though a great variety of system identification methods have been proposed during the last few decades, still adaptive identification of large dimension systems affected by noise remains a difficult task [1,2].

Besides an accurate system model, the state estimator also requires accurate statistical information about noises to function properly. For instance, the Kalman filter [3]—the most famous state estimator—needs both the process noise and the measurement noise covariances for state estimation. The process noise, which affects the dynamics of the system, arises from unknown excitations and/or the unmodeled parts associated with the model; while the measurement noise, which is usually additive in attribution (does not affect system dynamics), arises from imperfect instruments. In practice, the noise statistical information is hard to obtain. The measurement noise is relatively easier to deal with, because it can be studied by sending known signals through the instruments and analyzing the results; as for the process noise, accurate statistical properties are extremely difficult, if possible, to obtain. The process noise is usually associated with the states of the system, and in some cases the states may not have any physical meanings.

For flexible space structures, the process noise is usually significantly strong. The reason is that the system is quite non-linear because the structures are large and flexible, and the unmodeled dynamics is more significant because the model dimension can not be too large for practical consideration. Even the measurement noise could be changing due to material aging and/or the extreme working environment. Since system models have to be adaptively obtained on-orbit, adaptive state estimation is also needed in control of flexible space structures.

Traditionally adaptive system identification and adaptive state estimation are regarded as two related but different fields, because adaptive system identification estimates unknown system parameters while traditional adaptive state estimation, based on a known system model, estimates unknown noise parameters. Adaptive state estimation under unknown system model has seldom been addressed. However, recently it is found that adaptive system identification and adaptive state estimation can be treated at the same time [4-7]. Actually, the steady state optimal Kalman filter gain in a sense characterizes the stochastic property of a system associated with a set of input-output data.

The objective of this paper is to present an effective *integrated* adaptive state-space system identification and state estimation algorithm for applications in control of linear stochastic systems, and in particular for flexible space structures. The algorithm starts from knowing nothing about the system except the linearity assumption. From input and output data, a state-space model and a Kalman filter gain are identified. The identified model and the filter gain then can be used in controllers and state estimators design to fulfill control tasks.

Aiming at the ultimate objective, the problem is presented along the theme of adaptive state estimation, and is divided into three stages according to complexity. Section II investigates the first stage: *state estimation under full a priori information*. A projection filter is derived for this purpose based on parameter estimation theories and can be used as an alternative to the Kalman filter. The relation between these two filters provides a better understanding of the Kalman filter. In addition, the projection filter also provides a relation between state-space models and finite difference models of linear stochastic systems which will be used in Section IV.

Section III investigates the second stage: *state estimation under unknown noise covariances*. A method is introduced to estimate the steady-state optimal Kalman filter gain directly instead of estimating noise covariances first and then calculating the filter gain, as classic adaptive state estimation does. The idea is that with a known state-space model, the deterministic part of the output can be computed and subtracted out; the remaining signal represents the stochastic part and can be modeled by a Moving Average (MA) model. To identify the MA model, we first identify its inverse model—an AutoRegressive (AR) model—and then invert it. From the coefficients of the MA model, an estimated Kalman filter gain is calculated; thus, state estimation can be performed.

Section IV investigates the third and the most challenging stage: *state estimation under unknown system models and unknown noise statistics*. It is under this situation system identification comes in sight, and one faces a compound problem of system identification and state estimation. Two approaches are introduced. The first is called the simultaneous method, which is based on the relationship between a state-space model and an *infinite-termed* AutoRegressive with eXogeneous input (ARX) model *connected by the Kalman filter*. The coefficients of the ARX model are expressed in terms of the state-space parameters and the steady state Kalman filter gain; from these coefficients a state-space model and the corresponding Kalman filter gain can be obtained. The second approach is called the sequential method, which is based on the relationship between a state-space model and a *finite-termed* ARX model *connected by the projection filter*. It is similar to the first method in identifying a system model; however, it cannot identify the Kalman gain simultaneously.

Hence, after obtaining the state-space model the steady-state Kalman filter gain can be estimated using the technique developed in Section III. Compared to the simultaneous method, the sequential method has the advantage of being capable of using an ARX model of smaller order to reduce computation, but has the disadvantage of requiring an additional procedure for identifying the Kalman gain. To illustrate the feasibility of the approach, an experimental example of a ten-bay structure is given at the end of Section IV.

II. STATE ESTIMATION UNDER FULL A PRIORI INFORMATION

Estimation is a technique of deriving interested quantities based on relevant data when the interested quantities cannot be measured either directly or correctly [8]. It is an information extraction process, extracting useful information out from available data. Estimation theories also provide a strategy of combining information obtained from several different sources to yield more accurate information.

The idea of estimating unknown quantities from observed data can be traced back to Gauss in the early 1800's. In his astronomical studies, he invented the famous least-square technique [9]. In more recent times, the Kalman filter is a widely used technique in state estimation [3].

In this section, state estimation of linear time-invariant dynamic systems with sufficient a priori information (i.e., the system model, the noise covariances and the initial conditions) is investigated. Although under this situation the Kalman filter is already the optimal solution, and it seems no need to investigate this already solved problem, looking an old problem from different points of view usually could bring useful insight. Under such a belief an alternative method called *the projection filter* is developed. Compared with the Kalman filter, this alternative method has advantages and disadvantages. But at any rate, the derivation of the projection filter provides a good insight into the relationship between the classical estimation theory and the Kalman filter and paves a way to a new system identification method which will be introduced in Section IV.

In this section, first the optimal Kalman filter and some of its extreme cases are briefly discussed to provide a background for comparison with the projection filter; then, the projection filters for systems under various conditions are derived based on parameter estimation theories. Finally, the relation between the projection filter and the Kalman filter is discussed, and a numerical example is provided.

A. The Kalman Filter

A finite-dimensional, linear, discrete-time, time-invariant stochastic system can be modeled as:

$$x_{k+1} = Ax_k + Bu_k + w_k \tag{2.1}$$

$$y_k = Cx_k + Du_k + v_k \tag{2.2}$$

where $x \in R^{n\times 1}$, $u \in R^{m\times 1}$, $y \in R^{p\times 1}$ are state, input and output vectors, respectively; $\{w_k\}$ is the process noise, $\{v_k\}$ the measurement noise. Matrices A, B, C and D are respectively the state matrix, input matrix, output matrix and direct-influence matrix. The integer k is the sample indicator. Sequences $\{w_k\}$ and $\{v_k\}$ are assumed gaussian, white, zero-mean, stationary with covariances Q and R respectively, and are uncorrelated with each other.

The discrete Kalman filter can be summarized as follows [3]:

a. Initial Conditions:

$$E[x_0] = \hat{x}_0, \qquad E\left[(x_0 - \hat{x}_0)(x_0 - \hat{x}_0)^T\right] = P_o \tag{2.3}$$

b. Prediction (Extrapolation):

$$\hat{x}_k^- = A\hat{x}_{k-1}^+ + Bu_{k-1}, \tag{2.4}$$

$$P_k^- = AP_{k-1}^+A^T + Q \tag{2.5}$$

c. Measurement Update:

$$\hat{x}_k^+ = \hat{x}_k^- + K_k(y_k - C\hat{x}_k^- - Du_k) = (I_n - K_kC)\hat{x}_k^- + K_k(y_k - Du_k) \tag{2.6}$$

$$P_k^+ = (I_n - K_kC)P_k^- \tag{2.7}$$

$$K_k = P_k^-C^T(CP_k^-C^T + R)^{-1} \tag{2.8}$$

where $\hat{x}$ is the estimated state vector, P the corresponding estimation error covariance matrix, I_n the n-dimensional identity matrix, K_k the Kalman filter gain and the superscripts "$-$" and "+" distinguish the estimates before and after taking account of the current measurement, respectively. Three special cases of Kalman filtering are presented to provide a background for comparison with the projection filter derived later.

1. *Kalman Filter for Noise Free System*

If the system is noise free (Q = 0 and R = 0) with non-zero initial error covariance $(P_0 \neq 0)$, the Kalman filter gain becomes

$$K_k = P_k^-C^T(CP_k^-C^T)^{-1}. \tag{2.9}$$

Note that $CK_k = I_p$; therefore, K_k is actually a weighted pseudo-inverse of C and P_k^- is the weighting matrix. Premultiplying (2.6) by C gives

$$C\hat{x}_k^+ = C(I_n - K_kC)\hat{x}_k^- + CK_k(y_k - Du_k) = y_k - Du_k, \tag{2.10}$$

which implies that the estimate satisfies the measurement equation exactly for each time step. Based on (2.5) and (2.7) where A represents a stable system and Q = 0, the Kalman filter state error covariance will converge to zero in steady-state $(P_\infty^+ = P_\infty^- = 0)$.

2. *Kalman Filter for Systems without Process Noise*

If the system has no process noise but has measurement noise (Q = 0 and R $\neq$ 0), again based on (2.5), (2.7) and (2.8) the Kalman filter has a zero steady-state error covariance $\left(P_\infty^+ = P_\infty^- = 0\right)$ and a zero steady-state Kalman gain $\left(K_\infty = 0\right)$. In this case, the estimate is perfect in steady-state and it becomes

$$\hat{x}_k^+ = \hat{x}_k^- = A\hat{x}_{k-1}^- + Bu_{k-1},$$

which indicates that in steady-state the Kalman filter ignores the noise-corrupted measurement and relies solely on system model to predict state information and the prediction is the same as the estimation in steady-state.

B. The Projection Filter

The projection filter is a linear transformation which projects (transforms) a finite number of input/output data of a system into its current state space. The filter is designed such that the mean square estimation error is minimized; therefore, the image of the projection is an optimal estimate of the current state [4]. Here "filter" is a generic term referring to a data processing procedure which extracts desired information from data. To derive the projection filter, we start from a simple case and then gradually move to some more general cases.

1. *Projection Filter for Noise-Free Systems*

For a noise free system, (2.1) and (2.2) become

$$x_{k+1} = Ax_k + Bu_k \tag{2.11}$$

$$y_k = Cx_k + Du_k \tag{2.12}$$

It is easy to show that

$$\begin{bmatrix} y_k \\ y_{k-1} \\ \vdots \\ y_{k-q+1} \end{bmatrix} = \begin{bmatrix} C \\ CA^{-1} \\ \vdots \\ CA^{-q+1} \end{bmatrix} x_k - \begin{bmatrix} -D & 0 & \cdots & 0 \\ 0 & CA^{-1}B - D & \cdots & 0 \\ \vdots & \vdots & \ddots & \vdots \\ 0 & CA^{-q+1}B & \cdots & CA^{-1}B - D \end{bmatrix} \begin{bmatrix} u_k \\ u_{k-1} \\ \vdots \\ u_{k-q+1} \end{bmatrix}, \tag{2.13}$$

or in short,

$$Y_{q,k} = H_q x_k - G_q U_k, \tag{2.14}$$

or in a normal form

$$H_q x_k = Y_{q,k} + G_q U_k, \tag{2.15}$$

where q denotes the number of data stacking up to form the equation. The meanings of other notations are self-evident. If x_k is the unknown variable to solve, (2.15) contains n unknowns and $p \times q$ equations. However, there are only at most n independent equations; therefore, for a number q sufficiently large to make H_q full-column-ranked ($p \times q \geq n$), the unique solution of x_k is

$$\hat{x}_k = F_q\left(Y_{q,k} + G_q U_k\right), \tag{2.16}$$

where

$$F_q = \left(H_q^T H_q\right)^{-1} H_q^T \tag{2.17}$$

is the pseudo-inverse of H_q and the "projection filter" in this case. If $p \times q = n$ (i.e., H_q is square), F_q becomes H_q^{-1}. The number q can be any integer larger than or equal to the integer $q_{\min}$ which is the minimum number required to make H_q full-column-ranked. The solution $\hat{x}_k$ is identical with the true value x_k in this noise-free case.

Compared to the Kalman filter for the same noise-free case, the projection filter does not need the initial values of state and its error covariance to initiate the estimation. The projection filter can also achieve perfect estimation once the required minimum number of measurement becomes available, while the Kalman filter takes more steps to converge to zero-error estimation if the initial values are poorly estimated. However, the projection filter can not produce estimation until the matrix H_q becomes full-column-ranked, while the Kalman filter can.

2. *Projection Filter for Systems without Process Noise*

For a system without process noise but with additive measurement noise shown in (2.11) and (2.2), one can derive a matrix equation

$$H_q x_k = Y_{q,k} + G_q U_k - V_{q,k},$$

where $V_{q,k}^T = \left[v_k^T, v_{k-1}^T, \mathrm{L}\, , v_{k-q+1}^T\right]$. Note that the unknown x_k is still a deterministic variable in this case. Hence, by the theory of optimal parameter estimation for deterministic parameters from a linear equation with independent white noise [10], one can write the optimal estimate of x_k the same as (2.16), but the filter F_q becomes

$$F_q = \left(H_q^T \overline{R}^{-1} H_q\right)^{-1} H_q^T \overline{R}^{-1} \tag{2.18}$$

which is a weighted pseudo-inverse of H_q. In (2.18) $\overline{R} = R \otimes I_q$, where $\otimes$ is the Kronecker product, R the covariance of the measurement noise. The optimality is defined by the minimum variance of state estimation error. The corresponding state error covariance is a constant with

$$P_k = \left(H_q^T \overline{R}^{-1} H_q\right)^{-1}. \tag{2.19}$$

Compared to the Kalman filter under the same situation of no process noise, the projection filter with a small order q is less accurate because P_k in (2.19) is not zero for a small q, while the Kalman filter can achieve perfect estimation $\left(P_k^+ = 0\right)$. It can approach perfect estimation if the order q becomes sufficiently large. However, the computational load increases as the order increases. On the other hand, the projection filter does not need the initial values and its error covariance of the state to start the estimation.

3. *Projection Filter for Systems with both Process and Measurement Noises*

Next, consider a more general case of a system with both process and measurement noises shown in (2.1) and (2.2). Similarly, by writing the previous output in terms of the current state, one can derive

$$Y_{q,k} = H_q x_k - G_q U_k - M_q W_{q,k} + V_{q,k}, \tag{2.20}$$

where

$$M_q = \begin{bmatrix} 0 & \mathrm{L} & 0 \\ CA^{-1} & \mathrm{L} & 0 \\ \mathrm{M} & \mathrm{O} & \mathrm{M} \\ CA^{-q+1} & \mathrm{L} & CA^{-1} \end{bmatrix}, \qquad W_{q,k} = \begin{bmatrix} w_{k-1} \\ \mathrm{M} \\ w_{k-q+1} \end{bmatrix}.$$

It can be further simplified to

$$H_q x_k = Y'_{q,k} + \xi_{q,k}, \tag{2.21}$$

where

$$Y'_{q,k} = Y_{q,k} + G_q U_k \qquad \xi_{q,k} = M_q W_{q,k} - V_{q,k}.$$

Note that the unknown x_k is a random variable in this case. The overall noise vector $\xi_{q,k}$ is still gaussian and zero-mean because $W_{q,k}$ and $V_{q,k}$ are gaussian and zero-mean, but it is correlated with x_k because $W_{q,k}$ is correlated with x_k (current states are affected by previous process noise). Denote the covariance between x_k and $\xi_{q,k}$ by $P_{x\xi}$ and the auto-covariance of $\xi_{q,k}$ by R_ξ. For (2.21), given the mean of the current state $\bar{x}_k$ and its variance P_x, by the theory of random parameters estimation [4,10], the optimal estimate of x_k is

$$\hat{x}_k = \bar{x}_k + F_q\left(Y'_{q,k} - \bar{Y}'_{q,k}\right), \tag{2.22}$$

where the overbar "−" denotes the expectation value,

$$\bar{Y}'_{q,k} = H_q \bar{x}_k, \tag{2.23}$$

and

$$F_q = \left(P_x H_q^T + P_{x\xi}\right)\left(H_q P_x H_q^T + H_q P_{x\xi} + P_{x\xi}^T H_q^T + R_\xi\right)^{-1}. \tag{2.24}$$

The matrix F_q is the projection filter in this case. The optimality is defined by minimum variance of state estimation error.

C. Relation Between the Kalman Filter and the Projection Filter

The projection filter and the Kalman filter are closely related. In fact, a projection filter of order q ($q \geq 2$) can be transformed to have a Kalman filter structure, and the recursive projection filter of order one is identical to the Kalman filter. The identity can be proved by re-deriving the Kalman filter through the recursive projection filter of order one. Here the transformation of the projection filter to the Kalman filter structure is derived first.

The transformation is based on the concept of correlation cancellation [11]. From (2.21) and (2.23), one has

$$Y'_{q,k} - \bar{Y}'_{q,k} = Hx_k^* - \xi_{q,k} \tag{2.25}$$

where $x_k^* \equiv x_k - \bar{x}_k$. The noise term $\xi_{q,k}$ is correlated with x_k^*. According to the correlation cancellation theory, $-\xi_{q,k}$ can be divided into two parts: one is correlated with x_k^* and the other is not; namely,

$$-\xi_{q,k} = H'x_k^* + \zeta_k \tag{2.26}$$

where $H'x_k^*$ is the correlated part. The orthogonal projection [11], denoted by H', is

$$H' = P_{\xi_k x_k} P_{x_k}^{-1} \tag{2.27}$$

where $P_{\xi_k x_k} = E\left[\xi_{q,k} x_k^{*T}\right]$ and $P_{x_k} = Cov\left[x_k^*\right]$. Introducing (2.26) into (2.25) yields

$$\tilde{Y}_{q,k} \equiv Y'_{q,k} - \bar{Y}'_{q,k} = (H_q + H')x_k^* + \zeta_k = \tilde{H}_q x_k^* + \zeta_k \tag{2.28}$$

From (2.27) one can derive

$$P_{\xi_k x_k} = H'P_{x_k} \text{ and } P_{x_k \xi_k} = P_{x_k}(H')^T \tag{2.29}$$

The noise ζ_k in (2.28) is uncorrelated with x_k^*, and its covariance is [11]

$$\tilde{R}_k \equiv Cov[\zeta_k] = R_{\xi_k} - H'P_{x_k \xi_k} = R_{\xi_k} - H'P_{x_k}(H')^T \tag{2.30}$$

According to (2.22), (2.24) and (2.29), the projection filter is

$$\hat{x}_k = \bar{x}_k + F_{q,k}(Y'_{q,k} - \bar{Y}'_{q,k}) = \bar{x}_k + F_{q,k}\tilde{Y}_{q,k} \tag{2.31}$$

where

$$\begin{aligned} F_{q,k} &= (P_{x_k} H_q^T + P_{x_k \xi_k})(H_q P_{x_k} H_q^T + H_q P_{x_k \xi_k} + P_{x_k \xi_k}^T H_q^T + R_{\xi_k})^{-1} \\ &= \left(P_{x_k} H_q^T + P_{x_k}(H')^T\right)\left(H_q P_{x_k} H_q^T + H_q P_{x_k}(H')^T \right. \\ &\qquad \left. + H'P_{x_k} H_q^T + H'P_{x_k}(H')^T + \tilde{R}_k\right)^{-1} \\ &= P_{x_k} \tilde{H}_q^T \left(\tilde{H}_q P_{x_k} \tilde{H}_q + \tilde{R}_k\right)^{-1} \end{aligned} \tag{2.32}$$

The corresponding error covariance [11] is

$$\begin{aligned} P_k &= P_{x_k} - F_{q,k}\left(P_{x_k} H_q^T + P_{x_k \xi_k}\right)^T = P_{x_k} - F_{q,k}\left(H_q P_{x_k} + H'P_{x_k}\right) \\ &= \left(I - F_{q,k}\tilde{H}_q\right)P_{x_k}. \end{aligned} \tag{2.33}$$

A close examination of (2.31) to (2.33) shows that they have a Kalman filter format, where $F_{q,k}$ is equivalent to the Kalman filter gain K_k, and $\tilde{H}_q$, $\tilde{R}_k$ and P_{x_k} are equivalent to C, R and P_k^-, respectively. The a priori values of the estimates $\bar{x}_k$ and their corresponding error covariances P_{x_k} of the projection filter are either obtained by propagating from their initial values or set to be the steady-state values. However, the counterparts in the Kalman filter ($\hat{x}_k^-$ and P_k^-) are conditioned on all previous data. Using conditional means as the a priori estimates allows the Kalman filter to use all the data (from the beginning till the current moment) recursively in estimating the current state. On the contrary, a projection filter of order q uses only q's most recent data to do the same task. As a result, the Kalman filter is more accurate than a projection filter of small order in general. Besides, the Kalman filter treats only one measurement at each step, while the projection filter of order q needs to treat a batch of q data. Therefore, computationally the Kalman filter is more efficient. However, the projection filter can be modified to have the capability of using all the data available to produce the same result as the Kalman filter.

Based on (2.2) and the assumption that state and measurement noise are uncorrelated, the projection filter of order one and its corresponding estimation error covariance according to (2.31) to (2.33) are

$$\hat{x}_k = \bar{x}_k + P_{x_k} C^T \left(CP_{x_k} C^T + R\right)^{-1} \left(y_k - C\bar{x}_k - Du_k\right), \tag{2.34}$$

$$P_k^+ = P_{x_k} - P_{x_k} C^T \left(CP_{x_k} C^T + R\right)^{-1} CP_{x_k} \tag{2.35}$$

where $\bar{x}_k$ is the unconditional mean of x_k. In order to take advantage of previous estimations, conditional mean of state and conditional state covariance should be used. Suppose the optimal estimate of the state $\hat{x}_{k-1}$ at time $k-1$ and its corresponding error covariance P_{k-1} have been obtained using the projection filter of order $k-1$ based on all the data available. Because the projection filter also estimates the conditional mean of the parameters, conditioned on all the data used [8], $\hat{x}_{k-1}$ can be written as

$$\hat{x}_{k-1} = E\left[x_{k-1} | Y_{k-1}\right], \quad Y_{k-1} = \left\{y_0, \ldots, y_{k-1}, u_0, \ldots, u_{k-1}\right\}.$$

Using the estimate at $k-1$ and the system model, a prediction of the state at k and its corresponding error covariance can be made; that is,

$$\begin{aligned}\bar{x}_{k|Y_{k-1}} &= E\left[x_k | Y_{k-1}\right] = E\left[Ax_{k-1} + Bu_{k-1} + w_{k-1} | Y_{k-1}\right] \\ &= E\left[Ax_{k-1} | Y_{k-1}\right] + Bu_{k-1} = A\hat{x}_{k-1} + Bu_{k-1}.\end{aligned} \tag{2.36}$$

Since $\bar{x}_{k|Y_{k-1}}$ is the conditional mean conditioned on Y_{k-1}, it is also the a priori estimate of x_k. Similarly, the conditional state covariance can be calculated as

$$\begin{aligned}P_{x_k} = P_{x_k|Y_{k-1}} &= Cov\left[x_k - \bar{x}_{k|Y_{k-1}}\right] = Cov\left[A(x_{k-1} - \hat{x}_{k-1}) + w_{k-1}\right] \\ &= AP_{k-1}A^T + Q_{k-1} \equiv P_k^-\end{aligned} \tag{2.37}$$

Therefore, the recursive projection filter ((2.34) and (2.35)) becomes

$$\hat{x}_k = \bar{x}_k + K_k\left(y_k - C\bar{x}_k - Du_k\right) \tag{2.38}$$

$$K_k = P_k^- C^T \left(CP_k^- C^T + R\right)^{-1}, \tag{2.39}$$

$$P_k^+ = P_k^- - P_k^- C^T \left(CP_k^- C^T + R\right)^{-1} CP_k^- = \left(I_n - K_k C\right)P_k^-. \tag{2.40}$$

Equations (2.36) to (2.40) are exactly the same as the Kalman filter formations. This proves that *the recursive projection filter of order one is identical to the Kalman filter.*

Enlightened by the equivalence just derived, immediately we can find a modified Kalman filter for the case in which measurement noise is correlated with state. In this case the structure of the modified Kalman filter is the same as the conventional one, but the filter gain becomes

$$K_k = \left(P_k^- C^T + P_{xv}\right)\left(CP_k^- C^T + CP_{xv} + P_{xv}^T C^T + R\right)^{-1}, \tag{2.41}$$

where P_{xv} is the covariance of the state and measurement noise.

For recursive projection filters of orders greater than one in which the recursive feature is obtained by using conditional a priori mean and state covariance, the formulations are the same as (2.36) to (2.40), except C and R should be replaced by $\tilde{H}_q$ and $\tilde{R}_k$, respectively, and I_n in (2.40) replaced by the identity matrix of proper dimension. In this case, some measurements are used more than one time in estimating one single state, that is, some measurements are used both in calculating the a priori estimate and in calculating the filter part, or the modifying part. However, this does not help in improving the results. Since the projection filter seeks the conditional mean of the state, it makes no difference whether a measurement is conditioned once, twice, or more. Consequently, recursive projection filter of an order greater than one is computationally inefficient. Though there is no benefit in computation, the concept of projection filter is still valuable; it helps developing a new system identification method which will be introduced in Section IV.

D. Numerical Example

A lumped-mass beam-like system (see Fig. 1) is considered. This system has three masses and thus has three modes (six states). The state-space parameters $[A, B, C]$ in modal form are

$$A = diag\left\{ \begin{bmatrix} 0.9856 & 0.1628 \\ -0.1628 & 0.9856 \end{bmatrix}, \begin{bmatrix} 0.8976 & 0.4305 \\ -0.4305 & 0.8976 \end{bmatrix}, \begin{bmatrix} 0.8127 & 0.5690 \\ -0.5690 & 0.8127 \end{bmatrix} \right\} \tag{2.42}$$

$$B = \begin{bmatrix} 0.0011 & 0.0134 & -0.0016 & -0.0072 & 0.0011 & 0.0034 \end{bmatrix}^T \tag{2.43}$$

$$C = \begin{bmatrix} 1.5119 & 0.0000 & 2.0000 & 0.0000 & 1.5119 & 0.0000 \\ 1.3093 & 0.0000 & 0.0000 & 0.0000 & -1.3093 & 0.0000 \end{bmatrix}. \tag{2.44}$$

The sampling frequency is 10 Hz. The system starts from an initial condition x_0 = 0, and is excited by Gaussian white noise $\{w_k\}$ with covariance Q = 0.005 $\times I_6$; no external deterministic force is applied. The two output measurements are contaminated by additive Gaussian white noise with covariance $R = 0.3 \times I_2$. The noise-to-signal ratio in variance is about 10%.

The optimal Kalman filter and the projection filter in order 2, 10 and 20 are used to estimate the state using the same initial conditions ($\hat{x}_0 = 0$ and $P_0 = 4.0 \times I_6$) and noise covariances. The estimation of the first state is shown in Fig. 2, where the result of the optimal Kalman filter are compared with those of the projection filters with different order. Figure 3 shows the error variances of state 1 of the projection filter with different orders. The error variances are calculated by averaging 200 samples discarding the first 50 samples (not obtained from theory). As the order of the projection filter increases, the result is improved and approaches the performance of the optimal Kalman filter.

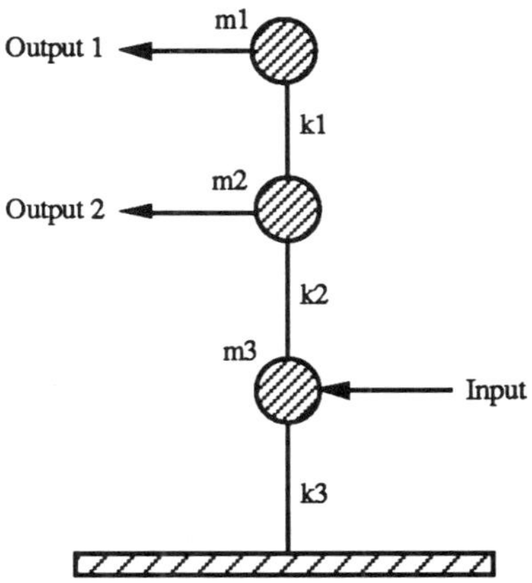

Table 1: Mass and Stiffness

Mass (kg)	
m1	50
m2	100
m2	100
Stiffness (N/cm)	
k1	10
k2	10
k3	10

Fig. 1 A simulated lumped-mass beam-like system.

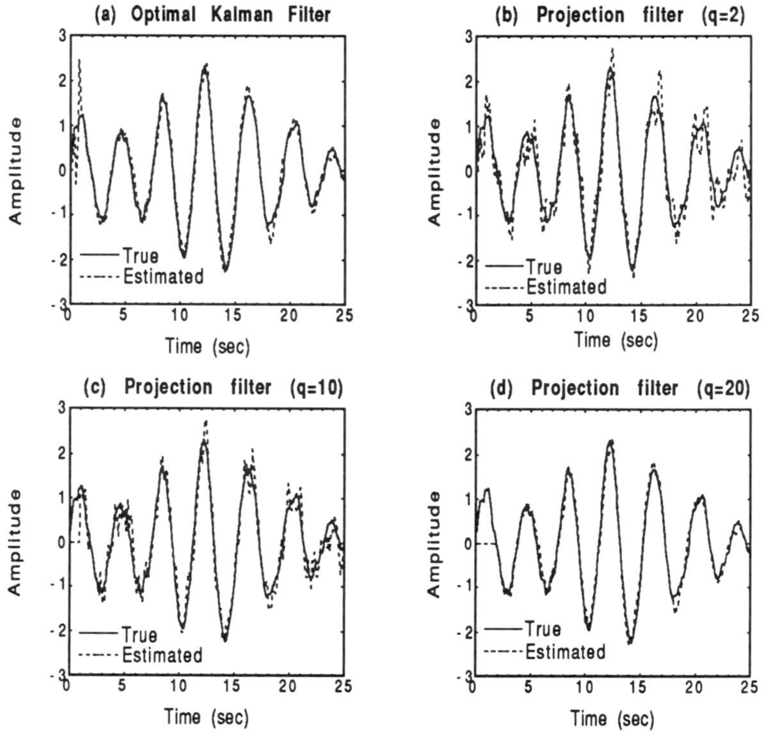

Fig. 2 Comparison of the state estimation of the first state by the Kalman filter and the projection filters with different order q.

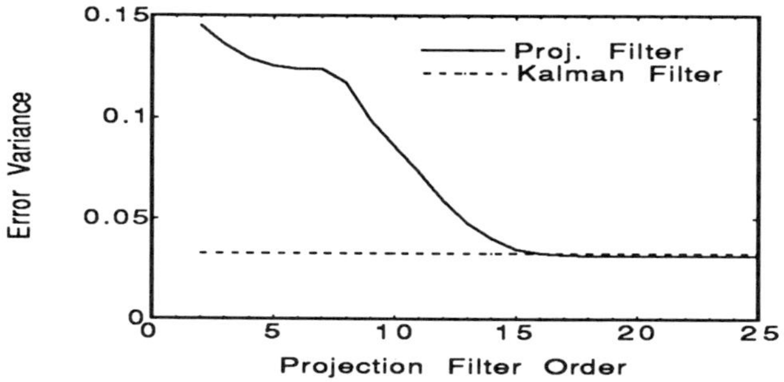

Fig. 3 The error variance of the first state by the projection filter with different orders compared to the Kalman filter.

In this section, a projection filter based on optimal parameter estimation theories is derived for state estimation of linear stochastic systems. The projection filter for noise-free systems is a least-squares filter, which is simply the pseudo-inverse of the measurement matrix. The projection filter for systems without process noise is a weighted least-squares filter, in which the weighting matrix is determined by the covariance of measurement noise. It does not need initial values to initiate the estimation but cannot yield estimates until it has accumulated a certain number of data. The estimation is less accurate as compared to the Kalman filter. The projection filter for systems with both process and measurement noises compute the conditional mean of the state based on all the measurements used. It needs the same amount of a priori information of system and noises as the Kalman filter does, but a small order projection filter is less accurate and requires more computation. However, the *recursive* projection filter is *equivalent* to the Kalman filter and can provide optimal state estimation even when the measurement noise is correlated with the state.

III. STATE ESTIMATION UNDER UNKNOWN NOISE COVARIANCES

A well-known restriction of the Kalman filter is that it requires the statistical information about the noises [3]. In practice, such information is difficult to obtain, especially for the process noise [12,13]. To use the Kalman filter without prior information about the noises, one should either guess the values based on engineering judgment, which is unreliable, or conduct some estimation procedures. Conducting estimation during operation to improve the performance of the Kalman filter is called adaptive Kalman filtering [13,14].

The methods of adaptive Kalman filtering have been divided into four categories [13]: Bayesian, maximum likelihood, correlation, and covariance matching. Each category has its own drawbacks: Bayesian approaches require statistical information of data which is seldom available in practice; maximum likelihood approaches are very time-consuming; correlation and covariance matching approaches base on the assumption of stationarity of the output or the residual and require a great amount of data.

The method introduced here does not have the above mentioned drawbacks. It estimates the filter gain directly without estimating noise covariances. Like other direct approaches [13,15], it is simple in theory, has fewer parameters to estimate, and can prevent the problem of non-uniqueness in estimating the process noise covariance. As compared to the other existing direct approaches, the method introduced here is simpler but yet is very effective.

In this section a state-space model is assumed given, but noise covariances are not available. First, an input-output relation is derived through the Kalman filter innovation model. This input-output model explicitly shows that the output is composed of two parts: the deterministic part caused by external deterministic input and the stochastic part caused by the stochastic input—noise. Second, with a known state-space model, the deterministic part of the output is computed and subtracted out from the output, and the remaining stochastic part is described by a Moving Average (MA) model. Third, in order to identify the MA model the method first identifies its inverse model—an AutoRegressive (AR) model–using a whitening filter to process on the remaining stochastic signal, and then invert it. The reason to employ this inverse filter approach is to avoid nonlinear optimization. Finally, from the coefficients of the estimated MA model an estimate of the Kalman gain is computed. One possible drawback of this method is that it could be suboptimal due to the approximation of the MA model by the inverse of an autoregressive (AR) model. However, as illustrated in the numerical example it can yield satisfactory results if the order of the AR model is sufficiently large.

A. Estimating a Steady-State Kalman Gain Using the Inverse Filter Method

One can write the Kalman filter in the following innovation format [16,17]:

$$\hat{x}_{k+1}^- = A\hat{x}_k^- + Bu_k + AK_k\varepsilon_k \tag{3.1}$$

$$y_k = C\hat{x}_k^- + Du_k + \varepsilon_k \tag{3.2}$$

where $\hat{x}_k^-$ is the optimal prediction of the state x_k based on the data $\{y_{k-1},\ldots,y_0,u_{k-1},\ldots,u_0\}$. The term ε_k, called the residual, is the difference between the real measurement y_k and the predicted measurement, $\hat{y}_k$ $(= C\hat{x}_k^- + Du_k)$. The quantity ε_k called the "innovation" in a sense contains new information that can not be obtained from previous data. A useful property of $\{\varepsilon_k\}$ is that it is a white sequence for an optimal Kalman filter [16].

From the innovation model (3.1) and (3.2) one can see that the Kalman filter is actually driven by the deterministic input u_k through B and the stochastic input ε_k through AK_k (in steady-state AK_k becomes AK). Consequently, the filter state and output can also be decomposed into two different parts—one caused by the deterministic external input and the other caused by the stochastic input. Accordingly, the innovation model can be divided into two models:

$$\hat{x}_{k+1,1}^- = A\hat{x}_{k,1}^- + Bu_k \tag{3.3}$$

$$y_{k,1} = C\hat{x}_{k,1}^- + Du_k \tag{3.4}$$

and

$$\hat{x}_{k+1,2}^- = A\hat{x}_{k,2}^- + AK\varepsilon_k \tag{3.5}$$

$$y_{k,2} = C\hat{x}_{k,2}^- + \varepsilon_k \tag{3.6}$$

where $\hat{x}_{k+1}^- = \hat{x}_{k+1,1}^- + \hat{x}_{k+1,2}^-$ and $y_k = y_{k,1} + y_{k,2}$. Expanding (3.4) and (3.6) based on (3.3) and (3.5), respectively, and assuming zero initial state, one can derive

$$y_{k,1} = Du_k + \sum_{i=1}^{k} CA^{i-1}Bu_{k-i}, \tag{3.7}$$

$$y_{k,2} = \varepsilon_k + \sum_{i=1}^{k} CA^{i}K\varepsilon_{k-i}. \tag{3.8}$$

Combining the above two equations yields

$$y_k = Du_k + \sum_{i=1}^{k} CA^{i-1}Bu_{k-i} + \varepsilon_k + \sum_{i=1}^{k} CA^{i}K\varepsilon_{k-i}. \tag{3.9}$$

Equation (3.9) explicitly shows the two parts of which the output is composed. Note that the sequence $CA^{i-1}B$ $(i=1,2,\mathrm{L}\ )$ is called the *system Markov parameters* and $CA^{i}K$ $(i=1,2,\mathrm{L}\ )$ is called the *stochastic Markov parameters*. If the state-space parameters $[A, B, C, D]$ are known, the deterministic part $y_{k,1}$ can be calculated and subtracted out from the output y_k; that is, by defining $s_k \equiv y_{k,2} = y_k - y_{k,1}$, then

$$s_k = \varepsilon_k + \sum_{i=1}^{k} CA^{i}K\varepsilon_{k-i} \equiv \sum_{i=0}^{k} C_i\varepsilon_{k-i}, \tag{3.10}$$

where $C_0 = I_p$, $C_i = CA^{i}K$ $(i=1,\mathrm{L}\ ,k)$. This remaining signal s_k represents the stochastic component driven by the sequence $\{\varepsilon_k\}$. For a stable system there exists a number q such that all the terms $CA^{i}K$ are negligibly small when $i > q$. Therefore, when k is large the upper limit of the summation of (3.10) can be replaced by q. In (3.10) s_k is described as a linear transformation of a white sequence $\{\varepsilon_k\}$. This equation is called a Moving Average (MA) model [18], and the coefficient matrices $C_1,\mathrm{L}\ ,C_q$ are called the MA parameters. The term "moving average" arises because s_k can be regarded as a weighted average of $\varepsilon_k,\mathrm{L}\ ,\varepsilon_{k-q}$. Note that the MA parameters are expressed in terms of state-space parameters A, C and the steady-state Kalman filter gain K. If the MA model is known, the filter gain can be computed from its coefficients.

The problem encountered in estimating the MA model in (3.10) is that the white sequence $\{\varepsilon_k\}$ is not available; therefore, the commonly used least-squares methods in identifying linear models can not be applied. Estimating an MA model is a non-linear problem, which can be solved by the maximum likelihood (ML) method [18]; however, the ML method is time-consuming and not guaranteed to converge and, in fact, may converge to an erroneous local minimum. An alternative is to estimate the MA model by inverting its inverse AR model [19,20]. To explain this point, taking z-transform of both sides of (3.10) one has

$$S(z) = \sum_{i=0}^{q} C_i z^{-i} E(z) = M(z^{-1})E(z), \tag{3.11}$$

where $M(z^{-1})$ is a polynomial matrix in z^{-1} (a matrix whose entries are polynomials in z^{-1}). Matrix $M(z^{-1})$ can be regarded as a filter which receives ε_k and its delayed versions as input and yields s_k as output. If the inverse filter $N(z^{-1})$ of $M(z^{-1})$ (i.e., $N(z^{-1})M(z^{-1}) = I_p$) can be found, pre-multiplying (3.19) by $N(z^{-1})$ one has

$$N(z^{-1})S(z) = E(z) \tag{3.12}$$

The inverse matrix $N(z^{-1})$ is in general an infinite-ordered polynomial matrix. In (3.12), $N(z^{-1})$ can be considered as a whitening filter which receives s_k and its delayed versions as input and yields white sequence $\{\varepsilon_k\}$ as output.

To obtain a whitening filter for the signal s_k, one can write an AutoRegressive model of s_k with order r in time domain as

$$\sum_{i=0}^{r} N_i s_{k-i} = \varepsilon_k \tag{3.13}$$

where $N_0 = I_p$, and estimate the AR parameters $N_1,\mathrm{L}\ ,N_r$ [21]. Comparing (3.12) with (3.13) it can be seen that the infinite-ordered polynomial matrix

$N(z^{-1})$ is approximated by a finite-ordered polynomial matrix $\sum_{i=0}^{r} N_i z^{-i}$. The parameter estimation of the AR model can be accomplished using the ordinary least-squares method.

After obtaining an identified $N(z^{-1})$, one can invert it to approximate $M(z^{-1})$. Inverting a square polynomial matrix is similar to inverting an ordinary scalar square matrix; the result is the adjoint matrix divided by the determinant of the matrix. In operation, polynomial multiplication is equivalent to convoluting the coefficient sequences of the two polynomials, and polynomial division is equivalent to deconvoluting the coefficient sequence of the numerator polynomial over that of the denominator polynomial, expanding to as long as desired. Note that another way of computing the MA model in the discrete-time domain is presented in [22].

After obtaining an estimated MA model, collecting $q1$ successive terms of its coefficients one can form a matrix

$$Z = \left[\hat{M}_1^T, \hat{M}_2^T, \ldots, \hat{M}_{q1}^T\right]^T \tag{3.14}$$

where

$$\hat{M}(z^{-1}) = I_p + \sum_{i=1}^{r} \hat{M}_i z^{-i}, \quad \hat{M}_i = \overline{CA^iK}$$

and $\overline{CA^iK}$ is an estimate of CA^iK. From the state-space model one can form another matrix:

$$H = \left[(CA)^T \quad (CA^2)^T \quad \cdots \quad (CA^{q1})^T\right]^T, \tag{3.15}$$

which is similar to the observability matrix. A least-squares estimate of K can be obtained by

$$\hat{K} = (H^T H)^{-1} H^T Z, \tag{3.16}$$

if H is full column-ranked. For an observable system, to make H full column-ranked, $q1$ must satisfies $(n/p) \leq q1 \leq r$ (n: number of states, p: number of outputs). However, if $q1$ is chosen barely larger than n/p, there will be too few data in the least-squares and the result is usually poor. Increasing $q1$ can improve the result in the beginning but will saturate in the long run. Choose $q1$ to be several times (say, 4 to 10) larger than n/p will be appropriate.

Because of the approximation used in the process, the estimated Kalman filter gain $\hat{K}$ is not optimal; however, it represents an identified state estimator gain (or suboptimal Kalman filter gain). The quality of the identified gain relies on the accuracy of the given state-space model and on the sufficiency of the order of the whitening filter r. If the state-space model is accurate and the order r chosen is large enough, the identified gain will converge to the optimal steady-state Kalman filter gain.

B. Estimating Measurement and Process Noise Covariances

Though a well-estimated steady-state Kalman filter gain is sufficient for a state estimation purpose, the covariances of process and measurement noises might be of interest to those who want to know the stochastic properties of the system and sensors. If this information can be derived using the estimated filter gain, the method will be useful. Indeed, one can derive the noise covariances after knowing the Kalman filter gain. In general, the covariance of measurement noise is comparatively easier to obtain than that of process noise. As one will see later, identifying the process noise covariance is difficult and the solution might not be unique.

From Kalman filter formulations, the steady-state filter gain is

$$K = P^- C^T \left(CP^- C^T + R\right)^{-1}, \tag{3.17}$$

where P^- is the a priori state error covariance. Without confusion, the superscript is omitted for simplicity. The covariance of the optimal residual sequence is

$$\Phi_k \equiv Cov\left[y_k - \hat{y}_k\right] = Cov\left[Ce_k^- + v_k\right] = CP_k C^T + R, \tag{3.18}$$

where e_k^- is the a priori state error. In steady-state subscript k of P_k can be dropped. From (3.17) and (3.18), it is obvious that

$$PC^T = K\Phi; \tag{3.19}$$

therefore,

$$R = \Phi - CK\Phi = \left(I_p - CK\right)\Phi. \tag{3.20}$$

The optimal residual can be estimated by running a Kalman filter using the estimated filter gain. For a well-estimated gain, the covariance of the estimated residual $\hat{\Phi}$ calculated by sample averaging should be very close to the true one. Replacing all the true values on the right hand side of (3.20) by the estimated versions, an estimate of measurement noise covariance, $\hat{R}$, is obtained.

Next, from Kalman filter formulations in steady-state, one has

$$P = AP^+A^T + Q, \tag{3.21}$$

$$P^+ = \left(I_n - KC\right)P, \tag{3.22}$$

where P^+ is the posterior state error covariance. Combining the above two equations yields

$$P = A\left(I_n - KC\right)PA^T + Q = APA^T - AKCPA^T + Q. \tag{3.23}$$

According to (3.23), suppose K is known, by assigning any value to P, Q is uniquely determined, and vice versa. However, P and Q cannot be chosen arbitrarily. First, both covariance matrices should be symmetric and positive definite. In addition, since the covariance of the optimal residual is fixed for a fixed gain, the value PC^T is fixed according to (3.19). Any symmetric positive definite matrix P' that satisfies (3.19) and produces a symmetric positive definite matrix Q' by (3.23) is a qualified candidate solution of P and Q. Using Q' thus found and R back to the Kalman filter will result in an identical steady-state filter gain K. Solving P and Q from (3.19) and (3.23) under the constraints of being both symmetric and positive definite is not a trivial task and the solution is not unique [4].

C. Numerical Example

The simulated dynamic system is the same as the one used in the last section (Fig. 1), except external input force is added this time. The system is excited by a known random force $\{u_k\}$ with a variance $\sigma_u^2 = 40$ at node 3, while

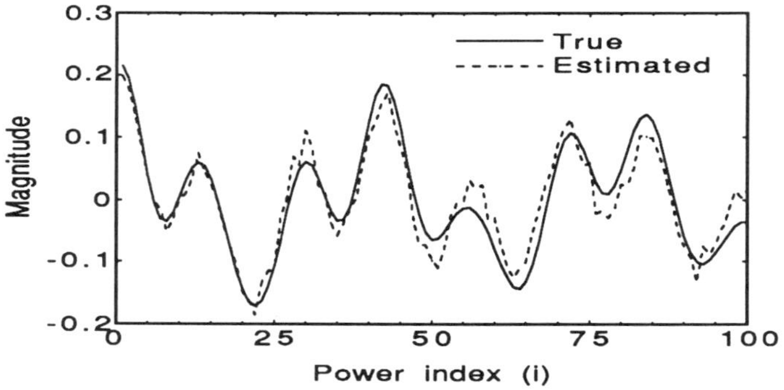

Fig. 4 Comparison of the true and the estimated stochastic Markov parameters CA^iK (the (1,1) element).

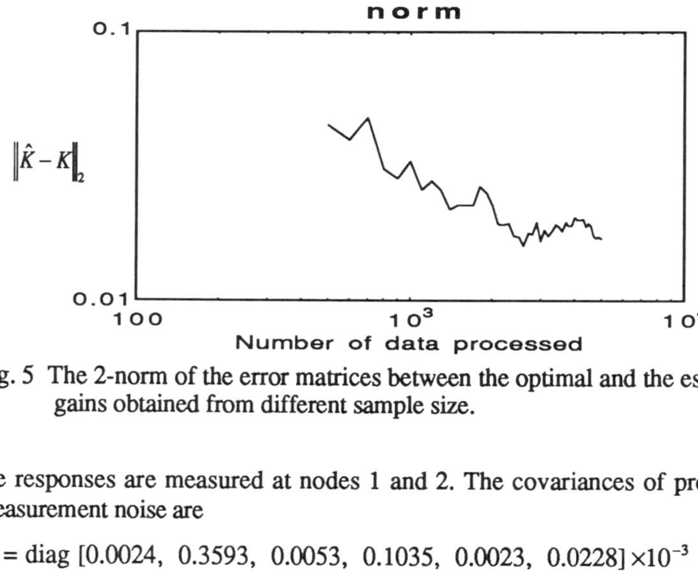

Fig. 5 The 2-norm of the error matrices between the optimal and the estimated gains obtained from different sample size.

the responses are measured at nodes 1 and 2. The covariances of process and measurement noise are

$$Q = \text{diag}\,[0.0024,\ 0.3593,\ 0.0053,\ 0.1035,\ 0.0023,\ 0.0228] \times 10^{-3} \tag{3.24}$$

$$R = 0.0279 \times I_2 . \tag{3.25}$$

The standard deviation of the process noise $\{w_k\}$ is about 23% with respect to $\{Bu_k\}$; the standard deviation of the measurement noise $\{v_k\}$ is about 10% with respect to $\{y_k\}$. Under these settings, the theoretical optimal steady-state Kalman filter gain is

$$K = \begin{bmatrix} 0.0604 & 0.0279 & 0.0471 & 0.0146 & 0.0132 & 0.0055 \\ 0.0648 & 0.0366 & -0.0059 & 0.0143 & -0.0162 & -0.0011 \end{bmatrix}^T . \tag{3.26}$$

The order of the AR model is set to 100, and the inverse of the polynomial matrix is also expanded to have 100 terms. For calculating the filter gain, the number ql in (3.15) is also set to 100. After processing 5,000 data, the (1,1) element of the estimated stochastic Markov parameters CA^iK, (i = 1,..., 100) is plotted along with their theoretical true values in Fig. 4. Figure 5 shows the 2-norm [23] of the difference matrix between the estimated and the optimal gain against the number of data processed. This figure shows the convergence of the estimation of the filter gain. The saturation occurred around 3,000 data is due to the order chosen for the whitening AR model being finite. The estimated gain after processing 5,000 data is

$$\hat{K} = \begin{bmatrix} 0.0580 & 0.0220 & 0.0457 & 0.0139 & 0.0122 & 0.0028 \\ 0.0666 & 0.0251 & -0.0090 & 0.0220 & -0.0234 & -0.0026 \end{bmatrix}^T \tag{3.27}$$

which can be compared with the theoretical optimal value (3.26).

The estimated gain is used in state estimation and the results are shown in Figs. 6–8. Figure 6 shows the comparison of the true and the estimated values of the first state; Fig. 7 is the comparison of the optimal residual (obtained using the optimal Kalman filter) and the estimated residual; Fig. 8 shows the normalized auto-correlation function of the estimated residual, from which one can tell that the residual is nearly white—an indication of good estimation. Though the estimated gain is still seemly different from the optimal one for some elements, the results show that the state estimation is near optimal already. The reason is that around the optimal point the gradients with respect to entry variables of K are not the same. Along some directions (variables) the slopes could be steeper or flatter than other directions. In other words, some variables are more sensitive than the others around the optimal point. Nevertheless, the method can estimate one that provides good results.

The Kalman filter using the estimated gain yields a residual having a covariance matrix (calculated by averaging 700 samples):

$$\hat{\Phi} = \begin{bmatrix} 0.0349 & 0.0031 \\ 0.0031 & 0.0306 \end{bmatrix}.$$

According to (3.20), the estimated measurement noise covariance $\hat{R}$ is very close to the theoretical one R shown in (3.25):

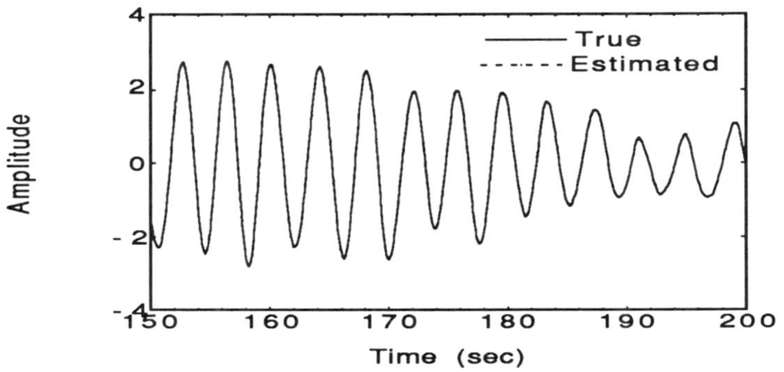

Fig. 6 Comparison of the true and the estimated values of the first state.

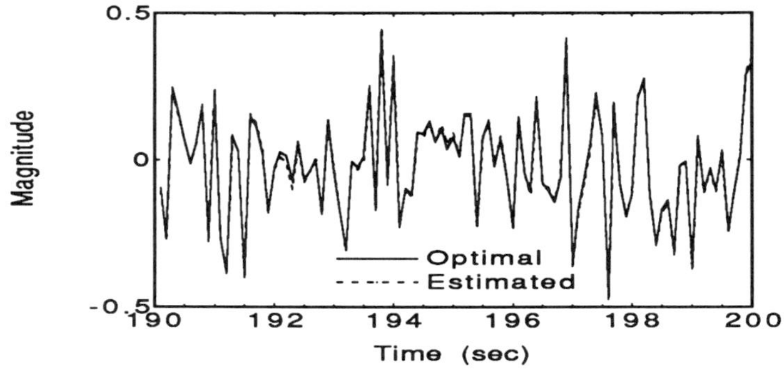

Fig. 7 Comparison of the optimal and the estimated values of the first residual.

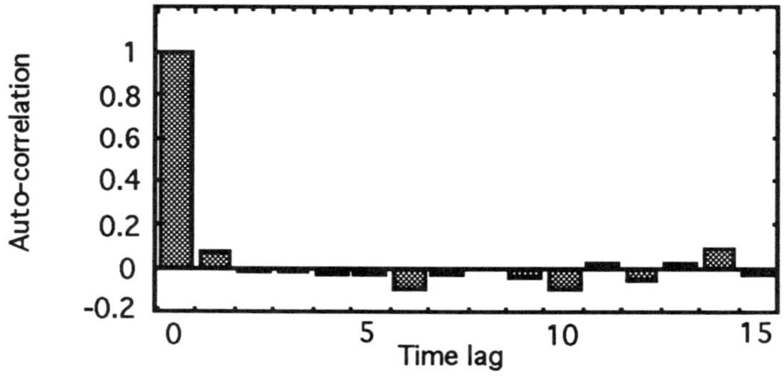

Fig. 8 The autocorrelation function of the estimated residual.

$$\hat{R} = \begin{bmatrix} 0.0278 & 0.0010 \\ 0.0006 & 0.0279 \end{bmatrix}.$$

If calculated from the optimal Kalman filter residual, the residual covariance and estimated measurement noise covariance are

$$\hat{\Phi}_{opt} = \begin{bmatrix} 0.0333 & 0.0024 \\ 0.0024 & 0.0312 \end{bmatrix}, \quad \hat{R}_{opt} = \begin{bmatrix} 0.0263 & 0.0000 \\ 0.0001 & 0.0278 \end{bmatrix}.$$

Therefore, even from the optimal Kalman filter, one cannot estimate the covariance exactly. The reason is that the number of data processed is finite.

In this section, a simple and effective method of estimating the optimal steady-state Kalman filter gain for a linear stochastic system is presented. With known state-space parameters, the method requires no information about the noises. Although the estimated gain is generally suboptimal, a numerical example shows that this method is capable of providing satisfactory state estimation results.

IV. STATE ESTIMATION UNDER UNKNOWN SYSTEM MODEL AND UNKNOWN NOISE—Integrated System Identification and State Estimation

In this section state estimation under unknown system model and unknown noise covariances is considered. In other words we know nothing about the system except the linearity assumption; all we have at hand is just input-output data. To conduct state estimation under this situation, one faces a compound problem of system identification and state estimation. Of course one possible approach is to identify a system model first, and then based on the identified model to develop the Kalman filter for state estimation as described in Section III. However, it is found [5], fortunately, that the Kalman filter gain and the system state-space parameters can be identified at the same time. Thus, an integrated method is proposed. In fact, the steady-state Kalman filter gain can be regarded as a parameter which characterizes the stochastic properties of a stochastic system. In other words, for a deterministic system, the state-space parameters $[A, B, C, D]$ are sufficient for characterizing the system; however, for a stochastic system one needs parameters $[A, B, C, D, K]$ to characterize the system. This fact can be understood by considering output prediction. For a deterministic system, given $[A, B, C, D]$ the output can be well predicted through using an observer [24], but for a stochastic system one needs $[A, B, C, D, K]$ to have good output prediction. Therefore, in conducting stochastic system identification one should have $[A, B, C, D, K]$ as the result.

Two adaptive system identification and state estimation algorithms for linear stochastic systems from input and output data are presented in this section. The first one, the simultaneous method, is derived based on the Kalman filter. The second one, the sequential method, is derived based on the projection filter.

A. The Simultaneous Method

The simultaneous method is based on the relation between the ARX model and the state-space model of a linear system connected through the Kalman filter. This relation enables one to obtain a state-space model and the corresponding Kalman filter gain from the ARX model. An ARX model has the advantage of being easy to identified from input-output data, because identifying an ARX model using the least-squares method is a linear problem and requires no initial information. It also can be identified recursively using the recursive least-squares method, which is the most desirable feature for on-line adaptive applications.

Traditionally a state-space model can be obtained from the input-output transfer function, such as using some canonical forms [23]; however, this approach suffers from several drawbacks. First, the input-output transfer function itself is not easy to be identified accurately, especially when process and measurement noises are involved. Second, the pole locations of a transfer function is known to be sensitive to the deviation of its parameters; consequently, small error in parameter identification can cause large error in the resulting model. Third, model reduction is usually needed to reduce the dimension of a canonical form to an observable and controllable model. Fourth, this approach can not identify the Kalman filter gain. In this section, though the ARX model is also a transfer function (from input and output to output) in time-domain format, the state-space model is obtained in a different way. Briefly speaking, the ARX model is used to calculate system pulse responses (the Markov parameters) first, and from the pulse responses a state-space model is obtained. This approach is more robust; in addition, the Kalman filter gain can be identified at the same time.

The relation between the ARX model and the state-space model of a linear system is derived first. Then the way of calculating the Markov parameters from the ARX model is introduced. Obtaining a state-space model using the Eigensystem Realization Algorithm (ERA) [25] is briefly discussed, and finally a numerical example using the same system as the last section is given to demonstrate the feasibility of the method.

1. *The Relation Between a State-Space Model and an ARX Model Through the Kalman Filter*

The linear system shown in (2.1) and (2.2) can also be represented by an ARMAX model [16]:

$$A(q^{-1})y_k = B(q^{-1})u_k + C(q^{-1})e_k \tag{4.1}$$

where

$$A(q^{-1}) = I_p + A_1q^{-1} + \ldots + A_{na}q^{-na} \tag{4.2}$$

$$B(q^{-1}) = D + B_1q^{-1} + \ldots + B_{nb}q^{-nb} \tag{4.3}$$

$$C(q^{-1}) = I_p + C_1q^{-1} + \ldots + C_{nc}q^{-nc} \tag{4.4}$$

q^{-1} is a backward shift operator (i.e., $q^{-1}y_k = y_{k-1}$), and na, nb, nc are the orders of the polynomials $A(q^{-1})$, $B(q^{-1})$, and $C(q^{-1})$, respectively. The sequence $\{e_k\}$ is a white noise process with zero mean. The term $A(q^{-1})y_k$ is the autoregressive (AR) part, $C(q^{-1})e_k$ the moving average (MA) part, and $B(q^{-1})u_k$ the exogenous variable (X) part.

The state-space model provides inner messages—states—about the system besides input-output information, whereas the ARMAX model gives the relation between input and output only. However, since these two different models describe the same system, they must be related somehow. Indeed, they are closely related and the relation can be connected through the Kalman filter.

From the innovation model of the Kalman filter introduced in Section III, A, one can derive

$$\varepsilon_k = y_k - C\hat{x}_k^- - Du_k. \tag{4.5}$$

Introducing (4.5) into (3.1) yields

$$\hat{x}_{k+1}^- = A(I_n - K_kC)\hat{x}_k^- + Bu_k + AK_k(y_k - Du_k) \tag{4.6}$$

The existence of the steady-state Kalman filter gain K is guaranteed if the system is detectable and $(A, Q^{1/2})$ is stablizable [16,8]. From (3.2) and (4.6) with the steady-state Kalman filter gain K, one can obtain the following input-output description:

$$y_k - Du_k = \sum_{i=1}^{q} C\bar{A}^{i-1}AK(y_{k-i} - Du_{k-i}) + \sum_{i=1}^{q} C\bar{A}^{i-1}Bu_{k-i} + \varepsilon_k + C\bar{A}^{q}\hat{x}_{k-q}^{-} \tag{4.7}$$

for some integer q and

$$\bar{A} = A(I_n - KC) \tag{4.8}$$

Note that, although the steady-state Kalman filter gain might not be known at the very beginning, it has already existed. This implies that (4.7) are a valid relation even for the very beginning data. In other words, once the value of every input/output term in (4.7) is known, this equation holds. Note that $\bar{A}$ in (4.8) is the system matrix of the Kalman filter dynamic equation (4.6), and is the system matrix of the filter error dynamic system [8]. For a stable filter, the matrix $\bar{A}$ is asymptotically stable; therefore, the last term of (4.7) is negligibly small and can be neglected for a sufficiently large number q. Moving all of the terms containing output y to the left, equation (4.7) becomes

$$y_k - \sum_{i=1}^{q} C\bar{A}^{i-1}AKy_{k-i} = Du_k + \sum_{i=1}^{q} C\bar{A}^{i-1}(B - AKD)u_{k-i} + \varepsilon_k \tag{4.9}$$

which is a special form of an ARMAX model (see (4.1) to (4.4)) with $C(q^{-1}) = I_p$. Hence it is called an ARX model because it has no moving average part. Note that all the coefficients of this model are expressed in terms of the state-space parameters A, B, C, D and the Kalman gain K, and the noise term is the residual of the optimal Kalman filter, which is white, zero-mean, and not correlated with previous input and output data according to the orthogonality principle in estimation.

Estimation of the ARX model given in (4.9) can be accomplished by the least-square method. A general ARX model of a linear system can be written as

$$y_k = \sum_{i=1}^{q1} A_i y_{k-i} + \sum_{i=0}^{q2} B_i u_{k-i} + \varepsilon_k \tag{4.10}$$

where the order of the model is denoted by ($q1$, $q2$). Given a set of input and output data $\{y_k, \ldots, y_0, u_k, \ldots, u_0\}$, the least-squares method can find a set of matrix coefficients $\{\hat{A}_1, \ldots, \hat{A}_{q1}, \hat{B}_0, \ldots, \hat{B}_{q2}\}$ which fits the data into the equation "best" in the sense of least-squares error of output prediction [21]. If the noise is white, the least-squares parameter estimation is unbiased and will converge to the true value as the number of data goes to infinity [26]. Note that the model in (4.9) is a special case of the general ARX model, having an order (q, q), or just q for simplicity.

2. *Estimation of the State-Space Parameters and the Kalman Filter Gain*

Note that (4.9) contains two groups of parameters: one is $C\bar{A}^{i-1}AK$ $(i = 1, \ldots, M)$ associated with the outputs; the other is $C\bar{A}^{i-1}(B - AKD)$ $(i = 1, \ldots, M)$ associated with the inputs; both of them have the same structure as the Markov parameters ($CA^{i-1}B$, $(i = 1, \ldots, M)$) [25]. Because the system matrices of these two parameter sets are the filter system $\bar{A}$ rather than the original system A, they are called *filter Markov parameters*. Due to the Markov structure, once being estimated they can be decomposed into the component parameters through the Eigensystem Realization Algorithm (ERA) or its variant ERA/DC [25,27]. The ERA is an algorithm for identification of linear systems from pulse response samples. It has been proved valuable for modal parameter identification from test data [28,29]. The algorithm uses the pulse response samples (i.e., the Markov parameters for a discrete-time system) to form a large block data matrix which is referred to as the general Hankel matrix. Then the technique of singular value decomposition is used to decompose the Hankel matrix into several matrices. The small singular values are attributed to noise and are truncated, and the system order is determined by counting the number of singular values retained. The state-space model can be computed from the decomposed matrices. For further details, the readers are referred to [25,27].

In [5] a method was derived to obtain a set of state-space parameters and the Kalman filter gain directly using this two parameters set through the ERA. However, the effectiveness of the method relies on the smoothness of the estimated ARX parameter sequence. Though the sequence will converge to a smooth sequence in the long run, the method usually requires a considerable large amount of data to result in a good model.

Another more effective method is introduced here. Denote $C\bar{A}^{i-1}AK$ by A_i and $C\bar{A}^{i-1}(B-AKD)$ by B_i, $B_0 = D$. One can calculate the system Markov parameters iteratively according to the following equation:

$$CA^jB = B_{j+1} + \sum_{i=1}^{j} A_i CA^{j-i}B + A_{j+1}D, \quad (j = 0, \ldots, q-1) \,. \tag{4.11}$$

Proof:

$$B_1 + A_1D = C(B - AKD) + CAKD = CB, \qquad (j = 0) \tag{4.12}$$

$$\begin{aligned}
&B_{j+1} + \sum_{i=1}^{j} A_i CA^{j-i}B + A_{j+1}D \\
&= C\bar{A}^j(B - AKD) + CAKCA^{j-1}B + \ldots + C\bar{A}^{j-2}AKCAB \\
&\quad + C\bar{A}^{j-1}AKCB + C\bar{A}^jAKD \\
&= C\left(\bar{A}^j + AKCA^{j-1} + \ldots + \bar{A}^{j-2}AKCA + \bar{A}^{j-1}AKC\right)B \\
&= C\left[\bar{A}^{j-1}(\bar{A} + AKC) + \bar{A}^{j-2}AKCA + \ldots + AKCA^{j-1}\right]B \\
&= C\left[\bar{A}^{j-2}(\bar{A} + AKC)A + \ldots + AKCA^{j-1}\right]B \\
&= \ldots \\
&= C\left[\bar{A}(\bar{A} + AKC)A^{j-2} + AKCA^{j-1}\right]B \\
&= C(\bar{A} + AKC)A^{j-1}B \\
&= CA^jB
\end{aligned} \qquad (j = 1, \ldots, q-1) \tag{4.13}$$

Δ Δ Δ

After obtaining the Markov parameters one can easily decompose them into a set of state-space parameters [A, B, C, D] using the ERA or ERA/DC. Though

Markov parameters of a system are unique, the realized state-space model through the ERA is not unique, because state-space models are not unique. However, all possible solutions are equivalent under equivalence transformation [23]. If the system contains no repeated eigenvalues, the solutions can be transformed to the modal coordinate and normalized so that the result is unique [5]. For a simulation case, only after doing model transformation and normalization on both the estimated and the theoretical models can one compare them.

Similarly, One can calculate the stochastic Markov parameters iteratively according to the following equation:

$$CA^{j+1}K = A_{j+1} + \sum_{i=1}^{j} A_{j-i+1}CA^{i}K, \quad (j = 0, \cdots, q-1). \tag{4.14}$$

<u>Proof :</u>

$$A_1 = CAK, \quad (j = 0) \tag{4.15}$$

$$\begin{aligned}
&A_{j+1} + \sum_{i=1}^{j} A_{j-i+1}CA^{i}K \\
&= C\bar{A}^{j}AK + C\bar{A}^{j-1}AKCAK + C\bar{A}^{j-2}AKCA^{2}K + \cdots + CAKCA^{j}K \\
&= C\left(\bar{A}^{j} + \bar{A}^{j-1}AKC + \bar{A}^{j-2}AKCA + \cdots + AKCA^{j-1}\right)AK \\
&= C\left[\bar{A}^{j-1}(\bar{A} + AKC) + \bar{A}^{j-2}AKCA + \cdots + AKCA^{j-1}\right]AK \\
&= C\left[\bar{A}^{j-2}(\bar{A} + AKC)A + \cdots + AKCA^{j-1}\right]AK \\
&= \cdots \\
&= C\left[\bar{A}(\bar{A} + AKC)A^{j-2} + AKCA^{j-1}\right]AK \\
&= C(\bar{A} + AKC)A^{j-1}AK \\
&= CA^{j+1}K
\end{aligned}$$

$$(j = 1, \cdots, q-1) \tag{4.16}$$

Δ Δ Δ

After obtaining this sequence, one can easily calculate an estimated Kalman filter gain as described in Section III, (3.14) to (3.16). Thus, the state estimation can be performed.

3. *Numerical Example*

The simulated dynamic system used in the last section (Fig. 1) is employed again. All the settings including external force and noise covariances are the same; therefore, the system state-space model and the optimal Kalman filter gain are identical to that of the last section. The order of the ARX model is set to 100. After processing 5,000 data, the estimated state-space parameters and the Kalman filter gain (after transforming to the modal coordinate and normalizing) are:

Table 2. Comparison of the true and the identified modal parameters

	Mode 1		Mode 2		Mode 3	
	Freq. (Hz)	Damp. (%)	Freq. (Hz)	Damp. (%)	Freq. (Hz)	Damp. (%)
Estimated	0.2611	0.86	0.7127	1.21	0.9727	1.39
True	0.2605	0.63	0.7117	1.01	0.9722	1.30

$$\hat{A} = \text{diag}\left\{\begin{bmatrix} 0.9852 & 0.1631 \\ -0.1631 & 0.9852 \end{bmatrix}, \begin{bmatrix} 0.8966 & 0.4306 \\ -0.4306 & 0.8966 \end{bmatrix}, \begin{bmatrix} 0.8121 & 0.5689 \\ -0.5689 & 0.8121 \end{bmatrix}\right\} \quad (4.17)$$

$$\hat{B} = \begin{bmatrix} 0.0011 & 0.0134 & -0.0016 & -0.0074 & 0.0010 & 0.0034 \end{bmatrix}^T \quad (4.18)$$

$$\hat{C} = \begin{bmatrix} 1.5113 & 0.0000 & 2.0000 & 0.0000 & 1.4999 & 0.0000 \\ 1.3099 & -0.0030 & -0.0107 & -0.0056 & -1.3227 & -0.0244 \end{bmatrix} \quad (4.19)$$

$$\hat{K} = \begin{bmatrix} 0.0578 & 0.0224 & 0.0449 & 0.0144 & 0.0124 & 0.0029 \\ 0.0669 & 0.0254 & -0.0093 & 0.0230 & -0.0235 & 0.0021 \end{bmatrix}^T, \quad (4.20)$$

which can be compared with equations (2.42)–(2.44) and (3.25). The identified modal parameters are listed in Table 2, compared with their true values. As one can see that the identified parameters are very close to the true ones. The results of state estimation using the identified state-space model and Kalman filter gain are very similar to Figs. 6–8 of the last section, which indicate the estimation is near optimal. They are omitted for conciseness.

B. The Sequential Method

The simultaneous method just introduced has a drawback of requiring an ARX model of considerably large order in general, which causes intensive computation in the embedded least-squares operation. Hence, efforts have been devoted to find the methods which can use an ARX model with smaller order or which can fasten computation. In [6] a method similar to the simultaneous method of obtaining a state-space model from input-output data using the notion of state observers is derived. This approach can use an ARX model with an order much smaller than that derived through the Kalman filter; however, the derivation is based on a *deterministic* approach, and the identified gain is an observer gain rather than a Kalman filter gain. In [7] it has been proved that as the order of the ARX model increase to infinity, the observer identification converges to the Kalman filter identification. However, for a stochastic system and an ARX model of a small order, to what the least-squares identification of the ARX model will converge in a stochastic sense is not clear. To addresses the above mentioned problems under stochastic setting, the following sequential method is presented.

This approach is primarily based on the relationship between the state-space model and the ARX model linked by the projection filter developed in Section II. It is similar to the simultaneous method and the method introduced in [6] in procedures, but the meaning of the parameters and the justification of the method are different. However, using this method one can not identify the Kalman gain at the same time; therefore, after obtaining the state-space model, the filter gain should be identified separately. Of course, one can use the technique developed in Section III. Nevertheless, because the effect of noise has been considered in the stochastic approach, the system model can be identified more accurately. An experimental example is given to illustrate the feasibility of the approach.

1. *The Relation Between a State-Space Model and an ARX Model Through the Projection Filter*

Similar to the development of the projection filter in Section II, we start deriving an ARX model using the relationship provided by the projection filter from a simple case and then gradually move to some more general cases.

a. An ARX Model for a Noise-Free System

To derive an ARX model of a noise-free system expressing the current output as a linear transformation of finite previous input/output data, one can combine (2.11), (2.12) and (2.16) to have:

$$\begin{aligned} y_k &= Cx_k + Du_k = CAx_{k-1} + CBu_{k-1} + Du_k \\ &= CA\left[F_q\left(Y_{q,k-1} + G_q U_{k-1}\right)\right] + CBu_{k-1} + Du_k \\ &= \sum_{i=1}^{q} CAF_{qi} y_{k-i} + Du_k + \left(CB - CAF_{q1}D\right)u_{k-1} + \sum_{i=2}^{q} CAF_q G_{qi} u_{k-i} \\ &= \sum_{i=1}^{q} A_i y_{k-i} + \sum_{i=0}^{q} B_i u_{k-i}, \end{aligned} \tag{4.21}$$

where F_{qi} ($\in R^{n\times p}$) and G_{qi} ($\in R^{(p\times q)\times m}$) are the i-th partitions of F_q and G_q, respectively, defined as

$$F_q = \left[F_{q1} \vdots F_{q2} \vdots \cdots \vdots F_{qq}\right], \qquad G_q = \left[G_{q1} \vdots G_{q2} \vdots \cdots \vdots G_{qq}\right], \tag{4.22}$$

and $A_i = CAF_{qi}$, $(i = 1, \ldots, q)$; $B_0 = D$, $B_1 = CB - CAF_{q1}D$, $B_i = CAF_q G_{qi}$ (i = 2, $\ldots$, q). Note that F_q is defined by (2.17). The model described by (4.21) is an ARX model.

b. An ARX Model for a System without Process Noise

To derive an ARX model for a system without process noise using the relationship provided by the projection filter, one can form a one-step-ahead output prediction using the estimated state of the last step, that is,

$$\hat{y}_k = CA\hat{x}_{k-1} + CBu_{k-1} + Du_k \tag{4.23}$$

and define

$$y_k = \hat{y}_k + \eta_k \tag{4.24}$$

where η_k is the prediction error. From (2.16) and (2.18), one has

$$\begin{aligned} y_k &= CA\hat{x}_{k-1} + CBu_{k-1} + Du_k + \eta_k \\ &= CA\left[F_q\left(Y_{q,k-1} + G_q U_{k-1}\right)\right] + CBu_{k-1} + Du_k + \eta_k \end{aligned}$$

$$= \sum_{i=1}^{q} CAF_{qi} y_{k-i} + Du_k + \left(CB - CAF_{q1}D\right)u_{k-1} + \sum_{i=2}^{q} CAF_q G_{qi} u_{k-i} + \eta_k$$
$$= \sum_{i=1}^{q} A_i y_{k-i} + \sum_{i=0}^{q} B_i u_{k-i} + \eta_k, \tag{4.25}$$

where F_{qi} and G_{qi} are defined in (4.22), but now F_q is defined by (2.18) in this case.

c. An ARX Model for a System with both Process and Measurement Noises

Similarly, to derive an ARX model for a system with both process and measurement noises, one can use one-step-ahead output prediction as (4.23) and (2.22) to (2.24) to have

$$\hat{y}_k = CA\hat{x}_{k-1} + CBu_{k-1} + Du_k$$
$$= CA\left[\bar{x}_{k-1} + F_q\left(Y'_{q,k-1} - \bar{Y}'_{q,k-1}\right)\right] + CBu_{k-1} + Du_k$$
$$= CAF_q Y_{q,k-1} + CBu_{k-1} + CAF_q G_q U_{k-1} + CA\left(I_n - F_q H_q\right)\bar{x}_{k-1} + Du_k$$
$$= \sum_{i=1}^{q} CAF_{qi} y_{k-i} + Du_k + \left(CB - CAF_{q1}D\right)u_{k-1}$$
$$+ \sum_{i=2}^{q} CAF_q G_{qi} u_{k-i} + CAL\bar{x}_{k-1}, \tag{4.26}$$

where

$$L = I_n - F_q H_q,$$

and F_{qi} and G_{qi} are defined the same way as in (4.22) but F_q is defined by (2.24) instead.

Equation (4.26) represents the best prediction of y_k one can make using q previous input/output data. If the prediction is made conclusively, namely, no prediction of previous state is made, the best value assigned to $\bar{x}_k$ is zero. However, if previous state estimation has been made, the best choice for $\bar{x}_k$ is the a priori Kalman filter estimate. Note that for a Kalman filter

$$\bar{x}_{k-1} = A\bar{x}_{k-2} + AK\left(y_{k-2} - C\hat{x}_{k-2} - Du_{k-2}\right) + Bu_{k-2} = \text{L}$$
$$= \sum_{i=1}^{q-1} \bar{A}^{i-1} AKy_{k-1-i} + \sum_{i=1}^{q-1} \bar{A}^{i-1}(B - AKD)u_{k-1-i} + \bar{A}^q \hat{x}_{k-q}, \tag{4.27}$$

where

$$\bar{A} = A\left(I_n - KC\right),$$

and K is the optimal steady-state Kalman filter gain. Based on the argument above, one can replace $\bar{x}_{k-1}$ in (4.26) by (4.27) and obtain

$$y_k = \hat{y}_k + \eta_k = CAF_{q1} y_{k-1} + \sum_{i=2}^{q} CA\left(F_{qi} + L\bar{A}^{i-2}AK\right)y_{k-i} + Du_k$$
$$+ \left(CB - CAF_{q1}D\right)u_{k-1} + \sum_{i=2}^{q} CA\left(F_q G_{qi} + L\bar{A}^{i-2}(B - AKD)\right)u_{k-i} + \eta'_k$$

$$= \sum_{i=1}^{q} A_i y_{k-i} + \sum_{i=0}^{q} B_i u_{k-i} + \eta'_k, \tag{4.28}$$

where $\eta'_k = \eta_k + C\bar{A}^q \hat{x}_{k-q}$. Note that if q is not large enough, $\{\eta'_k\}$ is not white. Equations (4.21), (4.25) and (4.28) represent the ARX models of linear systems in several different noise situations. The equation in each case provides a best prediction of the output measurement at time k in the sense of minimum state error at time k-1 using q previous input and output data.

2. *Identification of a State-Space Model and the Kalman Filter Gain*

There are some relationships between the system Markov parameters and the coefficient matrices of the ARX models described in (4.21), (4.25) or (4.28). Based on these relationships one can compute the system Markov parameters from the ARX models.

For a noise-free system, from (4.21) if the coefficient matrices of y_{k-j} and u_{k-j} are denoted by A_j and B_j, respectively, one can derive a relation:

$$CA^jB = B_{j+1} + \sum_{i=1}^{j} A_i CA^{j-i}B + A_{j+1}D \tag{4.29}$$

This equation allows one to calculate the system Markov parameters CA^jB $(j = 1, \cdots, q-1)$ iteratively from the coefficient matrices of an ARX model of order q (note that $B_0 = D$ and $CB = B_1 + A_1 D$).

Proof of Eq. (4.29):.

From (2.13), (2.14) and (4.22), $G_{q1} = \left[-D^T,\ 0^T,\ \cdots\ 0^T\right]^T$, and for $j \geq 2$

$$G_{qj} = \begin{bmatrix} 0 \\ \vdots \\ 0 \\ CA^{-1}B - D \\ \vdots \\ CA^{-q+j-1}B \end{bmatrix} = \begin{bmatrix} C \\ \vdots \\ CA^{-j+2} \\ CA^{-j+1} \\ \vdots \\ CA^{-q+1} \end{bmatrix} A^{j-2}B - \begin{bmatrix} CA^{j-2} \\ \vdots \\ C \\ 0 \\ \vdots \\ 0 \end{bmatrix} B - \begin{bmatrix} 0 \\ \vdots \\ 0 \\ D \\ \vdots \\ 0 \end{bmatrix}$$

$$= H_q A^{j-2}B - E^{(j-2)}B - D^{(j)} \tag{4.30}$$

where

$$E^{(j-2)} = \left[(CA^{j-2})^T,\ \cdots,\ C^T,\ 0^T,\ \cdots,\ 0^T\right]^T,$$

$$D^{(j)} = \left[0^T,\ \cdots\ 0^T,\ D^T,\ 0^T,\ \cdots\ 0^T\right]^T,$$

($D^{(j)}$ has D in the j-th block and zero elsewhere). Therefore, with $j \geq 1$

$$CAF_q G_{q(j+1)} = CAF_q H_q A^{j-1}B - CAF_q E^{(j-1)}B - CAF_{q(j+1)}D$$

$$= CAF_q H_q A^{j-1}B - \sum_{i=1}^{j} CAF_{qi} CA^{j-i}B - CAF_{q(j+1)}D \tag{4.31}$$

Now, because from (2.17)

$$F_q H_q = I_n \tag{4.32}$$

and from (4.21) and (4.30), one has

$$B_{j+1} = CAF_q G_{q(j+1)} = CA^j B - \sum_{i=1}^{j} A_i CA^{j-1} B - A_{j+1} D.$$

So, (4.29) follows.

$\Delta\ \Delta\ \Delta$

Note that $CA^j B$ $(j = q, q+1, \ldots)$ can also be iteratively calculated by

$$CA^j B = CA(F_q H_q) A^{j-1} B = \sum_{i=1}^{q} A_i CA^{j-i} B. \tag{4.33}$$

Though derived from a noise-free system, the above equations ((4.29) to (4.33)) also hold for systems with additive white measurement noise. Because for systems with white measurement noise the projection filter F_q is nothing but a weighted pseudo-inverse of H_q, hence (4.32) also holds.

It is interesting to see that (4.29) also holds for systems with both process and measurement noise even though (4.32) does not hold in this case. This can be proved as follows.

Proof:

By definition (see Eq. (4.28))

$$
\begin{aligned}
&B_{j+1} + \sum_{i=1}^{j} A_i CA^{j-i} B + A_{j+1} D \\
&= CA\left(F_q G_{q(j+1)} + L\bar{A}^{j-1}(B - AKD)\right) + CAF_{q1} CA^{j-1} B + CA\left(F_{q2} + LAK\right) CA^{j-2} B \\
&\quad + \cdots + CA(F_{qj} + L\bar{A}^{j-2} AK) CB + CA\left(F_{q(j+1)} + L\bar{A}^{j-1} AK\right) D \\
&= CAF_q G_{q(j+1)} + CAL\bar{A}^{j-1}(B - AKD) + \sum_{i=1}^{j} CAF_{qi} CA^{j-i} B \\
&\quad + \sum_{i=1}^{j-1} CAL\bar{A}^{i-1} AKCA^{j-i-1} B + CA\left(F_{q(j+1)} + L\bar{A}^{j-1} AK\right) D \\
&= CAF_q H_q A^{j-1} B - \sum_{i=1}^{j} CAF_{qi} CA^{j-i} B - CAF_{q(j+1)} D + CAL\bar{A}^{j-1} B - CAL\bar{A}^{j-1} AKD \\
&\quad + \sum_{i=1}^{j} CAF_{qi} CA^{j-i} B + \sum_{i=1}^{j-1} CAL\bar{A}^{i-1} AKCA^{j-i-1} B + CAF_{q(j+1)} D + CAL\bar{A}^{j-1} AKD \\
&= CAF_q H_q A^{j-1} B + CAL\bar{A}^{j-1} B + CAL\bar{A}^{j-2} AKCB + \sum_{i=1}^{j-2} CAL\bar{A}^{i-1} AKCA^{j-i-1} B \\
&= CAF_q H_q A^{j-1} B + CAL\bar{A}^{j-2}\left(\bar{A} + AKC\right) B + \sum_{i=1}^{j-2} CAL\bar{A}^{i-1} AKCA^{j-i-1} B \\
&= CAF_q H_q A^{j-1} B + CAL\bar{A}^{j-2} AB + \sum_{i=1}^{j-2} CAL\bar{A}^{i-1} AKCA^{j-i-1} B \\
&= \cdots \\
&= CAF_q H_q A^{j-1} B + CAL\bar{A} A^{j-2} B + CALAKCA^{j-2} AB
\end{aligned}
$$

$$= CAF_qH_qA^{j-1}B + CALA^{j-1}B$$
$$= CA\left(F_qH_q + L\right)A^{j-1}B = CA^jB \qquad (4.34)$$

where the relations $\bar{A} + AKC = A$, $F_qH_q + L = I_n$, and (4.31) are used.

Δ Δ Δ

However, (4.33) does not hold for a system with process noise. Hence, for an ARX model of order q, only q terms of the system Markov parameters can be obtained. A system of dimension n has only n independent Markov parameters; all the rests are the linear combination of these n independent ones. Therefore, the order chosen for the ARX model q, should be greater than or equal to n. Usually more Markov parameters can improve the accuracy of the identified state-space model in the later procedure. However, the trade-off is the increase in computation.

It can be proved that the least-squares identification of an ARX model of order q will converge to the model derived through the projection filter as the number of data goes to a large number, even when q is not large and the noise $\{\eta_k'\}$ is not white [30]. This fact and the relation described in (4.29) enable us to use an ARX model of a smaller order in identifying a system model.

Similar to the simultaneous method, the first step of the sequential method is to identify an ARX model of an order q using the least-squares method. Several least-squares techniques have been studied and compared [21,31]. However, due to the smaller order chosen for the ARX model, the identified ARX parameters should be interpreted by the projection filter rather than by the Kalman filter. Then the relation (4.29) is used to calculate the system Markov parameters, and the ERA is used to yield a state-space model. Up to this step, the procedures are actually the same as the simultaneous method. However, the interpretation through the projection filter prevents obtaining the Kalman filter from the ARX model. After obtaining the state-space parameters $[A, B, C, D]$, the state estimator gain K can be identified using the inverse filter method presented in Section III. Finally, the realized $[A, B, C, D, K]$ can be used for further state estimation or controller design.

3. *Experimental Validation*

An experimental example is used to demonstrate the feasibility of the sequential method developed above. A ten-bay structure as shown in Fig. 9 is considered [32]. The truss is one of the structures built in NASA Langley Research Center for experiments in studies of control and structure interaction (CSI). It is 100 inches long, with a square cross section of 10 in $\times$10 in. All the tubing (longerons, battens, and diagonals) and ball joints are made of aluminum. The structure is in a vertical configuration attached from the top using an L-shaped fixture to a backstop. Two cold air thrusters acting in the same direction are placed at the tip. The thrusters which are used for excitation and control have a maximum thrust of 2.2 lb each. A mass of approximately 20 lb is attached at the beam tip to lower the fundamental frequency of the truss. Two servo accelerometers located at a corner of the square cross section provide the in-plan tip acceleration.

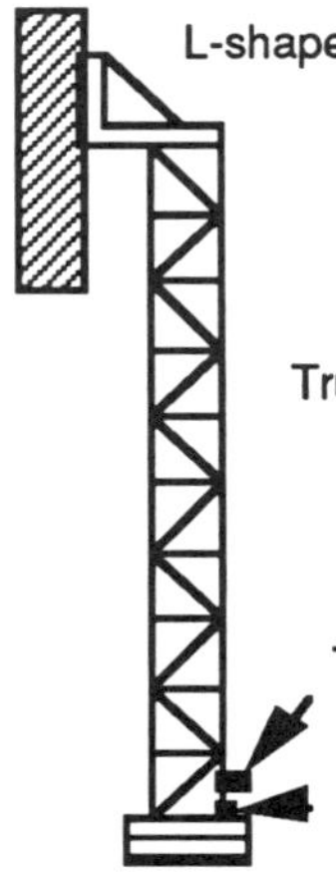

Fig. 9 Ten-bay truss structure test configuration.

The structure was excited using random inputs to both thrusters for 30 seconds. The input signals were filtered to concentrate the energy in the low frequency range. A total of 7499 data points at sampling rate 250 Hz is taken. The two output acceleration signals were filtered using a three-pole Bessel filter with a break frequency of 20 Hz.

From the output one can tell the dominant mode is about 5 to 6 Hz. To avoid using high order in the least-squares filter the sampling rate is reduced to 1/2 of the original one by choosing one out of every two samples. Hence, the sampling rate becomes 125 Hz and totals 3750 data. The order of the ARX model is set to 100. Figure 10 shows the identified system Markov parameters $CA^{i-1}B$ $(i = 1, \ldots, 100)$. By the ERA three modes are identified and the identified modal frequencies and dampings are listed in Table 3.

The corresponding state-space model matrices in normalized modal format are

$$\hat{A} = diag\left\{ \begin{bmatrix} 0.9555 & 0.2922 \\ -0.2922 & 0.9555 \end{bmatrix}, \begin{bmatrix} 0.9229 & 0.3568 \\ -0.3568 & 0.9229 \end{bmatrix}, \begin{bmatrix} -0.7538 & 0.6423 \\ -0.6423 & -0.7538 \end{bmatrix} \right\}, \tag{4.35}$$

$$\hat{B} = \begin{bmatrix} 0.1725 & -0.1789 \\ -0.1117 & 0.1267 \\ 0.1122 & -0.1522 \\ -0.0321 & 0.0556 \\ 0.3241 & 0.3309 \\ -0.1871 & -0.2725 \end{bmatrix}, \qquad \hat{C} = \begin{bmatrix} 1.7754 & 0.9201 \\ 0.0000 & 0.0362 \\ 1.0023 & -1.6909 \\ 0.0000 & -0.3692 \\ 1.3946 & -1.4287 \\ 0.0000 & 0.1185 \end{bmatrix}^T, \tag{4.36}$$

Table 3 The identified modal parameters of the ten-bay structure

Mode	Frequency (Hz)	Damping (%)
1	5.904	0.27
2	7.340	2.87
3	48.460	0.40

$$\hat{D} = \begin{bmatrix} -0.0042 & 0.0165 \\ 0.0215 & 0.0009 \end{bmatrix}. \tag{4.37}$$

The identified state-space model matrices are used to estimate the corresponding Kalman filter gain. The identified stochastic Markov parameters CA^iK $(i = 1, \ldots, 100)$ are shown in Fig. 11, and the estimated state estimator gain is

$$\hat{K} = \begin{bmatrix} 0.2787 & 0.1807 & 0.3437 & 0.1072 & 0.0065 & 0.0357 \\ 0.2277 & -0.0685 & 0.1236 & -0.0884 & -0.0874 & -0.0182 \end{bmatrix}^T. \tag{4.38}$$

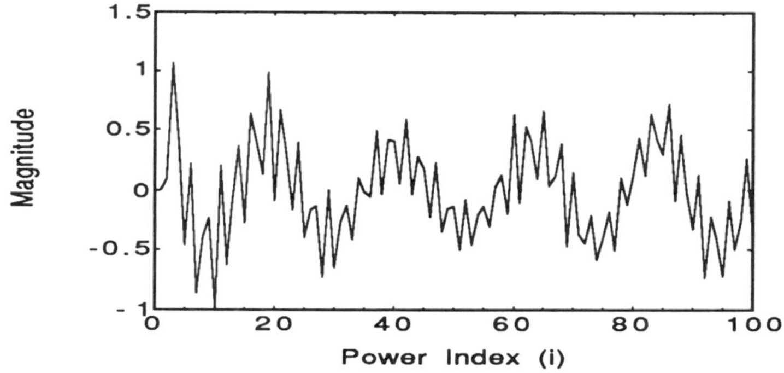

Fig. 10 Identified system Markov parameters $CA^{i-1}B$ (the (1,1) element).

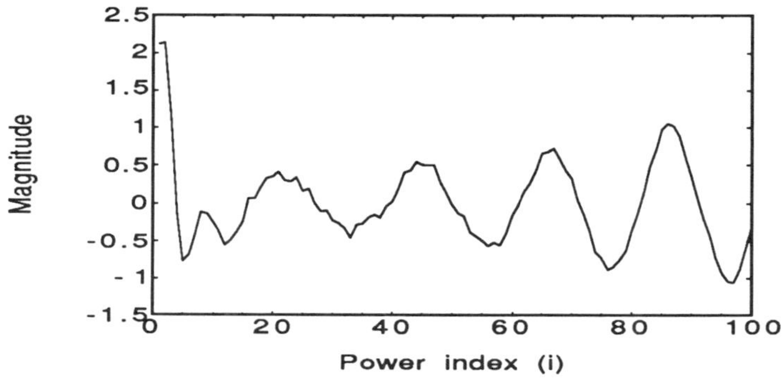

Fig. 11 Identified stochastic Markov parameter CA^iK (the (1,1) element).

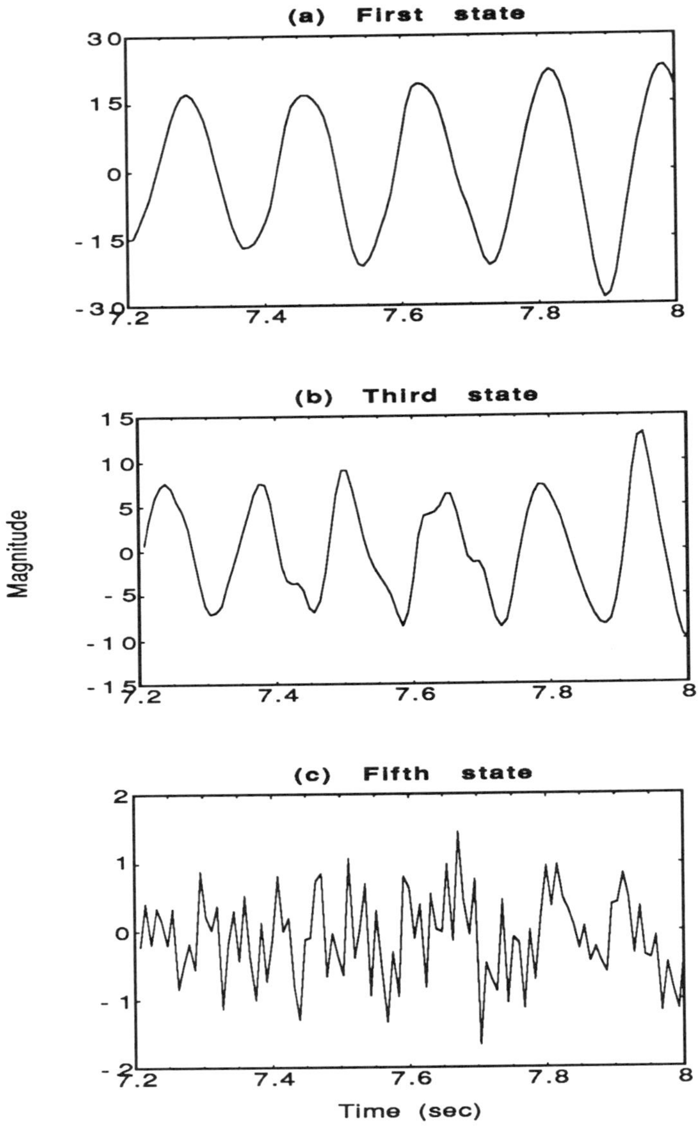

Fig. 12 Estimated modal states

Figure 12 shows the estimation of the first, third and fifth states. Since the modal model has been normalized, the amplitude of each modal state indicates the energy allocated in that specific mode. To evaluate the quality of the system identification and state estimation, the estimated outputs calculated based on the estimated state are compared to the true outputs.

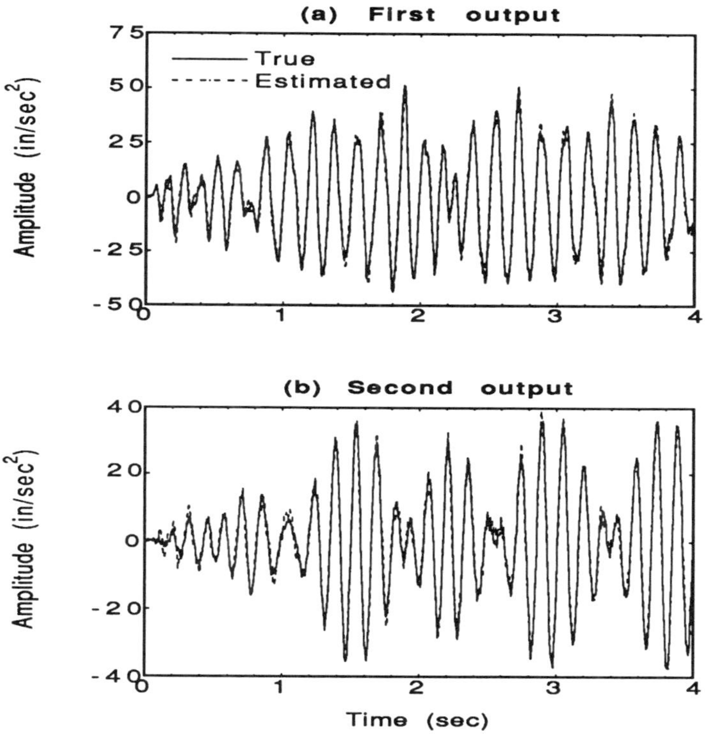

Fig. 13 Comparison of the true and estimated outputs

Because the true state is not available, output comparison is the only way to validate the results. Comparison of the outputs is shown in Fig. 13, where one can see the estimated and the true outputs are in good agreement. The covariance of the error is less than 1.5% of the covariance of the output.

V. CONCLUSIONS

Several approaches are presented here for identifying a state-space model and/or the corresponding Kalman filter gain for a linear stochastic system from input-output data: the inverse filter method identifies the Kalman filter gain based on a given system model; the simultaneous and the sequential methods identify both the system model and the Kalman gain. These methods are primarily based on the relationship between the state-space model and the difference models of linear stochastic systems, linked by the Kalman filter or by the projection filter. In contrast to most of the existing system identification methods—the great majority of them are deterministic approaches—these methods are derived under a stochastic framework, taking noise into consideration. Consequently, they are capable of reducing the effect of noise on parameter identification—especially the effect of process noise which is usually neglected in deterministic approaches—and yields better results. By using the recursive least-squares method, these algorithms can continually improve the system and filter models based on input-output data only, which is suitable for on-line applications. Furthermore, if forgetting factors are employed in the recursive least-squares, the methods can even trace a slowly time-varying linear system. The integrated identification of a state-space model and the corresponding Kalman filter gain of the methods is especially desirable for control applications because a controller and a state estimator are readily to be constructed. All these features indicate high potential of applications of these methods in the control of flexible space structures.

ACKNOWLEDGMENT

Financial support for this work through NASA grant NAG1-830 is gratefully acknowledged.

REFERENCES

1 Ljung, L., *System Identification —Theory for the User*, Prentice-Hall, Inc., Englewood Cliffs, New Jersey, 1987.

2 Juang, J.-N., and Pappa, R. S., "A Comparative Overview of Modal Testing and System Identification for Control of Structures," *The Shock and Vibration Digest*, Vol. 20, No. 6, June 1988, pp. 4-15.

3 Kalman, R. E., "A new approach to linear filtering and prediction problems," *Trans. ASME, J. Basic Eng.*, Vol 82, 1960, pp. 35-44.

4 Chen, C. W., "Integrated System Identification and State Estimation for Control of Flexible Structures," Ph.D. dissertation, August, 1991, Old Dominion University, Norfolk, Virginia.

5 Chen, C. W., Huang, J. K., Phan, M., and Juang, J. N., "Integrated System Identification and Modal State Estimation for Control of Flexible Space Structures," *Journal of Guidance, Control and Dynamics*, Vol. 15, No. 1, Jan.-Feb. 1992, pp. 88-95.

6 Phan, M., Horta, L. G., Juang, J.-N., and Longman, R. W., "Linear System Identification via an Asymptotically Stable Observer", Paper No. 91-2734, *Proceedings of the AIAA Guidance, Navigation and Control Conference*, New Orleans, LA, Aug. 12-14, 1991.

7 Juang, J.-N., Phan, M., Horta, L. G., and Longman, R. W., " Identification of Observer/Kalman Filter Markov Parameters: Theory and Experiments," Paper No. 91 2735, *Proceedings of AIAA Guidance, Navigation and Control Conference*, New Orleans, LA, Aug. 12-14, 1991, and NASA TM-104069, March 1991.

8 Lewis, F. L., *Optimal Estimation*, Wiley, New York, 1986.

9 Sorenson, H. W., "Least-squares estimation: from Gauss to Kalman," *IEEE Spectrum*, vol. 7, July 1970, pp. 63-68.

10 Elbert, T. F., *Estimation and Control of Systems*, Van Nostrand Reinhold Co., New York, 1984.

11 Orfanidis, S. J., *Optimal Signal Processing—An Introduction*, Macmillan Publishing Co., London, 1985, pp. 7-12.

12 Mehra, R. K., "On the identification of variances and adaptive Kalman filtering," *IEEE Transactions On Automatic Control*, AC-15, April 1970, pp. 175-184.

13 Mehra, R. K., "Approach to adaptive filtering," *IEEE Transactions On Automatic Control*, Oct., AC-17, 1972, pp. 693-698.

14 Carew, B. and Belanger, P. R., "Identification of optimal filter steady-state gain for system with unknown noise covariances," *IEEE Trans. Automatic Control*, vol. AC-18, 1973, pp. 582-587.

15 Tajima K., "Estimation of Steady-State Kalman Filter Gain," *IEEE Transactions On Automatic Control*, AC-23, Oct. 1978, pp. 944-945.

16 Goodwin, G. C., and Sin K. S., *Adaptive Filtering, Prediction and Control*, Prentice Hall, Englewood Cliffs, New Jersey 07632, 1984.

17 Kailath, T., "An innovations approach to least-square estimation - Part I: linear filtering in Additive White Noise," *IEEE Transactions on Automatic Control*, vol. AC-13, 1968, pp. 646-655.

18 Box, G. E. P., and Jenkins, G. M., *Time Series Analysis: Forecasting and Control*, Rev. ed. San Francisco: Holden Day, 1976,

19 Chen, C.-W., and Huang, J.-K. "Estimation of Optimal Kalman Filter Gain from Non optimal Filter Residuals," *Proceedings of SPIE's Conference on Controls for Optical Systems*, Orlando, Florida, April, 1991.

20 Graupe, D., *Time Series Analysis, Identification and Adaptive Filtering*, Robert E. Krieger Co., Florida 32950, 1989, pp. 125-160.

21 Haykin, S., *Adaptive Filter Theory*, Second Edition, Prentice-Hall, Englewood Cliffs, New Jersey 07632, 1991.

22 Juang, J.-N., Chen, C. W., and Phan, M., "Estimation of Kalman Filter Gain From Output Residual" *NASA Technical Memorandum 107603*, April, 1992.

23 Chen, C.-T., *Linear System Theory and Design*, CBS College Publishing, New York, 1984.

24 Luenberger, D. G., "Observing the State of a Linear System," *IEEE Transaction on Military Electronics*, Vol. MIL 8, April 1964.

25 Juang, J.-N., and Pappa, R. S., "An Eigensystem Realization Algorithm for Modal Parameter Identification and Model Reduction," *Journal of Guidance, Control, and Dynamics*, Vol. 8, Sept.-Oct. 1985, pp. 620-627.

26 Mayne, D. Q., and Firoozan, F., "Linear Identification of ARMA Process," *Automatica*, Vol. 18, N0.4, 1982, pp. 461–466.

27 Juang, J. N., Cooper, J. E., and Wright, J. R., " An Eigensystem Realization Algorithm Using Data Correlation (ERA/DC) for Modal Parameter Identification," *Journal of Control Theory and Advanced Technology*, Vol. 4, No. 1, pp. 5–14, March 1988.

28 Pappa, R. S. and Juang J.-N., "Some Experiences With the Eigensystem Realization Algorithm," *Sound and Vibration*, January 1988, pp. 30-34.

29 Pappa, R. S., and Schenk, A., "ERA Modal Identification Experiences with Mini Mast," Presented at *the 2nd USAF/NASA Workshop on System Identification and Health Monitoring of Precision Space Structures*, Pasadena, CA, March 27-29, 1990.

30 Chen, C.-W., Huang, J.-K., and Juang, J.-N., "Identification of Linear Stochastic Systems Through Projection Filters" Paper No. 92-2520, *Proceedings of AIAA Structures, Structural Dynamics & Materials Conference*, Dallas, Texas, April 1992, pp. 2330-2340.

31 Chen, C.-W., Lee, G., and Juang, J.-N., "Several Recursive Techniques for Observer/Kalman Filter System Identification From Data," *Proceedings of the AIAA Guidance, Navigation and Control Conference*, Hilton Head, South Carolina, Aug. 1992.

32 Horta, L. G., Phan, M., Juang, J.-N., Longman, R. W., and Sulla, J., "Frequency Weighted System Identification and Linear Quadratic Controller Design," *Proceedings of the AIAA Guidance, Navigation and Control Conference*, New Orleans, Louisiana, Aug. 1991.

Analysis and Robust Control Techniques for an Ideal Flexible Beam

Kathryn Lenz
Department of Mathematics and Statistics
University of Minnesota at Duluth
Duluth, MN 55812

Hitay Özbay
Department of Electrical Engineering
The Ohio State University
Columbus, OH 43210

I. INTRODUCTION

In this chapter we investigate problems of controlling the vibrations of a highly flexible beam subjected to perturbations and delays. Our investigation includes derivation and analysis of transfer functions for Euler-Bernoulli beams with Kelvin-Voigt damping in a variety of configurations including free-free, hinged ends, and clamped-free. We find that the transfer functions for two particular free-free configurations are qualitatively representative of all of the beam transfer functions we consider. We focus our control design and analysis discussion on these two beam systems. In the first system the displacement of one end of the beam is to be controlled by a force applied to the same end of the beam. In the second system, the displacement of one end of the beam is to be controlled by a force applied to the opposite end of the beam. We formulate a weighted mixed sensitivity robust control design problem for the beam systems, derive optimal and approximately optimal controllers, then analyze our control designs. Most of this chapter is based on the work described in [1].

The flexible beam systems we consider are examples of distributed parameter systems. They have transfer function representations as transcendental functions of a complex variable s. State-space representations of

CONTROL AND DYNAMIC SYSTEMS, VOL. 57

them would have an infinite number of states. One must use a finite order approximation of this kind of system in order to perform H^∞ optimization with standard state-space methods. In doing so one finds a controller that is optimal for an approximation of the system, but not for the distributed system itself. One does not find out how the design compares to an optimal design or what the best achievable performance is. One might expect that the number of states could be chosen sufficiently large so that a finite order system could be found to approximate a given distributed system as closely as one would like. However, even if this is the case, there are practical limits to the size of matrices that can be handled by any given computer. There are also numerical issues associated with state-space H^∞ optimization that depend on the conditioning and size of matrices.

We use skew Toeplitz theory, (see e.g. [2], [3], [4] and the references therein) to derive H^∞ optimal controllers for weighted mixed sensitivity problems for the two beam systems. Skew Toeplitz theory allows for the generalization of H^∞ control theory for finite order, linear, time-invariant systems to linear, time-invariant, distributed parameter systems. It does not use state-space representations of distributed system components.

We represent the transcendental beam transfer functions in the frequency domain as infinite products of second order terms. The infinite product representations facilitate factorizations of the plants as quotients of H^∞ functions. Such factorizations are central in solving weighted mixed sensitivity optimization problems. Truncation of the infinite products provides a straight forward way of generating finite order approximately optimal controllers.

For the simplest case we consider the measurement and the applied force are at the same end of the beam and there is no time delay in the system. The transfer function for the beam system is minimum phase (has no right half plane zeros) and stable (has no right half plane poles or singularities). We show that the optimal controller, C_{opt}, can be expressed as the product of a rational function and a transcendental function. The rational part is determined by the rational weighting functions in our design. The transcendental part is based on the transcendental transfer function for the beam system. We derive finite order, approximately optimal controllers by truncating the high frequency pole/zero pairs of a product representation for the transcendental part of C_{opt}. By preserving the exact poles and zeros of C_{opt} within a finite frequency band, we preserve a certain level of closed-loop system stability and performance. We show that these finite order controllers are also robust to a small, unmodelled feedback time delay.

The form of the optimal controller is much more complicated when the measurement and applied force are located at the opposite ends of the beam. In this case, we show that the beam system transfer function has infinitely many right half plane zeros that can be factored out of the plant in the form of an infinite Blaschke product. We measure the effect of these zeros on achievable performance in in terms of the numerical value of the H^∞ norm of the optimal weighted closed loop system. We derive finite order controllers for this system and compare their performance with that of the optimal controller. We also show that they are robust to a small, unmodelled feedback time delay.

We demonstrate mathematical techniques on a highly simplified and idealized beam model. Modifications would have to be made in order to make the results meaningful for a physical system. Our goal is to show what happens when a special partial differential equation model is used in a design problem of engineering interest. The weighted mixed sensitivity optimization technique we consider has been used in studying practical control problems for aerospace vehicles. However, we perform the design for a distributed parameter system, as opposed to a finite order system. We do not approximate the dynamics of a pure time delay or the effects of having actuators spacially separated from sensors. We determine directly their impact on the optimal design. The partial differential equation representation of the beam gives us an economical representation of the system dynamics as a transcendental function of s. By varying just a few parameters, stiffness and damping for example, one can make changes that affect all the structural modes simultaneously while maintaining the integrity of model assumptions. By varying one parameter that represents the duration of the time delay in the system, we can readily assess the impact of the length of the delay on achievable performance.

II. IDEAL BEAM TRANSFER FUNCTIONS

A variety of models have been developed to describe the motion of flexible beams. Ideal beam models have been used in the study of control problems for aerospace vehicles, large space structures, etc. The control design techniques we consider in this chapter are frequency domain based and therefore require transfer function representations of beam systems. In addition, we must be able to express beam transfer functions as quotients of H^∞ functions. In this section we derive transfer function representations of an ideal Euler-Bernoulli beam with Kelvin-Voigt damping subjected to several different sets of boundary conditions. We analyze in detail the two beam transfer functions studied in [1]. These two transfer functions are qualitatively representative of all of the transfer functions we derive here.

The basic Euler-Bernoulli beam equation is

$$\rho \frac{\partial^2 w}{\partial t^2} + \frac{\partial^2}{\partial x^2}\left(EI \frac{\partial^2 w}{\partial x^2}\right) = 0.$$

Here $w(x,t)$ denotes the deflection of the beam at time $t \geq 0$ and location x along the beam, $\rho(x)$ denotes the mass density per unit length of the beam, and $EI(x)$ denotes the second moment of the modulus of elasticity about the elastic axis. In this model, it is assumed that $w(x,t)$ is smooth and that no energy is lost or gained internally within the beam.

Damping terms may be incorporated to account for beam vibration attenuation due to dissipative mechanisms internal to the beam. In the Kelvin-Voigt damping mechanism, one assumes that the dissipative forces are obtained from the velocity in the same way that the restoring forces are obtained from the displacement. The dynamics for an Euler-Bernoulli beam with Kelvin-Voigt damping are described by [5]

$$\rho \frac{\partial^2 w}{\partial t^2} + 2\alpha \frac{\partial^2}{\partial x^2}\left(EI \frac{\partial^3 w}{\partial x^2 \partial t}\right) + \frac{\partial^2}{\partial x^2}\left(EI \frac{\partial^2 w}{\partial x^2}\right) = 0 \quad \text{for } \alpha > 0.$$

A. BEAM SYSTEMS ONE AND TWO

Consider an Euler-Bernoulli beam having free ends and Kelvin-Voigt damping with $2\alpha = \epsilon > 0$, length 1, $\rho = 1$, and $EI = 1$. Denote the deflection of the beam at time t and location x along the elastic axis of the beam by $w(x,t)$. Suppose that a transverse force, $-u(t)$, is applied at one end of the beam, say at $x = 1$. The dynamics of the beam ([1], [5]) are described by the linear partial differential equation

$$\frac{\partial^2 w}{\partial t^2} + \epsilon \frac{\partial^5 w}{\partial x^4 \partial t} + \frac{\partial^4 w}{\partial x^4} = 0 \tag{1}$$

with boundary conditions

$$\frac{\partial^2 w}{\partial x^2}(0,t) + \epsilon \frac{\partial^3 w}{\partial x^2 \partial t}(0,t) = 0, \quad \frac{\partial^2 w}{\partial x^2}(1,t) + \epsilon \frac{\partial^3 w}{\partial x^2 \partial t}(1,t) = 0, \tag{2}$$

$$\frac{\partial^3 w}{\partial x^3}(0,t) + \epsilon \frac{\partial^4 w}{\partial x^3 \partial t}(0,t) = 0, \quad \frac{\partial^3 w}{\partial x^3}(1,t) + \epsilon \frac{\partial^4 w}{\partial x^3 \partial t}(1,t) = u(t).$$

Consider the following two cases for the measurement of the beam. First consider the measurement to be $y(t) = w(1,t)$, where $w(1,t)$ is the deflection at the same end of the beam as the applied force $-u(t)$. Next consider $y(t) = w(0,t)$, where $w(1,t)$ is the deflection at the opposite end of the beam as the applied force $-u(t)$.

Taking the Laplace transform of (1) leads to the fourth order homogeneous, ordinary differential equation in x given by

$$\frac{d^4 w}{dx^4}(1 + \epsilon s) + s^2 w = 0.$$

The general solution to this differential equation is

$$w(x,s) = c_1 \sin(\beta x) + c_2 \cos(\beta x) + c_3 \sinh(\beta x) + c_4 \cosh(\beta x), \tag{3}$$

where $\beta^4 = \frac{-s^2}{(1+\epsilon s)}$. The boundary conditions given by (2) determine uniquely the values of the coefficients c_1, c_2, c_3, and c_4 in (3), and lead to the transfer function from $u(s)$ to $w(x,s)$, where $x \in [0,1]$, given by

$$\frac{w(x,s)}{u(s)} = \frac{1}{(1+\epsilon s)\beta^3} \times \tag{4}$$

$$\frac{(\cos\beta - \cosh\beta)(\sin\beta x + \sinh\beta x) + (\sinh\beta - \sin\beta)(\cos\beta x + \cosh\beta x)}{2(\cos\beta\cosh\beta - 1)}$$

In particular, we have transfer functions

$$P_1 = \frac{w(1,s)}{u(s)} = \frac{1}{(1+\epsilon s)\beta^3}\left(\frac{\sinh\beta\cos\beta - \cosh\beta\sin\beta}{\cos\beta\cosh\beta - 1}\right), \tag{5}$$

for beam system one, and

$$P_2 = \frac{w(0,s)}{u(s)} = \frac{1}{(1+\epsilon s)\beta^3}\left(\frac{\sinh\beta - \sin\beta}{\cos\beta\cosh\beta - 1}\right), \tag{6}$$

for beam system two.

One can show that both $P_1(s)$ and $P_2(s)$ are meromorphic in $C\backslash\left\{\frac{-1}{\epsilon}\right\}$ with no poles in the open right half plane. Also, P_1 and P_2 are strictly proper in the sense that $|P_1(s)| \to 0$ and $|P_2(s)| \to 0$ as $|s| \to \infty$ for $Re(s) \geq 0$. In Appendix.A we bound the magnitudes of $P_1(j\omega)$ and $P_2(j\omega)$ for all frequencies, ω, outside a certain frequency range $(-\omega_o, \omega_o)$, with decreasing functions that tend to 0 as $\omega \to \infty$. These bounding functions allow us to ascertain a finite frequency range, outside of which $|P_1(j\omega)|$ and $|P_2(j\omega)|$ are very small.

1. Product representations of P_1 and P_2

Both P_1 and P_2 can be expressed as infinite products of second order terms. These product representations display the poles and zeros of P_1 and P_2. They also facilitate inner/outer factorizations which are essential for solving the H^∞ optimization problems of Section IV.

Product representations for P_1 and P_2 can be derived as follows. We first express as infinite products the entire functions

$$\sinh\beta\cos\beta - \cosh\beta\sin\beta = \frac{-2}{3}\beta^3\prod_{n=1}^{\infty}\left(1 - \frac{\beta^4}{\alpha_n^4}\right), \tag{7}$$

$$\sinh\beta - \sin\beta = \frac{\beta^3}{3}\prod_{n=1}^{\infty}\left(1 + \frac{\beta^4}{4\alpha_n^4}\right),$$

$$\cos\beta\cosh\beta - 1 = \frac{-\beta^4}{6}\prod_{n=1}^{\infty}\left(1 - \frac{\beta^4}{\phi_n^4}\right),$$

where $\cos\alpha_n\sinh\alpha_n = \sin\alpha_n\cosh\alpha_n$ and $\cos\phi_n\cosh\phi_n = 1$, for $\alpha_n, \phi_n > 0$. Assuming the ordering $\phi_j < \phi_k$ and $\alpha_j < \alpha_k$ for $j < k$, one can establish that the locations of the $\phi_n' s$ alternate with the locations of the $\alpha_n' s$. In fact, the $\phi_n' s$ tend to $\pi/2 + n\pi$ and the $\alpha_n' s$ tend to $\pi/4 + n\pi$ as $n \to \infty$.

Thus, for values of s that are not roots of any of the denominator terms and such that the right hand side converges,

$$\begin{aligned} P_1(s) &= \frac{4}{(1+\epsilon s)\beta^4} \frac{\prod_{n=1}^{\infty}\left(1 - \frac{\beta^4}{\alpha_n^4}\right)}{\prod_{n=1}^{\infty}\left(1 - \frac{\beta^4}{\phi_n^4}\right)} \\ &= \frac{-4}{s^2} \frac{\prod_{n=1}^{\infty}\frac{1}{(1+\epsilon s)}\left(1 + \epsilon s + \frac{s^2}{\alpha_n^4}\right)}{\prod_{n=1}^{\infty}\frac{1}{(1+\epsilon s)}\left(1 + \epsilon s + \frac{s^2}{\phi_n^4}\right)}. \end{aligned}$$

The quotient of infinite products in the above expression for P_1 can be replaced with a single infinite product. For any finite, positive integer M,

$$\frac{\prod_{n=1}^{M}\frac{1}{(1+\epsilon s)}\left(1 + \epsilon s + \frac{s^2}{\alpha_n^4}\right)}{\prod_{n=1}^{M}\frac{1}{(1+\epsilon s)}\left(1 + \epsilon s + \frac{s^2}{\phi_n^4}\right)} = \prod_{n=1}^{M} \frac{\left(1 + \epsilon s + \frac{s^2}{\alpha_n^4}\right)}{\left(1 + \epsilon s + \frac{s^2}{\phi_n^4}\right)} \tag{8}$$

where $s \neq -1/\epsilon$ and s is not a pole of this product. If the limits as $M \to \infty$ of both sides of (8) exist, they must be equal. Thus,

$$P_1(s) = \frac{-4}{s^2}\prod_{n=1}^{\infty} f_n(s)\ , \quad \text{where} \quad f_n(s) = \frac{\left(1 + \epsilon s + \frac{s^2}{\alpha_n^4}\right)}{\left(1 + \epsilon s + \frac{s^2}{\phi_n^4}\right)} \tag{9}$$

for values of s where this infinite product converges.

By a similar argument,

$$P_2(s) = \frac{2}{s^2}\prod_{n=1}^{\infty} g_n(s)\ , \quad \text{where} \quad g_n(s) = \frac{\left(1 + \epsilon s - \frac{s^2}{4\alpha_n^4}\right)}{\left(1 + \epsilon s + \frac{s^2}{\phi_n^4}\right)} \tag{10}$$

for values of s where this infinite product converges.

In Appendix.B we establish that (9) and (10) converge everywhere in the closed right half plane and can be written as quotients of H^∞ functions.[1]

[1] A complex valued function $G(s)$ is in H^∞ if it is bounded and analytic in the right half plane Re $s > 0$. When $G \in H^\infty$, $||G||_\infty = \sup_{\text{Re } s>0} |G(s)| = \text{ess } \sup_\omega |G(j\omega)|$.

We show that P_1 admits a coprime factorization over H^∞ given by $P_1(s) = P_{1o}(s)(\frac{s^2}{(s+1)^2})^{-1}$, where

$$P_{1o}(s) = \frac{-4}{(s+1)^2} \prod_{n=1}^{\infty} \frac{\left(1 + \epsilon s + \frac{s^2}{\alpha_n^4}\right)}{\left(1 + \epsilon s + \frac{s^2}{\phi_n^4}\right)} .$$

We factor $P_2 = \frac{N_2}{D}$, where $D(s) = \frac{s^2}{(s+1)^2} \in H^\infty$, and N_2 has an inner/outer factorization given by $N_2(s) = B(s)P_{2o}(s)$, where

$$P_{2o}(s) = \frac{2}{(s+1)^2} \prod_{n=1}^{\infty} \frac{\left(1 + s\sqrt{\epsilon^2 + \frac{1}{\alpha_n^4}} + \frac{s^2}{4\alpha_n^4}\right)}{\left(1 + \epsilon s + \frac{s^2}{\phi_n^4}\right)}, \tag{11}$$

$$B(s) = \prod_{n=1}^{\infty} \left(\frac{2\alpha_n^4 \left(\epsilon + \sqrt{\epsilon^2 + \frac{1}{\alpha_n^4}}\right) - s}{2\alpha_n^4 \left(\epsilon + \sqrt{\epsilon^2 + \frac{1}{\alpha_n^4}}\right) + s} \right) . \tag{12}$$

We establish that $P_{1o}(s), P_{2o}(s) \in H^\infty$ and that the Blaschke product $B(s) \in H^\infty$ converges in the closed right half-plane.

Further analysis of (9) shows that the zeros of P_1 are at

$$s = \frac{-\alpha_n^4}{2} \left(\epsilon \pm \sqrt{\epsilon^2 - \frac{4}{\alpha_n^4}} \right) \qquad \text{for } n = 1, 2, \ldots \tag{13}$$

where $\cos(\alpha_n)\sinh(\alpha_n) = \sin(\alpha_n)\cosh(\alpha_n)$, for $\alpha_n > 0$.

From (10) one can establish that the zeros of P_2 are at

$$s = 2\alpha_n^4 \left(\epsilon \pm \sqrt{\epsilon^2 + \frac{1}{\alpha_n^4}} \right) \qquad \text{for } n = 1, 2, \ldots \tag{14}$$

where each α_n is the same as in (13). Also P_1 and P_2 each have a singularity at $-1/\epsilon$, a second order pole at $s = 0$, and poles at

$$s = \frac{-\phi_n^4}{2} \left(\epsilon \pm \sqrt{\epsilon^2 - \frac{4}{\phi_n^4}} \right) \qquad \text{for } n = 1, 2, \ldots$$

where $\cos(\phi_n)\cosh(\phi_n) = 1$, for $\phi_n > 0$.

For any $\epsilon > 0$, the finite number of complex poles and zeros of P_1 alternate along the circle $|s+1/\epsilon| = 1/\epsilon$. One branch of real poles and zeros tends to $-\infty$ and the other branch tends to $-1/\epsilon$. When $\epsilon = 0.001$, straight forward calculations show that P_1 has 13 pairs of complex poles and zeros. Examination of (14) shows that P_2 has all real zeros, one branch tending to $-1/\epsilon$ and the other to $+\infty$. Thus P_2 has infinitely many nonminimum phase, simple zeros, while its poles are the poles of P_1.

B. ADDITIONAL BEAM TRANSFER FUNCTIONS

In deriving P_1 and P_2 we first found the general solution (4). This general solution together with the boundary conditions specified in (2) lead to the beam transfer functions given in (5) and (6). Similarly, one can derive and analyze transfer functions for the beam systems described in this subsection. For each system we assume an Euler-Bernoulli beam with Kelvin-Voigt damping parameter $2\alpha = \epsilon > 0$, $\rho = 1$, and $EI = 1$. The function $w(x,t)$ denotes the deflection of the beam at time t and location x along the length of the beam.

Although these beam systems all have different configurations, their transfer functions are, from an H^∞ control design perspective, qualitatively either like P_1 or like P_2. Thus, to a certain extent one can consider the solutions to H^∞ optimal control design problems for P_1 and P_2 as representative of solutions to similar control problems for beam systems such as those we describe here.

1. A two-input/two-output transfer function

Consider a beam with free ends and length 2, with $x = -1$ and $x = 1$ denoting the locations of the ends of the beam. Suppose that the control inputs are $U(t) = \begin{bmatrix} f(t) \\ m(t) \end{bmatrix}$, where $f(t)$ is a point normal force and $m(t)$ is a point moment applied to the middle of the beam. This beam has been considered in previous studies (e.g. [6], [7], and [8]). The dynamics of the beam are described by

$$\frac{\partial^2 w}{\partial^2} + \epsilon \frac{\partial^5 w}{\partial x^4 \partial t} + \frac{\partial^4 w}{\partial x^4} = \delta(0) f(t) - \delta_x(0) m(t), \tag{15}$$

with boundary conditions

$$\frac{\partial^2 w}{\partial x^2}(-1,t) = \frac{\partial^2 w}{\partial x^2}(1,t) = 0, \quad \frac{\partial^3 w}{\partial x^3}(-1,t) = \frac{\partial^3 w}{\partial x^3}(1,t) = 0. \tag{16}$$

Let the measurements be $Y(t) = \begin{bmatrix} w(0,t) \\ \frac{\partial w}{\partial x}(0,t) \end{bmatrix}$, where $w(0,t)$ is the deflection and $\frac{\partial w}{\partial x}(0,t)$ is the angle of inclination of the middle of the beam. Taking the Laplace transform of (15) and (16) and solving the resulting fourth order ordinary inhomogeneous differential equation in s leads to the transfer function from $U(s)$ to $Y(s)$ given by

$$P(s) = \begin{bmatrix} P_3(s) & 0 \\ 0 & P_4(s) \end{bmatrix}$$

$$\text{where} \quad P_3(s) = \frac{\beta}{2s^2} \left(\frac{1 + \cosh\beta \cos\beta}{\cosh\beta \sin\beta + \sinh\beta \cos\beta} \right)$$

$$\text{and} \quad P_4(s) = \frac{\beta^3}{2s^2} \left(\frac{1 + \cosh\beta \cos\beta}{\cosh\beta \sin\beta - \sinh\beta \cos\beta} \right) \text{for } \beta^4 = \frac{-s^2}{(1+\epsilon s)}.$$

It has been shown, [7], that $P(s)$ is meromorphic in $C \backslash \left\{ \frac{-1}{\epsilon} \right\}$ with no poles or zeros in the open right half-plane and has an essential singularity at $-1/\epsilon$. Also $P(s)$ is strictly proper. That is, $|P(s)| \to 0$ as $|s| \to \infty$ for s in the closed right half-plane.

Infinite product representations for $P_3(s)$ and $P_4(s)$ can be derived similarly to how we derived product representations for $P_1(s)$ and $P_2(s)$. Begin with product representations (7) together with

$$\cosh\beta \sin\beta + \sinh\beta \cos\beta = 2\beta \prod_{n=1}^{\infty} \left(1 - \frac{\beta^4}{\mu_n^4}\right)$$

$$1 + \cosh\beta \cos\beta = 2 \prod_{n=1}^{\infty} \left(1 - \frac{\beta^4}{c_n^4}\right)$$

where $\cos(c_n)\cosh(c_n) = -1$, $\cos(\mu_n)\sinh(\mu_n) = -\sin(\mu_n)\cosh(\mu_n)$, and $\cos(\alpha_n)\sinh(\alpha_n) = \sin(\alpha_n)\cosh(\alpha_n)$ for c_n, μ_n, and $\alpha_n > 0$. Assume that for $j < l$, $c_j < c_l$, $\mu_j < \mu_l$, and $\alpha_j < \alpha_l$. The locations of the $c_n's$ tend to $\pi/2 + n\pi$, the locations of the $\mu_n's$ tend to $3\pi/4 + n\pi$, and the locations of the $\alpha_n's$ tend to $\pi/4 + n\pi$.

One can show that

$$P_3(s) = \frac{1}{2s^2}\prod_{n=1}^{\infty}\frac{\left(1+\epsilon s+\frac{s^2}{c_n^4}\right)}{\left(1+\epsilon s+\frac{s^2}{\mu_n^4}\right)}, \quad P_4(s) = \frac{-3}{2s^2}\prod_{n=1}^{\infty}\frac{\left(1+\epsilon s+\frac{s^2}{c_n^4}\right)}{\left(1+\epsilon s+\frac{s^2}{\alpha_n^4}\right)}.$$

The zeros of P_3 and P_4 occur at $s = \frac{-c_n^4}{2}\left(\epsilon \pm \sqrt{\epsilon^2 - \frac{4}{c_n^4}}\right)$. In addition to the poles at $s = 0$, the poles of P_3 and P_4 are at

$$s = \frac{-\mu_n^4}{2}\left(\epsilon \pm \sqrt{\epsilon^2 - \frac{4}{\mu_n^4}}\right) \quad \text{and at} \quad s = \frac{-\alpha_n^4}{2}\left(\epsilon \pm \sqrt{\epsilon^2 - \frac{4}{\alpha_n^4}}\right),$$

respectively.

Although $P(s)$ is matrix-valued, it consists of two uncoupled scalar-valued functions. Thus we may design controllers for P_3 and P_4 separately. Both P_3 and P_4 are, from an H^∞ control design perspective, like P_1. All three transfer functions are stable, have no right half plane zeros, and are strictly proper. One can factor P_3 and P_4 as the quotients of H^∞ functions analogously to how we have factored P_1. Furthermore, for any $\epsilon > 0$, as discussed in [7] for P_3 and P_4, the finite number of complex poles and zeros of each of these transfer functions lie on the circle $\left|s + \frac{1}{\epsilon}\right| = \frac{1}{\epsilon}$. For each of these transfer functions one branch of real poles and zeros tends to $-\infty$ and the other branch tends to $-1/\epsilon$.

2. A beam with hinged ends

Consider a beam with hinged ends and length 2 with $x = -1$ and $x = 1$ denoting the locations of the two ends of the beam. Suppose that a moment $m(t)$ is applied at one end of the beam while $-m(t)$ is applied simultaneously at the other end. Let the beam dynamics be described by (1) together with the boundary conditions

$$w(-1,t)=0;\quad \frac{\partial^2 w}{\partial x^2}(-1,t)+\epsilon\frac{\partial^3 w}{\partial x^2\partial t}(-1,t)=-m(t); \tag{17}$$

$$w(1,t)=0;\quad \frac{\partial^2 w}{\partial x^2}(1,t)+\epsilon\frac{\partial^3 w}{\partial x^2\partial t}(1,t)=m(t).$$

These boundary conditions determine uniquely the values of c_1, c_2, c_3, and c_4 in (5) and lead to the transfer function from $m(s)$ to $w(x,s)$, for $x\in[-1,1]$, given by

$$\frac{w(x,s)}{m(s)}=\frac{1}{2(1+\epsilon s)\beta^2}\left(\frac{(\sin\beta\sinh\beta x-\sinh\beta\sin\beta x)}{\sinh\beta\sin\beta}\right).$$

For $x=1/2$ we obtain the transfer function given by

$$P_5(s)=\frac{w(\frac{1}{2},s)}{m(s)}=\frac{1}{4(1+\epsilon s)\beta^2}\left(\frac{\cos\frac{\beta}{2}-\cosh\frac{\beta}{2}}{\cosh\frac{\beta}{2}\cos\frac{\beta}{2}}\right).$$

In terms of a product expansion

$$P_5(s)=\frac{-1}{16(1+\epsilon s)(1+\epsilon s+\frac{s^2}{\pi^4})}\prod_{n=1}^{\infty}\frac{\left(1+\epsilon s-\frac{s^2}{64n^4\pi^4}\right)}{\left(1+\epsilon s+\frac{s^2}{(2n+1)^4\pi^4}\right)}.$$

One can factor P_5 as a quotient of H^∞-functions analogously to how we factored P_2. Like P_2, P_5 is strictly proper, stable, and has infinitely many nonminimum phase simple zeros. All of the zeros of P_5 are real, with one branch tending to $-1/\epsilon$ and the other to $+\infty$. P_5 has a finite number of complex poles and they are located along the circle $\left|s+\frac{1}{\epsilon}\right|=\frac{1}{\epsilon}$. One branch of real poles tends to $-\infty$ and the other branch tends to $-1/\epsilon$.

3. A cantilever beam

Consider a beam in a cantilever configuration, with length 1. Let $x=0$ be the location of the clamped end of the beam and $x=1$ be the location of the free end of the beam. Suppose that a moment $m(t)$ and a lateral force $f(t)$ are applied at the free end. Let the beam dynamics be described by (1) together with the boundary conditions

$$w(0,t) = 0; \quad \frac{\partial^2 w}{\partial x^2}(1,t) + \epsilon \frac{\partial^3 w}{\partial x^2 \partial t}(1,t) = m(t);$$

$$\frac{\partial w}{\partial x}(0,t) = 0; \frac{\partial^3 w}{\partial x^3}(1,t) + \epsilon \frac{\partial^4 w}{\partial x^3 \partial t}(1,t) = -f(t)$$

These boundary conditions lead to the transfer function from $\begin{bmatrix} -f(s) \\ m(s) \end{bmatrix}$ to $w(x,s)$ given by $[P_{6x}, P_{7x}]$, for

$$P_{6x}(s) = \frac{1}{2(1+\epsilon s)\beta^3} P'_{6x}$$

where P'_{6x} is given by

$$\frac{(\cos\beta + \cosh\beta)(\sin\beta x - \sinh\beta x) - (\sinh\beta + \sin\beta)(\cos\beta x - \cosh\beta x)}{(\cos\beta\cosh\beta + 1)},$$

and

$$P_{7x}(s) = \frac{1}{2(1+\epsilon s)\beta^2} P'_{7x}(s)$$

where P'_{7x} is given by

$$\frac{(\sin\beta - \sinh\beta)(\sinh\beta x - \sin\beta x) + (\cosh\beta + \cos\beta)(\cosh\beta x - \cos\beta x)}{(\cos\beta\cosh\beta + 1)}.$$

In particular for $x = 1$ and $m(t) = 0$ we have

$$P_6(s) = \frac{w(1,s)}{f(s)} = \frac{1}{(1+\epsilon s)\beta^3}\left(\frac{(\cosh\beta\sin\beta - \sinh\beta\cos\beta)}{\cos\beta\cosh\beta + 1}\right). \quad (18)$$

For $x = 1$ and $f(t) = 0$ we have

$$P_7(s) = \frac{w(1,s)}{m(s)} = \frac{1}{(1+\epsilon s)\beta^2}\left(\frac{\sinh\beta\sin\beta}{\cos\beta\cosh\beta + 1}\right). \quad (19)$$

Infinite product representations can be derived for P_6 and P_7. One can show that P_6 and P_7 are stable, strictly proper, and have no right half-plane zeros. One can design and analyze single-input/single-output controllers for P_6 or P_7 similarly to how we design and analyze controllers for P_1 in the following sections.

III. FEEDBACK CONTROL PROBLEMS

When designing a feedback compensator one should take into account possible variations and unmodelled dynamics or uncertain dynamics that may affect system stability and performance. Otherwise, one may design a controller that intensifies system sensitivity, degrades performance, and even causes instabilities when these perturbations are present.

The H^∞-control design approach allows for the inclusion of perturbations to systems such as uncertain high frequency dynamics, sensor noise, unmodelled or uncertain delays, and external disturbances. In Sections IV, V, and VI we will account for the presence of these kinds of perturbations in feedback compensator design and analysis for beam systems one and two. In this section we define several types of feedback control problems including robust stabilization, sensitivity reduction, and disturbance attenuation. We show how these problems can be expressed within a two-block H^∞ control design framework.

Consider the feedback control system shown in Figure 1. Here P denotes the plant to be controlled and C represents the controller to be designed. The external signals are reference, r, actuator disturbance, d, output disturbance, v, and measurement noise, n. The internal signals of interest are measured error, e, command input, u, and plant output, y. We assume that the energy contents of the external signals are finite. Thus we model these signals as $L^2[0,\infty)$ functions.

Definition 3.1: Consider a transfer function $G(s)$ with input x and output $y = Gx$. We say that the system is stable if for every $x \in L^2[0,\infty)$ the output $y = Gx$ is in $L^2[0,\infty)$ and

$$\sup_{x \neq 0} \frac{||y||_2}{||x||_2} < \infty. \tag{20}$$

That is, we shall say that G is stable if and only if multiplication by G is a bounded linear operator from $L^2[0,\infty)$ to $L^2[0,\infty)$. One can interpret this definition as follows: A system is stable if all finite energy input signals u to the system give rise to finite energy responses y, and the maximum energy amplification from u to y is finite.

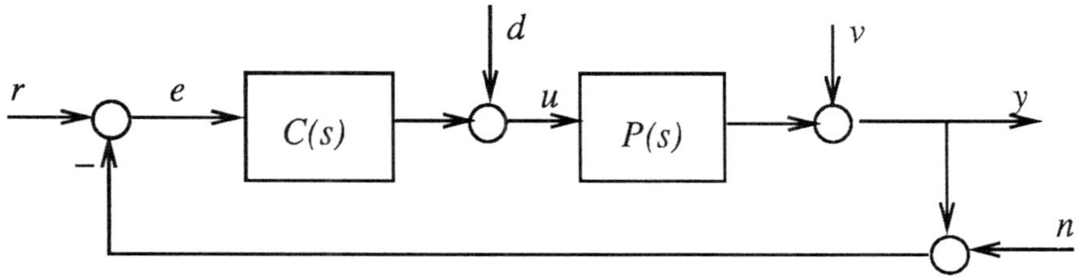

Figure 1: Closed Loop System

The stability of a system is equivalent to having its transfer function $G(s)$, in H^∞. In fact the maximum energy amplification given by (20) is the norm of this operator, which is equal to the H^∞ norm of the transfer function $G(s) = y(s)/x(s)$. That is,

$$\sup_{x \neq 0} \frac{||y||_2}{||x||_2} = \operatorname{ess\,sup}_{\omega \in \mathrm{R}} |G(j\omega)| = ||G||_\infty .$$

This definition of stability is an input/output stability concept in the sense of bounded energy amplification.

Definition 3.2: The *closed loop system* (C, P) *is stable* if all transfer functions from any external input to any internal signal are in H^∞.

Thus closed loop system stability is equivalent to the following: all finite energy external inputs give rise to finite energy internal signals and the maximum energy amplification in the system is finite.

One can deduce from Fig.1 that all of the closed loop system transfer functions can be expressed in terms of the following four functions:

$$\begin{aligned} S(s) &:= (1 + P(s)C(s))^{-1}, \\ T(s) &:= 1 - S(s) = P(s)C(s)(1 + P(s)C(s))^{-1}, \end{aligned}$$

$$\begin{aligned} C(s)S(s) &= C(s)(1+P(s)C(s))^{-1}, \\ P(s)S(s) &= P(s)(1+P(s)C(s))^{-1}. \end{aligned}$$

For example, the sensitivity function S is the transfer function from v to y, or from r to e; the complementary sensitivity function T is the transfer function from n to y, or from r to y; PS is the transfer function from d to y; CS is the transfer function from r to u; etc. Therefore, the closed loop system is stable if and only if all four transfer functions S, T, CS, PS are in H^∞

A. THE SET OF ALL STABILIZING CONTROLLERS

The most important requirement in control design is the stability of the closed loop system (C, P), with controller C and plant P. If (C, P) is a stable closed loop system, then we say that the controller C stabilizes P. The set of all stabilizing controllers for a plant P can be characterized as follows.

Assume that P can be expressed as a quotient $P(s) = N(s)/D(s)$, where $N, D \in H^\infty$ and such that there exist $X, Y \in H^\infty$ satisfying the Bezout identity

$$N(s)X(s) + D(s)Y(s) = 1 \ . \tag{21}$$

If such a factorization exists for P then N and D are called coprime factors of P. Existence of such a coprime factorization is necessary for the existence of stabilizing controllers [9].

Theorem 3.3([9], [10]): A controller C stabilizes the plant P if and only if C can be expressed in the form

$$C(s) = \frac{X(s) + D(s)Q(s)}{Y(s) - N(s)Q(s)}, \tag{22}$$

where $Q \in H^\infty$ is a free parameter to be chosen according to design specifications other than stabilization.

We are particularly interested in stabilization of the plants P_1 and P_2 derived in Section II.A. Recall that coprime factors of P_1 and P_2 are:

$$P_1(s) = \frac{N_1(s)}{D(s)} \quad , \qquad P_2(s) = \frac{N_2(s)}{D(s)}$$

where

$$N_1(s) = P_{1o}(s) = \frac{-4}{(s+1)^2} \prod_{n=1}^{\infty} f_n(s), \qquad N_2(s) = \frac{2}{(s+1)^2} \prod_{n=1}^{\infty} g_n(s),$$

with $D(s) = \frac{s^2}{(s+1)^2}$, and $f_n(s)$, and $g_n(s)$ as defined in Section II. In order to have

$$Y(s) = \frac{1 - N(s)X(s)}{D(s)} \in H^\infty$$

we need to choose $X \in H^\infty$ such that

$$X(0) = \frac{1}{N(0)}, \quad \text{and} \quad \frac{\partial}{\partial s} N(s)X(s)\bigg|_{s=0} = 0.$$

It can be shown that for N_1 and N_2 given above these conditions reduce to

$$X_1(0) = \frac{-1}{4} \quad \frac{\partial}{\partial s} X_1(s)\bigg|_{s=0} = \frac{-1}{2}, \quad X_2(0) = \frac{1}{2} \quad \frac{\partial}{\partial s} X_2(s)\bigg|_{s=0} = 1.$$

One can check that

$$X_1(s) = -\frac{0.25 + s}{1 + 2s} \quad \text{and} \quad X_2(s) = \frac{0.5 + 2s}{1 + 2s}$$

satisfy the above conditions and $Y_i = (1 - N_i X_i)/D \in H^\infty$ for $i = 1, 2$. Thus, a controller C_i stabilizing P_i can be obtained from

$$C_i(s) = D(s) \, \frac{X_i(s) + D(s)Q(s)}{1 - N_i(s)(X_i(s) + D(s)Q(s))},$$

where D, N_i, X_i are given above, and $Q \in H^\infty$ is the free parameter.

B. ROBUST STABILITY

In the above discussion we have characterized the set of all controllers that stabilize a fixed plant P. However, many practically motivated control problems require that controllers stabilize, in addition to a given plant P, plants that are close to P. This kind of requirement can be found in the classical control robustness measures of gain and phase margins. The gain and phase margins of the open loop transfer function PC indicate some of the ways in which P can be perturbed and yet still be stabilized by C. In H^∞ control analysis we measure the robustness of a controller to more general perturbations such as combinations of gain and phase variations, as well as norm bounded dynamic perturbations.

If one considers $P(s)$ to be merely an imperfect representation of the actual plant whose transfer function is denoted by $P_\Delta(s)$, the part of P_Δ which does not appear in P is called the *unmodeled dynamics*. The unmodeled dynamics can be represented in a number of ways. One typical way is in the form of a multiplicative perturbation to the plant:

$$P_\Delta(s) = P(s)\ (1 + \Delta(s)),$$

Here $P(s)$ is the nominal plant and $\Delta(s)$ is the unmodeled dynamics. We will assume that all we know about $\Delta(s)$ is that it is bounded by a given function $W_\Delta(s)$, called the uncertainty weight. Thus,

$$||\Delta(j\omega)|| < |W_\Delta(j\omega)| \quad \forall \ \omega \in \mathrm{R}, \tag{23}$$

where $W_\Delta(s)$ is known, stable, and minimum phase.

We will assume that $\Delta \in H^\infty$. This means that the unstable poles of the actual plant P_Δ coincide with those of P. In the case that $P = P_1$ or $P = P_2$ this would mean that P_Δ would have a double pole at $s = 0$ but no other poles in the closed right half plane. This assumption is quite reasonable if the actual plant is a model of a physical beam whose responses can be observed to die out with time. The true plant would not have unstable poles or any poles on the imaginary axis, except possibly at s = 0.

A pure time delay is an example of unmodelled dynamics that can be expressed as a multiplicative plant perturbation. The nominal plant model is $P(s)$ while the true plant has transfer function $e^{-hs}P(s)$. In this case, $\Delta(s) = e^{-hs} - 1$. We shall investigate stability and performance robustness to this kind of unmodelled dynamics for specific examples in Section VI.

In the presence of unmodeled plant dynamics we require that the controller C, which is fixed, stabilize not only the nominal plant P but also all possible plants P_Δ with $\Delta \in H^\infty$ satisfying (23) for a given weight W_Δ. If C meets this requirement then we say that C *robustly stabilizes* the plant. Necessary and sufficient conditions for a controller C to robustly stabilize P are stated in the following theorem.

Theorem 3.4: Consider the set of plants described by P_Δ, with the nominal plant P and $\Delta \in H^\infty$ satisfying (23) for a given $W_\Delta \in H^\infty$. Let C be of the form (22), and assume that $P(j\omega)$ and $C(j\omega)$ are continuous on the imaginary axis except at finitely many points. Then C robustly stabilizes P if and only if

$$||W_\Delta PC(1+PC)^{-1}||_\infty \leq 1, \tag{24}$$

Proof: To show that inequality (24) is a necessary requirement for robust stability we must establish that if

$$||W_\Delta PC(1+PC)^{-1}||_\infty > 1, \tag{25}$$

then there exists an admissible multiplicative perturbation Δ destabilizing the closed loop system. Such a perturbation can be constructed the same way as in the finite order case (see e.g. [11], see also [12] for the infinite order case) because of the assumption on the continuity of C and P. For sufficiency we note that

$$(1+P_\Delta C)^{-1} = (1+PC)^{-1}\frac{1}{1+\Delta PC(1+PC)^{-1}}. \tag{26}$$

Since C stabilizes P, $(1+PC)^{-1}$ and $PC(1+PC)^{-1}$ are in H^∞. Therefore, for $\Delta \in H^\infty$, (24) implies that $(1+\Delta PC(1+PC)^{-1})^{-1} \in H^\infty$. Hence $(1+P_\Delta C)^{-1} \in H^\infty$ for all admissible $\Delta \in H^\infty$. The identity (26) further implies that $C(1+P_\Delta C)^{-1}$ and $P_\Delta(1+P_\Delta C)^{-1}$ are also in H^∞. Hence the closed loop system (C, P_Δ) is stable for all admissible P_Δ. □

C. SENSITIVITY REDUCTION

The above discussion summarized by Theorem 3.4 gives conditions for a controller to robustly stabilize all plants of the form P_Δ. In addition to stability, a controller must provide satisfactory closed loop system performance. A typical H^∞ control performance objective is sensitivity reduction. Mathematically we state the following problem: Given a desired upper bound $W_d(s)$ for the sensitivity function, find a robustly stabilizing controller C such that

$$|(1 + P_\Delta(j\omega)C(j\omega))^{-1}| \leq |W_d(j\omega)| \qquad a.e. \quad \omega \in \mathrm{R}, \tag{27}$$

where P_Δ represents the actual plant, with nominal plant P and uncertainty weight W_Δ. The function $(1 + P_\Delta C)^{-1}$ is the sensitivity of the closed loop system with the actual plant P_Δ and controller C.

This problem is refered to as a robust performance problem because a controller solving it will provide acceptable performance in the presence of any Δ satisfying (23). Theorem 3.5 gives a sufficient condition for testing whether or not a given controller C solves this robust performance problem.

Theorem 3.5: Assume that the class of plants is described by multiplicative perturbations as defined in Theorem 3.4. Consider P and C satisfying the assumptions stated in Theorem 3.4. Then C solves the robust performance problem if it stabilizes the nominal plant P and satisfies the inequality

$$\left\| \begin{bmatrix} W_d^{-1}(1 + PC)^{-1} \\ W_\Delta PC(1 + PC)^{-1} \end{bmatrix} \right\|_\infty \leq \frac{1}{\sqrt{2}}. \tag{28}$$

This result is a consequence of Theorem 3.4, see e.g. [13] for details. We shall consider this kind of robust performance problem in designing feedback controllers for P_1 and P_2. In Section IV we outline how the problem of minimizing the left hand side of (28) over all controllers C stabilizing P is solved as a two-block H^∞-optimization problem.

D. DISTURBANCE ATTENUATION

One can express a disturbance attenuation control problem in the form of (28). Consider the closed loop system shown in Figure 1. Assume that $d = r = 0$, and

$$v \in \mathcal{D}_v := \{W_v(v_1) \ : \ v_1 \in L^2[0,\infty), \ \|v_1\|_2 \leq 1\}$$

$$n \in \mathcal{D}_n := \{W_n(n_1) \ : \ n_1 \in L^2[0,\infty), \ \|n_1\|_2 \leq 1\},$$

where W_v and W_n are linear, time-invarient weights shaping the frequency and magnitude of the output disturbance and the measurement noise, respectively.

One can formulate a disturbance attenuation problem within an H^∞ control design framework as follows. Find a controller C that stabilizes the plant P and minimizes

$$\gamma(C) = \sup_{v \in \mathcal{D}_v, \ n \in \mathcal{D}_n} \|y\|_2 .$$

This problem is equivalent to finding a stabilizing controller which minimizes

$$\gamma(C) = \left\| \begin{bmatrix} W_v(1+PC)^{-1} \\ W_n PC(1+PC)^{-1} \end{bmatrix} \right\|_\infty . \tag{29}$$

Note that this problem is similar to that of (28), provided the weights are chosen appropriately.

Several kinds of robust stabilization, robust performance, and disturbance attenuation problems can be solved by finding the H^∞ optimal controller C_{opt} for the weighted mixed sensitivity H^∞ optimal control problem

$$\gamma_{opt} = \inf_{C \text{ stabilizing } P} \left\| \begin{bmatrix} W_1(1+PC)^{-1} \\ W_2 PC(1+PC)^{-1} \end{bmatrix} \right\|_\infty , \tag{30}$$

where C_{opt} stabilizes the nominal plant P and achieves the H^∞ optimal performance γ_{opt}. That is,

$$\gamma_{opt} = \left\| \begin{bmatrix} W_1(1+PC_{opt})^{-1} \\ W_2PC_{opt}(1+PC_{opt})^{-1} \end{bmatrix} \right\|_\infty ,$$

where W_1 and W_2 are appropriate weights related to the control problems defined above. We will assume that $W_1 = W_d^{-1}$ (or W_v) is a proper rational minimum phase function and $W_2 = W_\Delta$ (or W_n) is a rational minimum phase function.

The solution to (30) depends upon the choice of W_1 and W_2. In general these weighting functions are low order rational functions. Since W_1 penalizes the sensitivity function, it is usually a low-pass filter. This means that v in the disturbance attenuation problem is a low frequency signal. On the other hand W_2 represents the magnitude of unmodeled dynamics. Typically one assumes that the nominal plant models the actual plant quite well at low frequencies, but not at high frequencies. Therefore, W_2 is usually a high pass filter. This also means that in the disturbance attenuation problem the measurement noise n is assumed to be largely a high frequency signal. This is typically a reasonable assumption to make when modelling measurement noises present in physical systems.

IV. H^∞-OPTIMAL CONTROL

We shall first discuss the general solution to the weighted mixed sensitivity H^∞ optimal control problem (30) defined in Section III. Then for some specific weights we will apply the solution to beam systems one and two. The solutions presented here are based on the formulae developed in [4]. As discussed in Section II, the transfer functions $P_1(s)$ and $P_2(s)$ can be written in the form $P(s) = N(s)/D(s)$, where D is outer, and N admits an inner/outer factorization $N = m_pN_o$ where m_p is inner and N_o is outer. This notation will be used throughout the rest of this chapter.

A. SOLUTIONS TO H^∞-OPTIMAL CONTROL PROBLEMS

Using the parametrization of Theorem 3.1, we know that all stabilizing controllers must be of the form

$$C = \frac{X + DQ_c}{Y - NQ_c} \quad Q_c \in H^\infty ,$$

where $X, Y \in H^{\infty}$ satisfy $NX + DY = 1$. Therefore, all admissible sensitivity functions $S = (1 + PC)^{-1}$ and complementary sensitivity functions $T = 1 - S$ in (30) are of the form:

$$S = 1 - NX - NDQ_c \qquad T = NX + NDQ_c, \quad Q_c \in H^{\infty}. \tag{31}$$

Thus,

$$\gamma_{opt} = \inf_C \gamma(C) = \inf_{Q_c \in H^{\infty}} \left\| \begin{bmatrix} W_1 \\ 0 \end{bmatrix} - \begin{bmatrix} W_1 \\ -W_2 \end{bmatrix} N(X + DQ_c) \right\|_{\infty}. \tag{32}$$

For P_1 or P_2 the numerator N is strictly proper so that $\gamma(C) \geq |W_1(\infty)|$ for all stabilizing controller C. Therefore $\gamma_{opt} \geq |W_1(\infty)|$. Similarly the properties that $D(0) = 0$ and $(1 - NX) = YD$ imply from (32) that $\gamma_{opt} \geq |W_2(0)|$. We will assume that $\gamma_{opt} > \gamma_o = \max\{|W_1(\infty)|, \ |W_2(0)|\}$.

By finding an inner outer factorization of $\begin{bmatrix} W_1 \\ -W_2 \end{bmatrix}$ we can reduce the problem (32) to a problem of the form

$$\gamma_{opt} = \inf_{Q_c \in H^{\infty}} \left\| \begin{bmatrix} W_1^* G^{-*} W_1 \\ W_1 W_2 G^{-1} \end{bmatrix} - \begin{bmatrix} 1 \\ 0 \end{bmatrix} GN(X + DQ_c) \right\|_{\infty}, \tag{33}$$

where G is a rational function such that $G^{-1} \in H^{\infty}$, and

$$G^*(W_1^* W_1 + W_2^* W_2)^{-1} G = 1.$$

This is a spectral factorization problem for a rational transfer function and as such can be solved by using a state space realization of $(W_1^* W_1 + W_2^* W_2)^{-1}$ as described in [14] pp. 90-94.

We now find an inner function m_w whose zeros are the right half plane poles of $W_1^* G^{-*} W_1$. By multiplying the first block of (33) by m_w and by defining $Q_1 := GN_o(X + DQ_c)$ we reduce the problem (32), (see [15] and [16] for the details), to the two-block general distance problem

$$\gamma_{opt} = \inf_{Q_1 \in H^{\infty}} \left\| \begin{bmatrix} W_o - mQ_1 \\ G_o \end{bmatrix} \right\|_{\infty}, \tag{34}$$

where $W_o = m_w W_1 W_1^* G^{-*}, \quad G_o = W_1 W_2 G^{-1}, \quad m = m_p m_w.$

After finding a $Q_1^{opt} \in H^\infty$ such that

$$\gamma_{opt} = \left\| \begin{bmatrix} W_o - mQ_1^{opt} \\ G_o \end{bmatrix} \right\|_\infty ,$$

we construct the optimal controller

$$C_{opt} = N_o^{-1} D_o \frac{Q_1^{opt} G^{-1}}{1 - mQ_1^{opt} G^{-1}}. \tag{35}$$

The H^∞ performance γ_{opt} can be found as follows. Let $F_o := m^* W_o$, then $\gamma_{opt} = ||\mathbf{A}||$, for $\mathbf{A} = \begin{bmatrix} \Gamma_{F_o} \\ \Upsilon_{G_o} \end{bmatrix}$, where Γ_{F_o} and Υ_{G_o} are the Hankel and Toeplitz operators with symbol F_o and G_o respectively.

By definition[2] $\Gamma_{F_o}\colon H^2 \to H^2_\perp$, and $\Upsilon_{G_o}\colon H^2 \to H^2$,

$$\Gamma_{F_o}(h) = \mathbf{\Pi}_- F_o h, \quad \text{and} \quad \Upsilon_{G_o}(h) = \mathbf{\Pi}_+ G_o h \quad (= G_o h \quad \text{since } G_o \in H^\infty)$$

for all $h \in H^2$, where $\mathbf{\Pi}_-$ and $\mathbf{\Pi}_+$ are the orthogonal projections on $H^2_\perp$ and H^2 respectively.

Since F_o is possibly irrational, $||\mathbf{A}||$ is given by $||\mathbf{A}|| = \max\{||\mathbf{A}||_e, \sigma_{max}\}$ where $||\mathbf{A}||_e$ is the essential norm of the operator $\mathbf{A}$ and σ_{max} is the largest singular value, of finite multiplicity, of $\mathbf{A}$.

Lemma: $||\mathbf{A}||_e = \max\{||G_o||_\infty \ , \ \eta\}$ where

$$\eta = \max \left\{ \left\| \begin{bmatrix} W_o(j\omega_i) \\ G_o(j\omega_i) \end{bmatrix} \right\| \quad i = 1, \ldots, k. \right\}$$

and the $j\omega_i$'s are the points on the extended imaginary axis $(\{j\omega \ : \ \omega \in [-\infty, +\infty]\})$ at which $m(j\omega)$ is not defined.

For a proof of this lemma see [17].

[2] L^2 denotes the Hilbert space of square integrable functions on the Im-axis, H^2 is the subspace of functions in L^2 that are analytic in the open right half plane, and $H^2_\perp$ is the orthogonal complement of H^2 in L^2, i.e. it is the subspace of functions in L^2 that are analytic in the open left half plane.

In summary we need to compute σ_{max} to find γ_{opt}. This problem has been first solved in [18] for the special case where G_o is constant. Then, in [17] a solution is given for arbitrary rational G_o. Finally, in [4] a more general case has been solved for G_o rational and $F_o = m^* W_o + E_o$, where W_o and m are as before and E_o is a rational function in L^∞. This problem corresponds to the situation where the plant to be controlled has unstable poles.

By definition $\sigma > ||\mathbf{A}||_e$ is a singular value of $\mathbf{A}$ if and only if there exists a nonzero $h \in H^2$ satisfying

$$\left(\sigma^2 \mathbf{I} - \mathbf{A}^* \mathbf{A}\right) h = 0, \qquad \text{that is,} \qquad \left(\sigma^2 \mathbf{I} - \Upsilon_{G_o}^* \Upsilon_{G_o} - \Gamma_{F_o}^* \Gamma_{F_o}\right) h = 0.$$

It can be shown [4] that h must be of the form $h(s) = \frac{R_\sigma(s)\Phi}{\lambda_\sigma(s)}$, where $R_\sigma(s)$ is a vector of finite size (say $n \times 1$) whose entries can be explicitly computed from the problem data W_o, G_o, E_o, m and σ; $\lambda_\sigma(s)$ is a polynomial whose coefficients can be explicitly computed; and Φ is an $n \times 1$ unknown complex vector whose entries are related to the unknown function $h(s)$. This observation leads to the following:

Theorem: The number $\sigma > ||\mathbf{A}||_e$ is a singular value of $\mathbf{A}$ if and only if there exists a nonzero constant complex vector Φ of size $n \times 1$, satisfying $\mathcal{R}_\sigma \Phi = 0$, where $\mathcal{R}_\sigma$ is an $n \times n$ complex matrix whose entries can be explicitly computed from the problem data W_o, G_o, E_o, m and σ.

See [4] for the proof of this result and the construction of the matrix $\mathcal{R}_\sigma$. The size n depends on the dimensions of W_o, G_o and E_o.

Corollary: The number γ_{opt} is the largest value of σ which makes $\mathcal{R}_\sigma$ singular.

Hence we can find γ_{opt} by plotting the minimum singular value of $\mathcal{R}_\sigma$ as a function of σ decreasing σ from an upper bound γ_{max} to a lower bound, e.g. $||\mathbf{A}||_e$. Once γ_{opt} is computed we can find a nonzero vector Φ_{opt} which makes

$$\mathcal{R}_{\gamma_{opt}} \Phi_{opt} = 0$$

and the optimal controller can be computed from the problem data together with γ_{opt} and Φ_{opt}, (see [19]).

In summary we can solve the two-block H^∞ optimal control problem by constructing the matrix $\mathcal{R}_\sigma$ and analyzing its singular values and vectors. A computer program, which uses Matlab commands, has been written by Handong Tu of University of Minnesota at Duluth, to construct the matrix $\mathcal{R}_\sigma$ and to find γ_{opt} and Φ_{opt}.

B. H^∞ OPTIMAL CONTROL OF AN IDEAL BEAM

We shall apply the H^∞ optimal control design procedure outlined in Section IV.A to solve weighted mixed sensitivity optimization problems for beam systems one and two. For this purpose consider

$$W_1(s) = \sqrt{b}\,\frac{as + 1/b}{as + 1} \qquad W_2(s) = \sqrt{b}\,\frac{as + 1/b}{abs + 1/b}, \tag{36}$$

where $a > 1 \gg b > 0$. The weight W_1 penalizes the sensitivity function mainly in the low frequency range up to $(1/ab)$ rd/sec. The weight W_2 penalizes the complementary sensitivity function mainly at high frequencies above $(1/ab)$ rd/sec. This means that the frequency content of the reference signals and output disturbances are mainly below $(1/ab)$ rd/sec, while the frequency content of the measurement noise and the unmodeled dynamics are mainly above $(1/ab)$ rd/sec.

For weights with this particular structure we see that

$$\gamma_o = \sqrt{b} \qquad \text{and} \qquad G^{-1} = \sqrt{\frac{b}{1+b^2}}\,W_1^{-1}W_2^{-1}. \tag{37}$$

This leads to

$$G_o(s) = \sqrt{\frac{b}{1+b^2}}\,, \quad W_o(s) = \sqrt{\frac{b}{1+b^2}}\left(\frac{1/b - abs}{1 + as}\right) \tag{38}$$

$$\text{and} \quad m_w(s) = \frac{1/b - as}{1/b + as}. \tag{39}$$

1. Minimum phase plant

Consider the plant $P_1 = N_1/D$, where $N_1 = P_{1o}$ is outer. In this case the H^∞ optimal performance and controller can the be computed as

$$\gamma_{opt} = \frac{\sqrt{2b}}{1+b} \quad \text{and} \quad C_{opt,1} = DN_1^{-1}W_1W_2^{-1}. \tag{40}$$

These lead to the following optimal sensitivity and complementary sensitivity, respectively:

$$S_{opt,1} = \frac{\sqrt{b}}{1+b}W_1^{-1}, \qquad T_{opt,1} = \frac{\sqrt{b}}{1+b}W_2^{-1}.$$

Note that we need to have $b < \sqrt{2} - 1$, in order to have $\gamma_{opt} > \gamma_o$.

2. Non-minimum phase plant

Suppose that the plant is $P(s) = e^{-hs}P_2(s)$, where $h \geq 0$ is a time delay in the system. The general form for the inner factor of the beam transfer function is then

$$m_p(s) = e^{-hs}\prod_{n=1}^{\infty}\frac{z_n - s}{z_n + s} = e^{-hs}B(s)$$

where $h \geq 0$ and $z_n > 0$ is given in (14), and $z_k > z_n$, for $k > n$, with $z_n \to \infty$ as $n \to \infty$.

In this case the optimum controller can be computed from (35) where

$$\begin{aligned} Q_1^{opt}G^{-1}(s) &= \left(\frac{b}{1+b^2}\right)\left(\frac{1}{(as+1)(as+1/b)}\right)\times \\ &\quad \left(\frac{a^2(\gamma_1^2 - b^2)(s^2+\omega_o^2)}{\gamma_1 + m_w(s)m_p(s)W_1(s)W_2^{-1}(s)}\right) \end{aligned} \tag{41}$$

where γ_1 is the largest root of the equation

$$\gamma_1 + m_w(j\omega_o)m_p(j\omega_o)W_1(j\omega_o)W_2(j\omega_o)^{-1} = 0, \tag{42}$$

for $b < \gamma_1 < 1/b$ and

$$\omega_o = \frac{1}{a}\sqrt{\frac{1/b^2 - \gamma_1^2}{\gamma_1^2 - b^2}} = \frac{1}{a}\sqrt{\frac{1 - \gamma_{opt}^2 b}{\gamma_{opt}^2 b - b^2}}. \tag{43}$$

The optimal performance γ_{opt} can be computed from the equations (42) and (43). It can be shown that (42) is equivalent to

$$\begin{aligned} \pi \quad = \quad & h\omega_o + \tan^{-1}(\frac{\omega_o}{a}) + 2\tan^{-1}(\frac{b\omega_o}{a}) - \tan^{-1}(\frac{b^2\omega_o}{a}) + \\ & \sum_{n=1}^{\infty} 2\tan^{-1}(\frac{\omega_o}{z_n}). \end{aligned} \tag{44}$$

Thus, when the inner factor of the plant is nontrivial, the form of the controller given in (35) does not reduce to a simple expression like that of (40). From (44) we see that for the plant P_2 dependence of the optimal performance level on ϵ enters via the zeros z_n of the Blaschke product $B(s)$.

Another interesting point about (44) is that for a fixed ω_o such that $0 < \omega_o < \infty$, the infinite sum in the right hand side of (44) converges if and only if the sum $\sum_{n=1}^{\infty} \frac{1}{z_n}$ converges. This fact can be seen easily from the following estimates which are obtained by formal power series expansions of sine and cosine:

$$\sum_{n=N}^{\infty} \left(\frac{\omega_o}{z_n} - \frac{1/3(\omega_o/z_n)^3}{1 - 1/2(\omega_o/z_n)^2}\right) \leq \sum_{n=N}^{\infty} \tan^{-1}\left(\frac{\omega_o}{z_n}\right) \leq \sum_{n=N}^{\infty} \left(\frac{\omega_o}{z_n}\right)$$

for all fixed ω_o and positive integers N such that $z_N > \omega_o$. On the other hand convergence of $\sum_{n=1}^{\infty} 1/z_n$ is equivalent to the convergence of the Blaschke product $B(s)$ in the closed right half plane. In summary (44) is well defined if and only if the infinite Blaschke product given by (12) converges in the closed right half plane. As discussed at the end of Appendix.B, $B(s)$ does converge in the closed right half plane.

One can observe that the optimal controllers for P_1 and P_2 have infinite number of poles and zeros, and are improper. Also, the formula for the optimal controller suggest that the double pole at $s = 0$ is cancelled by introducing a double zero in the controller at $s = 0$. This violates internal stability in the sense that $P(1 + PC_{opt})^{-1}$ does not belong to H^∞. We can fix this problem by replacing the double zero of the controller, at $s = 0$, by a double zero at $s = -\zeta$ where ζ is a small positive real number. This will be discussed in more detail in Section V.

We see from the optimal controller expression (35) that for the above choice of weights C_{opt} cancels the left half plane poles of the plant, (i.e. N_o^{-1} appears as a factor of C_{opt}). From a practical perspective this is not a desirable situation when the cancelled poles are close to the imaginary axis. We could avoid this situation by choosing a W_2 whose zeros are the poles of the plant that we don't want to cancel. On the other hand, such a choice would increase the dimension of the weights and hence increase the complexity of the solutions. For the weights given by (36) the formulae (41) and (44) were derived by hand, without using a computer. For other choices of weights these computations would best be done by machine.

V. FINITE ORDER ROBUST CONTROLLERS

Typically, a linear time-invariant controller designed for implementation purposes is proper and low order. However, for certain problems H^∞ optimization leads to improper controllers that are high or even infinite order. We observed this in Section IV.B in the cases of beam systems one and two. The optimal controllers have an infinite number of poles and zeros, and are improper. We can see, more generally, from the optimal controller expression (35) that if $N_o(s)$ is strictly proper then the optimal controller will be improper, and hence not physically realizable. Thus, practical considerations lead us to investigate the problem of constructing low order, proper controllers that provide stability and an acceptable level of performance.

We shall say that an Nth order controller C_N provides acceptable performance if the closed loop system (C_N, P) is stable and the H^∞ performance measure $\gamma(C_N)$ is less than a certain value. This value indicates a level above which performance is considered inadequate. It depends on specifications of the problem and the weighting functions. What ever the level of acceptable performance is, one can say that in general the closer $\gamma(C_N)$ is to the optimal performance level γ_{opt}, the closer C_N is to being optimal. We shall say that if the difference between the numbers $\gamma(C_N)$ and γ_{opt} is relatively small, then C_N is approximately optimal.

We shall consider the following two basic approaches to obtaining finite order approximately optimal controllers. The first approach is to approximate the optimal controller C_{opt} by a controller of finite order N, say C_N, in such a way that C_N is approximately optimal for plant P. Although at first this approach may seem straight forward, it may not be. The controller appears in a linear-fractional way in the closed loop system transfer function. Thus, a controller that is close to C_{opt} in H^∞ norm, may result in a closed loop system transfer function with a much higher H^∞ norm. Worse yet, the close controller may not even stabilize the closed loop system. Also, because the set of stabilizing controllers may contain controllers quite different from C_{opt}, a straight forward model reduction procedure applied to C_{opt} may miss the most desirable low order approximately optimal controllers. A good way to implement the first approach is to approximate the coprime factors of C_{opt} separately, so that the approximate controller is close to the optimal controller in the gap metric. The approximate controller will be guaranteed to stabilize the nominal plant and achieve an H^∞ performance level close to the optimum. One can find detailed discussions of this approach in [20], [21], and the references therein. The second approach is to approximate the plant transfer function in an appropriate fashion by a plant of finite order N, say P_N, so that the controller C_N that is optimal for plant P_N is approximately optimal for the original plant P. There are a variety of ways to implement this approach also depending on how one chooses to approximate P. For details of the second approach and further references see [22].

Generally speaking, the success one has with either the first or the second approach depends on several factors such as the plant dynamics, the amount of performance degradation that can be tolerated, and how the approach is implemented. In this section we will consider the first approach for constructing Nth order controllers for P_1 and the second approach for constructing Nth order controllers for P_2.

A. FINITE ORDER CONTROLLERS FOR P_1

We will construct finite order controllers for P_1 which, within a certain frequency range, have the same poles and zeros as $C_{opt,1}$, and approximately the same gain. By preserving the poles, zeros and the gain of $P_1 C_{opt,1}$

within a frequency band containing the crossover region we are able to maintain the closed loop stability and certain level of H^∞ performance.

Define finite order proper controllers C_{N1} of the form

$$C_{N1}(s) = \frac{W_1(s)}{W_2(s)} P_{N1}^{-1}(s) \tag{45}$$

where

$$P_{N1}(s) = \frac{-4(1+\epsilon s + s^2/\alpha_N^4)}{(s+\zeta)^2} \prod_{n=1}^{N-1} \frac{(1+\epsilon s + s^2/\alpha_n^4)}{(1+\epsilon s + s^2/\phi_n^4)} \tag{46}$$

for some given positive integer N and some small positive real number ζ. Notice that C_{N1} is of order $2N+1$. We obtain C_{N1} by truncating the high frequency terms of the infinite product part of $C_{opt,1}$ given in (40) and replacing the s^2 term of $C_{opt,1}$ with $(s+\zeta)^2$. We keep fewer zeros than poles in this product to make $C_{N1}(s)$ proper. The added high frequency roll–off of $P_1 C_{N1}$ diminishes the effects of the high frequency modes of P_1. Replacing the s^2 term by $(s+\zeta)^2$ allows C_{N1} to stabilize P_1 in the H^∞ sense for N sufficiently large. In this case the open loop transfer function $P_1(s)C_{N1}(s)$ is given by

$$P_1(s)C_{N1}(s) = \frac{W_1(s)}{W_2(s)} \left(\frac{s+\zeta}{s}\right)^2 \frac{1}{1+\epsilon s + s^2/\phi_N^4} \prod_{n=N+1}^{\infty} \frac{1+\epsilon s + s^2/\alpha_n^4}{1+\epsilon s + s^2/\phi_n^4}.$$

By referring to the derivations in Section IV one can show that C_{N1} is the H^∞ optimal controller for the plant P_{N1}. Thus, had we approximated P_1 by P_{N1} at the start, then performed H^∞ optimization, we would have obtained the same finite order controllers C_{1N} as we did designing directly with P_1 then approximating $C_{opt,1}$. In other words the two approaches to obtaining low order approximately optimal controllers yield the same finite order controllers in this case. We would also have this special situation for other plants which, like P_1, can be expressed as a ratio of two outer functions. However, for plants such as P_2 that have a nontrivial inner factor this implementation of approach one would in general give different finite order controllers than a corresponding implementation of approach two.

B. FINITE ORDER CONTROLLERS FOR P_2

We shall consider an indirect scheme for constructing finite order controllers for P_2. Each finite order controller will be defined as the H^∞ optimal controller for a finite order approximation of P_2. The method of approximating P_2 will then determine the finite order controllers that we select.

Consider finite order plants of the form

$$P_{N2}(s) = \frac{2(1+\epsilon s - s^2/4\alpha_N^4)}{(s+\zeta)^2} \prod_{n=1}^{N-1} \frac{(1+\epsilon s - s^2/4\alpha_n^4)}{(1+\epsilon s + s^2/\phi_n^4)}. \tag{47}$$

Let C_{N2} be the H^∞ optimal controller for plant P_{N2}.

Since the optimal controller for P_2 is unique, C_{N2} is not optimal for P_2. However, it seems intuitively reasonable that for N large enough, C_{N2} will be approximately optimal for P_2. Likewise, one would expect that for N large enough C_{N1} will be approximately optimal for P_1. This must be established, along with estimates of what value of N is large enough to ensure that C_N stabilizes and provides adequate performance for P. In the next subsection we consider the problems of determining whether or not a given finite order controller stabilizes a system and if so, assessing the performance that the finite order controller provides.

C. ASSESSING APPROXIMATE OPTIMALITY

We would like to determine conditions under which C_{N1} and C_{N2} are approximately optimal for plants P_1 and P_2. To do this we investigate the sensitivity function $S = (1+PC_N)^{-1}$, and the complementary sensitivity function $T = PC_N(1+PC_N)^{-1}$ when the controller C_N is used in the feedback loop for plant P. Here P may denote P_1, or P_2 and C_N will denote the corresponding finite order controller C_{N1} or C_{N2}.

Let $S_N = (1+P_NC_N)^{-1}$ and $T_N = P_NC_N(1+P_NC_N)^{-1}$ be the optimal sensitivity and the optimal complementary sensitivity functions when the plant is P_N. One can express S and T in terms of S_N and T_N as follows:

$$S(s) \;=\; \frac{1}{1+PC_N} = \frac{1}{1+P_N(s)C_N(s)+C_N(s)(P(s)-P_N(s))}$$

$$= \frac{S_N(s)}{1 + T_N(s)((\frac{s+\zeta}{s})^2 \tau_N(s) - 1)}$$

$$= \frac{s^2}{(s+\zeta)^2} \frac{S_N(s)}{1 + T_N(s)(\tau_N(s) - 1) + S_N(s)(\frac{s^2}{(s+\zeta)^2} - 1)},$$

$$\text{where } \tau_N(s) = \frac{s^2}{(s+\zeta)^2} P_N^{-1}(s) P(s).$$

By the small gain theorem C_N is guaranteed to stabilize P if the magnitude of

$$\Delta_N(j\omega) = T_N(j\omega)(\tau_N(j\omega) - 1) + S_N(j\omega)(\frac{-\omega^2}{(j\omega+\zeta)^2} - 1)$$

is less than 1 everywhere on the Imaginary axis. That is if $||\Delta_N||_\infty < 1$, then we conclude that C_N stabilizes the plant P. We shall measure the performance of C_N according to the value of

$$\gamma(C_N) = \left\| \begin{bmatrix} W_1(1+PC_N)^{-1} \\ W_2 PC_N(1+PC_N)^{-1} \end{bmatrix} \right\|_\infty . \tag{48}$$

Recall that C_N achieves the optimal performance for plant P_N, denoted by

$$\gamma_N = \left\| \begin{bmatrix} W_1 S_N \\ W_2 T_N \end{bmatrix} \right\|_\infty .$$

We will investigate the relationship between $\gamma(C_N)$, γ_N and γ_{opt}. Algebraic manipulations show that

$$\begin{bmatrix} W_1(1+PC_N)^{-1} \\ W_2 PC_N(1+PC_N)^{-1} \end{bmatrix} = \frac{1}{1+\Delta_N} \begin{bmatrix} W_1 S_N \frac{s^2}{(s+\zeta)^2} \\ W_2 T_N \tau_N \end{bmatrix} .$$

Therefore,

$$\gamma(C_N) \le \frac{\lambda}{1 - ||\Delta_N||_\infty} \gamma_N, \tag{49}$$

where

$$\lambda = \max\{|| \frac{s^2}{(s+\zeta)^2} ||_\infty \ , \ ||\tau_N||_\infty\} = \max\{1 \ , \ ||\tau_N||_\infty\}.$$

It can be seen from (44) that $\gamma_N \to \gamma_{opt}$, as $N \to \infty$, because the term $\sum_{n=N+1}^{\infty} 2\tan^{-1}(\frac{\omega_o}{z_n}) \to 0$, as $N \to \infty$, for every finite ω_o. Thus, in order to find an estimate of $\gamma(C_N)$ we will bound $\|\Delta_N\|_\infty$ and $\|\tau_N\|_\infty$.

In Appendix.C we show that for both plants P_1 and P_2, $\|\tau_{Ni}\|_\infty \to 1$ as $N \to \infty$, for $i = 1, 2$. In fact, $|\tau_{Ni}(j\omega) - 1|$ is very small up to a certain frequency, say ω_N, and as $N \to \infty$, $\omega_N \to \infty$ for $i = 1, 2$. But for each fixed N, $|\tau(j\omega) - 1|$ converges to 1 as $\omega \to \infty$. We further establish that

$$\lim_{N\to\infty,\ \zeta\to 0} \|\Delta_N\|_\infty \leq 2\gamma_{opt}\sqrt{b}, . \tag{50}$$

This leads to

$$\gamma_{opt} \leq \lim_{N\to\infty,\zeta\to 0} \gamma(C_N) \leq \frac{\gamma_{opt}}{1 - 2\gamma_{opt}\sqrt{b}}. \tag{51}$$

Although this bound is not tight, for the typical case where $1 \gg 2\gamma_{opt}\sqrt{b}$ we have that $\gamma(C_N) \approx \gamma_{opt}$. In fact it is possible to make this bound tight by choosing the weighting functions in such a way that $|W_2(\infty)|^{-1} = 0$ and $|W_1(0)|^{-1} = 0$. Such a choice means that W_2 has to be improper, and W_1 has to have a pole at $s = 0$.

VI. COMPUTATIONAL RESULTS

In this section we compute optimal performance and optimal controllers for specific examples of the weighted mixed sensitivity problem (30) described in Section III for beam systems one and two developed in Section II. We quantify the effects on optimal achievable performance of a pure feedback time delay. We derive and quantify the approximate optimality of finite order controllers in the manner described in Section V. We compute H^∞ optimal performance, bound suboptimal performance, and analyze finite order proper controllers achieving suboptimal performance.

As discussed in Section II.A, for any given $\epsilon > 0$, P_1 and P_2 have the same poles, a finite number of which are complex. P_1 has a finite number of complex zeros, and P_2 has all real zeros. When $\epsilon = 0.001$, for example, P_1 has 13 pairs of complex poles and zeros. Table I lists the first five pairs of complex poles of P_1 and P_2, the first five complex pairs of zeros of P_1, and the corresponding ten real zeros of P_2 for $\epsilon = 0.001$.

Table I. Five pairs of poles of P_1 and P_2, zeros of P_1 and 10 real zeros of P_2.

	Zeros of P_1	Poles of P_1, P_2	Zeros of P_2
Magnitude	15.418	22.374	-30.364
Damping	0.0077091	0.011187	31.315
Magnitude	49.965	61.673	-95.062
Damping	0.024983	0.030837	105.05
Magnitude	104.25	120.90	-187.89
Damping	0.052124	0.060452	231.36
Magnitude	178.27	199.86	-298.60
Damping	0.089135	0.099930	425.72
Magnitude	272.03	298.55	-415.83
Damping	0.13602	0.14928	711.84

Figure 2 shows Bode gain plots of P_1 and P_2 for $\epsilon = 0.001$. Figure 3 shows a phase plot of P_2. Plots of P_1 and P_2 for other small, positive values of ϵ would be qualitatively similar.

Letting $a = 33$ and $b = 0.001$ for the weights W_1 and W_2 given by (36),

$$W_1(s) = \sqrt{.001}\,\frac{33s + 1000}{33s + 1}, \qquad W_2 = \sqrt{.001}\,\frac{33s + 1000}{0.033s + 1000}. \tag{52}$$

This choice of a and b reflects an assumption that low frequency disturbances occur mainly below 30.3 $\frac{rad}{sec}$, while high frequency noises and unmodelled dynamics are present mainly above 30.3 $\frac{rad}{sec}$. We shall refer to the weighted mixed sensitivity design problem (30) with the specific weights given by (52) as design $\gamma(W_1, W_2)$.

For P_1 the results of Section IV.B show that the optimal open loop system for $\gamma(W_1, W_2)$ is $P_1 C_{opt,1} = \frac{0.033s+1000}{33s+1}$. Note that $P_1 C_{opt,1}$ behaves like the function $\frac{30.3}{s}$ in the sense that $P_1 C_{opt1}$ is high at low frequencies, low at high frequencies, and crosses over at about 30.3 $\frac{rad}{sec}$ with a log-log slope of -1. One could raise or lower the very low and very high frequency magnitudes of $P_1 C_{opt1}$ while maintaining the midrange behavior by adjusting the weighting parameters a and b accordingly.

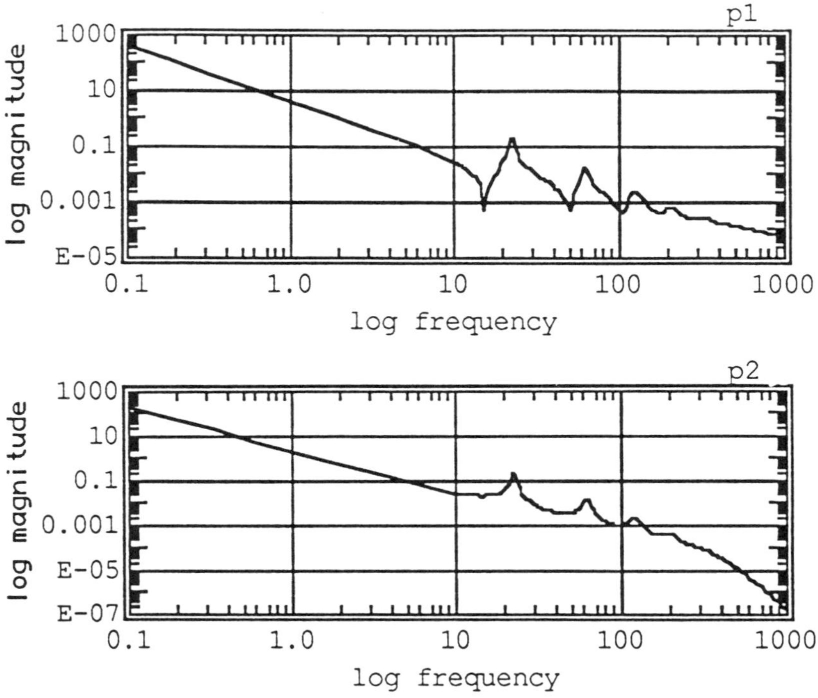

Figure 2: (a) Gain plot of P_1, (b) gain plot of P_2.

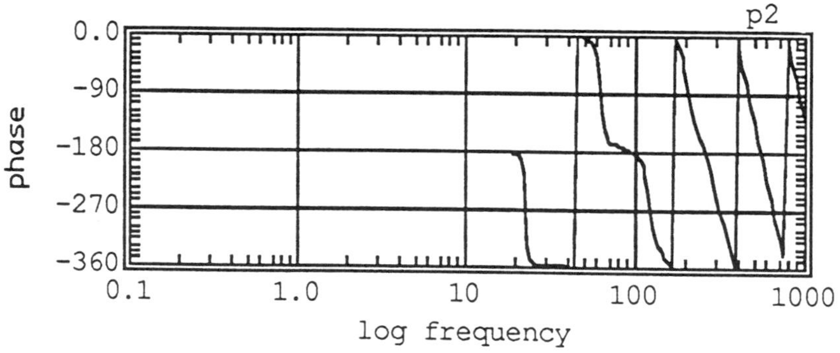

Figure 3: Phase plot of P_2.

It may seem surprising at first observation that $P_1 C_{opt,1}$ crosses over between the first and second pair of lightly damped poles of P_1. One might imagine situations in which this optimal design would leave the closed loop system vulnerable to perturbations exciting the low frequency lightly damped poles of P_1. However, the possibility of such a situation is not incorporated into the problem formulation of $\gamma(W_1, W_2)$.

Using the stability analysis discussed in Section V.C, we find that for $N \geq 4$, C_{N1} given by (45) stabilizes P_1. We determine this by estimating $||\Delta_{N1}||_\infty$. Our estimates for $||\Delta_{N1}||_\infty$ are obtained by computing $|\Delta_{N1}(j\omega)|$ directly for values of ω ranging from 10^{-2} to 10^4. Our analysis in Appendix.A indicates that $|\Delta_{N1}(j\omega)|$ assumes its maximum value within this finite frequency range. Table II shows computed estimates of $||\Delta_{N1}||_\infty$ for $N = 3$ through $N = 10$.

We found that for $N = 10$, the finite order controller C_{N1} provides a performance level $\gamma(C_{N1})$ of less than 0.0484. We determine the estimate 0.0484 by computing the right hand side of (49). Table III shows numerical estimates of (49) for $N = 4$ through $N = 10$. Here the optimal performance level is $\gamma_{opt} = \frac{\sqrt{0.002}}{1.001}$, approximately 0.0447. By inequality (51), as $N \rightarrow \infty$, we have that $\gamma(C_{N1}) \rightarrow \tilde{\gamma} < 0.04481$.

The situation changes when we incorporate a time delay into our nominal model of the system. Table IV shows values of γ_{opt} given values of h for the plant $e^{-hs}P_1$. The optimal norm is much larger for $h \geq 0.1$ than it is for $h = 0$. This is not surprising considering that when $h \geq 0.1$, the delay adds significant phase variation in the desired crossover region. This causes a trade-off between the magnitude of PC_{opt} below 30.3 $\frac{rad}{sec}$ and the magnitudes of S_{opt} and T_{opt} near 30.3 $\frac{rad}{sec}$. If the delay is small enough, say $h \leq 0.01$, it has little effect on γ_{opt}. At frequencies where the delay adds significant phase, $|PC_{opt}(j\omega)|$ is well below 1.

The finite order controllers that we designed for the nondelay case are robust to a small, unmodelled delay of $h \leq 0.01$. Because e^{-hs}, for $h \leq 0.01$, adds no gain variation to $P_1 C_{N1}$ and very little phase variation near 30.3 $\frac{rad}{sec}$, if C_{N1} stabilizes P_1 and provides a sufficient phase margin, C_{N1} also stabilizes $e^{-hs}P_1$. In terms of our stability bound, this can be verified by replacing τ_{N1} by $e^{-hs}\tau_{N1}$. As indicated in Table V, when $h = 0.01$,

$||\Delta_N||_\infty < 1$ for $N \geq 4$.

Unlike $e^{-0.01s}$, the delay term $e^{-0.1s}$ changes the phase of the open loop transfer function by more than 90 deg at 30.3 $\frac{rad}{sec}$. This delay will destabilize the nondelayed system since $P_1 C_{opt,1}$ and $P_1 C_{N1}$ for $N \geq 4$ have a phase of approximately -90 deg at 30.3 $\frac{rad}{sec}$ where their gain is 1. Hence, if a delay of $h \geq 0.1$ is present, one must either lower the bandwidth of $P_1 C_{N1}$ or incorporate the presence of the delay into the design process.

Consider the problem of controlling P_2 with the finite order controller C_{N2}, where C_{N2} is the H^∞ optimal controller for the finite order plant P_{N2} given by (47). The optimal open loop system is

$$P_2 C_{opt,2} = \frac{B Q_1^{opt} G^{-1}}{1 - m_w B Q_1^{opt} G^{-1}},$$

where $m_w = \frac{1000-33s}{1000+33s}$, B is the Blaschke product given by (12), and

$$Q_1^{opt}(s) G^{-1}(s) = \frac{1}{1 + (0.001)^2} \left(\frac{0.003s + 1000}{33s + 1000} \right) \times$$

$$\frac{33^2 \left(\gamma_1^2 - (.001)^2\right)\left(s^2 + \omega_o^2\right)}{\gamma_1 (33s + 1000)(1 + 33s) - (33s - 1000)(0.033s + 1000) B(s)},$$

where γ_1 is approximately 3.22 and $\omega_o = \frac{311}{33}$.

Computing $||\Delta_{N2}||_\infty$ for various values of N, we find that N must be at least 9 before $||\Delta_{N2}||_\infty < 1$. Table VI shows numerical estimates of $||\Delta_{N2}||_\infty$ for $N = 9$ through $N = 16$. In general, we find that in order to obtain values for $||\Delta_{N2}||_\infty$ comparable to those we obtained in the case of P_1, we have to include at least 6 additional product terms in P_{N2}.

Table VII shows numerical estimates for the upper bound on $\gamma(C_{N2})$. For brevity, only even values of N from 14 through 26 are listed. The optimal performance level γ_{opt} is about 0.1065. By inequality (51), as $N \to \infty$, $\gamma(C_{N2}) \to \tilde{\gamma} < 0.1073$. Our numerical results indicate that the convergence of $\gamma(C_{N2})$ is much slower than the convergence of $\gamma(C_{N1})$. This slower convergence is also indicated by the factors δ_{N1} and δ_{N2} given in Appendix.C. The ratio $\frac{\delta_{N1}}{\delta_{N2}} = \frac{\epsilon}{N} e^{-2(2/\pi^2 \epsilon - 1)/N}$ predicts a much faster convergence rate for $\gamma(C_{N1})$ than for $\gamma(C_{N2})$.

Table II. Stability bound $||\Delta_{N1}||_\infty$

Value of N	3	4	5	6	7	8	9	10
$\|\|\Delta_{N1}\|\|_\infty$	1.61	0.589	0.270	0.147	0.0906	0.0615	0.0451	0.0356

Table III. Upper bound on $\gamma(C_{N1})$ for P_1 and C_{N1}

Value of N	4	5	6	7	8	9	10
Upper bound on $\gamma(C_{N1})$	0.427	0.165	0.106	0.0787	0.0632	0.0538	0.0481

Table IV. Values of γ_{opt} given values of h for $P = e^{-hs}P_1$

Value of h	0	0.001	0.005	0.01	0.05	0.1	0.5	1
γ_{opt}	.0447	.0452	.0481	.0512	.0749	0.105	0.345	0.612

Table V. Stability bound $||\Delta_N||_\infty$ for $e^{-0.01s}P_1$ and C_{N1}

Value of N	3	4	5	6	7	8	9	10
$\|\|\Delta_N\|\|_\infty$	1.84	0.758	0.387	0.311	0.307	0.305	0.304	0.303

Table VI. Stability bound $||\Delta_{N2}||_\infty$ for P_2 and C_{N2}

Value of N	9	10	11	12	13	14	15	16
$\|\|\Delta_{N2}\|\|_\infty$	0.904	0.566	0.383	0.278	0.214	0.172	0.145	0.126

Table VII. Upper bound on $\gamma(C_{N2})$ for P_2 and C_{N2}

Value of N	14	16	18	20	22	24	26
Upper bound on $\gamma(C_{N2})$	0.191	0.138	0.136	0.123	0.117	0.115	0.114

Table VIII. Values of γ_{opt} given values of h for $P = e^{-hs}P_2$

Value of h	0	0.001	0.005	0.01	0.05	0.1	0.5	1
γ_{opt}	0.1065	0.107	0.109	0.113	0.137	0.167	0.408	0.705

Table IX. Stability bound $||\Delta_N||_\infty$ for $e^{-0.005s}P_2$ and C_{N2}

Value of N	9	10	11	12	13	14	15	16
$\|\|\Delta_N\|\|_\infty$	0.910	0.617	0.583	0.574	0.570	0.567	0.565	0.564

Finding that significantly more states are needed in this approximation scheme to satisfy the stability bound for P_2 than for P_1 suggests that P_2 is harder to control. The fact that γ_{opt} is much larger in the case of P_2 than in the case of P_1 indicates that our design objectives are more easily met when the plant is P_1 than when it is P_2. We also see in Table IV that the presence of a delay makes our design objectives more difficult to satisfy. In general, as the formulas in Section IV for the optimal solution indicate, this H^∞ optimization problem becomes difficult when the inner factor of the plant is nontrivial.

Consider the addition of a pure time delay to beam system two. Table VIII shows the values of γ_{opt} given values of h for a plant of the form $e^{-hs}P_2$. As h increases, the optimal norm increases comparably to how the optimal norm increases for plants $e^{-hs}P_1$. Also, as for plants $e^{-hs}P_1$, we find that if the delay is small, say $h \leq 0.005$, C_{N2} stabilizing P_2 stabilizes $e^{-hs}P_2$. One can verify that the stability bound, $||\Delta_N||_\infty < 1$, is satisfied for $\tau_N = e^{-hs}\tau_{N2}$, when $h \leq 0.005$ and $N \geq 9$. Table IX lists values of $||\Delta_N||_\infty$ for $N \geq 9$ and $h = 0.005$.

It is interesting to note that for plants of the form $e^{-hs}P_1$, the value of γ_{opt} is independent of the damping parameter ϵ, assumed to be a known, positive constant. However, for plants of the form $e^{-hs}P_2$, the value of γ_{opt} is affected by the value of ϵ. Computing γ_{opt} as a function of ϵ for ϵ between 0 and 1 using (34), we find that γ_{opt} ranges from about 0.1154 for $\epsilon = 0.0$ to about 0.0459 for $\epsilon = 1.0$.

VII. DISCUSSION AND CONCLUSIONS

The flexible beam control design and analysis techniques that we described in this chapter exemplify a robust control design approach that can be applied to a variety of problems. Using skew Toeplitz theory we computed the H^∞ optimal performance and H^∞ optimal controllers for flexible beam systems with and without a pure time delay. We investigated finite order, realizable controllers that are indicated by the form of the optimal controllers. We demonstrated how the value of the optimal performance γ_{opt} can be used as a precise measure by which to judge the approximate optimality of given finite order controllers.

To carry out the H^∞ optimization we needed to factor the beam transfer functions as quotients of H^∞ functions. To facilitate this we derived infinite product representations of the beam transfer functions. We found that when the displacement is measured at the opposite end of the beam from the applied control force, the beam transfer function, P_2, has infinitely many right half-plane zeros. We factored these zeros out of the product representation for P_2 as an infinite Blaschke product. Even so, we were able to solve H^∞ optimal control problems for this plant directly, without approximation.

We used weighting functions to model system uncertainty and to specify desired sensitivity reduction at each frequency. We chose the specific form of the weights W_1 and W_2 as defined in (36) to make the quantities of Section IV as simple as possible to derive by hand. The hand computations allowed us to see the forms of intermediate quantities not readily viewed when H^∞ optimization is carried out by computer. In general, for arbitrary weights, the computations outlined in Section III are done numerically.

Our computations of Section VI showed that the weights W_1 and W_2 do not impose a substantial trade-off between S and T when the plant is P_1. Thus, we solved an easy design problem for P_1; one that could be interpreted as not limited by actuator bandwidth constraints and having performance requirements well within the frequency range where the design model is accurately represented. In Section IV we showed that, in fact, these weights lead to a reduction of our two-block general distance problem to a one-block general distance problem, in the sense that G_o is constant.

We showed how the optimal design is degraded when the actuator and sensor locations are changed from both being located at one end of the beam to being located at opposite ends of the beam. We quantified the effect of a pure time delay on the optimal designs for both P_1 and P_2, as exemplified in Table IV and Table VIII. We showed how the optimal H^∞ performance degrades as the length of the assumed delay increases. We also analyzed the tolerance of approximately optimal designs to an unmodelled feedback delay.

APPENDIX

This appendix consists of three parts. In Appendix.A we derive functions that bound $|P_1(j\omega)|$ and $|P_2(j\omega)|$ for all ω such that $|\omega| > \omega_o$, for ω_o sufficiently large. The bounding functions tend to 0 as $\omega \to \infty$. In Appendix.B we establish that the product representations for P_1 and P_2 introduced in Section II.A converge in the closed right half-plane and admit coprime factorizations over H^∞ given by $P_1(s) = \frac{(s+1)^2}{s^2} P_{1o}(s)$ and $P_2(s) = \frac{(s+1)^2}{s^2} B(s) P_{2o}(s)$. In Appendix.C we derive Section V.C performance bounds. The notation and terminology are consistent with that of Sections II through V.

A. UPPER BOUNDS ON THE MAGNITUDES OF P_1 AND P_2

Theorem A.1

For $|\omega| > \omega_o$, for ω_o sufficiently large,

$$|P_1(j\omega)| \le \left| \frac{1}{(1+j\epsilon\omega)\beta(j\omega)^3} \right| \left| \frac{\sqrt{2}(1 + e^{-2(x+y)} + e^{-2y} + e^{-2x})}{1 - e^{-2y} - e^{-2x} - e^{-2(x+y)} - 4e^{-x-y}} \right|$$

where $x = |Re(\beta(j\omega))|$ and $y = |Im(\beta(j\omega))|$.

Proof: Recall that $\beta^4 = \frac{-s^2}{(1+\epsilon s)}$. Thus for $s = j\omega$

$$\beta^4 = \frac{\omega^2}{(1+j\epsilon\omega)} = \frac{\omega^2}{\sqrt{(1+j\epsilon^2\omega^2)}} e^{j\tan^{-1}(-\epsilon\omega)}$$

and the fourth roots of β^4 are, for $k = 0, 1, 2, 3$,

$$\beta_k = \frac{\sqrt{|\omega|}}{(1+\epsilon^2\omega^2)^{1/8}} e^{j\left(\frac{\tan^{-1}(-\epsilon\omega)}{4} + \frac{k}{2}\pi\right)}.$$

$$\text{As } \omega \to \infty, \quad \frac{\sqrt{|\omega|}}{(1+\epsilon^2\omega^2)^{1/8}} \to \infty \text{ and } \frac{\tan^{-1}(-\epsilon\omega)}{4} \to \frac{-\pi}{8}.$$

$$\text{As} \quad \omega \to -\infty, \quad \frac{\sqrt{|\omega|}}{(1+\epsilon^2\omega^2)^{1/8}} \to \infty \quad \text{and} \quad \frac{\tan^{-1}(-\epsilon\omega)}{4} \to \frac{\pi}{8}.$$

Moreover, there is a positive number ω_o such that $|Re(\beta_k)|$ and $|Im(\beta_k)|$ are increasing functions of ω for $\omega > \omega_o$, and $|Re(\beta_k)| \to \infty$ and $|Im(\beta_k)| \to \infty$ as $\omega \to \infty$. Also, as $|\omega| \to \infty$, $\frac{1}{(1+\epsilon j\omega)\beta_k{}^3} \to 0$.

Note that $P_1(s) = \frac{1}{(1+\epsilon s)\beta^3} F_1(s)$, for

$$F_1 = \left(\frac{\sinh \beta_k \cos \beta_k - \cosh \beta_k \sin \beta_k}{\cos \beta_k \cosh \beta_k - 1} \right) =$$

$$\frac{(e^{\beta_k} - e^{-\beta_k})(e^{j\beta_k} + e^{-j\beta_k}) + j(e^{\beta_k} + e^{-\beta_k})(e^{j\beta_k} - e^{-j\beta_k})}{(e^{\beta_k} + e^{-\beta_k})(e^{j\beta_k} + e^{-j\beta_k}) - 4}.$$

Suppose that $Re(\beta_k) = x > 0$ and $Im(\beta_k) = y > 0$. Factoring $e^{(1-j)\beta_k}$ out of both the numerator and denominator of the right hand side,

$$F_1(j\omega) = \frac{(1 - e^{-2\beta_k})(e^{j2\beta_k} + 1) + j(1 + e^{-2\beta_k})(e^{j2\beta_k} - 1)}{(1 + e^{-2\beta_k})(e^{j2\beta_k} + 1) - 4e^{(j-1)\beta_k}}$$

where $|e^{j2\beta_k}|, |e^{-2\beta_k}|, |e^{(j-1)\beta_k}|$ are bounded, decreasing functions of ω, for $|\omega| > \omega_o$. Replacing β_k by $(x + jy)$, algebraic manipulations establish that for $|\omega| > \omega_o$, for ω_o sufficiently large,

$$|F_1(j\omega)| \leq \left| \frac{\sqrt{2}(1 + e^{-2(x+y)} + e^{-2y} + e^{-2x})}{1 - e^{-2y} - e^{-2x} - e^{-2(x+y)} - 4e^{-x-y}} \right|,$$

where $x = |Re(\beta_k)|$ and $y = |Im(\beta_k)|$. As $|\omega| \to \infty, x \to \infty, y \to \infty$, and the right hand side tends to $\sqrt{2}$.

For the case that $Re(\beta_k) > 0$ and $Im(\beta_k) < 0$, factoring $e^{(1+j)\beta_k}$ out of the numerator and denominator of our first expression for F_1 leads again to our expression for $|F_1(j\omega)|$. Since $|F_1(j\omega)|$ is symmetric in e^{β_k} and $e^{-\beta_k}$, we also obtain equivalent results for $Re(\beta_k) < 0, Im(\beta_k) < 0$ and $Re(\beta_k) < 0$, $Im(\beta_k) > 0$. □

Theorem A.2 For $|\omega| > \omega_o$, for ω_o sufficiently large,

$$|P_2(j\omega)| \leq \left|\frac{1}{(1+j\epsilon\omega)\beta(j\omega)^3}\right| \left|\frac{2(e^{-2x-y}+e^{-y}+e^{-x}+e^{-x-2y})}{1-e^{-2y}-e^{-2x}-e^{-2(x+y)}-4e^{-x-y}}\right|$$

where $x = |Re(\beta(j\omega))|$ and $y = |Im(\beta(j\omega))|$.

Proof: Note that $P_2(s) = \frac{1}{(1+\epsilon s)\beta^3}F_2(s)$, where

$$F_2(j\omega) = \left(\frac{\sinh\beta_k - \sin\beta_k}{\cos\beta_k \cosh\beta_k - 1}\right) = \frac{2(e^{\beta_k} - e^{-\beta_k} + j(e^{j\beta_k} - e^{-j\beta_k}))}{(e^{\beta_k}+e^{-\beta_k})(e^{j\beta_k}+e^{-j\beta_k}) - 4}$$

Suppose that $Re(\beta_k) = x > 0$ and $Im(\beta_k) = y > 0$. Factoring $e^{(1-j)\beta_k}$ out of both the numerator and denominator we obtain

$$F_2 = 2\frac{(1-e^{-2\beta_k})e^{j\beta_k} + j(e^{j2\beta_k}-1)e^{-\beta_k}}{(1+e^{-2\beta_k})(e^{j2\beta_k}+1) - 4e^{-\beta_k(1-j)}}.$$

One can show that for $|\omega| > \omega_o$, for ω_o sufficiently large,

$$|F_2(j\omega)| \leq 2\left|\frac{e^{-2x-y}+e^{-y}+e^{-x}+e^{-x-2y}}{1-e^{-2y}-e^{-2x}-e^{-2(x+y)}-4e^{-x-y}}\right|,$$

where $x = |Re(\beta_k)|$ and $y = |Im(\beta_k)|$. As $|\omega| \to \infty$, $x \to \infty$, $y \to \infty$, and the right hand side tends to 0. For the remaining three cases given by $Re(\beta_k) < 0$ and $Im(\beta_k) < 0$, $Re(\beta_k) > 0$ and $Im(\beta_k) < 0$, and $Re(\beta_k) < 0$ and $Im(\beta_k) > 0$ one obtains this bound in an analogous way. □

B. CONVERGENCE AND FACTORIZATIONS OF P_1 AND P_2

1. Factorization of P_1

Consider the sequence $\{F_{M,N}\}_{M=N}^{\infty}$, for N fixed and

$$F_{M,N}(s) = \left(\frac{1}{s+1}\right)\prod_{n=N}^{M} f_n(s), \quad \text{where} \quad f_n(s) = \frac{\left(1+\epsilon s+\frac{s^2}{\alpha_n^4}\right)}{\left(1+\epsilon s+\frac{s^2}{\phi_n^4}\right)}$$

Theorem A.3 The sequence $\{F_{M,N}\}$ is a Cauchy sequence in H^∞ that converges to

$$F_{\infty,N}(s) = \frac{1}{(s+1)} \prod_{n=N}^{\infty} \frac{\left(1+\epsilon s+\frac{s^2}{\alpha_n^4}\right)}{\left(1+\epsilon s+\frac{s^2}{\phi_n^4}\right)} \in H^\infty .$$

Proof: One can show that $|f_n(j\omega)|^2 = \frac{\left(1-\frac{\omega^2}{\alpha_n^4}\right)^2+(\epsilon\omega)^2}{\left(1-\frac{\omega^2}{\phi_n^4}\right)^2+(\epsilon\omega)^2}$. Therefore, $\frac{\partial}{\partial\omega}|f_n(j\omega)|^2 = 0$ if and only if $\frac{\phi_n^8(\alpha_n^8-\omega^4)}{\alpha_n^8(\phi_n^8-\omega^4)} = |f_n(j\omega)|^2$. Hence, at the points ω_n where $\frac{\partial}{\partial\omega}|f_n(j\omega)|^2$ vanishes, either $|\omega| < \alpha_n^2 < \phi_n^2$ or $|\omega| > \phi_n^2 > \alpha_n^2$. Moreover, if $|\omega| < \alpha_n^2$, then $|f_n(j\omega)|^2 \leq \frac{\phi_n^8}{\alpha_n^8}$.

Solving $\frac{\partial}{\partial\omega}|f_n(j\omega_n)|^2 = 0$, one finds that

$$\omega_n^2 = \left(\frac{1}{\epsilon^2 - \frac{2}{\phi_n^4+\alpha_n^4}}\right)\left(-1+\sqrt{1+\frac{\epsilon^2(\phi_n^4+\alpha_n^4-2)}{1+\frac{1}{2}\left(\frac{\phi_n^4}{\alpha_n^4}+\frac{\alpha_n^4}{\phi_n^4}\right)}}\right).$$

Thus, as $n \to \infty$, $\omega_n^2 \to \frac{\sqrt{\phi_n^4+\alpha_n^4}}{\sqrt{2}\epsilon}$. Hence, as $n \to \infty$, ω_n^2 is $O(\alpha_n^2) \ll \alpha_n^4$. So at the local extrema, $|f_n(j\omega_n)| \leq \frac{\phi_n^4}{\alpha_n^4}$, while as $\omega \to \infty$, $|f_n(j\omega)| \to \frac{\phi_n^4}{\alpha_n^4}$. Therefore, $||f_n||_\infty \leq \frac{\phi_n^4}{\alpha_n^4}$ for n sufficiently large.

To establish that $\{F_{M,N}\}$ is a Cauchy sequence, observe that

$$\left|\frac{-\omega^2}{(1+j\omega)\left(1-\frac{\omega^2}{\phi_n^4}+j\epsilon\omega\right)}\right|^2 = \left|\frac{\omega^4}{(1+\omega^2)\left(\left(1-\frac{\omega^2}{\phi_n^4}\right)^2+\epsilon^2\omega^2\right)}\right|$$

$$\leq \left|\frac{\omega^4}{(1+\omega^2)\epsilon^2\omega^2}\right| \leq \frac{1}{\epsilon^2}.$$

Therefore $\left|\left|\frac{s^2}{(1+s)\left(1+\epsilon s+\frac{s^2}{\phi_n^4}\right)}\right|\right|_\infty \leq \frac{1}{\epsilon}$. Thus,

$$||F_{M,N} - F_{M-1,N}||_\infty \le \left\| \prod_{n=N}^{M-1} f_n \right\|_\infty \left\| \frac{1}{1+s}(f_M - 1) \right\|_\infty$$

$$\le \left(\prod_{n=N}^{M-1} ||f_n||_\infty \right) \left\| \frac{s^2 \left(\frac{1}{\alpha_M^4} - \frac{1}{\phi_M^4} \right)}{(1+s)(1+\epsilon s + \frac{s^2}{\phi_M^4})} \right\|_\infty \le$$

$$\frac{1}{\epsilon} \left(\frac{1}{\alpha_M^4} - \frac{1}{\phi_M^4} \right) \prod_{n=N}^{M-1} \left(\frac{\phi_n}{\alpha_n} \right)^4 \le \left(1 - \frac{\alpha_M}{\phi_M} \right) \frac{1}{\epsilon \alpha_M^4} \prod_{n=N}^{M} \left(\frac{\phi_n}{\alpha_n} \right)^4 .$$

Recall that $\phi_n \to \frac{\pi}{2} + n\pi$ and $\alpha_n \to \frac{\pi}{4} + n\pi$ as $n \to \infty$. One can show that for each n, $\frac{\pi}{4} + n\pi \ge \alpha_n \ge \frac{\pi}{4} + n\pi - 2e^{-2n\pi}$. Also, $\phi_n \le \frac{\pi}{2} + n\pi + 4e^{-n\pi}$.

$$\text{Thus,} \quad \prod_{n=N}^{M} \left(\frac{\phi_n}{\alpha_n} \right)^4 \le \prod_{n=N}^{M} \left(\frac{\frac{\pi}{2} + n\pi + 4e^{-n\pi}}{\frac{\pi}{4} + n\pi - 2e^{-2n\pi}} \right)^4 \le \frac{M}{N} e^{2/N}.$$

Therefore,

$$\left(1 - \frac{\alpha_M}{\phi_M} \right) \frac{1}{\epsilon \alpha_M^4} \prod_{n=N}^{M} \left(\frac{\phi_n}{\alpha_n} \right)^4 \le \left(\frac{\frac{\pi}{4} + 2e^{-M\pi} + e^{-2N\pi}}{\frac{\pi}{2} + M\pi} \right) \frac{M e^{2/N}}{\epsilon N M^4 \pi^4}$$

$$\le \frac{5}{2\epsilon(1+2M) N M^3 \pi^4} e^{2/N} \le \frac{5}{4\epsilon N M^4 \pi^4} e^{2/N}.$$

Thus, $||F_{M,N} - F_{M-1,N}||_\infty \le \frac{5e^{2/N}}{4\pi^4 \epsilon M^4 N}$. It follows that for any M, M' with $M' > M > N$,

$$||F_{M,N} - F_{M',N}||_\infty \le \frac{5e^{2/N}}{4\pi^4 \epsilon N} \sum_{n=M}^{M'} \frac{1}{n^4} \le \frac{5e^{2/N}}{2\pi^4 \epsilon M^3 N}.$$

Therefore $\{F_{M,N}\}$ is a Cauchy sequence. Since H^∞ is complete, $\{F_{M,N}\}$ converges to $F_{\infty,N} \in H^\infty$. □

$$\text{Thus,} \quad P_{1o} = \left(\frac{-4}{(s+1)^2} \right) \prod_{n=N}^{\infty} \frac{\left(1 + \epsilon s + \frac{s^2}{\alpha_{n^4}} \right)}{\left(1 + \epsilon s + \frac{s^2}{\phi_n^4} \right)} \in H^\infty.$$

2. Factorization of P_2

Consider the H^∞ sequence $\{H_{M,N}\}$, where

$$H_{M,N} = \frac{1}{s+1} \prod_{n=N}^{M} \left(\frac{1 + s\sqrt{\epsilon^2 + \frac{1}{\alpha_n^4}} + \frac{s^2}{4\alpha_n^4}}{1 + \epsilon s + \frac{s^2}{\phi_n^4}} \right).$$

Theorem A.4 The sequence $\{H_{M,N}\}_{M=N}^{\infty}$ is Cauchy and converges to

$$H_{\infty,N}(s) = \frac{1}{(s+1)} \prod_{n=N}^{\infty} \left(\frac{1 + s\sqrt{\epsilon^2 + \frac{1}{\alpha_n^4}} + \frac{s^2}{4\alpha_n^4}}{1 + \epsilon s + \frac{s^2}{\phi_n^4}} \right) \in H^\infty.$$

Proof: Let $g_n(s) = \frac{1+\epsilon s - \frac{s^2}{4\alpha_n^4}}{1+\epsilon s + \frac{s^2}{\phi_n^4}}$. For $\omega \neq 0$, $|g_n(j\omega)|^2 = \frac{1+\frac{1}{\epsilon^2}\left(\frac{1}{\omega}+\frac{\omega}{4\alpha_n^4}\right)^2}{1+\frac{1}{\epsilon^2}\left(\frac{1}{\omega}-\frac{\omega}{\phi_n^4}\right)^2}$.

Therefore, $\frac{\partial}{\partial\omega}|g_n(j\omega)|^2 = 0$ if and only if

$$|g_n(j\omega)|^2 = \frac{\left(\frac{1}{\omega} + \frac{\omega}{4\alpha_n^4}\right)\left(\frac{-1}{\omega^2} + \frac{1}{4\alpha_n^4}\right)}{\left(\frac{1}{\omega} - \frac{\omega}{\phi_n^4}\right)\left(\frac{-1}{\omega^2} - \frac{1}{\phi_n^4}\right)} = \frac{1 - \frac{\omega^4}{16\alpha_n^8}}{1 - \frac{\omega^4}{\phi_n^8}}.$$

Solving for $\omega = \omega_n$ we have

$$\omega_n^2 = \frac{-1 + \sqrt{1 + \frac{16\alpha_n^4\phi_n^4}{4\alpha_n^4 - \phi_n^4}\left(\frac{\epsilon^2}{2} + \frac{1}{\phi_n^4 + 4\alpha_n^4}\right)}}{\epsilon^2 + \frac{2}{4\alpha_n^4 - \phi_n^4}}$$

As $n \to \infty$, ω_n^2 is $O(\phi_n^2) << \phi_n^4$. So, for n sufficiently large, local extremum of $|g_n(j\omega)|$ occur for $\omega << \phi_n^2$. As $\omega \to \infty$, $|g_n(j\omega)| \to \frac{1}{4}\frac{\phi_n^4}{\alpha_n^4}$.

For $\omega < \phi_n^2$, we have

$$|g_n(j\omega)| = \left| 1 + \frac{\omega^2\left(\frac{1}{4\alpha_n^4} + \frac{1}{\phi_n^4}\right)}{1 - \frac{\omega^2}{\phi_n^4 + j\omega}} \right| \leq 1 + \left| \frac{\omega}{\epsilon}\left(\frac{1}{4\alpha_n^4} + \frac{1}{\phi_n^4}\right) \right| \leq 1 + \frac{2}{\epsilon\phi_n^2}.$$

Therefore, for n sufficiently large, $||g_n||_\infty \leq 1 + \frac{2}{\epsilon\phi_n^2}$.

Note that $\left| \frac{1}{j\omega+1} \left(\frac{j\omega\left(\epsilon - \sqrt{\epsilon^2 + \frac{1}{\alpha_M^4}}\right) + \omega^2\left(\frac{1}{4\alpha_M^4} - \frac{1}{\phi_M^4}\right)}{1 + j\epsilon\omega - \frac{\omega^2}{\phi_M^4}} \right) \right|^2$

$$\leq \frac{1}{\omega^2+1} \left(\frac{\omega^2\left(\epsilon - \sqrt{\epsilon^2 + \frac{1}{\alpha_M^4}}\right)^2 + \omega^4\left(\frac{1}{4\alpha_M^4} - \frac{1}{\phi_M^4}\right)^2}{\epsilon^2\omega^2} \right)$$

$$\leq \left(1 - \sqrt{1 + \frac{1}{\epsilon^2\alpha_M^4}}\right)^2 + \frac{1}{\epsilon^2}\left(\frac{1}{4\alpha_M^4} - \frac{1}{\phi_M^4}\right)^2 \leq$$

$$\left(\frac{1}{\epsilon^2\alpha_M^4}\right)^2 + \frac{1}{\epsilon^2}\left(\frac{1}{4\alpha_M^4} + \frac{1}{\phi_M^4}\right)^2 \leq \frac{4}{\epsilon^2\alpha_M^8}$$

assuming that $\epsilon^2 \leq \frac{3}{4}$.

Then, $\left\| \frac{1}{s+1} \left(\frac{s\left(\epsilon - \sqrt{\epsilon^2 + \frac{1}{\alpha_M^4}}\right) + s^2\left(\frac{1}{\phi_M^4} - \frac{1}{4\alpha_M^4}\right)}{1 + \epsilon s + \frac{s^2}{\phi_M^4}} \right) \right\| \leq \frac{2}{\epsilon^2\alpha_M^4}.$

It follows that for $M > N$, $||H_{M,N} - H_{M-1,N}||_\infty$

$$\leq \frac{2}{\epsilon^2\alpha_M^4} \prod_{n=N}^{M-1} \left(1 + \frac{2}{\epsilon\phi_n^2}\right) \leq \frac{2}{\epsilon^2\alpha_M^4} \prod_{n=N}^{M-1} \left(1 + \frac{2}{\epsilon\alpha_n^2}\right).$$

Defining $r_M = \prod_{n=N}^{M}\left(1 + \frac{2}{\epsilon\alpha_n^2}\right)$, we have that

$$\ln(r_M) \leq \frac{2}{\epsilon}\sum_{n=N}^{M} \frac{1}{\alpha_n^2} \leq \frac{2}{\epsilon}\sum_{n=N}^{M} \frac{1}{\pi^2 n} \leq \frac{2}{\epsilon\pi^2}\left(\frac{1}{N} + \frac{1}{N^2}\right) \leq \frac{4}{\epsilon N\pi^2}.$$

Therefore, $r_M \leq e^{4/(\pi^2\epsilon N)}$. Thus for $M > N$,

$$||H_{M,N} - H_{M-1,N}||_\infty \leq \frac{2}{\epsilon^2\alpha_M^4} e^{4/(\pi^2\epsilon N)} \leq \frac{2}{\epsilon^2\pi^4 M^4} e^{4/(\pi^2\epsilon N)}.$$

It follows that for any M, M' with $M' > M > N$,

$$||H_{M,N} - H_{M',N}||_\infty \leq \frac{2}{\epsilon^2 \pi^2 M^3} e^{4/(\pi \epsilon N)} .$$

Therefore $\{H_{M,N}\}$ is Cauchy and converges to $H_{\infty,N} \in H^\infty$. □ This implies that the sequence $\frac{2}{s+1}\{H_{M,1}\}_{M=1}^{\infty}$ converges in H^∞ to P_{2o}.

To show that the Blaschke product $B(s)$ converges in the closed right half-plane we use the following result [23]: A necessary and sufficient condition for the convergence of a Blaschke product $\prod_{n=1}^{\infty} \frac{z_n - s}{z_n + s}$, where $z_n > 0$ is that $\sum_{n=1}^{\infty} \frac{z_n}{1+z_n^2} < \infty$. For $B(s)$, $z_n = 2\alpha_n^4 \left(\epsilon + \sqrt{\epsilon^2 + \frac{1}{\alpha_n^4}}\right)$. One can establish that $\frac{z_n}{1+z_n^2} < \frac{1}{z_n} < \frac{1}{2\alpha_n^2}$. Therefore $B(s)$ converges.

C. BOUNDS ON SUBOPTIMAL PERFORMANCE

1. Bounding $||\tau_N||_\infty$

Let τ_N be defined by $\frac{(s+\zeta)^2}{s^2} P_N \tau_N = P$. For $P = P_1$, τ_N is

$$\tau_{N1}(s) = \left(\frac{1}{1 + \epsilon s + \frac{s^2}{\phi_N^4}}\right) \prod_{n=N+1}^{\infty} \left(\frac{1 + \epsilon s + \frac{s^2}{\alpha_n^4}}{1 + \epsilon s + \frac{s^2}{\phi_n^4}}\right) .$$

For P_2, τ_N is

$$\tau_{N2}(s) = \left(\frac{1}{1 + \epsilon s + \frac{s^2}{\phi_N^4}}\right) \prod_{n=N+1}^{\infty} \left(\frac{1 + \epsilon s - \frac{s^2}{4\alpha_n^4}}{1 + \epsilon s + \frac{s^2}{\phi_n^4}}\right) .$$

Define the quantities

$$\tau_{N1}^a(s) = \left(\frac{1}{1 + \epsilon s + \frac{s^2}{\phi_N^4}}\right) \left(\frac{1 + \epsilon s + \frac{s^2}{\alpha_{N+1}^4}}{1 + \epsilon s + \frac{s^2}{\phi_{N+1}^4}}\right) .$$

$$\tau_{N2}^a(s) = \left(\frac{1}{1 + \epsilon s + \frac{s^2}{\phi_N^4}}\right) \left(\frac{1 + \epsilon s - \frac{s^2}{4\alpha_{N+1}^4}}{1 + \epsilon s + \frac{s^2}{\phi_{N+1}^4}}\right) .$$

Define the quantities

$$\delta_{N1} = \frac{5e^{2/N}}{2\pi^4 \epsilon N^4} \quad \text{and} \quad \delta_{N2} = \frac{5e^{4/\pi^2 \epsilon N}}{2\pi^4 \epsilon^2 N^3}.$$

Following the ideas used in the proof of Theorem A.3, one can show that for N sufficiently large, $||\tau_{N1}^a - \tau_{N1}||_\infty \leq \delta_{N1}$ and

$$1 \leq ||\tau_{N1}||_\infty \leq \frac{\phi_{N+1}^4}{\alpha_{N+1}^4} + \delta_{N1} \leq \left(1 + \frac{1}{N+1}\right)^4 + \delta_{N1}.$$

Hence $||\tau_{N1}||_\infty \to 1$ and $||\tau_{N1}^a - \tau_{N1}||_\infty \to 0$ as $N \to \infty$.

Similarly, following the ideas of the proof of Theorem A.4 one can show that for N sufficiently large, $||\tau_{N2}^a - \tau_{N2}||_\infty \leq \delta_{N2} \to 0$ as $N \to \infty$ and

$$1 \leq ||\tau_{N2}||_\infty \leq \prod_{n=N+1}^{\infty} \left(1 + \frac{2}{\epsilon \phi_n^2}\right) \leq \frac{2\pi^4 \epsilon^2 N^3}{5} \delta_{N2} \to 1 \text{ as } N \to \infty.$$

2. Bounding $\gamma(C_N)$

In the cases that $P = P_1$ or $P = P_2$, since $\left|\left|\frac{s^2}{(s+\zeta)^2}\right|\right|_\infty = 1$ and $|\tau_{Ni}(0)| = 1 \leq ||\tau_{Ni}||_\infty$ for $i = 1, 2$, it follows that $\lambda = ||\tau_{Ni}||_\infty$. Thus we have $\gamma_N \leq \gamma_{opt} \leq \gamma(C_N) \leq \frac{||\tau_N||_\infty}{1 - ||\Delta_N||_\infty} \gamma_N$, where $\gamma_N \to \gamma_{opt}$ and $||\tau_N||_\infty \to 1$ as $N \to \infty$.

Recalling $\Delta_N(j\omega)$ as defined in Section V.C, one can see that

$$|\Delta_N(j\omega)| \leq |S_N(j\omega)| \left| \frac{-\omega^2}{(j\omega + \zeta)^2} - 1 \right| + |T_N(j\omega)||\tau_N(j\omega) - 1|.$$

From the H^∞-optimality of S_N and T_N for the finite order plant P_N,

$$\gamma_N^2 = |W_1(j\omega)|^2 |S_N(j\omega)|^2 + |W_2(j\omega)|^2 |T_N(j\omega)|^2 \quad \text{for all } \omega.$$

Thus,

$$|S_N(j\omega)| = |W_1(j\omega)|^{-1} \sqrt{\gamma_N^2 - |W_2(j\omega)|^2 |T_N(j\omega)|^2} < \gamma_N |W_1(j\omega)|^{-1},$$

$$|T_N(j\omega)| = |W_2(j\omega)|^{-1}\sqrt{\gamma_N^2 - |W_1(j\omega)|^2|S_N(j\omega)|^2} < \gamma_N|W_2(j\omega)|^{-1}$$

for all ω. Therefore $||\Delta_N||_\infty$

$$< \gamma_N \sup_\omega \left(|W_1(j\omega)|^{-1} \left| \frac{-\omega^2}{(j\omega+\zeta)^2} - 1 \right| + |W_2(j\omega)|^{-1}|\tau_N(j\omega) - 1| \right).$$

Recalling that $W_1 = \sqrt{b}\frac{as+\frac{1}{b}}{as+1}$ and $W_2 = \sqrt{b}\frac{as+\frac{1}{b}}{abs+\frac{1}{b}}$ where $a > 1 >> b > 0$, and assuming that $\frac{1}{a} >> \zeta$, one can establish that

$$\sup_\omega |W_1(j\omega)|^{-1} \left| \frac{-\omega^2}{(j\omega+\zeta)^2} - 1 \right| \leq \sqrt{b}\sqrt{\frac{1+a^2\zeta^2}{1+a^2b^2\zeta^2}}.$$

Therefore, as $\zeta \to 0$, $\sup_\omega |W_1(j\omega)|^{-1} \left| \frac{-\omega^2}{(j\omega+\zeta)^2} - 1 \right| \to \sqrt{b}$.

For $i = 1, 2$, we have $|\tau_{Ni}(j\omega) - 1| \leq |\tau_{Ni}^a(jw) - 1| + \delta_{Ni}$. From this one can show that for N large, $|\tau_{Ni}(j\omega) - 1|$ is very small up to a certain frequency, say ω_N, and that as $N \to \infty$, $\omega_N \to \infty$. Further, given any large number Ω, one can find N sufficiently large so that $|\tau_{Ni}(j\omega) - 1|$ is arbitrarily small for all $\omega \in [0, \Omega]$. Hence, as $N \to \infty$,

$$\sup_\omega |W_2(j\omega)|^{-1} |\tau_{Ni}(j\omega) - 1| \to |W_2(\infty)|^{-1} = \sqrt{b}.$$

Thus,

$$\lim_{N\to\infty,\zeta\to 0} ||\Delta_N||_\infty \leq 2\gamma_{opt}\sqrt{2}.$$

This leads to the bound

$$\gamma_{opt} \leq \lim_{N\to\infty,\zeta\to 0} \gamma(C_N) \leq \frac{\gamma_{opt}}{1 - 2\gamma_{opt}\sqrt{b}}.$$

References

[1] Lenz, K., H. Özbay, A. Tannenbaum, J. Turi, B. Morton, "Frequency domain analysis and robust control design for an ideal flexible beam," *Automatica.* **27** 1991, pp.947-961.

[2] Bercovici, H., C. Foias, A. Tannenbaum, "On skew Toeplitz operators," *Operator Theory: Adv. and Appl.* **32** 1988, pp.21-43.

[3] Foias, C., A. Tannenbaum, "On the four-block problem II: the singular system" *Operator Theory and Int. Equ.* **11** 1988, pp.726-767.

[4] Özbay, H., M. C. Smith, A. Tannenbaum, "Mixed sensitivity optimization for a class of unstable infinite dimensional systems," to appear *Lin. Alg. and its Appl.*.

[5] Russel, D. L. "Mathematical models for elastic beams and their control-theoretic implications", (F. Kappel ed.), *Proc. 1984 Autumn College on Semigroups and their Appl.*, Birkhauser, Stuttgart.

[6] Bontsema, J., R. Curtain, J. Schumacher, "Robust Control of Flexible Structures: a case study," *Automatica*, **24** 1988, pp.177-186.

[7] Bontsema, J., and S. A. de Vries "Robustness of flexible structures against small time delays", *Proc. 27th IEEE Conf. Decision and Contr.*, 1988 pp.1647-1648.

[8] Lenz, K., H. Özbay, A. Tannenbaum, J. Turi, B. Morton, "Robust control design for a flexible beam using a distributed parameter H^∞ method," *Proc. 28th IEEE Conf. Decision and Contr.* 1989, pp.2673-2678.

[9] Smith, M. C., "On stabilization and existence of coprime factorizations," *IEEE Trans. Aut. Contr.*, 1989, pp.1005-1007.

[10] Youla, D. C., H. A. Jabr and J. J. Bongiorno Jr., "Modern Wiener Hopf design of optimal controllers: part II," *IEEE Trans. Aut. Contr.*, **AC-21** 1976, pp.319-338.

[11] Doyle, J. C. and G. Stein, "Multivariable feedback design: concepts for a classical/modern synthesis," *IEEE Trans. Aut. Contr.*, **26** 1981, pp.4-16.

[12] Chen, M. J. and C. A. Desoer, "Necessary and sufficient condition for robust stability of linear distributed feedback systems," *Int. J. Contr.*, **35** 1982, pp.255-267.

[13] Doyle, J., B. Francis, A. Tannenbaum, *Feedback Control Theory*, Macmillan Pub., New York, 1992.

[14] Francis, B., *A Course in H^∞ Control Theory*, Lecture Notes Contr. and Information Sci., **88**, Springer Verlag, 1987.

[15] Flamm, D. S., "Outer factor 'Absorption' for H^∞ control problems," to appear *Int. J. Robust and Nonlinear Contr.*.

[16] Francis, B., G. Zames, "On H^∞ optimal sensitivity theory for SISO feedback systems," *IEEE Trans. Aut. Contr.*, **29** 1984, pp.9-16.

[17] Zames, G. and S. K. Mitter, "A note on essential spectrum and norms of mixed Hankel-Toeplitz operators," *Systems and Contr. Letters* **10** 1988, pp.159-165.

[18] Foias, C., A. Tannenbaum, G. Zames, "Some explicit formulae for the singular values of a certain Hankel operators with factorizable symbol," *SIAM J. Math. Anal.* **19** 1988, pp.1081-1091.

[19] Özbay, H., M. C. Smith, A. Tannenbaum, "On the optimal two-block H^∞ compensators for distributed unstable plants," to appear *Proc. 1992 American Contr. Conf.* 1992.

[20] Özbay, H., "Controller reduction in the two-block H^∞ optimal design for distributed plants," *Int. J. Contr.*, **54** 1991, pp.1291-1308.

[21] Özbay, H. and A. Tannenbaum, "On the structure of suboptimal H^∞ controllers in the sensitivity minimization problem for distributed stable systems," *Automatica*, **27** 1991, pp.293-305.

[22] Rodriguez, A., M. Dahleh, "Weighted H^∞ optimization for stable infinite dimensional systems using finite dimensional techniques," *Proc. 29th IEEE Conf. Decision and Contr.* 1990, pp.1814-1819.

[23] Hoffman, K. *Banach Spaces of Analytic Functions*, Prentice Hall, Engelwood Cliffs NJ, 1962.

Simple Adaptive Controller with Applications to Wing Flutter

Rafael Livneh
Aerospace Research Center
College of Engineering and Applied Sciences
Arizona State University, Tempe, AZ 85287-8006

Gary L. Slater
Department of Aerospace Engineering and Engineering Mechanics
University of Cincinnati, Cincinnati, OH 45221-0070

I. INTRODUCTION

For any plant with a completely defined dynamics model and a fixed set of parameters, the construction of a feedback control system which will yield a stable well managed response is generally a straight forward task. The design problem becomes significantly more difficult when there is an intrinsic uncertainty about the validity of the dynamic model of the plant. Such an uncertainty may be the result of a non-random noise to the input or output, poor knowledge of some plant parameters (which might also be time dependent), and in many cases a lack of knowledge of both the structure and the order of the plant.

Examples of such "uncertain plants" which are of current interest to the aerospace and control community are large flexible space structures and the

CONTROL AND DYNAMIC SYSTEMS, VOL. 57

dynamics of flexible aircraft. In particular, for these types of flexible systems the large number of degrees of freedom preclude the possibility of exact modeling in an on-line controller. The problem is exacerbated by the difficulty of modeling accurately the dynamics of the vehicle. For example, many proposed large space structures cannot be tested prior to deployment, and might be subjected to significant structural changes during their operational life. The uncertainty in the aircraft problem is due to mass and aerodynamic changes, resulting from different wing and body store configurations. Also the large Mach number and dynamic pressure variations typical of modern aircraft cause significant changes in the aerodynamic forces. The aeroelastic parameters of the wing are thus subjected to large variations in the operational flight envelope of the aircraft.

An attractive possibility for the control of an uncertain plant is the "model reference adaptive control" (MRAC), in which the dynamics of the uncertain plant are compared and driven to a "user specified" reference model. The MRAC approach was introduced in the late 50's, and has led to the development of numerous adaptive algorithms (see e.g. Osborn *et al.* [1], Parks [2], and Butchart and Shakcloth [3]). More recent research by Landau [4], [5] utilized Popov's hyperstability theory to unify the designs and the stability proofs of numerous MRAC algorithms. To prove stability, most hyperstable MRAC require perfect modeling and a noise-free environment. Controllers which violate any of these restriction were shown to exhibit unsatisfactory performance (see Rohrs *et al* [6], Åström [7], and Slater and Livneh [8]).

These problems led to the introduction of the augmented error by Monopoi [9], and to an extensive study of the sensitivity of adaptive algorithms to the "amplitude/frequency" content of unmodeled disturbances (see Ioannou and Kokotovic [10] and Narendra and Annaswamy [11]).

A design methodology, termed "simple adaptive control" (SAC), which successfully addressed the problem of multi-input-multi-output (MIMO) systems with unmodeled dynamics, was initiated by Sobel *et al* [12]. This approach uses the concept of the "ideal trajectory" and the "command generator tracker" (CGT) developed by O'Brien and Broussard [13] to define the state error. The norm of this error is then reduced by the application of Lyapunov's direct method. The algorithm generates an adaptive input which causes the output of the plant to track the output of the reference model, provided that the plant transfer function can be made strictly positive real by some constant output feedback gain matrix. The algorithm uses only the plant output, and the plant and the reference model may be of different orders. The stability analysis of the SAC guarantees asymptotic stability only for reference model input excitation, which can be described as a state of a homogenous linear system. The algorithm is sensitive, however, to process and measurement noises. Those noises introduce a drift in the adaptive gains and lead eventually to either a physical or a numerical divergence.

The SAC algorithm was extended and simulated in a series of papers by Bar-Kana and Kaufman [14], [15] and Bar-Kana *et al.* [16]. They used the "σ - modification" of Ioannou and Kokotovic [10], [17] to address the drift in the adaptive gains, and provided simulations of the algorithm for the control of a simply supported plate, unstable helicopter dynamics, and a MIMO missile model.

A "branch-filtering" method to reduce the drift problem was introduced by Bayard *et al* [18]. They filtered the output vector $\mathbf{e}_y$ to obtain the filtered error $\widetilde{\mathbf{e}}_y$, and performed time integration on matrices of the form $[\widetilde{\mathbf{e}}_y \; \mathbf{e}_y^T]$ instead of $[\mathbf{e}_y \; \mathbf{e}_y^T]$ to generate the adaptive gains. Ih *et al.* [19], [20] also reported a successful experimental application of the SAC algorithm to the problem of a large, flexible-antenna-like structure. Both the "σ - modification" and the "branch-filtering" techniques result in ultimate boundedness rather than asymptotic stability. Also, the current form of the SAC does not fully address the problem of independent plant input/output excitations.

This paper addresses the problem of measurable and unmeasurable plant excitations. We begin the analysis in § II by adjoining a class of measurable and unmeasuarable plant excitations into the ideal trajectory. The results are general algebraic conditions for the existence of the ideal trajectory. Section III introduces the adaptive law, which is extended to include the measurable plant excitation. The stability analysis of the adaptive law utilizes Lyapunov's direct method and is summarized in Theorem 1 of § IV. Theorem 1 uses the time derivative of the Lyapunov function, which is brought into a simplified form for an "almost strictly positive real" (ASPR) plant. Section IV also addresses the question: "When is a plant ASPR?", which is discussed in Theorem 2-4. Section V concludes with a numerical simulation of the controller, applied to the 23rd-order wing flutter model of Abel [21]. The numerical results are in good agreement with the theoretical analysis and demonstrate the requirement for minimum phase plant, and the effects of the unmeasurable excitation and the filtering parameter on the algorithm performance. The algorithm exhibits good robustness over a wide range of variations in simulation parameters such as plant and reference model excitation, dynamic pressures on the wing, and sensor actuator configurations.

II. PROBLEM FORMULATION

Consider the plant given by

$$\dot{\mathbf{x}}_p = \mathbf{A}_p \mathbf{x}_p + \mathbf{B}_p \mathbf{u}_p + \mathbf{B}_{pv} \mathbf{v}_p + \mathbf{B}_{pz} \mathbf{z}_p \tag{1}$$

$$\mathbf{y}_p = \mathbf{C}_p \mathbf{x}_p + \mathbf{D}_p \mathbf{u}_p + \mathbf{D}_{pv} \mathbf{v}_p + \mathbf{D}_{pz} \mathbf{z}_p \tag{2}$$

$\mathbf{x_p} \in R^n$ is the plant state, $\mathbf{u_p} \in R^m$ is the adaptive input, and $\mathbf{y_p} \in R^m$ is the plant output. $\mathbf{v_p} \in R^\beta$ and $\mathbf{z_p} \in R^s$ are extraneous plant excitation functions with $\mathbf{v_p}$ corresponding to a measurable excitation, while $\mathbf{z_p}$ is unmeasurable. With only slight restrictions, $\mathbf{v_p}$ and $\mathbf{z_p}$ are confined to the class of excitations which are the solution of the homogenous linear differential equations given by

$$\dot{\mathbf{v}}_p = \mathbf{M}_p \mathbf{v}_p \tag{3}$$

$$\dot{\mathbf{z}}_p = \mathbf{N}_p \mathbf{z}_p \tag{4}$$

$\mathbf{A_p}$, $\mathbf{M_p}$, $\mathbf{N_p}$, $\mathbf{B_p}$, $\mathbf{B_{pv}}$, $\mathbf{B_{pz}}$, $\mathbf{C_p}$, $\mathbf{D_p}$, $\mathbf{D_{pv}}$, and $\mathbf{D_{pz}}$, are matrices of appropriate dimensions which are assumed constant but unknown. The reference model is given by

$$\dot{\mathbf{x}}_m = \mathbf{A}_m \mathbf{x}_m + \mathbf{B}_{mv} \mathbf{v}_m \tag{5}$$

$$\mathbf{y}_m = \mathbf{C}_m \mathbf{x}_m + \mathbf{D}_{mv} \mathbf{v}_m \tag{6}$$

The model is driven by the independent reference model excitation $\mathbf{v_m} \in R^\eta$, which is assumed to be generated by

$$\dot{\mathbf{v}}_m = \mathbf{M}_m \mathbf{v}_m \tag{7}$$

$\mathbf{x_m} \in R^\nu$ is the reference model state, $\mathbf{y_m} \in R^m$, is the plant output, and $\mathbf{A_m}$, $\mathbf{M_m}$, $\mathbf{B_{mv}}$,, $\mathbf{C_m}$, and $\mathbf{D_{mv}}$, are constant matrices of appropriate dimensions.

The plant and the reference model can be of different orders ($\nu \neq n$), but are both square matrices with an identical number of *m* inputs and outputs. Clearly, the case of unmodeled dynamics in the plant corresponds to $n > \nu$.

A. IDEAL TRAJECTORY

Since the plant and the reference model orders are different, the classical model following requirements are inapplicable. Nevertheless in many cases it is possible for the output of a higher order plant to track the model output. The purpose of the adaptive algorithm studied in this paper is to determine an appropriate adaptive input $\mathbf{u}_p(t)$ to obtain the steady-state output tracking condition

$$\lim_{t \to \infty} \mathbf{y}_p(t) = \lim_{t \to \infty} \mathbf{y}_m(t) \tag{8}$$

The plant input which achieves output tracking for all $t \geq 0$, and the plant state and output associated with it are termed the ideal trajectory and are denoted by $\mathbf{u}_p^*$, $\mathbf{x}_p^*$ and $\mathbf{y}_p^*$. The ideal trajectory was defined and studied by O'Brien and Broussard [13]. Their expressions for $\mathbf{x}_p^*$ and $\mathbf{u}_p^*$ can be written in terms of the current notation to give

$$\mathbf{x}_p^* = \mathbf{X}_{11}\,\mathbf{x}_m + \mathbf{X}_{12}\,\mathbf{v}_m + \mathbf{X}_{13}\,\mathbf{v}_p + \mathbf{X}_{14}\,\mathbf{z}_p \tag{9}$$

$$\mathbf{u}_p^* = \mathbf{K}_{xm\infty}\,\mathbf{x}_m + \mathbf{K}_{vm\infty}\,\mathbf{v}_m + \mathbf{K}_{vp\infty}\,\mathbf{v}_p + \mathbf{K}_{zp\infty}\,\mathbf{z}_p \tag{10}$$

where $\mathbf{X}_{11}$, $\mathbf{X}_{12}$, $\mathbf{X}_{13}$, $\mathbf{X}_{14}$, $\mathbf{K}_{xm\infty}$, $\mathbf{K}_{vm\infty}$, $\mathbf{K}_{vp\infty}$ and $\mathbf{K}_{zp\infty}$ are constant matrices of appropriate dimensions. The ideal trajectory satisfies, by definition, the plant Eqs. (1), (2) and the output tracking condition of Eq. (8), for all $t \geq 0$, i.e.

$$\dot{\mathbf{x}}_p^* = \mathbf{A}_p \mathbf{x}_p^* + \mathbf{B}_p \mathbf{u}_p^* + \mathbf{B}_{pv} \mathbf{v}_p + \mathbf{B}_{pz} \mathbf{z}_p \tag{11}$$

$$\mathbf{y}_p^* = \mathbf{C}_p \mathbf{x}_p^* + \mathbf{D}_p \mathbf{u}_p^* + \mathbf{D}_{pv} \mathbf{v}_p + \mathbf{D}_{pz} \mathbf{z}_p \tag{12}$$

$$\mathbf{y}_p^* = \mathbf{y}_m \tag{13}$$

Substitution of Eqs. (9), (10) into Eqs. (11), (12) enables one to express $\dot{\mathbf{x}}_p^*$ and $\mathbf{y}_p^*$ in terms of $\mathbf{x}_m, \mathbf{v}_m, \mathbf{v}_p$, and $\mathbf{z}_p$. Comparison of these to the time derivative of Eq. (9) and to the output of the reference model $\mathbf{y}_m$ yields

$$\begin{bmatrix} \mathbf{A}_p & \mathbf{B}_p \\ \mathbf{C}_p & \mathbf{D}_p \end{bmatrix} \begin{bmatrix} \mathbf{X}_{11} & \mathbf{X}_{12} & \mathbf{X}_{13} & \mathbf{X}_{14} \\ \mathbf{K}_{xm\infty} & \mathbf{K}_{vm\infty} & \mathbf{K}_{vp\infty} & \mathbf{K}_{zp\infty} \end{bmatrix} =$$

$$\begin{bmatrix} \mathbf{X}_{11}\mathbf{A}_m & (\mathbf{X}_{11}\mathbf{B}_{mv}+\mathbf{X}_{12}\mathbf{M}_m) & (\mathbf{X}_{11}\mathbf{B}_{mv}+\mathbf{X}_{12}\mathbf{M}_m) & (\mathbf{X}_{11}\mathbf{B}_{mv}+\mathbf{X}_{12}\mathbf{M}_m) \\ \mathbf{C}_m & (\mathbf{X}_{11}\mathbf{B}_{mv}+\mathbf{X}_{12}\mathbf{M}_m) & (\mathbf{X}_{11}\mathbf{B}_{mv}+\mathbf{X}_{12}\mathbf{M}_m) & (\mathbf{X}_{11}\mathbf{B}_{mv}+\mathbf{X}_{12}\mathbf{M}_m) \end{bmatrix} \tag{14}$$

The ideal trajectory will exist only if $\mathbf{X}_{11}$, $\mathbf{X}_{12}$, $\mathbf{X}_{13}$, $\mathbf{X}_{14}$, $\mathbf{K}_{xm\infty}$, $\mathbf{K}_{vm\infty}$, $\mathbf{K}_{vp\infty}$ and $\mathbf{K}_{zp\infty}$ satisfy Eq. (14). Conditions for this existence are based on some properties of the transmission zeros of the plant $(\mathbf{A}_p, \mathbf{B}_p, \mathbf{C}_p, \mathbf{D}_p)$ of Eqs. (1), (2), outlined below. Using the definition given by Davison and Wang [22], the transmission zeros of the system $(\mathbf{A}_p, \mathbf{B}_p, \mathbf{C}_p, \mathbf{D}_p)$ are the values of λ satisfying

$$\begin{vmatrix} (\mathbf{A}_p - \lambda \mathbf{I}) & \mathbf{B}_p \\ \mathbf{C}_p & \mathbf{D}_p \end{vmatrix} = 0 \tag{15}$$

Clearly if the plant $(\mathbf{A}_p, \mathbf{B}_p, \mathbf{C}_p, \mathbf{D}_p)$ has no transmission zeros at the origin

$$\begin{vmatrix} \mathbf{A_p} & \mathbf{B_p} \\ \mathbf{C_p} & \mathbf{D_p} \end{vmatrix} \neq 0 \tag{16}$$

and the matrices Ω_{11}, Ω_{12}, Ω_{21} and Ω_{22} can be defined by

$$\begin{bmatrix} \Omega_{11} & \Omega_{12} \\ \Omega_{21} & \Omega_{22} \end{bmatrix} \equiv \begin{bmatrix} \mathbf{A_p} & \mathbf{B_p} \\ \mathbf{C_p} & \mathbf{D_p} \end{bmatrix}^{-1} \tag{17}$$

in this case

$$\begin{aligned} \begin{vmatrix} (\mathbf{A_p} - \lambda \mathbf{I_n}) & \mathbf{B_p} \\ \mathbf{C_p} & \mathbf{D_p} \end{vmatrix} &= \begin{vmatrix} \mathbf{A_p} & \mathbf{B_p} \\ \mathbf{C_p} & \mathbf{D_p} \end{vmatrix} \left| \begin{bmatrix} \mathbf{I_n} & 0 \\ 0 & \mathbf{I_m} \end{bmatrix} - \lambda \begin{bmatrix} \Omega_{11} & \Omega_{12} \\ \Omega_{21} & \Omega_{22} \end{bmatrix} \begin{bmatrix} \mathbf{I_n} & 0 \\ 0 & 0 \end{bmatrix} \right| \\ &= \begin{vmatrix} \mathbf{A_p} & \mathbf{B_p} \\ \mathbf{C_p} & \mathbf{D_p} \end{vmatrix} \left| \begin{bmatrix} (\mathbf{I_n} - \lambda \Omega_{11}) & 0 \\ -\lambda \Omega_{21} & \mathbf{I_m} \end{bmatrix} \right| = \lambda^n \begin{vmatrix} \mathbf{A_p} & \mathbf{B_p} \\ \mathbf{C_p} & \mathbf{D_p} \end{vmatrix} \left| \frac{1}{\lambda} \mathbf{I_n} - \Omega_{11} \right| \end{aligned} \tag{18}$$

and the transmission zeros λ are equal to the inverse of the eigenvalues of Ω_{11}. Equations (14) and (18) lead to the ideal trajectory algebraic equations

$$\Omega_{11} \mathbf{X}_{11} \mathbf{A}_m - \mathbf{X}_{11} = -\Omega_{12} \mathbf{C}_m \tag{19}$$

$$\Omega_{11} \mathbf{X}_{12} \mathbf{M}_m - \mathbf{X}_{12} = -\Omega_{11} \mathbf{X}_{11} \mathbf{B}_{mv} - \Omega_{12} \mathbf{D}_{mv} \tag{20}$$

$$\Omega_{11} \mathbf{X}_{13} \mathbf{M}_p - \mathbf{X}_{13} = \Omega_{11} \mathbf{B}_{pv} + \Omega_{12} \mathbf{D}_{pv} \tag{21}$$

$$\Omega_{11} \mathbf{X}_{14} \mathbf{N}_p - \mathbf{X}_{14} = -\Omega_{11} \mathbf{B}_{pz} + \Omega_{12} \mathbf{D}_{pz} \tag{22}$$

$$\mathbf{K}_{\mathbf{xm}\infty} = \Omega_{21}\ \mathbf{X}_{11}\ \mathbf{A}_{\mathbf{m}} + \Omega_{22}\ \mathbf{C}_{\mathbf{m}} \tag{23}$$

$$\mathbf{K}_{\mathbf{vm}\infty} = \Omega_{21}\ \left(\mathbf{X}_{11}\ \mathbf{B}_{\mathbf{mv}} + \mathbf{X}_{12}\ \mathbf{M}_{\mathbf{m}}\right) + \Omega_{22}\ \mathbf{D}_{\mathbf{mv}} \tag{24}$$

$$\mathbf{K}_{\mathbf{vp}\infty} = \Omega_{21}\ \left(\mathbf{X}_{13}\mathbf{M}_{\mathbf{p}}\ - \mathbf{B}_{\mathbf{pv}}\right) - \Omega_{22}\ \mathbf{D}_{\mathbf{pv}} \tag{25}$$

$$\mathbf{K}_{\mathbf{zp}\infty} = \Omega_{21}\ \left(\mathbf{X}_{14}\mathbf{N}_{\mathbf{p}}\ - \mathbf{B}_{\mathbf{pz}}\right) - \Omega_{22}\ \mathbf{D}_{\mathbf{pz}} \tag{26}$$

Equations (19)-(22) are generalized discrete Lyapunov equations of the form $\mathbf{A}\ \mathbf{X}\ \mathbf{B} - \mathbf{X} = \mathbf{C}$. This equation is known to have a unique solution if and only if no eigenvalue of $\mathbf{A}$ is a reciprocal of an eigenvalue of $\mathbf{B}$ (LaSalle [23]). In other words, when the plant has no transmission zeros at the origin of the s-plane, each of Eqs. (19)–(22) will have a unique solution if and only if none of the products of any eigenvalue of Ω_{11} with any eigenvalue of $\mathbf{A}_{\mathbf{m}}, \mathbf{M}_{\mathbf{m}}, \mathbf{A}_{\mathbf{p}},$ and $\mathbf{N}_{\mathbf{p}}$ is equal to 1. Each one of the non-zero eigenvalues of Ω_{11} was shown (see Eq. (18)) to be the inverse of one of the finite transmission zeros of the plant. Hence one arrives at the conclusion that, for a plant with no transmission zeros at the origin, the ideal trajectory (corresponding to Eqs. (1)–(7)) will exist and be unique if and only if the plant does not have any transmission zeros in common with any eigenvalues of $\mathbf{A}_{\mathbf{m}}$, $\mathbf{M}_{\mathbf{m}}$, $\mathbf{A}_{\mathbf{p}}$, and $\mathbf{N}_{\mathbf{p}}$.

III. ADAPTIVE ALGORITHM

In the development of the adaptive algorithm we will assume that the ideal trajectory exists and is unique, even though none of its matrices $\left(\mathbf{X}_{11}, \mathbf{X}_{12}, \ldots, \mathbf{K}_{\mathbf{zp}\infty}\right)$ needs to be explicitly computed. Upon comparing the ideal trajectory $\left(\mathbf{u}_{\mathbf{p}}^{*}, \mathbf{x}_{\mathbf{p}}^{*}, \mathbf{y}_{\mathbf{p}}^{*}\right)$ with the actual plant trajectory $(\mathbf{u}_{\mathbf{p}}, \mathbf{x}_{\mathbf{p}}, \mathbf{y}_{\mathbf{p}})$, we can define the state error $\mathbf{e}_{\mathbf{x}}$ and the output error $\mathbf{e}_{\mathbf{y}}$ by

$$\mathbf{e_x} \equiv \mathbf{x_p}^* - \mathbf{x_p} \tag{27}$$

$$\mathbf{e_y} \equiv \mathbf{y_p}^* - \mathbf{y_p} \tag{28}$$

and restate the purpose of the algorithm, given by Eq. (8), as: generate an input $\mathbf{u_p}$ which will achieve

$$\lim_{t \to \infty} \mathbf{e_y} = \mathbf{0} \tag{29}$$

The form of the ideal input $\mathbf{u_p^*}$, given by Eq. (10), is the motivation for generating the adaptive input $\mathbf{u_p}$ by

$$\mathbf{u_p} = \mathbf{K_e}\,\mathbf{e_y} + \mathbf{K_{xm}}\,\mathbf{x_m} + \mathbf{K_{vm}}\,\mathbf{v_m} + \mathbf{K_{vp}}\,\mathbf{v_p} \tag{30}$$

where $\mathbf{K_e}, \mathbf{K_{xm}}, \mathbf{K_{vm}}$ and $\mathbf{K_{vp}}$ are time dependent gain matrices which are generated by the adaptive controller. The term $\mathbf{K_{zp}}\,\mathbf{z_p}$ does not appear in the adaptive input because $\mathbf{z_p}$ is unmeasurable. Also, the term $\mathbf{K_e}\,\mathbf{e_y}$ does not appear in the ideal input of Eq. (10), because in the ideal case $\mathbf{e_y} = \mathbf{y_m} - \mathbf{y_p^*} = \mathbf{0}$. Defining the extended state $\mathbf{r}$ by

$$\mathbf{r} \equiv \left[\mathbf{e_y^T}\ \ \mathbf{x_m}^T\ \ \mathbf{v_m}^T\ \ \mathbf{v_p}^T\right]^T \tag{31}$$

and the adaptive gain matrix $\mathbf{K}$ by

$$\mathbf{K} = \left[\ \mathbf{K_e}\ \ \ \mathbf{K_{xm}}\ \ \mathbf{K_{vm}}\ \ \mathbf{K_{vp}}\ \right] \tag{32}$$

one can write Eq. (30), for the adaptive input $\mathbf{u_p}$, as

$$\mathbf{u_p} = \mathbf{K}\,\mathbf{r} \tag{33}$$

The adaptive gains are generated by the "proportional + integral" form, suggested by Bar-Kana and Kaufman [24], [15], and are given by

$$\mathbf{K} = \mathbf{K_p} + \mathbf{K_i} \tag{34}$$

$$\mathbf{K_p} = \mathbf{e_y}\,\mathbf{r}^T\,\mathbf{T_p} \tag{35}$$

$$\dot{\mathbf{K}}_\mathbf{i} = \left[\mathbf{e_y}\,\mathbf{r}^T - \sigma\,\mathbf{K_i}\right]\mathbf{T_i} \tag{36}$$

The weight matrices $\mathbf{T_p}$, and $\mathbf{T_i}$ are positive definite and $\sigma \geq 0$ is a filtering parameter, introduced by Ioannou and Kokotovic [17], which is used to guarantee the boundedness of $\mathbf{K_i}$ in the presence of the unmeasurable excitation $\mathbf{z_p}$. Figure 1 depicts the block diagram of the plant, the reference model, the excitation dynamics and the adaptive law of Eqs. (1)-(7), (30)-(33). The ideal input $\mathbf{u_p^*}$ of Eq. (10) can be expressed in terms of the extended state vector $\mathbf{r}$ by

$$\mathbf{u_p^*} = \mathbf{K_\infty}\,\mathbf{r} - \mathbf{K_{e\infty}}\,\mathbf{e_y} + \mathbf{K_{zp\infty}}\,\mathbf{z_p} \tag{37}$$

where $$\mathbf{K_\infty} = \left[\begin{array}{cccc} \mathbf{K_{e\infty}} & \mathbf{K_{xm\infty}} & \mathbf{K_{vm\infty}} & \mathbf{K_{vp\infty}} \end{array}\right] \tag{38}$$

The reference gain matrices $\mathbf{K_{xm\infty}}, \mathbf{K_{vm\infty}}$ and $\mathbf{K_{vp\infty}}$ which achieve perfect tracking when $\mathbf{z_p} = 0$ are the unique solutions to Eqs. (23)-(26), while $\mathbf{K_{e\infty}}$ is determined from stability considerations, as shown later. The reference adaptive gain $\mathbf{K_{e\infty}}$, the adaptive gain $\mathbf{K}$ and the integral part of the adaptive gain $\mathbf{K_i}$ are next used to define the gain error $\Delta\mathbf{K}$ and the integral gain error $\Delta\mathbf{K_i}$, by

$$\Delta K \equiv K_{\infty} - K \tag{39}$$

$$\Delta K_i \equiv K_{\infty} - K_i \tag{40}$$

Definitions (39) and (40) are utilized in the stability proof of the next section.

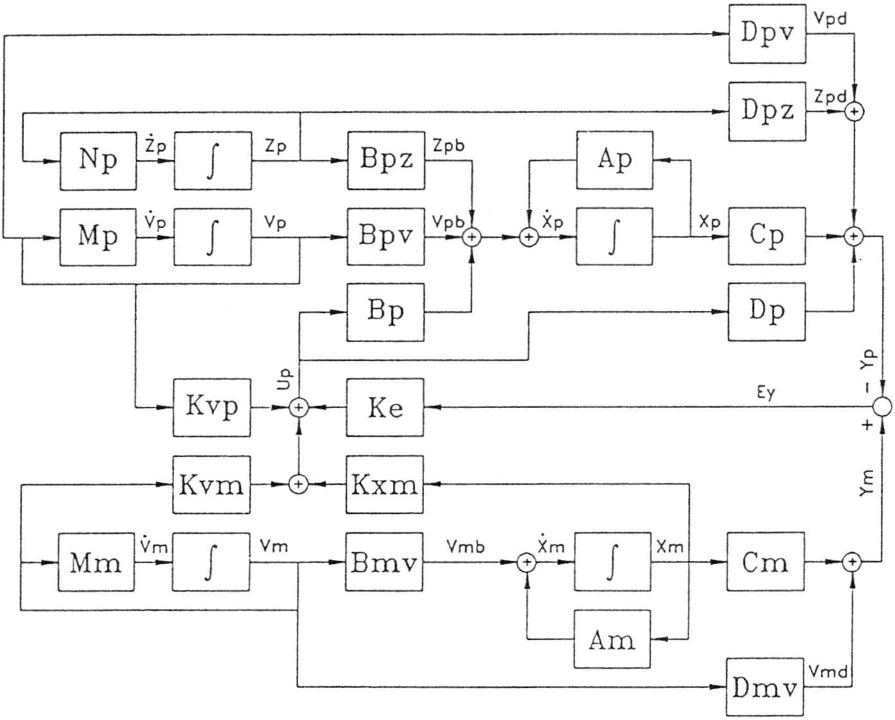

Figure 1: Block Diagram of the adaptive Controller

IV. MAIN STABILITY RESULT

The stability of the adaptive rule of Eqs. (34)-(36) is examined by Lyapunov's direct method (LaSalle and Lefschetz [25]). This method is applied using the quadratic Lyapunov function $\boldsymbol{\nu}$, defined by

$$\boldsymbol{\nu} \equiv \tfrac{1}{2}\mathbf{e}_x^T \mathbf{P}\, \mathbf{e}_x + \tfrac{1}{2}\mathrm{tr}\left(\Delta\mathbf{K}_i^T \mathbf{T}_i^{-1} \Delta\mathbf{K}_i\right) \tag{41}$$

The time derivatives of $\boldsymbol{\nu}$ involve $\dot{\mathbf{e}}_x$, $\Delta\dot{\mathbf{K}}_i$, and $\mathbf{e}_y$. These entities are derived in appendix C by manipulation of Eqs. (27), (28), (36). The results are

$$\dot{\mathbf{e}}_x = \widetilde{\mathbf{A}}_p \mathbf{e}_x + \widetilde{\mathbf{B}}_p \left(\Delta\mathbf{K}\, \mathbf{r} + \mathbf{K}_{zp\infty} \mathbf{z}_p\right) \tag{42}$$

$$\mathbf{e}_y = \widetilde{\mathbf{C}}_p \mathbf{e}_x + \widetilde{\mathbf{D}}_p \left(\Delta\mathbf{K}\, \mathbf{r} + \mathbf{K}_{zp\infty} \mathbf{z}_p\right) \tag{43}$$

$$\Delta\dot{\mathbf{K}}_i = \left(-\sigma\, \Delta\mathbf{K}_i - \mathbf{e}_y \mathbf{r}^T + \sigma\, \mathbf{K}_\infty\right) \mathbf{T}_i \tag{44}$$

The matrices $\widetilde{\mathbf{A}}_p, \widetilde{\mathbf{B}}_p, \widetilde{\mathbf{C}}_p$ and $\widetilde{\mathbf{D}}_p$ represent the closed-loop dynamics which occur when the plant $\left(\mathbf{A}_p, \mathbf{B}_p, \mathbf{C}_p, \mathbf{D}_p\right)$, is subjected to the output feedback gain $\mathbf{K}_{e\infty}$. These matrices are given by

$$\widetilde{\mathbf{A}}_p = \mathbf{A}_p - \mathbf{B}_p \mathbf{K}_{e\infty} \left(\mathbf{I} + \mathbf{D}_p \mathbf{K}_{e\infty}\right)^{-1} \mathbf{C}_p \tag{45}$$

$$\widetilde{\mathbf{B}}_p = \mathbf{B}_p \left[\mathbf{I} - \mathbf{K}_{e\infty} \left(\mathbf{I} + \mathbf{D}_p \mathbf{K}_{e\infty}\right)^{-1} \mathbf{D}_p\right] \tag{46}$$

$$\widetilde{\mathbf{C}}_p = \left(\mathbf{I} + \mathbf{D}_p \mathbf{K}_{e\infty}\right)^{-1} \mathbf{C}_p \tag{47}$$

$$\widetilde{\mathbf{D}}_p = \left(\mathbf{I} + \mathbf{D}_p \mathbf{K}_{e\infty}\right)^{-1} \mathbf{D}_p \tag{48}$$

The main assumptions for the current algorithm is that there exists an output feedback gain $\mathbf{K}_{e\infty}$ such that the closed-loop dynamics $\left(\widetilde{\mathbf{A}}_p, \widetilde{\mathbf{B}}_p, \widetilde{\mathbf{C}}_p, \widetilde{\mathbf{D}}_p\right)$ satisfy the matrix conditions for a strictly positive real transfer function. A useful form of these conditions, given by Anderson [26], is

$$\mathbf{P}\,\widetilde{\mathbf{A}}_p + \widetilde{\mathbf{A}}_p^T\,\mathbf{P} = -\mathbf{L}\,\mathbf{L}^T - 2\,\mu\,\mathbf{P} = -\mathbf{Q} < 0, \quad \mu > 0 \tag{49}$$

$$\mathbf{P}\,\widetilde{\mathbf{B}}_p - \widetilde{\mathbf{C}}_p^T = -\mathbf{L}\,\mathbf{W} \tag{50}$$

$$\widetilde{\mathbf{D}}_p + \widetilde{\mathbf{D}}_p^T = \mathbf{W}^T\,\mathbf{W} \tag{51}$$

If such a $\mathbf{K}_{e\infty}$ exists then the open-loop transfer function $\mathbf{G}(s)$, corresponding to $(\mathbf{A}_p, \mathbf{B}_p, \mathbf{C}_p, \mathbf{D}_p)$, is termed "almost strictly positive real" (ASPR) (see Bar-Kana and Kaufman [24], [15]). For an ASPR plant one can show (see Appendix – A) that

$$\begin{aligned}\dot{\boldsymbol{\nu}} \equiv &-\mu\,\mathbf{e}_x^T\,\mathbf{P}\,\mathbf{e}_x - \sigma\,\mathrm{tr}\left(\Delta\mathbf{K}_i\,\Delta\mathbf{K}_i^T\right) - \tfrac{1}{2}\mathbf{z}^T\,\mathbf{z} \\ &- \mathbf{e}_y^T\,\mathbf{e}_y\,\mathbf{r}^T\,\mathbf{T}_p\,\mathbf{r} + \mathbf{e}_y^T\,\mathbf{K}_{zp\infty}\,\mathbf{z}_p + \sigma\,\mathrm{tr}\left(\mathbf{K}_\infty\Delta\mathbf{K}_i^T\right)\end{aligned} \tag{52}$$

where $$\mathbf{z} \equiv \mathbf{L}^T\mathbf{e}_x + \mathbf{W}\left(\Delta\mathbf{K}\,\mathbf{r} + \mathbf{K}_{zp}\,\mathbf{z}_p\right) \tag{53}$$

Note that $\dot{\boldsymbol{\nu}}$ is sign-indefinite, owing to the last two terms in Eq. (52), hence in general stability cannot be assumed. However for sufficiently large values of $\mathbf{e}_x$, $\Delta\mathbf{K}_i$, and $\mathbf{e}_y$, $\dot{\boldsymbol{\nu}}$ is dominated by the first four negative quadratic terms on the right hand side of Eq. (52). Such a system was shown by LaSalle and Lefschetz [25] to be ultimately bounded; hence the state error and the adaptive gains will remain finite. If $\sigma = 0$, $\dot{\boldsymbol{\nu}}$ is negative definite for $\mathbf{z}_p = \mathbf{0}$, and sign indefinite for $\mathbf{z}_p \neq \mathbf{0}$. These cases correspond to asymptotic stability and to a likely divergence, respectively, as is demonstrated below. The stability properties of the adaptive algorithm are summarized in the following theorem.

Theorem 1

Assume that the ideal trajectory associated with Eqs. (1)–(7) exists and is unique, and that the plant is ASPR. Then the stability properties of the adaptive controller of Eqs. (34)–(36), are dependent upon the presence of the filtering gain σ and the unmeasurable plant input $\mathbf{z}_p$ according to the following Table

Table 1: Stability properties of the adaptive algorithm and their dependence on σ and $\mathbf{z}_p$

Case	σ	$\mathbf{z}_p$	stability result for $\mathbf{e}_x$, $\Delta\mathbf{K}_i$, and $\mathbf{e}_y$
1	$\sigma > 0$	bounded	ultimate boundedness
2	$\sigma = 0$	$\mathbf{z}_p \neq \mathbf{0}$	possible divergence
3	$\sigma = 0$	$\mathbf{z}_p = \mathbf{0}$	asymptotic stability

Proof

The proof is given separately for each case of the Table-1.

Case No. 1 $\sigma > 0$

The ultimate boundedness of case 1 in Table-1 is proved by using Theorem B.1 from Appendix B. We start by defining the m×r order adaptive gain vectors $\mathbf{v}_{\Delta\mathbf{K}_i}$ and $\mathbf{v}_{\Delta\mathbf{K}_\infty}$ by

$$\mathbf{v}_{\Delta\mathbf{K}_i}(m \times r) \equiv \left[\Delta K_i(1,1),\ \Delta K_i(2,1),\ \ldots\ \Delta K_i(m,r)\ \right]^T \tag{54}$$

$$\mathbf{v}_{\Delta\mathbf{K}_\infty}(m \times r) \equiv \left[\Delta K_\infty(1,1),\ \Delta K_\infty(2,1),\ \ldots\ \Delta K_\infty(m,r)\ \right]^T \tag{55}$$

Where r - the number of columns in $\Delta \mathbf{K_i}$ is given by

$$r = m + \nu + \eta + 1 \tag{56}$$

The state error vector $\mathbf{e_x}$ and the adaptive gain vector $\mathbf{v}_{\Delta K_i}$ are next used to define the $(n + m \times r)$ order phase space vector $\mathbf{x}$ by

$$\mathbf{x} = \left(\mathbf{e}_x^T,\ \mathbf{v}_{\Delta K_i}{}^T\right)^T \tag{57}$$

Equation (52) for the time derivative of the Lyapunov function can be expressed in terms of the vectors $\mathbf{e_x}$ and $\mathbf{v}_{\Delta K_i}$ by

$$\begin{aligned} \dot{\boldsymbol{\nu}} = &- \mu\, \mathbf{e}_x^T \mathbf{P}\, \mathbf{e}_x - \sigma\, \mathbf{v}_{\Delta K_i}^T\, \mathbf{v}_{\Delta K_i} - \tfrac{1}{2}\mathbf{z}^T \mathbf{z} \\ &- \mathbf{e}_y^T \mathbf{e}_y\, \mathbf{r}^T \mathbf{T}_p\, \mathbf{r} + \mathbf{e}_y^T \mathbf{K}_{zp\infty}\, \mathbf{z}_p + \sigma\, \mathbf{v}_{\Delta K_i}^T\, \mathbf{v}_{\Delta K_\infty} \end{aligned} \tag{58}$$

According to Theorem B.1 one can prove the ultimate boundedness of $\mathbf{x}$ (and consequently the ultimate boundedness of $\mathbf{e_x}$ and $\mathbf{v}_{\Delta K_i}$) if $\dot{\boldsymbol{\nu}}$ satisfies

$$\dot{\boldsymbol{\nu}}(\mathbf{x}) < -|\varepsilon\ | < 0 \tag{59}$$

for all $\mathbf{x}$ outside some closed and bounded set M. Sufficient conditions for inequality (59) are

$$\mathbf{v}_{\Delta K_i}^T\, \mathbf{v}_{\Delta K_i} \geq \mathbf{v}_{\Delta K_\infty}^T\, \mathbf{v}_{\Delta K_i} \tag{60}$$

and

$$\mu\, \mathbf{e}_x^T \mathbf{P}\, \mathbf{e}_x + \mathbf{e}_y^T \mathbf{e}_y\, \mathbf{r}^T \mathbf{T}_p\, \mathbf{r} \geq \mathbf{e}_y^T \mathbf{K}_{zp\infty}\, \mathbf{z}_p \tag{61}$$

Inequality (60) will be satisfied whenever

$$\| v_{\Delta K i} \|_2 \geq \| v_{\Delta K\infty} \|_2 \tag{62}$$

To determine a closed and bounded region in the phase plane outside which inequality (61) is satisfied, consider the two possible cases

$$\mathbf{e}_y^T \mathbf{e}_y \mathbf{r}^T \mathbf{T}_p \mathbf{r} \geq \mathbf{e}_y^T \mathbf{K}_{zp\infty} \mathbf{z}_p \tag{63}$$

or

$$\mathbf{e}_y^T \mathbf{e}_y \mathbf{r}^T \mathbf{T}_p \mathbf{r} < \mathbf{e}_y^T \mathbf{K}_{zp\infty} \mathbf{z}_p \tag{64}$$

We analyze these two cases separately.

Analysis of Inequality (61) combined with inequality (63)

To find the lower bound of $\| \mathbf{e}_x \|$ which satisfies inequality (63), denote by λ_1 the smallest eigenvalue of the symmetric positive definite matrix $\mathbf{T}_p$ and bound the left hand side of inequality (63) from below by

$$\mathbf{e}_y^T \mathbf{e}_y \mathbf{r}^T \mathbf{T}_p \mathbf{r} \geq \mathbf{e}_y^T \mathbf{e}_y \mathbf{e}_y^T \lambda_1 \mathbf{e}_y = \lambda_1 \| \mathbf{e}_y \|_2^4 \tag{65}$$

and the right hand-side of this inequality from above by

$$\mathbf{e}_y^T \mathbf{K}_{zp\infty} \mathbf{z}_p \leq \| \mathbf{e}_y^T \mathbf{K}_{zp\infty} \mathbf{z}_p \|_2 \leq \| \mathbf{e}_y^T \|_2 \ \| \mathbf{K}_{zp\infty} \mathbf{z}_p \|_2 \tag{66}$$

Inequalities (65), (66) imply that a sufficient condition for inequality (63) is

$$\lambda_1 \| \mathbf{e}_y \|_2^4 \geq \| \mathbf{e}_y^T \|_2 \| \mathbf{K}_{zp\infty} \mathbf{z}_p \|_2 \tag{67}$$

Inequality (67) will be satisfied whenever

$$\left\| e_y^T \right\|_2 \geq \left(\frac{\left\| K_{zp\infty} \; z_p \right\|_2}{\lambda_1} \right)^{1/3} \tag{68}$$

Upon satisfaction of inequality (68) we are guaranteed the negativity of $\dot{\nu}$ for all **x** outside the closed and bounded set

$$M_1 = \left\{ \left(e_x, \; v_{\Delta K_i} \right) : \left\| e_x \right\| = 0, \left\| v_{\Delta K_i} \right\| \leq \left\| v_{\Delta K_\infty} \right\| \right\} \tag{69}$$

The closed and bounded set M_1 of Eq. (69) is based on inequality (63). We next derive the closed and bounded set M_2 outside of which one can satisfy inequality (64).

Analysis of inequality (61) combined with inequality (64)

We start the analysis of this case by proving the following lemma

Lemma 1 Inequality (64) implies

$$\left\| e_y \right\| < \left(\frac{\left\| K_{zp\infty} \; z_p \right\|}{\lambda_1} \right)^{1/3} \tag{70}$$

Proof (by contradiction): Assume that inequality (64) is valid and that inequality (70) is not. If inequality (70) is not valid then

$$\left\| e_y \right\| \geq \left(\frac{\left\| K_{zp\infty} \; z_p \right\|}{\lambda_1} \right)^{1/3} \tag{71}$$

But inequality (71) is identical to inequality (68) which implies inequality (63) and thus contradicts inequality (64). Q.E.D.

If inequality (64) is valid, then a sufficient for inequality (61) is

$$\mu\, \mathbf{e}_x^T \mathbf{P}\, \mathbf{e}_x \geq \mathbf{e}_y^T \mathbf{K}_{zp\infty}\, \mathbf{z}_p \tag{72}$$

Denoting by λ_2 the smallest eigenvalue of the symmetric positive definite matrix $\mathbf{P}$ one can bound from below the left hand-side of inequality (72) by

$$\mu\, \mathbf{e}_x^T \mathbf{P}\, \mathbf{e}_x \geq \mu\, \mathbf{e}_x^T \lambda_2\, \mathbf{e}_x = \mu\, \lambda_2 \| \mathbf{e}_x \|_2^2 \tag{73}$$

The right-hand side of inequality (72) can be bounded from above by

$$\begin{aligned} \mathbf{e}_y^T \mathbf{K}_{zp\infty}\, \mathbf{z}_p &\leq \| \mathbf{e}_y^T \mathbf{K}_{zp\infty}\, \mathbf{z}_p \|_2 \leq \| \mathbf{e}_y^T \|_2 \| \mathbf{K}_{zp\infty}\, \mathbf{z}_p \|_2 \\ &\leq \left(\frac{\| \mathbf{K}_{zp\infty}\, \mathbf{z}_p \|_2}{\lambda_1} \right)^{1/3} \| \mathbf{K}_{zp\infty}\, \mathbf{z}_p \|_2 \leq \left(\frac{\| \mathbf{K}_{zp\infty}\, \mathbf{z}_p \|_2^4}{\lambda_1} \right)^{1/3} \end{aligned} \tag{74}$$

Inequalities (73), (74) imply that a sufficient condition for inequality (72) is

$$\mu\, \lambda_2 \| \mathbf{e}_x \|_2^2 \geq \left(\frac{\| \mathbf{K}_{zp\infty}\, \mathbf{z}_p \|_2^4}{\lambda_1} \right)^{1/3} \tag{75}$$

which will be satisfied whenever

$$\| \mathbf{e}_x \|_2^2 \geq \frac{\| \mathbf{K}_{zp\infty}\, \mathbf{z}_p \|_2^{2/3}}{\mu^{1/2}\, \lambda_1^{1/6}\, \lambda_2^{1/2}} \tag{76}$$

The above derivation means that upon satisfaction of inequality (64) one can guarantee the negativity of $\dot{\mathcal{V}}$ for all $\mathbf{x}$ outside the closed and bounded set

$$M_2 = \left\{ (\mathbf{e}_x, \mathbf{v}_{\Delta K_i}) : \ \| \mathbf{e}_x \| \le \frac{\| \mathbf{K}_{zp\infty} \mathbf{z}_p \|_2^{2/3}}{\mu^{1/2} \lambda_1^{1/6} \lambda_2^{1/2}}, \ \| \mathbf{v}_{\Delta K_i} \| \le \| \mathbf{v}_{\Delta K_\infty} \| \right\} \tag{77}$$

Since the set M_1 includes the set M_2 of Eq. (69), we arrive at the conclusion that inequality (59) can be satisfied for any $\mathbf{x}$ outside the closed and bounded set M_2 of Eq. (77). Since all the conditions of Theorem B.1 have been met, the ultimate boundedness of $\mathbf{x}$ and consequently of $\Delta\mathbf{K}_i$, $\mathbf{e}_x$ and $\mathbf{e}_y$ is thus established. This completes the boundedness proof of Case 1 in Table 1.

Case 2 $\sigma = 0$

Setting $\sigma = 0$ in Eq. (52) results in

$$\dot{\nu} = -\mu\, \mathbf{e}_x^T \mathbf{P}\, \mathbf{e}_x - \tfrac{1}{2}\mathbf{z}^T \mathbf{z} - \mathbf{e}_y^T \mathbf{e}_y\, \mathbf{r}^T \mathbf{T}_p \mathbf{r} + \mathbf{e}_y^T \mathbf{K}_{zp_\infty} \mathbf{z}_p \tag{78}$$

The elimination of the quadratic term $-\sigma \operatorname{tr}\left(\Delta\mathbf{K}_i \Delta\mathbf{K}_i^T\right)$ from $\dot{\nu}$ means that $\dot{\nu}$ can remain sign-indefinite even for large values of the entries of $\Delta\mathbf{K}_i$. For example, if $\mathbf{W} = \mathbf{0}$ (no feed forward), and $\dot{\nu}$ is positive for some $\mathbf{e}_x$ then

$$\dot{\nu} = \mathbf{e}_x^T \left(-\mu\, \mathbf{P}\, \mathbf{e}_x - \tfrac{1}{2}\mathbf{L}\,\mathbf{L}^T \mathbf{e}_x - \mathbf{C}_p^T \mathbf{C}_p\, \mathbf{e}_x\, \mathbf{r}^T \mathbf{T}_p \mathbf{r} + \mathbf{C}_p^T \mathbf{K}_{zp_\infty} \mathbf{z}_p \right) > 0 \tag{79}$$

and no increase in $\Delta\mathbf{K}_i$ is going to make it negative. In this case one can not guarantee the ultimate boundedness of $\mathbf{e}_x$ and $\Delta\mathbf{K}_i$.

Case 3 $\sigma = 0$ $\mathbf{z}_p = \mathbf{0}$

Substitution of $\sigma = 0$ and $\mathbf{z}_p = \mathbf{0}$ in Eq. (52) yields

$$\dot{\nu} = -\mu\, \mathbf{e}_x^T \mathbf{P}\, \mathbf{e}_x - \tfrac{1}{2}\mathbf{z}^T \mathbf{z} - \mathbf{e}_y^T \mathbf{e}_y\, \mathbf{r}^T \mathbf{T}_p \mathbf{r} \tag{80}$$

$\dot{\boldsymbol{\nu}}$ is clearly negative definite, while $\boldsymbol{\nu}$ is autonomous and positive definite. Hence this case results in asymptotic stability. Q.E.D.

The main assumption underlying Theorem 1 is that the transfer function $\mathbf{G}(s)$ of the plant $(\mathbf{A}_p, \mathbf{B}_p, \mathbf{C}_p, \mathbf{D}_p)$ is ASPR. Some conditions for $\mathbf{G}(s)$ to be ASPR are given here, to supplement Theorem 1.

Theorem 2

A necessary condition for $(\mathbf{A}_p, \mathbf{B}_p, \mathbf{C}_p, \mathbf{D}_p)$ to be ASPR is that it is minimum phase.

Proof (by contradiction)

Suppose that the transfer function $\mathbf{G}(s)$ corresponding to $(\mathbf{A}_p, \mathbf{B}_p, \mathbf{C}_p, \mathbf{D}_p)$ is almost strictly positive real and at the same time has one or more transmission zeros with positive real part. Then there exists an output feedback gain matrix $\mathbf{K}_{e_\infty}$ such that the closed loop transfer function

$$\widetilde{\mathbf{G}}(s) \equiv \widetilde{\mathbf{C}}_p \left(s\mathbf{I} - \widetilde{\mathbf{A}}_p\right)^{-1} \widetilde{\mathbf{B}}_p + \widetilde{\mathbf{D}}_p \tag{81}$$

is strictly positive real. The strict positive realness of $\widetilde{\mathbf{G}}(s)$ implies the strict positive realness of $\widetilde{\mathbf{G}}(s)^{-1}$ as was shown by Newcomb ([27] - Theorem 5.8 and Lemma 5.3). The characteristic polynomial $\widetilde{P}(s)$ of the transfer function $\widetilde{\mathbf{G}}(s)$ is given by

$$\widetilde{P}(s) = \left|s\mathbf{I} - \widetilde{\mathbf{A}}_p\right| \tag{82}$$

and the zero polynomial $\widetilde{Z}(s)$ of the transfer function $\widetilde{\mathbf{G}}(s)$ is given by

$$\widetilde{Z}(s) = \begin{vmatrix} (\widetilde{\mathbf{A}}_p - s\,\mathbf{I}) & \widetilde{\mathbf{B}}_p \\ \widetilde{\mathbf{C}}_p & \widetilde{\mathbf{D}}_p \end{vmatrix} = \left| s\,\mathbf{I} - \widetilde{\mathbf{A}}_p \right| \left| \widetilde{\mathbf{C}}_p \left(s\,\mathbf{I} - \widetilde{\mathbf{A}}_p \right)^{-1} \widetilde{\mathbf{B}}_p + \widetilde{\mathbf{D}}_p \right| \quad (83)$$

Manipulation of Eqs. (82), (83) yields

$$\left| \widetilde{\mathbf{G}}(s) \right| = \left| \widetilde{\mathbf{C}}_p \left((s\,\mathbf{I} - \widetilde{\mathbf{A}}_p \right)^{-1} \widetilde{\mathbf{B}}_p + \widetilde{\mathbf{D}}_p \right| = \frac{\widetilde{Z}(s)}{\widetilde{P}(s)} \quad (84)$$

The invariance of the transmission zeros under output feedback implies that the open loop zero polynomial Z(s) is related to the closed loop zero polynomial $\widetilde{Z}(s)$ by

$$\widetilde{Z}(s) = c\, Z(s) \quad (85)$$

where c is a scalar constant. By substitution of Eq. (85) into Eq. (84) one can obtain

$$\left| \left[\widetilde{\mathbf{G}}(s) \right]^{-1} \right| = \frac{1}{\left| \widetilde{\mathbf{G}}(s) \right|} = \frac{\widetilde{P}(s)}{c\, Z(s)} \quad (86)$$

Equation (86) demonstrates that the poles of $\widetilde{\mathbf{G}}(s)^{-1}$ are also the transmission zeros of $\mathbf{G}(s)$. The assumption that $\mathbf{G}(s)$ has some unstable transmission zeros has led to the conclusion that $\widetilde{\mathbf{G}}(s)^{-1}$ is strictly positive real with some unstable poles, which is the required contradiction. Q.E.D.

Theorem 2 provides a necessary condition for a system to be ASPR. Two sufficient conditions for a system to be ASPR, which are naturally more restrictive than Theorem 2, are given by Theorems 3 and 4.

Theorem 3

Consider the open-loop plant $(\mathbf{A}_p,\mathbf{B}_p,\mathbf{C}_p,\mathbf{D}_{p1})$. If the system $(\mathbf{A}_p,\mathbf{B}_p,\mathbf{C}_p,\mathbf{0})$ is output-stabilizable by some positive definite output feedback gain $\mathbf{K}_0$, then augmenting the plant $(\mathbf{A}_p,\mathbf{B}_p,\mathbf{C}_p,\mathbf{D}_{p1})$ with a feedforward gain

$$\mathbf{D}_{p2} = \mathbf{K}_0^{-1} - \mathbf{D}_{p1} \tag{87}$$

results in a "feedforward augmented plant" $\left(\mathbf{A}_p,\mathbf{B}_p,\mathbf{C}_p,\mathbf{D}_p = \mathbf{D}_{p1} + \mathbf{D}_{p2} = \mathbf{K}_0^{-1}\right)$ which is ASPR.

Proof

To show that the system $\left(\mathbf{A}_p,\mathbf{B}_p,\mathbf{C}_p,\mathbf{D}_p = \mathbf{D}_{p1} + \mathbf{D}_{p2} = \mathbf{K}_0^{-1}\right)$ is ASPR, it should be shown that there exists an output feedback gain matrix $\mathbf{K}_{e_\infty}$ such that the closed loop transfer function $\widetilde{\mathbf{G}}(s)$ of Eq. (81) is strictly positive real (SPR). Consider applying output feedback with

$$\mathbf{K}_{e_\infty} = k\,\mathbf{K}_0 \tag{88}$$

where k is some positive finite scalar. Figure 2 depicts the block diagram resulting from Eqs. (87), (88) for the feedforward and the feedback gains. Substitution of the particular choice of $\mathbf{K}_{e_\infty}$ of Eq. (88) into Eqs. (45)-(48) yields

$$\widetilde{\mathbf{A}}_p = \mathbf{A}_p - \frac{k}{k+1}\mathbf{B}_p\,\mathbf{K}_0\,\mathbf{C}_p \tag{89}$$

$$\widetilde{\mathbf{B}}_p = \frac{1}{k+1}\mathbf{B}_p \tag{90}$$

$$\widetilde{\mathbf{C}}_p = \frac{1}{k+1}\mathbf{C}_p \tag{91}$$

$$\widetilde{D}_p = \frac{1}{k+1} K_0^{-1} \tag{92}$$

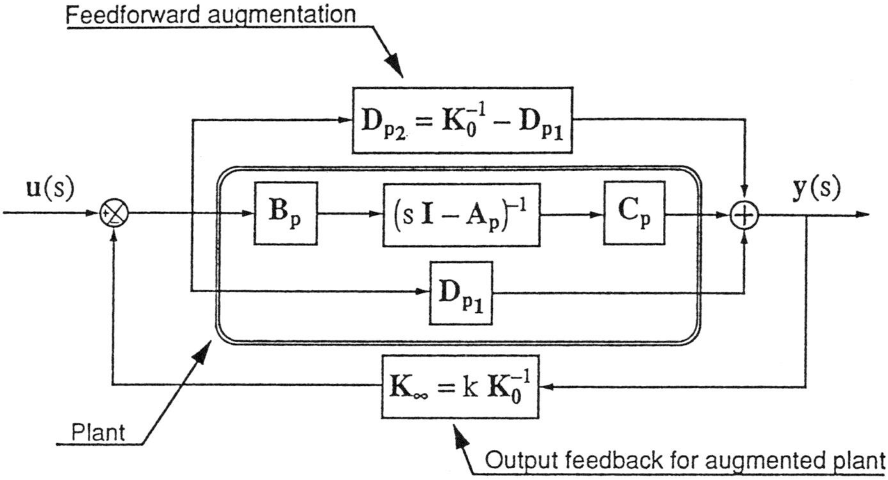

Figure 2: Feedforward augmented plant with output feedback

Equations (89) - (92) can be combined to get

$$\widetilde{G}(s) = \frac{1}{k+1}\left(K_0^{-1} + \frac{1}{k+1} F(s)\right) \tag{93}$$

where $$F(s) \equiv C_p \left(sI - \widetilde{A}_p\right)^{-1} B_p \tag{94}$$

To show that there exists a k such that $\widetilde{G}(s)$ of Eq. (93) is SPR, $\widetilde{G}(s)$ should be analytic for all $\Re(s) \geq 0$, and its hermitian part defined by

$$\widetilde{G}_H(s) \equiv \frac{1}{2}\left[\widetilde{G}^H(s) + \widetilde{G}(s)\right] \tag{95}$$

should satisfy $\quad \widetilde{\mathbf{G}}_H(s) > 0 \quad$ for all $\quad \Re(s) \geq 0$ (96)

Since $\mathbf{K_0}$ stabilizes the system then Eq. (89) implies that there exists a finite positive L_1 such that for any $k > L_1$ $\widetilde{\mathbf{A}}_\mathbf{p}$ is stable. This means that for any $k > L_1$ $\mathbf{F}(s)$ is analytic throughout $\Re(s) \geq 0$.

Substitution of Eq. (93) into the definition (95) of $\widetilde{\mathbf{G}}_H(s)$ yields

$$\widetilde{\mathbf{G}}_H(s) = \frac{1}{k+1}\left[\left(\mathbf{K}_0^{-1}\right)_H + \mathbf{F}_H(s)\right] \tag{97}$$

where
$$\mathbf{F}_H(s) \equiv \frac{1}{2}\left[\mathbf{F}^H(s) + \mathbf{F}(s)\right] \tag{98}$$

and
$$\left(\mathbf{K}_0^{-1}\right)_H \equiv \frac{1}{2}\left[\left(\mathbf{K}_0^{-1}\right)^H + \mathbf{K}_0^{-1}\right] \tag{99}$$

Examination of Eq. (94) reveals that for any $k > L_1 > 0$ $\mathbf{F}(s)$ is defined throughout $\Re(s) \geq 0$, is finite for finite s, and satisfies

$$\lim_{\substack{s \to \infty \\ \infty > k > M_1 > 0}} \mathbf{F}(s) = \mathbf{0} \tag{100}$$

These observations are also valid for $\mathbf{F}^H(s)$ and $\mathbf{F}_H(s)$. In other words, all the entries of $\mathbf{F}(s)$, $\mathbf{F}^H(s)$ and $\mathbf{F}_H(s)$, and consequently all the eigenvalues of these matrices, have finite lower and upper bounds in the domain $s \geq 0$ for any finite $k > L_1 > 0$. To abbreviate the rest of the proof it is assumed that $k > L_1 > 0$, and that any function of s is evaluated only in the closed right s-plane $(s \geq 0)$.

Note that for any hermitian matrix $\mathbf{S}$ $(\mathbf{S} = \mathbf{S}^H)$ and any hermitian positive definite $\mathbf{T}$ $(\mathbf{T} = \mathbf{T}^H > \mathbf{0})$, there exists a non-singular transformation $\mathbf{P}$ such that $\mathbf{P}^H\ \mathbf{S}\ \mathbf{P} = \Lambda$ and $\mathbf{P}^H\ \mathbf{T}\ \mathbf{P} = \mathbf{I}$, where Λ is a real diagonal matrix (see e.g.

Noble and Daniel [28], pp. 428). Therefore since $\left(\mathbf{K}_0^{-1}\right)_H$ is positive definite hermitian, and $\mathbf{F}_H(s)$ is hermitian and bounded, one can write Eq. (97) as

$$\widetilde{\mathbf{G}}_H(s) = \frac{1}{k+1}\left[\mathbf{P}^H(s)\right]^{-1}\left[\mathbf{I} + \frac{1}{k+1}\Lambda_H(s)\right]\left[\mathbf{P}(s)\right]^{-1} \tag{101}$$

where $\mathbf{P}(s)$ is a bounded non-singular matrix, and $\Lambda_H(s)$ is a real diagonal matrix, given by

$$\Lambda_H(s) = \mathbf{P}^H(s)\,\mathbf{F}_H(s)\,\mathbf{P}(s) \tag{102}$$

Clearly, all the eigenvalues of $\Lambda_H(s)$ are bounded. Let L_2 and L_3 be two finite scalars satisfying

$$L_2 < \Lambda_{H_{ii}}(s) < L_3, \qquad i = 1,2, \quad \ldots, m \tag{103}$$

and define L_4 by

$$L_4 = \max\left(|L_2|, |L_3|\right) \tag{104}$$

Then choosing k so that

$$L_4 > \max\left[L_1, \left(L_4 - 1\right)\right] \tag{105}$$

guarantees that all the eigenvalues of $\widetilde{\mathbf{A}}_p$ are stable and that all the eigenvalues of $\widetilde{\mathbf{G}}_H(s)$ are real and positive. This means that $\widetilde{\mathbf{G}}_H(s)$ is both analytic and positive definite in $\Re(s) \geq 0$, which means that $\widetilde{\mathbf{G}}(s)$ to be strictly positive real. Q.E.D.

Two special cases of Theorem 2 are: a) system whose feedforward can not be augmented and b) system with no feedforward. The conditions for such systems to be ASPR are given in the following corollaries

Corollary 1

Consider the open-loop plant $(\mathbf{A}_p,\mathbf{B}_p,\mathbf{C}_p,\mathbf{D}_{p1})$. If $\mathbf{D}_{p1} > \mathbf{0}$ and the output feedback gain $\mathbf{K}_0 = \mathbf{D}_{p1}^{-1}$ stabilizes $(\mathbf{A}_p,\mathbf{B}_p,\mathbf{C}_p,\mathbf{0})$, then the system $(\mathbf{A}_p,\mathbf{B}_p,\mathbf{C}_p,\mathbf{D}_{p1})$ is ASPR.

Proof

This is the special case $\mathbf{D}_{p2} = \mathbf{0}$ in Theorem 3. Q.E.D.

Corollary 2

If the plant $(\mathbf{A}_p,\mathbf{B}_p,\mathbf{C}_p,\mathbf{0})$ can be stabilized by an output feedback gain of the form $\mathbf{K}_0 = \lim_{r \to \infty} r\,\mathbf{K}_0^*$ for some $\mathbf{K}_0^* > \mathbf{0}$ then it is ASPR.

Proof

This case is a result of choosing $\mathbf{D}_{p2} = -\mathbf{D}_{p1}$ in Theorem 2 to get an ASPR $(\mathbf{A}_p, \mathbf{B}_p, \mathbf{C}_p, \mathbf{D}_p = \mathbf{D}_{p1} + \mathbf{D}_{p1} = \mathbf{0})$. Q.E.D.

Theorem 2 demonstrates the relation between the almost strict positive realness and the output stabilizability of a given system. The drawbacks of this approach are the lack of a unified criterion to check output stabilizability, and the need for a feedforward augmentation. The first drawback can be alleviated by using an algebraic approach which relates the almost strict positive realness of a system to the steady state Riccati equation and to the existence conditions for its positive definite symmetric solution **P**. The algebraic approach consists of augmenting $\mathbf{D}_{p1}$, and then adjusting $\mathbf{C}_p$ to achieve an ASPR system. This approach is summarized in the following theorem.

Theorem 4

Consider a plant with $\mathbf{A}_p$, $\mathbf{B}_p$, and $\mathbf{D}_{p1}$ given, and assume that $(\mathbf{A}_p, \mathbf{B}_p)$ is controllable. If one augments $\mathbf{D}_{p1}$ by

$$\mathbf{D}_{p2} = k\,\mathbf{K}_{e\infty}^{-1} - \mathbf{D}_{p1} \tag{106}$$

then for any positive definite output feedback gain $\mathbf{K}_{e\infty} > \mathbf{0}$, an output matrix $\mathbf{C}_p$ can be chosen such that the open-loop transfer function corresponding to $\left(\mathbf{A}_p, \mathbf{B}_p, \mathbf{C}_p, \mathbf{D}_p = \mathbf{D}_{p1} + \mathbf{D}_{p2} = k\,\mathbf{K}_{e\infty}^{-1}\right)$ is almost strictly positive real, and the closed-loop transfer function corresponding to $\left(\widetilde{\mathbf{A}}_p, \widetilde{\mathbf{B}}_p, \widetilde{\mathbf{C}}_p, \widetilde{\mathbf{D}}_p\right)$ of Eqs. (45)-(48), (89)-(92) is strictly positive real. The output matrix $\mathbf{C}_p$ is given by

$$\widetilde{\mathbf{C}}_p = \left(\mathbf{I} + \mathbf{D}_p\,\mathbf{K}_{e\infty}\right)\left\{\left[\left(\widetilde{\mathbf{D}}_p + \widetilde{\mathbf{D}}_p^T\right)\mathbf{K}_{e\infty}^T\,\mathbf{B}_p^T + \widetilde{\mathbf{B}}_p^T\right]\mathbf{P} - \mathbf{W}^T\mathbf{H}\right\} \tag{107}$$

where $\mathbf{H}$ is an arbitrary matrix chosen such that $(\mathbf{A}_p, \mathbf{H})$ is observable, and $\mathbf{P}$ is the symmetric positive definite solution to the steady-state Riccati equation

$$\mathbf{H}^T\mathbf{H} - \mathbf{P}\,\mathbf{B}_p\left(\mathbf{K}_{e\infty} + \mathbf{K}_{e\infty}^T\right)\mathbf{B}_p^T\,\mathbf{P} + \mathbf{A}_p^{\,T}\,\mathbf{P} + \mathbf{P}\,\mathbf{A}_p = \mathbf{0} \tag{108}$$

Proof

As is shown in Appendix D, the proof is by direct substitution of the closed loop dynamics, Eqs. (45)–(48), into the matrix conditions for strict positive realness, Eqs. (49)–(51). Substitution of Eq. (45) into Eq. (50) and use of Eqs. (50), (51), (45) yields Eq. (108) where

$$\mathbf{H} = \mathbf{W}\,\mathbf{K}_{e\infty}^{T}\,\mathbf{B}_p^T\,\mathbf{P} - \mathbf{L}^T \tag{109}$$

It is shown by Kwakernaak and Sivan [29] that if $\mathbf{K}_{e\infty} > \mathbf{0}$, $(\mathbf{A}_p, \mathbf{B}_p)$ is controllable, and $(\mathbf{A}_p, \mathbf{H})$ is observable, then there exists a unique symmetric positive definite solution $\mathbf{P}$ to the steady state Riccati equation (108). Equation (107) for $\mathbf{C}_p$ is obtained by pre-multiplication of Eq. (109) by $\mathbf{W}^T$ and use of Eqs. (50), (51), (91) for $\mathbf{W}^T\mathbf{W}$, $-\mathbf{L}\mathbf{W}$ and $\widetilde{\mathbf{C}}_p$, respectively. Q.E.D.

Theorem 4 lends itself to the following interpretation. Consider the case when $(\mathbf{A}_p, \mathbf{B}_p)$ is controllable and the feedforward $\mathbf{D}_{p1}$ is augmented by $\mathbf{D}_{p2}$ of Eq. (106). Then Eqs. (107), (108) can be viewed as a non-linear operator Σ which maps the set of all $\mathbf{K}_{e\infty} > \mathbf{0}$, and the set of all $\mathbf{H}$ such that $(\mathbf{A}_p, \mathbf{H})$ is observable into some set Γ which is a subset of all (m,n) real matrices ($\Gamma^{m,n} \subset \Re^{m,n}$). If $\mathbf{C}_p$ belongs to the range Γ of the operator Σ, then the open loop system $(\mathbf{A}_p, \mathbf{B}_p, \mathbf{C}_p, \mathbf{D}_p = \mathbf{D}_{p1} + \mathbf{D}_{p2} = k\,\mathbf{K}_{e\infty}^{-1})$ is ASPR, i.e. it would result in a SPR closed loop system $(\widetilde{\mathbf{A}}_p, \widetilde{\mathbf{B}}_p, \widetilde{\mathbf{C}}_p, \widetilde{\mathbf{D}}_p)$ upon application of some finite feedback gain $\mathbf{K}_{e\infty} > \mathbf{0}$.

Significant simplification of the formulation of Theorem 4 occurs when one chooses k=0 in Eq. (106) to obtain $\mathbf{D}_p = \mathbf{D}_{p1} + \mathbf{D}_{p2} = \mathbf{0}$. In this case $\widetilde{\mathbf{D}}_p = \mathbf{D}_p = \mathbf{0}$, $\mathbf{W}^T = \mathbf{0}$, $\mathbf{H} = \mathbf{L}^T$, $\widetilde{\mathbf{A}}_p = \mathbf{A}_p - \mathbf{B}_p\,\mathbf{K}_{e\infty}\,\mathbf{C}_p$, $\widetilde{\mathbf{B}}_p = \mathbf{B}_p$ and $\widetilde{\mathbf{C}}_p = \mathbf{C}_p$ which reduces Eqs. (107), (108) to the form

$$\mathbf{L}\,\mathbf{L}^T - \mathbf{P}\,\mathbf{B}_p\left(\mathbf{K}_{e\infty} + \mathbf{K}_{e\infty}^T\right)\mathbf{B}_p^T\,\mathbf{P} + \mathbf{A}_p^T\,\mathbf{P} + \mathbf{P}\,\mathbf{A}_p = \mathbf{0} \tag{110}$$

$$\mathbf{C}_p = \mathbf{B}_p^T\,\mathbf{P} \tag{111}$$

For this case Theorem 4 states that if $(\mathbf{Ap}, \mathbf{Bp})$ is controllable and $\mathbf{D}_p = \mathbf{D}_{p1} + \mathbf{D}_{p2} = \mathbf{0}$, then for any $\mathbf{K}_{e\infty} > \mathbf{0}$ and any $\mathbf{L}$ chosen such that

$(\mathbf{A}_p, \mathbf{L}^T)$ is observable , one can use the symmetric positive definite solution of Eq. (110) to calculate $\mathbf{C}_p$ from Eq. (111). The resulting system $(\mathbf{A}_p, \mathbf{B}_p, \mathbf{C}_p, \mathbf{0})$ is ASPR, and the closed loop system $(\mathbf{A}_p - \mathbf{B}_p \mathbf{K}_{e\infty} \mathbf{C}_p, \mathbf{B}_p, \mathbf{C}_p, \mathbf{0})$ is SPR. Given a controllable pair $(\mathbf{A}_p, \mathbf{B}_p)$ Theorem 4 provides a procedure to find $\mathbf{C}_p^*$ so that $(\mathbf{A}_p, \mathbf{B}_p, \mathbf{C}_p^*, \mathbf{0})$ is ASPR. The theorem fails, however to determine whether a system $(\mathbf{A}_p, \mathbf{B}_p, \mathbf{C}_p, \mathbf{0})$ (with a pre-specified $\mathbf{C}_p$) is ASPR. This problem can be addressed by incorporation of Theorem 4 with an optimization procedure (see Livneh *et al* [30]).

Theorem 2 provides a necessary condition for a plant to be ASPR. It demonstrates the requirement for a minimum phase plant, and is in excellent agreement with the numerical simulations. Theorems 3 and 4 provide sufficiency conditions for a plant to be ASPR. Both theorems use feedforward augmentation to achieve a minimum phase plant. Theorem 4 is more general than Theorem 3 in the sense that it requires only controllability rather than output stabilizability. However from an implementation point of view it is usually easier to adjust $\mathbf{D}_{p1}$ than $\mathbf{C}_p$. It is important to realize that while the plant matrices $(\mathbf{A}_p, \mathbf{B}_p, \mathbf{C}_p, \mathbf{D}_p)$ and the reference gain $\mathbf{K}_{e\infty}$, are utilized in the stability proof of the adaptive algorithm, these matrices need not be computed or even known for the successful application of the adaptive algorithm. The same is true for μ, $\mathbf{P}$, $\mathbf{L}$, $\mathbf{W}$ and $\mathbf{z}$.

The main stability result (Theorem 1) was proved for a minimum phase ASPR plant. Our simulation results indicate that a non-minimum phase plant is likely to result in divergence; however if the plant is minimum phase one can replace the requirement for almost strict positive realness by that of output stabilizability, which is less restrictive. The stable behavior of such a non-minimum phase output stabilizable plant is discussed below.

V. NUMERICAL EXAMPLES

To demonstrate the validity of the theoretical results of this work we provide numerical simulation of the SAC for the following two cases: a) adaptive regulation of a fourth order plant with one unstable mode, and b) tracking the output of a second order model by the output of a 23rd order wing flutter model with one unstable mode.

A. A FOURTH ORDER PLANT WITH ONE UNSTABLE MODE

An example of regulation of fourth order plant is used to demonstrate the uniqueness of the adaptive gains. The plant consists of two vibration modes with a single observed variable which consists of a linear combination of position and velocity collocated with single actuator. The plant is given by

$$\mathbf{A_p} = \begin{bmatrix} 0 & 1 & 0 & 0 \\ -16 & 2 & 0 & 0 \\ 0 & 0 & 0 & 1 \\ 0 & 0 & -49 & -10 \end{bmatrix}, \quad \mathbf{B_p} = \begin{bmatrix} 0 \\ 4 \\ 0 \\ 7 \end{bmatrix}, \quad \mathbf{C_p} = \begin{bmatrix} 2 \\ 4 \\ 3.5 \\ 7 \end{bmatrix}^T, \quad \mathbf{D_p} = [0] \qquad (112)$$

and its transfer function is

$$G_p(s) = \frac{(s+2)(65s^2 + 62s + 1568)}{2\,(s^2 - 2s + 16)\,(s^2 + 10s + 49)} \qquad (113)$$

The plant has a pair of stable poles at $1.0 \pm 3.873i$, a pair of unstable poles at $-5.0 \pm 4.899i$ and three stable zeros at $-0.477 \pm 4.888i$, and -2.0. It is minimum phase and, thus, meets the necessary condition of Theorem 2 for being ASPR. In the current example we restrict ourselves to measurable noise v_p and unmeasurable noise z_p applied only at the plant actuators. This is achieved by choosing

$$\mathbf{B}_{pv} = \mathbf{B}_p \mathbf{H}_{pv}, \quad \mathbf{B}_{pz} = \mathbf{B}_p \mathbf{H}_{pz}$$
$$\mathbf{D}_{pv} = \mathbf{D}_p \mathbf{H}_{pv}, \quad \mathbf{D}_{pz} = \mathbf{D}_p \mathbf{H}_{pz} \tag{114}$$

since in that case we can define

$$\mathbf{w}_{pv} = \mathbf{H}_{pv} \mathbf{v}_p, \quad \mathbf{w}_{pz} = \mathbf{H}_{pz} \mathbf{z}_p \tag{115}$$

and write Eqs.(1), (2) as

$$\dot{\mathbf{x}}_p = \mathbf{A}_p \mathbf{x}_p + \mathbf{B}_p (\mathbf{u}_p + \mathbf{w}_{vp} + \mathbf{w}_{zp})$$
$$\mathbf{y}_p = \mathbf{C}_p \mathbf{x}_p + \mathbf{D}_p (\mathbf{w}_p + \mathbf{w}_{vp} + \mathbf{w}_{zp}) \tag{116}$$

$\mathbf{H}_{pv}$ and $\mathbf{H}_{pz}$ are constant matrices of appropriate dimension which determine the phase and the magnitude of each frequency of $\mathbf{v}_p$ and $\mathbf{z}_p$ of in the combined signals $\mathbf{w}_{pv}$ and $\mathbf{w}_{pz}$, respectively.

Choosing

$$\mathbf{M}_p = \begin{bmatrix} 0 & 1 & 0 & 0 \\ -4.5^2 & 0 & 0 & 0 \\ 0 & 0 & 0 & 1 \\ 0 & 0 & -7.3^2 & 0 \end{bmatrix}, \quad \mathbf{v}_p(0) = \begin{bmatrix} 1 \\ 0 \\ 1 \\ 0 \end{bmatrix}, \quad \mathbf{H}_{pv} = \begin{bmatrix} 1 \\ 0 \\ 1 \\ 0 \end{bmatrix}^T \tag{117}$$

yields

$$\mathbf{B}_{pv} = \begin{bmatrix} 0 & 0 & 0 & 0 \\ 4 & 0 & 4 & 0 \\ 0 & 0 & 0 & 0 \\ 7 & 0 & 7 & 0 \end{bmatrix}, \quad \mathbf{D}_{pv} = \begin{bmatrix} 0 \\ 0 \\ 0 \\ 0 \end{bmatrix}^T, \quad \mathbf{w}_{pv} = \cos(4.4t) + \cos(7.3t) \tag{118}$$

for the measurable plant excitation, while choosing

$$\mathbf{N}_{\mathbf{p}} = \begin{bmatrix} 0 & 1 \\ -4.4^2 & 0 \end{bmatrix}, \quad z_p(0) = \begin{bmatrix} 1 \\ 0 \end{bmatrix}, \quad \mathbf{H}_{pz} = \begin{bmatrix} 1 \\ 0 \end{bmatrix}^T \tag{119}$$

yields

$$\mathbf{B}_{pz} = \begin{bmatrix} 0 & 0 \\ 4 & 0 \\ 0 & 0 \\ 7 & 0 \end{bmatrix}, \quad \mathbf{D}_{pz} = \begin{bmatrix} 0 \\ 0 \end{bmatrix}^T, \quad \mathbf{w}_{pz} = \cos(4.4t) \tag{120}$$

for the unmeasurable plant excitation.

The weight matrices used to determine the adaptive gains are given by $\mathbf{T}_p = 0.01\,\mathbf{I}$, and $\mathbf{T}_i = \mathbf{I}$, where I is the identity matrix of appropriate order. The filtering parameter is $\sigma = 0$, and the initial conditions to the state of the plant are given by $x_p(0) = [\ 0.1 \quad 0 \quad 0.1 \quad 0\ ]$.

Substitution of the above values for $\mathbf{A}_p, \mathbf{B}_p, \mathbf{C}_p, \mathbf{D}_p, \mathbf{M}_p, \mathbf{B}_{pv}$, and $\mathbf{D}_{pv}$ into Eqs. (15), (21), (25) yields

$$\mathbf{K}_{vp\infty} = \begin{bmatrix} -1.00 & 1.75\times10^{-17} & -1.00 & -2.16\times10^{-17} \end{bmatrix} \tag{121}$$

Figure 3 depicts the time variations of the adaptive gains corresponding to regulation of the plant of Eq.(112) which is subjected to a two frequency measurable excitation W_{vp} of Eq. (118). The algorithm results in five adaptive gains

$$\begin{aligned} \mathbf{K} &= [\ \mathbf{K}_e \quad \mathbf{K}_{vp}\ ] = [\ K_e(1) \quad K_{vp}(1) \quad K_{vp}(2) \quad K_{vp}(3) \quad K_{vp}(4)\ ] \\ &= [\ 1.19 \quad -1.0 \quad 2.1\times10^{-9} \quad -1.0 \quad -2.4\times10^{-10}\ ] \\ &= [\ K(1,1) \quad K(1,4) \quad K(1,5) \quad K(1,6) \quad K(1,7)\ ] \end{aligned} \tag{122}$$

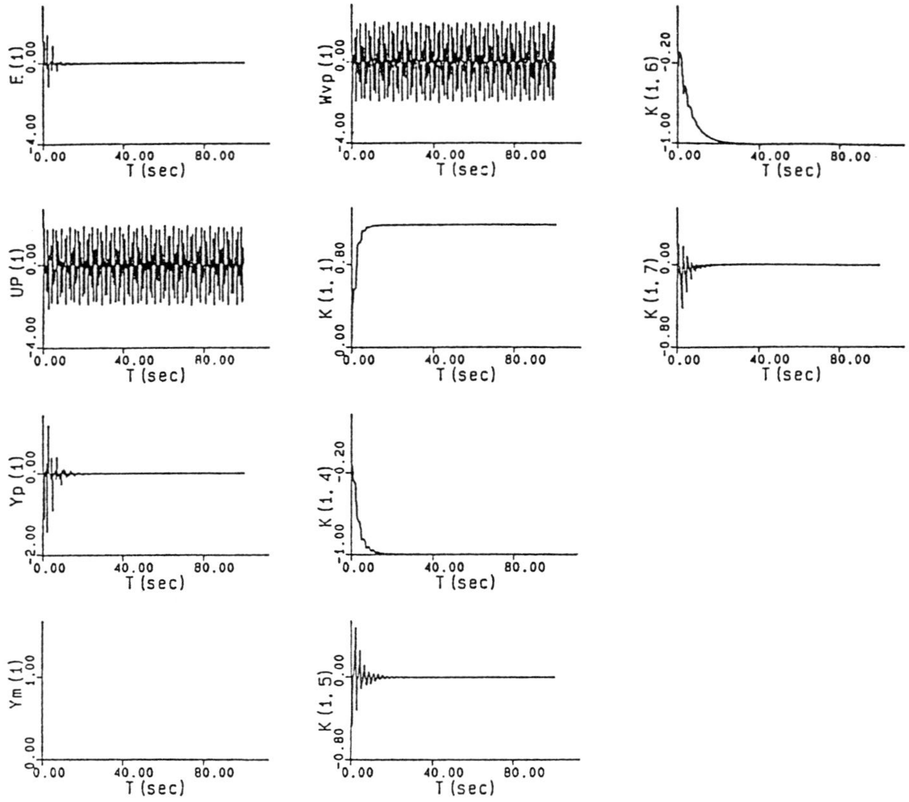

Figure 3: Regulation of $y_p(w_{vp})$, $\sigma = 0$, $t = 100$ sec

The last row of Eq. (122) corresponds to the notation used in Fig. 3. It is evident that the adaptive gains of Eq. (122) converge to the reference values of Eq. (121) neutralizing the independent excitation to the plant and achieving

$$\mathbf{K}_{vp} \Big|_{t=100} \cong - \mathbf{H}_{pv} \tag{123}$$

The effects of unmeasurable excitation z_p on the above regulation problem is depicted in Fig. 4. The unmeasurable excitation w_{pz} causes an increase in both $K_e(1)$ [=K(1,1)] and the steady state error $e_y(1)$ [=E(1)]. The monotonic increase in $K_e(1)$ is clearly non acceptable and should be bounded by choosing a positive filtering gain $\sigma > 0$. The effects of $\sigma > 0$ are discussed below for the wing flutter problem.

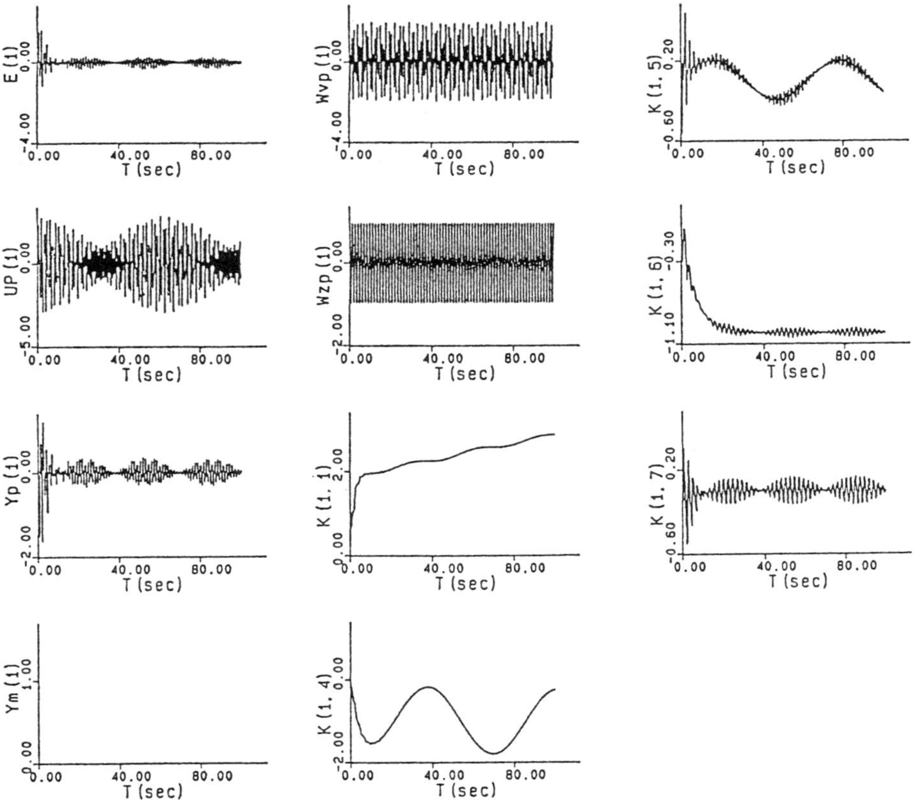

Figure 4: Regulation of $y_p(w_{vp}, w_{zp})$, $\sigma = 0$, $t = 100$ sec

B. FLUTTER CONTROL OF A CANTILEVER SWEPT BACK WING

The flutter model of Abel [21] is utilized to demonstrate the application of the adaptive algorithm to a realistic problem. The plant, depicted in Fig. 5 is an aeroelastic wind-tunnel model of a cantilever swept-back wing. This wing model was evaluated at Mach number M=0.9 for eight dynamic pressures spaced evenly between 20 and 160 lbf/ft^2 . The wing model for each dynamic pressure consists of five structural modes, ten aerodynamic lags (two for each mode) and a third-order actuator. The input to the plant is the actuator deflection command in degrees, and its output is the acceleration as measured by one of two accelerometers. In the application of the current algorithm it is assumed that the output measured by either the inboard or the outboard sensor consists of a linear combination of the position and velocity at the sensor location. The variation (with dynamic pressure) of the poles and zeros associated with the five aeroelastic modes is shown in Fig. 6 for the inboard sensor configuration. The first mode becomes unstable at dynamic pressures exceeding 100 lbf/ft^2 while the poles and zeros of the other modes become more stable with an increase in the dynamic pressure.

Figure 7 shows the root locus corresponding to the inboard sensor at q=160 lbf/ft^2. The ten branches near the imaginary axis correspond to the five structural modes. The thirteen branches to their left consist of ten branches of the aerodynamic lags and three branches of the actuator dynamics. The source of instability in the closed-loop configuration can be either the first structural mode at low feedback gains, or the third mode at high gains. However, for an output feedback gain in the range $0.22<k<1.82$ all the closed-loop poles are stable.

The system has a relative degree of two and therefore is clearly not ASPR. It is, however, minimum phase and output-stabilizable, and as such conforms quite well with the stability results for the algorithm (Theorem 1).

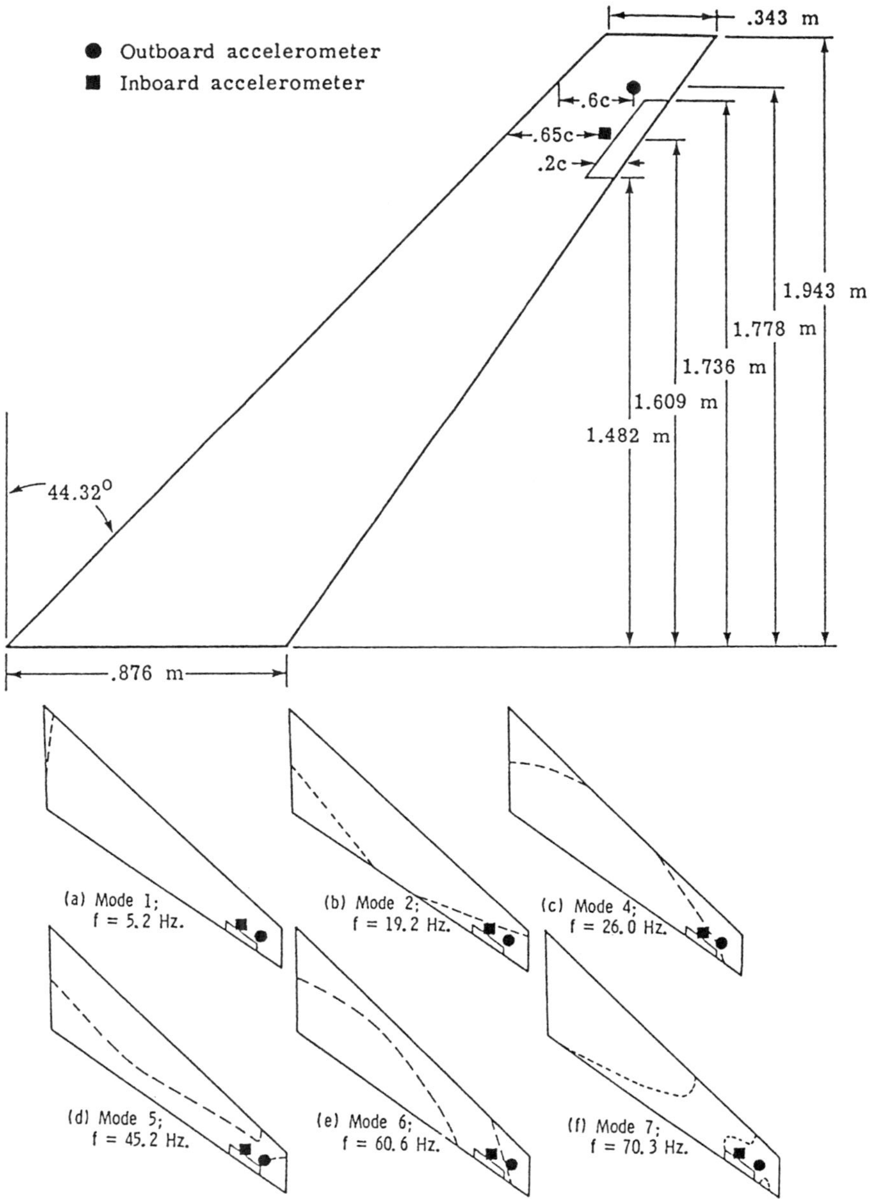

Figure 5: Wing geometry and the Inboard/Outboard sensor actuator configuration

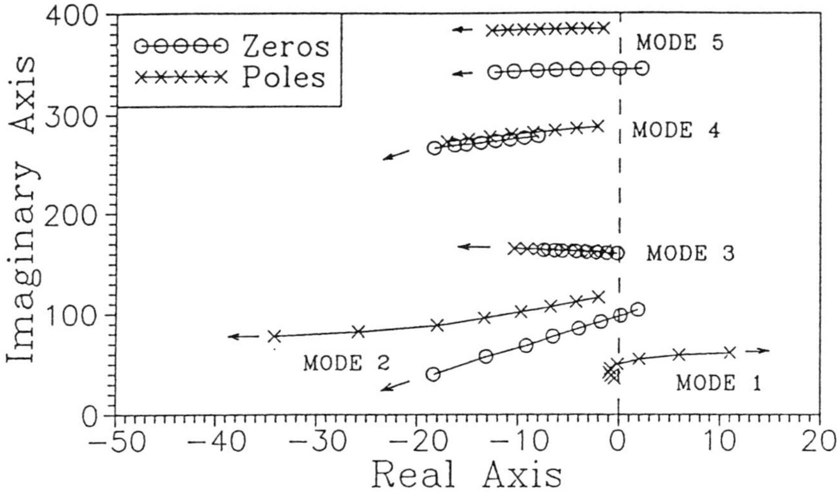

Figure 6: Variation with dynamic pressure of the poles and zeros of the structural modes for the 23rd order wing model with inboard sensor

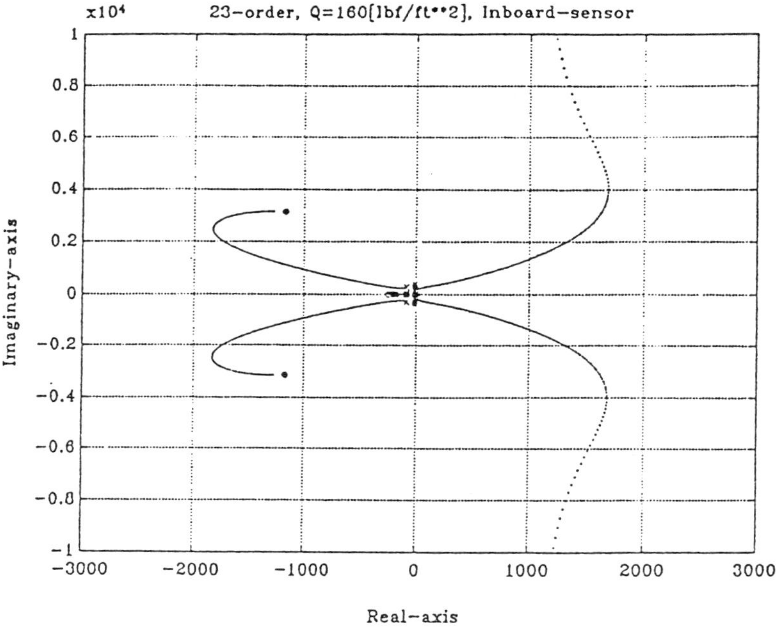

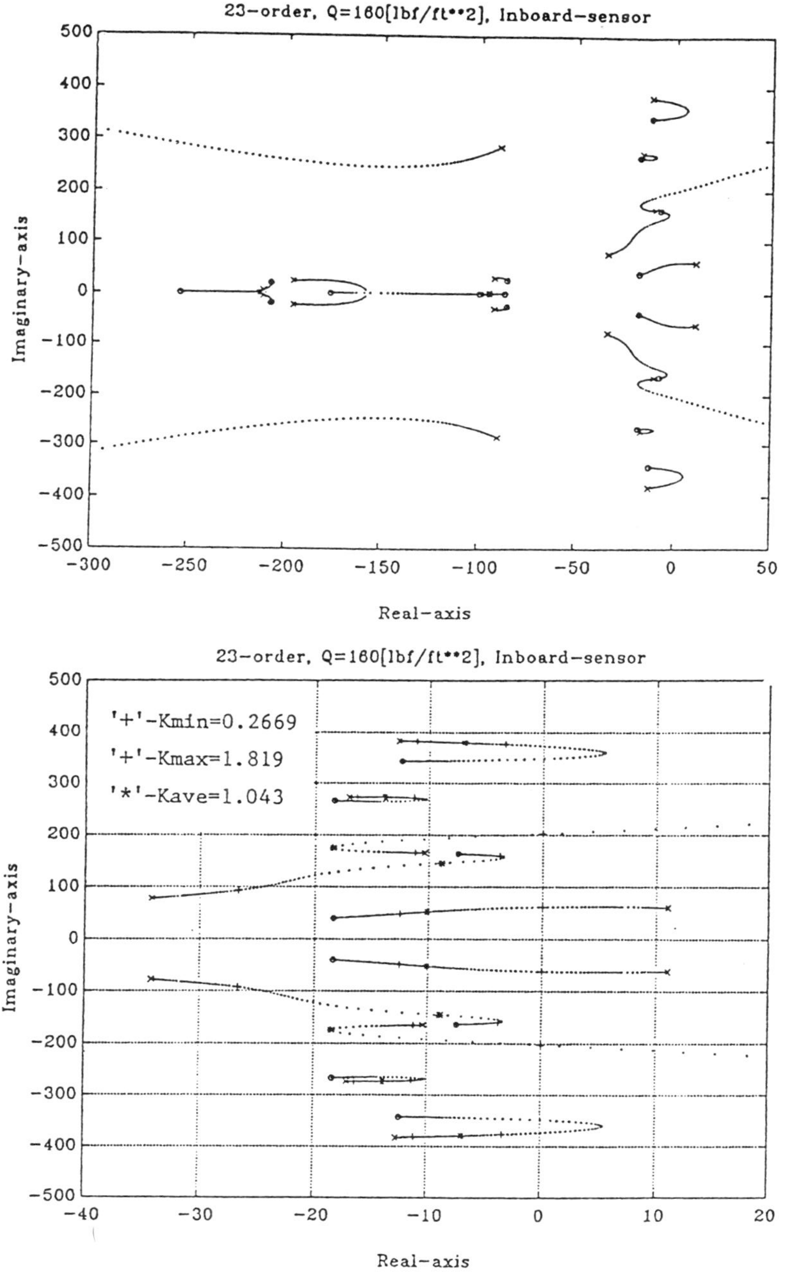

Figure 7: Root locus of the 23rd order wing model with inboard sensor at dynamic pressure $q = 160\ \text{lbf/ft}^2$.

The root locus depicted in Fig. 7 is characteristic of the dynamic pressure in the range 60-140 lbf/ft^2, which are all in good agreement with the main stability result. The 20 and 40 lbf/ft^2, inboard sensor configurations are non-minimum phase owing to the second and fifth mode zeros (see Fig. 6), and result in instability.

Another example of instability due to a non-minimum phase plant occurs when one uses the outboard rather than the inboard sensor. The root locus associated with the outboard sensor is very similar to that of Fig. 7. The main difference is in its two asymptotes about -465 (rather than 1000) and its two large zeros at 315±230i (rather than −1160±3150i). The fast divergence associated with this case is thus both expected and understood.

The main purpose of the numerical simulations presented for the wing flutter problem is to examine Theorem 1, i.e. to study the effect of the filtering parameter σ and the unmeasurable excitation $\mathbf{z_p}$ on the algorithm performance. Unless otherwise stated, all simulations correspond to a 23rd order wing model with the inboard sensor, evaluated at q=160 lbf/ft^2 (which corresponds to an unstable first mode). The reference model consists of one mode with natural frequency $\omega_m = 50$ rad/sec and damping $\zeta_m = 0.02$. The wing model is subjected to non-zero initial conditions of 0.1 inch in each of its structural modes, and the adaptation weight matrices are given by $\mathbf{T}_i = 0.1\ \mathbf{I}$ and $\mathbf{T}_p = 0.001\ \mathbf{I}$, where $\mathbf{I}$ is an identity matrix of appropriate order. The filtering parameter is either $\sigma = 0.0$ or $\sigma = 0.5$, as is discussed below. Figure 8 demonstrates the time responses resulting only from the non-zero initial conditions for q = 60 lbf/ft^2 and q = 160 lbf/ft^2. The faster convergence rate at q = 160 lbf/ft^2 is due to the position of its closed-loop poles which (as a result of $K_e(1)$ [=K (1,1)] which is larger for q = 160 lbf/ft^2) are further to the left of the closed-loop poles corresponding to q = 60 lbf/ft^2. This result gains further significance when one recalls that $q = 60$ lbf/ft^2 corresponds to more stable open-loop dynamics than $q = 160$ lbf/ft^2 (which is in fact unstable).

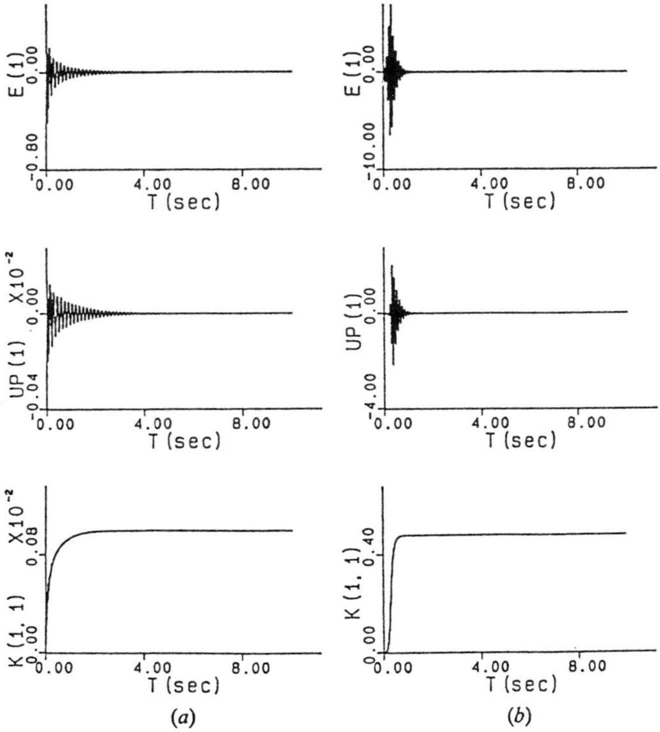

Figure 8: Time responses of the 23rd order plant with inboard sensor to non-zero initial conditions (no I/O excitations). a) q=60 lbf/ft^2, b) q=160 lbf/ft^2.

To demonstrate numerically the validity of Theorem 1 we provide Fig. 9 (a-c) which correspond to cases 3, 2 and 1 in the Table, respectively. For the SISO plant at hand, the adaptive input $\mathbf{u}_p$, the plant and reference model outputs y_p and y_m and the output error e_y are all scalar signals. Following the notation in Fig. 1, it is assumed that both the measurable and unmeasurable input excitation v_{pb}, and z_{pb}, are added directly to the actuator deflection command.

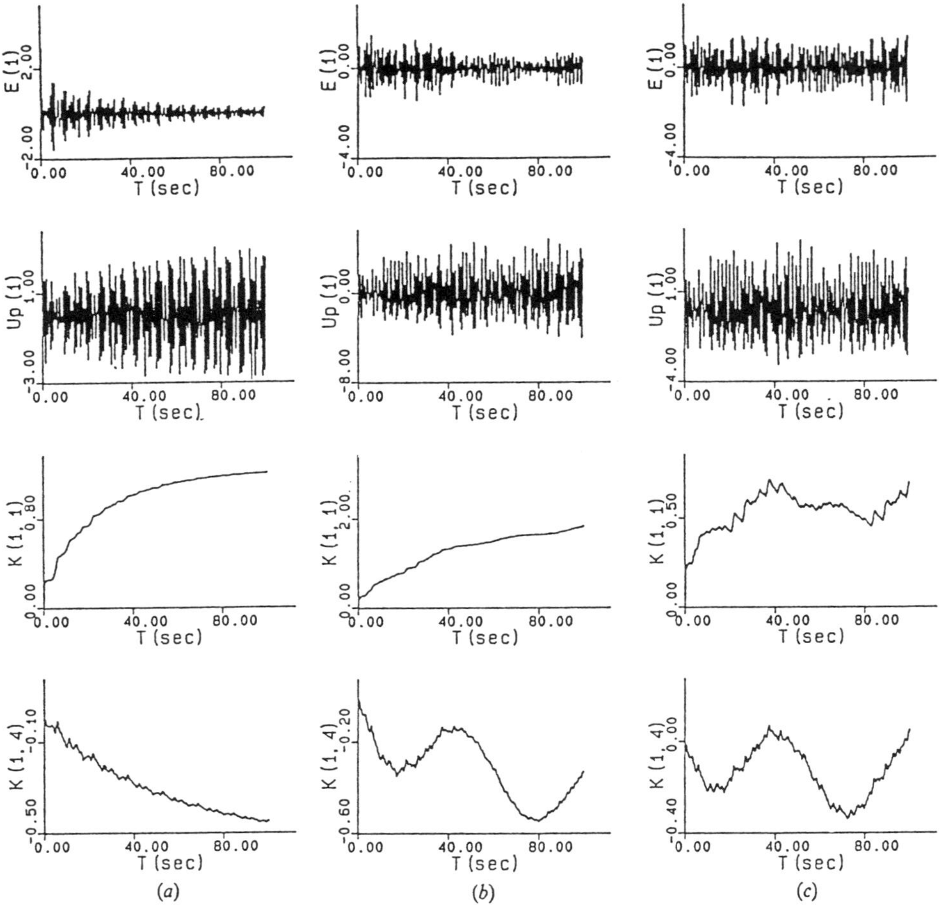

Figure 9: Time responses of the 23rd order wing model subjected to non-zero initial conditions and to $v_p \neq 0$ and $v_m \neq 0$. (a) $z_p = 0$, $\sigma = 0$, (b) $z_p \neq 0$, $\sigma = 0$, (c) $z_p \neq 0$, $\sigma \neq 0$.

Also, both the measurable and the unmeasurable output excitations v_{pd}, and z_{pd} are assumed to be added directly to the inboard sensor output measurement. These assumptions imply that, for the case at hand, v_{mb}, v_{pb}, z_{pb}, v_{mb}, v_{pd}, and z_{pd} are all scalar signals. These signals can include, however, an arbitrary number of frequency components with arbitrary amplitudes and phases. This is achieved by a proper choice of the excitation dynamics $(\mathbf{M}_m, \mathbf{M}_p, \mathbf{N}_p)$ the excitation I/O matrices $(\mathbf{B}_{mv}, \mathbf{B}_{pv}, \mathbf{B}_{pz} \cdots, \mathbf{D}_{pz})$ and the initial conditions for $\mathbf{v}_m$ $\mathbf{v}_p$ and $\mathbf{z}_m$. The current simulations were conducted for

$$\mathbf{v}_{pb} = \mathbf{B}_{pv}\, \mathbf{v}_p = 0.7 \cos(4.5t) - 1.2 \cos(7.3\, t) \tag{124}$$

$$\mathbf{v}_{pd} = \mathbf{D}_{pv}\, \mathbf{v}_p = 0.5 \cos(5.3t) \tag{125}$$

$$\mathbf{v}_{mb} = \mathbf{D}_{mv}\, \mathbf{v}_m = 0.8 \cos(6.1t) \tag{126}$$

$$\mathbf{v}_{md} = \mathbf{D}_{mv}\, \mathbf{v}_m = -0.3 \cos(3.7t) \tag{127}$$

The measurable excitations, given by (124)-(127), result in 13 scalar adaptive gains. Figure 9 depicts only two of these gains, namely $K_e(1)\,[= K(1,1)]$ and $K_{vm}(1)\,[= K(1,4)]$ together with the output error $\mathbf{e}_y\,[= E(1)]$ and the adaptive input $\mathbf{u}_p\,[= UP(1)]$ where the quantities enclosed in parentheses corresponds to the notation used in Figures 8 and 9.

Figure 9(a) corresponds to case 3 in Theorem 1 (σ=0, $\mathbf{z}_p$=0) and exhibits the asymptotic stability of the output error $\mathbf{e}_y$ and the convergence of the adaptive gains. Figures 9(b) and 9(c) are generated by adding to Fig. 9(a) the unmeasurable plant input/output excitation $\mathbf{z}_{pb}$ and $\mathbf{z}_{pd}$ (see Fig. 1) given by

$$\mathbf{z}_{pb} = \mathbf{B}_{pz}\, \mathbf{z}_p = 0.7 \cos(3.6t) \tag{128}$$

$$\mathbf{z}_{pd} = \mathbf{D}_{pz}\, \mathbf{z}_p = 0.7 \cos(7.4t) \tag{129}$$

Figure 9(b) corresponds to case 2 in Theorem 1 ($\sigma = 0$, $\mathbf{z}_p \neq 0$). The divergent tendency, expected in this case, is most evident in the plot of $K_e(1)$, which is monotonically increasing. Figure 9(c) corresponds to Case 3 in the Table ($\sigma \neq 0$, $z_p = 0$). The bounding effects of $\sigma=0.5$ are demonstrated in $K_e(1)$, which is non-monotonic and is kept at a lower level than its value in Fig. 9(b). Also the output error $\mathbf{e_y}$ is larger for $\sigma = 0.5$ than for $\sigma = 0$ This is due to the "corrupting effect" of σ, which moves the adaptive gains off their unique reference values.

Figures 9(b) and (c) also exhibit the effects of unmeasurable input/output excitations that have frequencies which are approximated by the measurable excitations. This was achieved by choosing the frequency of the input signal $\mathbf{z_{pb}}$ (3.6 rad/sec) to be 0.1 rad/sec off the frequency in $\mathbf{v_{md}}$ (3.7 rad/sec). Also the frequency of $\mathbf{z_{pd}}$ (7.4 rad/sec) was chosen to be 0.1 rad/sec off the second frequency in $\mathbf{v_{pb}}$ (7.3 rad/sec). The result of this approximation is a 0.1 rad/sec oscillation of $\mathbf{K_{vm}}(1)$, $\mathbf{K_{vm}}(2)$, $\mathbf{K_{vp}}(3)$, and $\mathbf{K_{vp}}(4)$,. Figures 9(b) and (c) depict this oscillation for $\mathbf{K_{vm}}(1)$ [=K (1,4)]. The 0.1 rad/sec frequency component is absent from K (1,4) in Fig. 9(a), since in that case all input/output excitations are measurable.

So far we have addressed differences in frequencies between the measurable and the unmeasurable excitations. Differences in amplitude and phase between those excitations are corrected by the adaptive adjustment of two gains for each frequency and do not affect significantly the performance and robustness of the algorithm.

VI. CONCLUSIONS AND RECOMMENDATIONS

The problem of designing a robust simplified controller for the control of a wing flutter vibration has been studied. The adaptive control algorithm developed by Bar-Kana *et al* was extended to include independent excitations to both the input and the output of the plant. These excitations were partitioned into measurable and unmeasurable parts, and incorporated into the ideal trajectory and the adaptive law. The stability of the adaptive law was proved by using ultimate stability results for non-autonomous systems., rather than the LaSalle invariance principle and the Gronwall-Bellman lemma which were used by Bar-Kana et al. The property of "almost strict positive realness" introduced by Bar-Kana was examined and related to minimum phase systems, output stabilizability, and the solution to the steady state Riccati equation. Numerical simulations were performed on a 23rd order wing flutter model. The simulation results are in good agreement with the theoretical analysis. The requirement for minimum phase plants should be strictly enforced, whereas the requirement for ASPR plant can be replaced by output stabilizability. The algorithm exhibits a good robustness over a large range of variations in the simulation parameters, such as plant and reference model excitations, dynamic pressure on the wing and sensor actuator configurations.

APPENDIX-A: EVALUATION OF $\dot{\nu}$ FOR AN ASPR PLANT $(\mathbf{A}_p, \mathbf{B}_p, \mathbf{C}_p, \mathbf{D}_p)$

Differentiation of (41) yields

$$\dot{\nu} = \frac{1}{2}\mathbf{e}_x^T \mathbf{P}\ \dot{\mathbf{e}}_x + \frac{1}{2}\dot{\mathbf{e}}_x^T \mathbf{P}\,\mathbf{e}_x + \mathrm{tr}\left(\Delta\mathbf{K}_i\,\mathbf{T}_i^{-1}\,\Delta\,\dot{\mathbf{K}}_i^T\right) \tag{A.1}$$

next define the vector s by

$$\mathbf{s} = \Delta\mathbf{K}\,\mathbf{r} + \mathbf{K}_{vp\infty}\,\mathbf{z}_p \tag{A.2}$$

which enables us to write equations (42), (43) and (53) in the form

$$\dot{\mathbf{e}}_x = \widetilde{\mathbf{A}}_p\,\mathbf{e}_x + \widetilde{\mathbf{B}}_p\,\mathbf{s} \tag{A.3}$$

$$\mathbf{e}_y = \widetilde{\mathbf{C}}_p\,\mathbf{e}_x + \widetilde{\mathbf{D}}_p\,\mathbf{s} \tag{A.4}$$

$$\mathbf{z} = \mathbf{L}^{T}\mathbf{e}_{x} + \mathbf{W}\,\mathbf{s} \tag{A.5}$$

The substitution of Eq. (A.3) for $\dot{\mathbf{e}}_{\mathbf{x}}$ and Eq. (44) for $\Delta\mathbf{K}_{\mathbf{i}}$ into Eq. (A.1) yields

$$\begin{aligned}\dot{\nu} = {} & \frac{1}{2}\mathbf{e}_x^T\left(\mathbf{P}\,\widetilde{\mathbf{A}}_p + \widetilde{\mathbf{A}}_p^T\,\mathbf{P}\right)\mathbf{e}_x + \mathbf{e}_x^{\,T}\,\mathbf{P}\,\widetilde{\mathbf{B}}_p\,\mathbf{s} - \mathbf{e}_y^{\,T}\,\Delta\mathbf{K}_i\,\mathbf{r} \\ & - \sigma\,\mathrm{tr}\,\left(\Delta\mathbf{K}_i\,\Delta\mathbf{K}_i^{\,T}\right) + \sigma\,\mathrm{tr}\,\left(\Delta\mathbf{K}_\infty\,\Delta\mathbf{K}_i^{\,T}\right)\end{aligned} \tag{A.6}$$

the third term on the right of Eq. (A.6) can be manipulated to give

$$\begin{aligned}\mathbf{e}_y^T\,\Delta\mathbf{K}_i\,\mathbf{r} &= \mathbf{e}_y^T\,\Delta\mathbf{K}\,\mathbf{r} + \mathbf{e}_y^T\,\mathbf{K}_p\,\mathbf{r} \\ &= \left(\widetilde{\mathbf{C}}_p\,\mathbf{e}_x + \widetilde{\mathbf{D}}_p\,\mathbf{s}\right)^T\left(\mathbf{s} - \mathbf{K}_{zp\infty}\,\mathbf{z}_p\right) + \mathbf{e}_y^T\,\mathbf{e}_y\,\mathbf{r}^T\,\mathbf{T}_p\,\mathbf{r}\end{aligned} \tag{A.7}$$

The substitution of Eq. (A.7) into Eq. (A.6) and some rearrangement yields

$$\begin{aligned}\dot{\nu} = {} & \frac{1}{2}\mathbf{e}_x^T\left(\mathbf{P}\,\widetilde{\mathbf{A}}_p + \widetilde{\mathbf{A}}_p^T\,\mathbf{P}\right)\mathbf{e}_x + \mathbf{e}_x^T\left(\mathbf{P}\,\widetilde{\mathbf{B}}_p - \widetilde{\mathbf{C}}_p^T\right)\mathbf{s} - \mathbf{s}^T\,\widetilde{\mathbf{D}}_p\,\mathbf{s} + \mathbf{e}_x^{\,T}\,\widetilde{\mathbf{C}}_p^T\,\mathbf{K}_{zp\infty}\,\mathbf{z}_p \\ & + \mathbf{s}^T\,\widetilde{\mathbf{D}}_p\,\mathbf{K}_{zp\infty}\,\mathbf{z}_p - \mathbf{e}_y^T\,\mathbf{e}_y\,\mathbf{r}^T\,\mathbf{T}_p\,\mathbf{r} - \sigma\,\mathrm{tr}\left(\Delta\mathbf{K}_i\Delta\mathbf{K}_i^T\right) + \sigma\,\mathrm{tr}\left(\Delta\mathbf{K}_\infty\,\Delta\mathbf{K}_i^T\right)\end{aligned} \tag{A.8}$$

Since the plant $(\mathbf{A}_p, \mathbf{B}_p, \mathbf{C}_p, \mathbf{D}_p)$ is ASPR, one can substitute Eqs. (49)–(51), (A.4) for the first four terms in the right-hand side of Eq. (A.7), respectively, to obtain

$$\begin{aligned}\dot{\nu} = {} & -\frac{1}{2}\mathbf{e}_x^T\,\mathbf{L}\,\mathbf{L}^T\,\mathbf{e}_x - \mathbf{e}_x^T\,\mathbf{L}\,\mathbf{W}\,\mathbf{s} - \mathbf{s}^T\,\mathbf{W}^T\,\mathbf{W}\,\mathbf{s} \\ & + \left(\mathbf{e}_y^T - \mathbf{s}^T\widetilde{\mathbf{D}}_p^T\right)\mathbf{K}_{zp\infty}\,\mathbf{z}_p + \mathbf{s}^T\,\widetilde{\mathbf{D}}_p^T\,\mathbf{K}_{zp\infty}\,\mathbf{z}_p - \mathbf{e}_y^T\,\mathbf{e}_y\,\mathbf{r}^T\,\mathbf{T}_p\,\mathbf{r} \\ & - \mu\,\mathbf{e}_x^T\,\mathbf{P}\,\mathbf{e}_x - \sigma\,\mathrm{tr}\left(\Delta\mathbf{K}_i\Delta\mathbf{K}_i^T\right) + \sigma\,\mathrm{tr}\left(\Delta\mathbf{K}_\infty\,\Delta\mathbf{K}_i^T\right)\end{aligned} \tag{A.9}$$

Equation (A.9) can be simplified to

$$\dot{\nu} = -\frac{1}{2}\left(\mathbf{L}^T \mathbf{e}_x + \mathbf{W}\,\mathbf{s}\right)^T\left(\mathbf{L}^T \mathbf{e}_x + \mathbf{W}\,\mathbf{s}\right) + \mathbf{e}_y^T \mathbf{K}_{zp\infty}\,\mathbf{z}_p - \mathbf{e}_y^T \mathbf{e}_y\,\mathbf{r}^T \mathbf{T}_p\,\mathbf{r} - \mu\,\mathbf{e}_x^T \mathbf{P}\,\mathbf{e}_x - \sigma\,\mathrm{tr}\left(\Delta\mathbf{K}_i\,\Delta\mathbf{K}_i^T\right) + \sigma\,\mathrm{tr}\left(\Delta\mathbf{K}_\infty\,\Delta\mathbf{K}_i^T\right) \tag{A.10}$$

The substitution of (A.5) for **z** and some rearrangement of the right-hand side of Eq. (A.10) yields the final result

$$\dot{\nu} \equiv -\mu\,\mathbf{e}_x^T \mathbf{P}\,\mathbf{e}_x - \sigma\,\mathrm{tr}\left(\Delta\mathbf{K}_i\,\Delta\mathbf{K}_i^T\right) - \tfrac{1}{2}\mathbf{z}^T\mathbf{z} - \mathbf{e}_y^T \mathbf{e}_y\,\mathbf{r}^T \mathbf{T}_p\,\mathbf{r} + \mathbf{e}_y^T \mathbf{K}_{zp\infty}\,\mathbf{z}_p + \sigma\,\mathrm{tr}\left(\mathbf{K}_\infty\,\Delta\mathbf{K}_i\right) \tag{A.11}$$

which is identical to Eq. (52).

APPENDIX-B: BACKGROUND FOR THE STABILITY PROOF OF THEOREM 1

The proof of Theorem 1 is based upon the following definition and Theorem from LaSalle and Lefschetz [23].

Definition B.1

The non-autonomous system given by

$$\dot{\mathbf{x}} = \mathbf{f}(\mathbf{x},t), \quad t>0 \tag{B.1}$$

is ultimately bounded if there is a finite $b > 0$ such that corresponding to each solution $\mathbf{x}(t)$ of Eq. (B.1) there is a $t_b > 0$ with the property that $\|\mathbf{x}(t)\| < b$ for all $t > t_b$. For the non-autonomous system of Eq. (B.1) and an autonomous Lyapunov function $\nu(\mathbf{x})$, we have the following ultimate boundedness result.

Theorem B.1

Let $\boldsymbol{\nu}(\mathbf{x})$ be a scalar function which for all $\mathbf{x}$ has continuous first partial derivatives with the property that

$$\lim_{\|\mathbf{x}\| \to \infty} \boldsymbol{\nu}(\mathbf{x}) = \infty$$

if $\dot{\boldsymbol{\nu}}(\mathbf{x}) \leq -\|\varepsilon\| < 0$ for all $\mathbf{x}$ outside some closed and bounded set M, then every solution of (B.1) is ultimately bounded.

APPENDIX-C: DERIVATION OF EQS. (42)-(44) FOR $\dot{\mathbf{e}}_\mathbf{x}$, $\mathbf{e}_\mathbf{y}$, AND $\Delta\dot{\mathbf{K}}_\mathbf{i}$

We start by noting that based on Eqs. (37), (33), and (39)for $\mathbf{u}_\mathbf{p}^*$, $\mathbf{u}_\mathbf{p}$, and $\Delta\mathbf{K}$, respectively

$$\mathbf{u}_\mathbf{p}^* - \mathbf{u}_\mathbf{p} = \left(\mathbf{K}_\infty - \mathbf{K}\right)\mathbf{r} - \mathbf{K}_{\mathbf{e}_\infty}\mathbf{e}_\mathbf{y} + \mathbf{K}_{\mathbf{zp}_\infty}\mathbf{z}_\mathbf{p} = \Delta\mathbf{K}\,\mathbf{r} - \mathbf{K}_{\mathbf{e}_\infty}\mathbf{e}_\mathbf{y} + \mathbf{K}_{\mathbf{zp}_\infty}\mathbf{z}_\mathbf{p} \tag{C.1}$$

A. DERIVATION OF EQ. (43) FOR $\mathbf{e}_\mathbf{y}$

Based on Eqs. (28), (12), (2), (27), (C.1) for $\mathbf{e}_\mathbf{y}$, $\mathbf{y}_\mathbf{p}^*$, $\mathbf{y}_\mathbf{p}$, $\mathbf{e}_\mathbf{x}$, and $\left(\mathbf{u}_\mathbf{p}^* - \mathbf{u}_\mathbf{p}\right)$

$$\begin{aligned}\mathbf{e}_\mathbf{y} &= \mathbf{y}_\mathbf{p}^* - \mathbf{y}_\mathbf{p} = \mathbf{C}_\mathbf{p}\left(\mathbf{x}_\mathbf{p}^* - \mathbf{x}_\mathbf{p}\right) + \mathbf{C}_\mathbf{p}\left(\mathbf{u}_\mathbf{p}^* - \mathbf{u}_\mathbf{p}\right) \\ &= \mathbf{C}_\mathbf{p}\,\mathbf{e}_\mathbf{x} + \mathbf{D}_\mathbf{p}\left(\Delta\mathbf{K}\,\mathbf{r} + \mathbf{K}_{\mathbf{zp}_\infty}\mathbf{z}_\mathbf{p}\right) - \mathbf{D}_\mathbf{p}\,\mathbf{K}_{\mathbf{e}_\infty}\mathbf{e}_\mathbf{y}\end{aligned} \tag{C.2}$$

Isolation of $\mathbf{e}_\mathbf{y}$ from Eq. (D.2) and use of Eq. (47), (48) for $\widetilde{\mathbf{C}}_\mathbf{p}$, and $\widetilde{\mathbf{D}}_\mathbf{p}$ yields

$$\begin{aligned}\mathbf{e}_\mathbf{y} &= \left(\mathbf{I} + \mathbf{D}_\mathbf{p}\,\mathbf{K}_{\mathbf{e}_\infty}\right)^{-1}\left[\mathbf{C}_\mathbf{p}\,\mathbf{e}_\mathbf{x} + \mathbf{D}_\mathbf{p}\left(\Delta\mathbf{K}\,\mathbf{r} + \mathbf{K}_{\mathbf{zp}_\infty}\mathbf{z}_\mathbf{p}\right)\right] \\ &= \widetilde{\mathbf{C}}_\mathbf{p}\,\mathbf{e}_\mathbf{x} + \widetilde{\mathbf{D}}_\mathbf{p}\left(\Delta\mathbf{K}\,\mathbf{r} + \mathbf{K}_{\mathbf{zp}_\infty}\mathbf{z}_\mathbf{p}\right)\end{aligned} \tag{C.3}$$

which is identical to Eq. (43).

B. DERIVATION OF EQ. (42) FOR $\dot{e}_x$

Based on Eqs. (27), (11), (2), (1), (C.1), and (C.3), for e_x, $\dot{x}^*_p$, $\dot{x}_p$, $(u^*_p - u_p)$, and e_y, , respectively

$$\begin{aligned}
\dot{e}_x &= \dot{x}^*_p - \dot{x}_p = A_p \left(x^*_p - x_p\right) + B_p\left(u^*_p - u_p\right) \\
&= A_p e_x + B_p\left(\Delta K r + K_{zp_\infty} z_p - K_{e_\infty} e_y\right) \\
&= A_p e_x + B_p\left(\Delta K r + K_{zp_\infty} z_p\right) \\
&\quad - B_p K_{e_\infty}\left[\widetilde{C}_p e_x + \widetilde{D}_p\left(\Delta K r + K_{zp_\infty} z_p\right)\right] \\
&= \left(A_p - B_p K_{e_\infty} \widetilde{C}_p\right) e_x + B_p\left(I - K_{e_\infty} \widetilde{D}_p\right)\left(\Delta K r + K_{zp_\infty} z_p\right)
\end{aligned} \tag{C.4}$$

Use of Eqs. (45)-(48) yields

$$\dot{e}_x = \widetilde{A}_p e_x + \widetilde{B}_p\left(\Delta K r + K_{zp_\infty} z_p\right) \tag{C.5}$$

which is identical to Eq. (42).

C. DERIVATION OF EQ. (44) FOR $\Delta\dot{K}_i$

Differentiation of Eq. (40) for ΔK_i and use of Eq. (36) for $\dot{K}_i$, yields

$$\begin{aligned}
\Delta\dot{K}_i &= -\dot{K}_i = \left[- e_y r^T - \sigma\left(K_\infty - K_i - K_\infty\right)\right] T_i \\
&= \left[- \sigma \Delta K_i - e_y r^T + \sigma K_\infty\right] T_i
\end{aligned} \tag{C.6}$$

which is identical to Eq. (44).

APPENDIX-D: THE RELATION BETWEEN ALMOST STRICY POSITIVE REALNESS AND THE STEADY STATE RICCATI EQUATION

The plant $(\mathbf{A}_p,\mathbf{B}_p,\mathbf{C}_p,\mathbf{D}_p)$ "almost strictly positive real" (ASPR) if there exists an output feedback gain such that the closed loop system $(\widetilde{\mathbf{A}}_p,\widetilde{\mathbf{B}}_p,\widetilde{\mathbf{C}}_p,\widetilde{\mathbf{D}}_p)$ given by Eqs. (45)-(48), is strictly positive real. A common version of the matrix condtions for the stricy positive realness of $(\widetilde{\mathbf{A}}_p,\widetilde{\mathbf{B}}_p,\widetilde{\mathbf{C}}_p,\widetilde{\mathbf{D}}_p)$ is (see e.g. [5] and [26]) is given by

$$\mathbf{P}\,\widetilde{\mathbf{A}}_p + \widetilde{\mathbf{A}}_p^T\,\mathbf{P} = -\mathbf{L}\,\mathbf{L}^T = -\mathbf{Q} < 0 \tag{D.1}$$

$$\mathbf{P}\,\widetilde{\mathbf{B}}_p - \widetilde{\mathbf{C}}_p^T = -\mathbf{L}\,\mathbf{W} \tag{D.2}$$

$$\widetilde{\mathbf{D}}_p + \widetilde{\mathbf{D}}_p^T = \mathbf{W}^T\mathbf{W} \tag{D.3}$$

We begin the analysis by noting that one can use Eqs. (45)-(48), to get

$$\widetilde{\mathbf{A}}_p = \mathbf{A}_p - \mathbf{B}_p\,\mathbf{K}_{e_\infty}\,\widetilde{\mathbf{C}}_p \tag{D.4}$$

$$\widetilde{\mathbf{B}}_p = \mathbf{B}_p\left(\mathbf{I} - \mathbf{K}_{e_\infty}\,\widetilde{\mathbf{D}}_p\right) \tag{D.5}$$

Next substitute Eq. (D.4) into Eq. (D.1) to get

$$\mathbf{P}\left(\mathbf{A}_p - \mathbf{B}_p\,\mathbf{K}_{e_\infty}\,\widetilde{\mathbf{C}}_p\right) + \left(\mathbf{A}_p^T - \widetilde{\mathbf{C}}_p^T\,\mathbf{B}_p\,\mathbf{K}_{e_\infty}^T\,\mathbf{B}_p^T\right)\mathbf{P} = -\mathbf{L}\,\mathbf{L}^T \tag{D.6}$$

Substitution of Eq. (D.2) into Eq. (D.6) and some rearrangement yields

$$\mathbf{P}\,\mathbf{A}_p + \mathbf{A}_p^T\mathbf{P} - \mathbf{P}\,\mathbf{B}_p\;\mathbf{K}_{e\infty}\left(\widetilde{\mathbf{B}}_p^T\mathbf{P} + \mathbf{W}^T\mathbf{L}^T\right)$$
$$- \left(\mathbf{P}\;\widetilde{\mathbf{B}}_p + \mathbf{L}\,\mathbf{W}\right)\mathbf{K}_{e\infty}^T\;\mathbf{B}_p\,\mathbf{P} = -\mathbf{L}\,\mathbf{L}^T \qquad \text{(D.7)}$$

Further rearrangement of Eq. (D.7) yields

$$\begin{aligned}&\mathbf{P}\,\mathbf{A}_p + \mathbf{A}_p^T\mathbf{P} - \mathbf{P}\left(\mathbf{B}_p\,\mathbf{K}_{e\infty}\,\widetilde{\mathbf{B}}_p^T + \widetilde{\mathbf{B}}_p\,\mathbf{K}_{e\infty}^T\,\mathbf{B}_p^T\right)\mathbf{P}\\ &- \mathbf{P}\,\mathbf{B}_p\;\mathbf{K}_{e\infty}\,\mathbf{W}^T\mathbf{L}^T - \mathbf{L}\,\mathbf{W}\,\mathbf{K}_{e\infty}^T\;\mathbf{B}_p^T\mathbf{P} = -\mathbf{L}\,\mathbf{L}^T\end{aligned} \qquad \text{(D.8)}$$

Evaluation of the third term on th left of Eq. (D.8) using Eqs. (D.5) and (D.3) yields

$$\begin{aligned}&\mathbf{B}_p\,\mathbf{K}_{e\infty}\,\widetilde{\mathbf{B}}_p^T + \widetilde{\mathbf{B}}_p\,\mathbf{K}_{e\infty}^T\,\mathbf{B}_p^T\\ &= \mathbf{B}_p\,\mathbf{K}_{e\infty}\left(\mathbf{I} - \widetilde{\mathbf{D}}_p^T\,\mathbf{K}_{e\infty}^T\right)\mathbf{B}_p^T + \mathbf{B}_p\left(\mathbf{I} - \mathbf{K}_{e\infty}\,\widetilde{\mathbf{D}}_p\right)\mathbf{K}_{e\infty}^T\,\mathbf{B}_p^T\\ &= \mathbf{B}_p\left(\mathbf{K}_{e\infty} + \mathbf{K}_{e\infty}^T\right)\mathbf{B}_p^T - \mathbf{B}_p\,\mathbf{K}_{e\infty}\left(\widetilde{\mathbf{D}}_p^T + \widetilde{\mathbf{D}}_p\right)\mathbf{K}_{e\infty}^T\,\mathbf{B}_p^T\\ &= \mathbf{B}_p\left(\mathbf{K}_{e\infty} + \mathbf{K}_{e\infty}^T\right)\mathbf{B}_p^T - \mathbf{B}_p\,\mathbf{K}_{e\infty}\,\mathbf{W}^T\mathbf{W}\,\mathbf{K}_{e\infty}^T\;\mathbf{B}_p^T\end{aligned} \qquad \text{(D.9)}$$

Substitution of Eq. (D.9) into Eq. (D.8) yields

$$\begin{aligned}&\mathbf{P}\,\mathbf{A}_p + \mathbf{A}_p^T\mathbf{P} - \mathbf{P}\,\mathbf{B}_p\left(\mathbf{K}_{e\infty} + \mathbf{K}_{e\infty}^T\right)\mathbf{B}_p^T\mathbf{P}\\ &+ \mathbf{P}\,\mathbf{B}_p\;\mathbf{K}_{e\infty}\,\mathbf{W}^T\mathbf{W}\,\mathbf{K}_{e\infty}^T\;\mathbf{B}_p^T\mathbf{P}\\ &- \mathbf{P}\,\mathbf{B}_p\;\mathbf{K}_{e\infty}\,\mathbf{W}^T\mathbf{L}^T - \mathbf{L}\,\mathbf{W}\,\mathbf{K}_{e\infty}^T\;\mathbf{B}_p^T\mathbf{P} + \mathbf{L}\,\mathbf{L}^T = 0\end{aligned} \qquad \text{(D.10)}$$

The last four terms on the left hand-side of Eq. (D.10) can be simplifies to

$$\begin{aligned}&\mathbf{P}\,\mathbf{B}_p\;\mathbf{K}_{e\infty}\,\mathbf{W}\,\mathbf{W}^T\,\mathbf{K}_{e\infty}^T\,\mathbf{B}_p^T\mathbf{P} - \mathbf{P}\,\mathbf{B}_p\;\mathbf{K}_{e\infty}\,\mathbf{W}^T\mathbf{L}^T\\ &- \mathbf{L}\,\mathbf{W}\,\mathbf{K}_{e\infty}^T\;\mathbf{B}_p^T\mathbf{P} + \mathbf{L}\,\mathbf{L}^T = \left(\mathbf{P}\;\mathbf{B}_p\,\mathbf{K}_{e\infty}\,\mathbf{W}^T - \mathbf{L}\right)\left(\mathbf{W}\,\mathbf{K}_{e\infty}^T\;\mathbf{B}_p^T\mathbf{P} - \mathbf{L}^T\right)\end{aligned}$$

(D.11)

By defining **H** by

$$\mathbf{H} = \mathbf{W}\,\mathbf{K}_{e\infty}^{T}\;\mathbf{B}_{p}^{T}\mathbf{P} - \;\mathbf{L}^{T} \tag{D.12}$$

one can write Eq. (D.10) as

$$\mathbf{H}^{T}\mathbf{H} - \mathbf{P}\,\mathbf{B}_{p}\;\left(\mathbf{K}_{e\infty} + \mathbf{K}_{e\infty}^{T}\right)\mathbf{B}_{p}^{T}\mathbf{P} + \mathbf{P}\,\mathbf{A}_{p}\;\; + \mathbf{A}_{p}^{T}\mathbf{P} = 0 \tag{D.13}$$

Pre-multiplication of Eq. (D.12) by $\mathbf{W}^{T}$ and manipulation of the results by using Eqs. (D.3), (D.2), and (47) for $\mathbf{W}^{T}\mathbf{W}$, $-\mathbf{L}\,\mathbf{W}$, and $\widetilde{\mathbf{C}}_{p}$ respectively, yields

$$\widetilde{\mathbf{C}}_{p} = \;\left(\mathbf{I} + \;\mathbf{D}_{p}\,\mathbf{K}_{e\infty}\right)\left\{\left[\left(\widetilde{\mathbf{D}}_{p} + \widetilde{\mathbf{D}}_{p}^{T}\right)\mathbf{K}_{e\infty}^{T}\;\mathbf{B}_{p}^{T} + \widetilde{\mathbf{B}}_{p}^{T}\right]\mathbf{P} - \;\mathbf{W}^{T}\mathbf{H}\right\} \tag{D.14}$$

which us identical to eq. (107).

REFERENCES

1. Osborn, P. V., Whitaker, H. P., and Kezer, A., "New Developments in the Design of Adaptive Control Systems", Institute of Aeronautical Science, Paper -61-43,.1961.

2. Parks, P. C., "Lyapunov Redesign of the Model Reference Adaptive Control Systems," *I.E.E.E. Transactions on Automatic Control,* **11,** pp. 362-367, 1966.

3. Butchart, R. L., and Shakcloth, B., "Synthesis of MRACS by Lyapunov's Second Method," *Proceeding of the 2nd IFAC Symposium on the Theory on Self-Adaptive Control Systems,* Teddington (Plenum Press), pp. 145-152, 1966.

4. Landau, Y. D., "A Hyperstability Criterion for Model Reference Adaptive Control Systems," *I.E.E.E. Transactions on Automatic Control,* **15,** pp. 552-555; 1969.

5. Landau, Y. D., 1969, *"Adaptive Control – the Model Reference Approach "*, New York: Marcel Decker, 1979.

6. Rohrs, C. E., Valvani, L., Athans, M. and Stein, G., "Robustness of Continuous Time Adaptive Control Algorithm in the Presence of Unmodeled Dynamics," *I.E.E.E. Transactions on Automatic Control,* **30,** pp. 881-889, 1985.

7. Åström, K. J., "A Commentary on the Rohrs *et al* Paper,". *I.E.E.E. Transaction on Automatic Control,* **30**, pp. 889, 1985.

8. Slater, G. L., and Livneh, R., "A Second Order Adaptive Controller for Wing Flutter Control," Paper No. 85.1965CP, *AIAA Guidance and Control Conference,* Snowmass, Colorado, 1985.

9. Monopoli, R. V., "Model Reference Adaptive Control with an Augmented Error Signal," *I.E.E.E. Transactions on Automatic Control,* **19,** pp. 474-485, 1974.

10. Ioannou, P. A., and Kokotovic, P. V., "An Asymptotic Error Analysis of Identifiers and Adaptive Observers in the Presence of Parasites," *I.E.E.E. Transactions on Automatic Control,* **27,** pp. 921-927, 1982.

11. Narendra, K. S., and Annaswamy, A. M., "Robust Adaptive Control in The Presence of Bounded Disturbances," *I.E.E.E. Transactions on Automatic Control,* **31,** pp. 306-315, 1986.

12. Sobel, K., Kaufman, H., and Mabius, L., "Implicit Adaptive Control of a Class of MIMO Systems," *I.E.E.E. Transactions on Aerospace and Electronic Systems,* **18,** pp. 576-590, 1982.

13. O'Brien, M. J., and Broussard, J. R., 1978, "Feedforward Control to Track the Output of a Forced Model," *Proceedings of the 17th I.E.E.E. Conference on Decision and Control,* pp. 1149-1154.

14. Bar-Kana, I., and Kaufman, H., "Some Applications of Direct Adaptive Control to Large Structural Systems," *A.I.A.A. Journal of Guidance Dynamics and Control,* **7,** pp. 717-724, 1984.

15. Bar-Kana, I., and Kaufman, H., "Global Stability and Performance of a Simplifies Adaptive Algorithm," *International Journal of Control,* **46,** pp. 1491-1505, 1985

16. Bar-Kana, I., and Kaufman, H., and Balas M., "Model Reference Adaptive Control of Large Structural Systems," *A.I.A..A. Journal of Guidance, Dynamics and Control,* **6,** pp. 112-118, 1983.

17. Ioannou, P. A., and Kokotovic, P. V., "Singular Perturbations and Robust Redesign of Adaptive Control," *Proceedings of the 21st I.E.E.E. Conference on Decision and Control,* Orlando, pp. 24-29, 1982 .

18. Bayard, D. S., Ih, C.-H. C., and Wang, S. J., "Adaptive Control for Flexible Structures with Measurement Noise," *Proceedings of the American Control Conference,* Minneapolis, pp. 368-379, 1987.

19. Ih C.-H. C., Bayard, D. S., Wang, S. J., and Eldred, D. B., "Adaptive Control Experiment with a Large Flexible Structure," *A.I.A.A. Guidance, Navigation and Control Conference,* Paper AIAA-88-4153-CP, Minneapolis, 1988.

20. Ih C.-H. C., Bayard, D. S., Ahmed, A., and Wang, S. J., "Experiments in Multivariable Adaptive Control of a Large Flexible Structure," *A.I.A.A. Guidance, Navigation and Control Conference,* Paper AIAA-89-3571-CP, Boston, Massachusetts, 1989.

21. Abel, I., "An Analytical Technique for Predicting the Characteristics of a Flexible Wing Equipped with an Active flutter Suppression System and Comparison with Wind Tunnel Data," NASA/TP-1367, 1979.

22. Davison, E. J., Wang, S. H., "Properties and Calculation of Transmission Zeros of Linear Multivariable Systems," *Automatica,* **10,** pp. 643-658, 1974.

23. LaSalle, J. P., *"The Stability and Control of Discrete Processes,"* Berlin: Springer-Verlag, 1986.

24. Bar-Kana, I., and Kaufman, H., "Robust Simplified Adaptive Control for a Class of Multivariable Continuous Time Systems," *Proceedings of the 24th Conference on Decision and Control,* Ft Lauderdale, Florida, pp. 141-146, 1985.

25. LaSalle, J. P., and Lefschetz, S., *"Stability by Lyapunov's Direct Methods with Applications,"* New York: Academic Press, 1961.

26. Anderson, B. D. O., "A Simplified Viewpoint of Hyperstability," *I.E.E.E. Transaction on Automatic Control,* **13**, pp. 292-294, 1968.

27. Newcomb, Y. D., *"Linear Multiport Synthesis,"* McGraw-Hill Book Company, 1966.

28. Noble, B., Daniel, J. W., *"Applied Linear Algebra,"* Prentice-Hall, 1977.

29. Kwakernaak, H., and Sivan, R., *"Linear Optimal Control Systems,"* Wiley-Interscience, 1972.

30. Livneh, R., McLaren, M. D., and Slater, G. L., "Some Conditions for Almost Strict Positive Realness," *Proceedings of the 29th IEEE Conference on Decision and Control,* Honolulu, Hawaii, Dec. 5-7, 1990, pp. 2512-2517.

INDEX

F

G

H

M

P

ISBN 0-12-012757-1
90065
9 780120 127573